23 Problems in Systems Neuroscience

23 Problems in Systems Neuroscience

Edited by

J. Leo van Hemmen
Terrence J. Sejnowski

OXFORD
UNIVERSITY PRESS

2006

OXFORD
UNIVERSITY PRESS

Oxford University Press, Inc., publishes works that further
Oxford University's objective of excellence
in research, scholarship, and education.

Oxford   New York
Auckland   Cape Town   Dar es Salaam   Hong Kong   Karachi
Kuala Lumpur   Madrid   Melbourne   Mexico City   Nairobi
New Delhi   Shanghai   Taipei   Toronto

With offices in
Argentina   Austria   Brazil   Chile   Czech Republic   France   Greece
Guatemala   Hungary   Italy   Japan   Poland   Portugal   Singapore
South Korea   Switzerland   Thailand   Turkey   Ukraine   Vietnam

Published by Oxford University Press, Inc.
198 Madison Avenue, New York, New York 10016

www.oup.com

Oxford is a registered trademark of Oxford University Press

Library of Congress Cataloging-in-Publication Data
23 problems in systems neuroscience / edited by Leo Van Hemmen and Terrence Sejnowski.
p. cm.
Includes bibliographical references and index.
ISBN-13: 978-0-19-514822-0
ISBN-10: 0-19-514822-3
1. Neurobiology.   2. Biological systems.   I. Title: Twenty three problems in systems
neuroscience.   II. Hemmen, J. L. van (Jan Leonard).   III. Sejnowski, Terrence J.
(Terrence Joseph)
QP355.2.A14 2005
573.8'6—dc22   2005003411

9 8 7 6 5 4 3 2

Printed in the United States of America
on acid-free paper

Preface

> As long as a branch of science offers an abundance of problems, so long is it alive; a lack of problems foreshadows extinction or the cessation of independent development.
>
> —David Hilbert in Paris, 8 August 1900

When invited in 1899 to give the opening address at the second International Congress of Mathematicians (ICM) in Paris the following year, David Hilbert (1862–1943), one of the greatest mathematicians of the last century and the father of Hilbert space, was facing the beginning of a new century. Henri Poincaré, who had a similar task in addressing the first ICM in Zurich, extolled the virtues of what had previously been accomplished in pure mathematics. Rather than focus on the glories of the past, Hilbert chose to make a break with tradition and instead emphasize what was *not* known. At the ICM on August 8, 1900, he outlined twenty-three key open mathematical problems as challenges to mathematicians for the twentieth century. These problems were influential in shaping the direction of mathematics.

To understand Hilbert's motivation, let us retrace his reasoning in the introduction to his lecture entitled *Mathematical Problems*[1]:

---

[1] The German text appeared in 1900 directly after the Paris meeting in *Nachr. Kgl. Ges. d. Wiss. zu Göttingen, math.-phys. Klasse* 3, 253–297. It can also be found in volume 3 of Hilbert's collected works (1935, 290–329). The translation above is by Dr. Mary Winston Newson and was published in 1902 in *Bull. Amer. Math. Soc.* 8: 437–479; it reappeared in 2000 in *Bull. Amer. Math. Soc.* (new series) 37: 407–436.

History teaches the <u>continuity of the development of science</u>. We know that every age has its own problems, which the following age either solves or casts aside as profitless and replaces by new ones. If we would obtain an idea of the probable development of mathematical knowledge in the immediate future, we must let the unsettled questions pass before our minds and look over the problems which the science of to-day sets and whose solution we expect from the future. To such a review of problems the present day, lying at the meeting of the centuries, seems to me well adapted. For the close of a great epoch not only invites us to look back into the past but also directs our thoughts to the unknown future.

The deep significance of certain problems for the advance of mathematical science in general and the important role that they play in the work of the individual investigator are not to be denied. As long as a branch of science offers an abundance of problems, so long is it alive; a lack of problems foreshadows extinction or the cessation of independent development. Just as every human undertaking pursues certain objects, so also mathematical research requires its problems. It is by the solution of problems that the investigator tests the temper of his steel; he finds new methods and new outlooks, and gains a wider and freer horizon.

It is difficult and often impossible to judge the value of a problem correctly in advance; for the final award depends upon the grain which science obtains from the problem. Nevertheless we can ask whether there are general criteria which mark a good mathematical problem. An old French mathematician said: "A mathematical theory is not to be considered complete until you have made it so clear that you can explain it to the first man whom you meet on the street." This <u>clearness and ease of comprehension</u>, here insisted on for a mathematical theory, I should still more demand for a mathematical problem if it is to be perfect; for what is clear and easily comprehended attracts, the complicated repels us.

As for systems neuroscience, why add more, if—*mutatis mutandis*—it has all been said so clearly a century ago? The same text, with "mathematical science" replaced by "systems neuroscience," applies to the twenty-first century. The complexity of the brain and the protean nature of behavior remain the most elusive area of science and also the most important. There is no single brilliant individual who could serve as the David Hilbert of systems neuroscience, so we have invited twenty-three experts from the many areas of systems neuroscience to formulate one problem each.

We thank the Max Planck Society for hosting a wonderful meeting on "Problems in Systems Neuroscience" in Dresden on September 4 to 8, 2000, that formed the core of the authors in this collection. We have asked several other colleagues to contribute to the twenty-three challenging questions in systems neuroscience that are included. It took Hilbert two years to get his problems published in English. It took us even longer to get the twenty-three different essays collected together in this volume. Although it is not comprehensive, and it is, like Hilbert's original list of problems, idiosyncratic, nonetheless it may serve as a source of inspirations for future explorers of the brain.

We have organized the chapters into five themes that range from evolution to qualia, each theme posed as a general question. Although each of the chapters was independently written and can be read separately, there are some common issues that reflect contemporary concerns.

J. Leo van Hemmen
Terrence J. Sejnowski

Contents

Contributors

L. F. Abbott
Volen Center
Brandeis University
Waltham, MA 02454-9110
abbott@brandeis.edu

Amos Arieli
Department of Neurobiology
The Weizmann Institute of
  Science
P.O. Box 26
Rehovot 76100
Israel

André Brechmann
Leibniz-Institut für Neurobiologie
Brenneckestrasse 6
39118 Magdeburg
Germany
andre.brechmann@
  ifn-magdeburg.de

Jean Bullier
Centre de Recherche Cerveau et
  Cognition
CNRS-UPS UMR 5549
Université Paul Sabatier
133, route de Narbonne

F-31062 Toulouse Cedex
France

C. E. Carr
Department of Biology
University of Maryland
College Park, MD 20742-4415
cecarr@umd.edu

Francis C. Crick
The Salk Institute
10010 North Torrey Pines Road
La Jolla, CA 92037

Günter Ehret
Abt. Neurobiologie
Universität Ulm
89069 Ulm
Germany
guenter.ehret@
  biologie.uni-ulm.de

David J. Field
Department of Psychology
Uris Hall
Cornell University
Ithaca, NY 14853-7601
djf3@cornell.edu

Bernhard Gaese
Zoologisches Institut
Universität Frankfurt
Siesmayerstrasse 70
60323 Frankfurt am Main
Germany
gaese@zoology.uni-frankfurt.de

Vittorio Gallese
Institute of Human Physiology
University of Parma
Italy
vittorio.gallese@unipr.it

Wulfram Gerstner
Ecole Polytechnique Fédérale de
   Lausanne
School of Computer and
   Communication Sciences
and
Brain-Mind Institute
CH-1015 Lausanne
Switzerland
Wulfram.Gerstner@epfl.ch

Amiram Grinvald
Department of Neurobiology
The Weizmann Institute of Science
P.O. Box 26
Rehovot 76100
Israel
Amiram.Grinvald@weizmann.ac.il

J. Leo van Hemmen
Physik Department
TU München
85747 Garching bei München
Germany
LvH@ph.tum.de

Andreas V. M. Herz
Institut für Theoretische Biologie
Humboldt Universität zu Berlin
Invalidenstrasse 43

10115 Berlin
Germany
a.herz@biologie.hu-berlin.de

Andreas Hess
Leibniz-Institut für Neurobiologie
Department of Auditory Plasticity
   and Speech
Brenneckestrasse 6
39118 Magdeburg
Germany
andreas.hessnospam@nospamucd.ie

Edward M. Hubbard
Brain and Perception Laboratory
University of California, San Diego
9500 Gilman Dr. 0109
La Jolla, CA 92093-0109
and
SNL-B
Salk Institute for Biological Studies
10010 North Torrey Pines Road
La Jolla, California 92037-1099
edhubbard@psy.ucsd.edu

S. Iyer
Department of Biology
University of Maryland
College Park, MD 20742-4415

S. Kalluri
Department of Biology
University of Maryland
College Park, MD, 20742-4415
and
Institute for Systems Research
University of Maryland
College Park, MD 20742

Tal Kenet
Department of Neurobiology
The Weizmann Institute of Science
P.O. Box 26
Rehovot 76100
Israel

Georg M. Klump
AG Zoophysiologie & Verhalten
IBU, Fakultät V
Carl von Ossietzky Universität
Oldenburg
26111 Oldenburg
Germany
Georg.Klump@uni-oldenburg.de

Christof Koch
Division of Biology, 139-74
California Institute of Technology
1200 E. California Blvd.
Pasadena, CA 91125
koch@klab.caltech.edu

Gilles Laurent
Division of Biology, 139-74
Caltech
1200 E. California Blvd.
Pasadena, CA 91125
laurentg@caltech.edu

David McAlpine
Department of Physiology
University College London
Gower Street
London WC1E 6BT
United Kingdom
d.mcalpine@ucl.ac.uk

Frank W. Ohl
Leibniz-Institut für Neurobiologie
Department of Auditory Plasticity
   and Speech
Brenneckestrasse 6
39118 Magdeburg
Germany
frank.ohl@ifn-magdeburg.de

Bruno A. Olshausen
Redwood Center for
   Theoretical Neuroscience
University of California, Berkeley
Helen Wills Neuroscience Institute
132 Barker, MC #3190
Berkeley, CA 94720-3190
baolshausen@ucdavis.edu

Alan R. Palmer
Department of Physiology
University College London
Gower Street
London WC1E 6BT
United Kingdom

V. S. Ramachandran
Department of Psychology
Univerersity of California,
   San Diego
La Jolla, CA 92093
vramacha@ucsd.edu

John H. Reynolds
The Salk Institute
10010 N. Torrey Pines Road
P.O. Box 85800
La Jolla, CA 92037
reynolds@salk.edu

Giacomo Rizzolatti
Institute of Human Physiology
University of Parma
Parma
Italy
giacomo.rizzolatti@unipr.it

Henning Scheich
Leibniz-Institut für Neurobiologie
Department of Auditory Plasticity
   and Speech
Brenneckestrasse 6
39118 Magdeburg
Germany
scheich@ifn-magdeburg.de

Holger Schulze
Leibniz-Institut für Neurobiologie
Department of Auditory Plasticity
   and Speech
Brenneckestrasse 6
39118 Magdeburg
Germany

Terrence J. Sejnowski
Howard Hughes Medical Institute
Computational Neurobiology
  Laboratory
Salk Institute for Biological
  Studies
La Jolla, CA 92037
and
Division of Biological Science
University of California, San Diego
La Jolla, CA 92093
terry@salk.edu

S. Murray Sherman
Department of Neurobiology,
  Pharmacology, and Physiology
The University of Chicago
947 E. 58th Street
MC 0926, 316 Abbott
Chicago, IL 60637
msherman@bsd.uchicago.edu

J. Z. Simon
Institute for Systems Research
University of Maryland
College Park, MD 20742

D. Soares
Department of Biology
University of Maryland
College Park, MD, 20742-4415
daph@wam.umd.edu

Misha Tsodyks
Department of Neurobiology
The Weizmann Institute of Science
P.O. Box 26
Rehovot 76100
Israel
misha@weizmann.ac.il

C. van Vreeswijk
Neurophysique et Physiologie du
  Système Moteur
UMR 8119 CNRS
Universit René Descartes
45 rue des Saints Pères
75270 Paris, Cedex 06
France
Carl.Van-Vreeswijk@biomedicale.
univ-paris5.fr

Hermann Wagner
Institut für Biologie II
RWTH Aachen
Lehrstuhl für Zoologie/
  Tierphysiologie
Kopernikusstrasse 16
52074 Aachen
Germany
wagner@bio2.rwth-aachen.de

Laurenz Wiskott
Institut für Theoretische Biologie
Humboldt Universität zu Berlin
Invalidenstrasse 43
10115 Berlin
Germany
wiskott@itb.biologie.hu-berlin.de

Steven W. Zucker
Department of Computer Science
Yale University
P.O. Box 208285
New Haven, CT 06520
zucker@cs.yale.edu

PART I

HOW HAVE BRAINS EVOLVED?

•

1

Shall We Even Understand the Fly's Brain?

Gilles Laurent

Why This Obsession with Cortex?

Integrative neuroscience is an odd biological science. Whereas most biologists would now agree that living organisms share a common evolutionary heritage and that, as a consequence, much can be learned about complex systems by studying simpler ones, systems neuroscientists seem generally quite resistant to this empirical approach when it is applied to brain function. Of course, no one now disputes the similarities between squid and macaque action potentials or between chemical synaptic mechanisms in flies and rats. In fact, much of what we know about the molecular biology of transmitter release comes from work carried out in yeast, which obviously possesses neither neurons nor brain. When it comes to computation, integrative principles, or "cognitive" issues such as perception, however, most neuroscientists act as if King Cortex appeared one bright morning out of nowhere, leaving in the mud a zoo of robotic critters, prisoners of their flawed designs and obviously incapable of perception, feeling, pain, sleep, or emotions, to name but a few of their deficiencies. How nineteenth century!

I do not dispute that large, complex systems such as mammalian cerebral cortices have their own idiosyncrasies, both worthy of intensive study and critical for mental life. Everything in nature is worth studying, including, obviously, the id and *le moi*. Yet considering our obsession with things cortical, can we say that we have, in over forty years, figured out how the visual cortex shapes the responses of simple and complex cells? Do we really understand the cerebellum? Do we even know what a memory is? Do we understand the simplest forms and mechanisms of pattern recognition, classification, or generalization? I believe that our hurried drive to tackle these immensely complicated problems using the most complex neuronal systems that evolution produced—have you ever looked down a microscope at a small section of Golgi-stained cerebral cortex?—makes little sense.

Why this resistance to the use of simpler systems? I see two main reasons. The first is subjective: we are dealing with the brain, which defines us humans; there is something in us that probably resists the idea of shared principles with "lower creatures." (Consider how surprised gene sequencers claim to have been when they found that Craig Venter has only twice as many genes as *Drosophila*!) The second is a misinterpretation of objective results or more precisely a lack of appropriate abstraction. We are, at this time, still in the midst of cataloging brain parts. These parts range between atoms within proteins and brain areas involved in so-called higher cognitive processes. We sequence, we look for homologies, we give names, we compare, but in the area of systems and computation we are still pretty poor at defining an appropriate level of description that transcends those physical details. For example, the main excitatory neurotransmitter in the central nervous system (CNS) of mammals is glutamate. In insects, it is acetylcholine. Does this matter? No, because either mechanism provides equivalent means for postsynaptic excitation over many timescales, using ionotropic and metabotropic receptors. The flow of activity within and across networks is, as far as we know, similar with either combination of neurotransmitter and receptors. On the other hand, the fact that glutamate ended up being used in the mammalian CNS enabled the exploitation of receptor subtypes with interesting nonlinear properties (e.g., N-methyl-D-aspartate [NMDA] receptors) for a variety of fundamental tasks such as rhythm generation or learning. We have no evidence, so far, for an insect cholinergic receptor whose voltage sensitivity and permeability could make it a functional equivalent of the vertebrate NMDA receptor. It could still be that the functions played by the NMDA receptor in the vertebrate CNS are implemented differently in other animals (by mechanisms that may not be as compact, that may have a different molecular identity, but that accomplish a similar input/output transformation).

In short, we are still much too attached to empirical descriptions, names, and molecular details to abstract, from the study of particular examples, principles of function that can then be compared across systems or used as models for the study of more complex brains. This situation, however, can easily be improved. First, we should strive to identify, within the raw complexity and idiosyncrasy of experimental data, the underlying functional principles. Second, we should be open to the possibility that such principles may, through different implementations, be at work equally in small and large brains. The remainder of this chapter focuses on work on olfactory systems. I hope to illustrate two main things: the first is that small systems, and particularly small olfactory systems, seem to use mechanisms and strategies that are not unique to them. The second is that small systems are not at all that simple; this reinforces my view that we may be better off starting with the modest goal of understanding flies first. Will the next century be sufficient? I am not so sure.

The Olfactory Brain as a System to Identify Rules of Potentially General Relevance

Olfactory systems have evolved over millions of years to solve a variety of object identification problems. Some of these are relatively simple (e.g., identification of CO_2 or oligomolecular pheromones). If these signals and their receptors coevolved, as is likely, tight recognition by individual receptors might enable the animal to identify the molecule, measure its local abundance, and track it through the use of olfactory "labeled lines" (Hansson 1999; Kauer and White 2001). Many olfactory problems, however, are more complex: they involve odors composed of multimolecular mixtures (sometimes containing hundreds of volatile components; Knudsen, Tollsten, and Bergstrom 1993). Odor perception tends to bind together rather than segment the elements of a mixture (Laing 1991; Chandra and Smith 1998; Livermore and Laing 1996): the olfactory system thus recognizes odors as patterns. Added complexity arises because an odor's precise composition often varies along the lifetime of the odor source; fluctuations can be due to noise or to processes such as oxidation or differential volatility of the analytes in a mixture. In addition, the biological chemistry of odor formation (e.g., flower scents) leads to the formation of mixtures of chemically related elements (e.g., citrus essences). Odor clusters can be defined qualitatively (with many degrees of resolution: e.g., "aromatic" → "minty" → spearmint) or quantitatively (Duchamp-Viret and Duchamp 1997; Friedrich and Laurent 2001), noting that concentration changes can also lead to changes in perceived quality. Human psychophysics reveals that such clusters can be identified perceptually (Laing 1991). I will therefore assume that through evolution the generalist olfactory system found solutions to these pattern recognition tasks in which the space of possible signals (perceptually definable odors) is immense and not smoothly occupied. The magic of olfactory perception is that the brain can achieve cluster separation at (seemingly) many levels of resolution, enabling both gross classification and precise identification. How does it do it?

The olfactory bulb (OB) and its insect analog, the antennal lobe (AL), are highly interconnected circuits in which inhibition is physically widespread owing to the projections of either principal or local neurons. Electrophysiological experiments have revealed many forms of temporal patterning of activity of their output elements, the mitral cells (MCs, in the OB) or projection neurons (PNs, in the AL) (Friedrich and Laurent 2001; Wellis, Scott, and Harrison 1989; Burrows, Boeckh, and Esslen 1982; Meredith 1986; Buonviso, Chaput, and Berthommier 1992; Laurent, Wehr, and Davidowitz 1996; Macrides and Chorover 1972; Yokoi, Mori, and Nakanishi 1995; Motokizawa 1996; Adrian 1942; Laurent and Davidowitz 1994). The function of this patterning, seen equally in species in which MCs and PNs are multiglomerular (e.g., lower vertebrates and some insects; Kauer and Moulton 1974; Friedrich and Laurent 2001; Laurent, Wehr, and Davidowitz 1996) and in ones where output neurons are mainly uniglomerular (mammals and other insects; Wellis, Scott, and Harrison 1989; Burrows, Boeckh, and Esslen 1982; Meredith 1986; Buonviso, Chaput, and Berthommier 1992; Laurent, Wehr, and Davidowitz 1996; Macrides and Chorover 1972; Yokoi, Mori, and Nakanishi 1995; Motokizawa 1996; Spors and Grinvald 2002), is unclear. Exploiting the small sizes of insect (locust, honeybee) and zebrafish brains, we have tried to address this issue from a systems perspective rather than a purely cellular or anatomical one; this approach reveals interesting computations that might otherwise remain undetected.

Working Hypothesis

We hypothesize that, because of the complexity of the olfactory pattern recognition problem (size and landscape of odor space, noisiness of odors), the brain exploits circuit dynamics to accomplish at least two objectives. The first is to create, through spatiotemporal patterns of neuronal activation, a very large coding space in which to spread representation clusters. The large size of this representation space is a consequence of the number of possible spatiotemporal combinations. The goal, though, is not to enable the storage of an infinite number of items; rather, it is to ease the handling of a smaller number of often unpredictable items that the animal will, in its lifetime, need to store and recall. The second objective is to use distributed dynamics both to confer stability on each representation in the face of noise and to optimize the filling of the representation space. In this chapter, I will present evidence that the first olfactory relay can, in two parallel operations, increase the separation between the representations of chemically related odors (decorrelation through slow dynamics) and format those representations so that they can be sparsened in the next station (exploiting oscillatory synchronization). I propose, therefore, that the OB and AL are "encoding machines" that actively transform a distributed, multidimensional afferent input to enable the formation of compact and easily recalled memories.

The OB and AL as Decorrelators

Slow Patterns

Single or multiple simultaneous recordings from zebrafish MCs (Friedrich and Laurent 2001) and insect PNs (Laurent, Wehr, and Davidowitz 1996) indicate, as shown in other species (Burrows, Boeckh, and Esslen 1982; Meredith

1986; Buonviso, Chaput, and Berthommier 1992; Laurent, Wehr, and Davidowitz 1996; Macrides and Chorover 1972; Stopfer et al. 1997), that principal neuron responses are not static. Rather, individual neurons respond with characteristic epochs of increased and decreased firing that are both neuron- and odor-specific. An example is shown (figure 1-1) of a locust PN and its response patterns to sixteen different airborne odors (from Perez-Orive et al. 2002). Similar patterning was seen across the responses of zebrafish MCs to many amino acids (Friedrich and Laurent 2001). Because not all responding neurons express the same patterns at the same time, the population representation is dynamic, carried by an assembly of neurons (MCs or PNs) that evolves in a stimulus-specific manner over time (figure 1-1b). In locusts, this evolution can be tracked along a periodic local field potential (20–30 Hz) caused by the synchronized periodic firing and updating of the participating neurons (Wehr and Laurent 1996). Local field potential oscillations in the same frequency range are also seen in fish, although their development during a response generally lags behind peak MC activity (Friedrich and Laurent 2001).

Decorrelation

Successive time epochs in a sustained response contain different but stimulus-specific assemblies of active projection cells. What are the functional consequences? In zebrafish, we found that spatiotemporal patterning results in a rapid decorrelation of odor representations (Friedrich and Laurent 2001); for this to be uncovered, however, representations must be considered across MC assemblies. Decorrelation means that the overlap between the representations of related odors (e.g., several aromatic amino acids) decreases over time, corresponding to divergent redistributions of activity over time across the MC array (Friedrich and Laurent 2001). Representation size remained constant on average. The trajectory followed by each stimulus-evoked evolution was reliable from trial to trial, and short response segments late in a response were more reliable than early ones for stimulus identification (Friedrich and Laurent 2001). Early epochs offered reliable clues for odor classification; later ones allowed for precise stimulus identification.

Mechanisms and Possible Formal Principles

The mechanisms underlying this slow population patterning involve at least interactions within the OB (or the AL) because afferent output shows no odor- or ORN-specific patterning and no decorrelation over time (Friedrich and Korsching 1997; Friedrich and Laurent 2001). In locusts, these mechanisms appear to be independent of fast inhibition in the AL (MacLeod and Laurent 1996) and do not involve feedback from downstream areas. Slow patterning

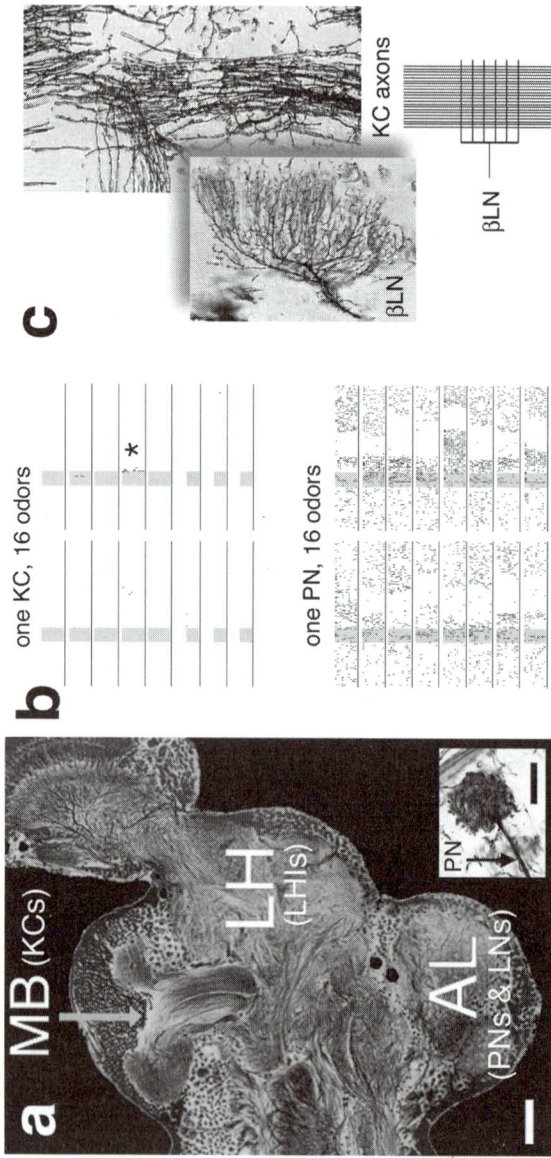

Figure 1-1. The locust olfactory circuits and the transformation of response properties between first and second relay. See color insert.

in the locust AL thus results from both slow inhibition (the mediation of which is still not understood) and, possibly, lateral excitatory interactions within the AL. A computational model of the AL and its connections was used to explore the minimum cellular, synaptic, and network requirements for generating realistic population dynamics (Bazhenov et al. 2001b; Bazhenov et al. 2001a). This revealed that distributed dynamics similar to those observed arise naturally in networks with realistic slow synapses and distributed lateral connections. A more abstract approach with smaller networks was used to explore fundamental aspects of these dynamic phenomena (Rabinovich et al. 2001). This chapter proposes, within the framework of nonlinear dynamical systems theory, a chaotic regime called "winner-less competition" (WLC; Rabinovich et al. 2001), in which the responding population follows an orbit that links unstable states. Orbits are highly sensitive to input, explaining the am- ORBITS? plification over time of small input differences; critically, however, they are stable, such that the population trajectory is resistant to noise in the participating neurons, explaining in part the trial-to-trial reliability of population patterns despite probabilistic responses in each neuron and epoch (Rabinovich et al. 2001). Qualitatively, each odor is represented by a constantly changing assembly in which each active neuron both participates in the dynamics of the others and benefits from the global stability of the assembly, preventing large individual response deviations. The global fate of the system is determined by the input pattern. Although the link between WLC dynamics and experimental observations needs to be strengthened, this approach provides a simplified framework for exploring olfactory responses, their causes (e.g., asymmetric inhibitory coupling), and their computational consequences (e.g., representation optimization and stability).

Oscillatory Synchronization and Sparse Representations

Oscillatory synchronization in the olfactory system was first described with electroencephalograph and local field potential (LFP) recordings in mammals (Adrian 1942). It has since been found in most other systems (e.g., visual, auditory, somatosensory, motor; Gray 1994; Engel, Fries, and Singer 2001), including other olfactory systems (Gray 1994; Gelperin and Tank 1990; Laurent and Naraghi 1994; Laurent and Davidowitz 1994). We are attempting to provide a high-resolution description of the cellular, synaptic, and circuit events underlying these oscillatory LFPs (Laurent, Wehr, and Davidowitz 1996; Wehr and Laurent 1996; MacLeod and Laurent 1996; Bazhenov et al. 2001a, 2001b) in insects and to test their functional relevance (Stopfer et al. 1997; MacLeod, Bäcker, and Laurent 1998). This system offers the prospect of understanding encoding, decoding, and functional/behavioral aspects of periodic and synchronized activity in a brain area.

Causes and Behavioral Relevance

Oscillatory synchronization in the locust AL arises through the action of local inhibitory (GABAergic) neurons (LNs) with widespread output to other LNs and to the PNs. Synchronization can be blocked by Cl^- channel blockers infused locally into the AL (MacLeod and Laurent 1996). Blocking fast inhibition, however, leaves untouched the slow inhibition that is important for generating slow response patterning (MacLeod and Laurent 1996). Consequently, the global, patterned PN population output can be maintained while disrupting periodic synchronization (MacLeod and Laurent 1996). This dichotomy of inhibitory actions allowed us to establish the relevance of oscillatory synchronization, using both behavioral (in honeybees; Stopfer et al. 1997) and physiological (in locusts; MacLeod and Laurent 1996; Laurent et al. 2001) assays.

Hidden Activity

If oscillations are functionally relevant, how and why are they useful? Let us first examine what an oscillatory LFP indicates. An LFP is a weighted average of local potential fluctuations caused by events (in our case, mainly synaptic currents) occurring in the vicinity of the sampling site. Oscillations in the LFP suggest that groups of neurons with outputs close to the electrode tend to fire in common periodic epochs. The LFP does not, however, reveal any of the detail of the activity that causes it. For example, in the simplest of cases it could be caused by a subgroup of neurons that all fire together, periodically and precisely in phase throughout the epoch of oscillation. Alternatively, it could be caused by a neuronal group whose members change as the response progresses and, when they fire, do so in the proper phase. It could be caused also by a dynamic group of cells among which some, but not all, phase-lock to each other. Because the LFP is a mean, the influence of the phase-locked neurons (even if they are few) on the LFP waveform can be greater than that of the neurons that fire independently. Finally, LFP oscillations could also result from a dynamic assembly in which the active neurons can produce both locked and nonlocked spikes at different times of the response. Although seemingly baroque, this is what occurs in the locust AL (Laurent, Wehr, and Davidowitz 1996; Wehr and Laurent 1996; Laurent et al. 2001): because PN output is temporally patterned (see the section "The Olfactory Brain as a System to Identify Rules of Potentially General Relevance"), not all PNs fire in the same epochs; when individual PN firing events collected over many trials are compared with the LFP, one observes that the spikes produced by individual PNs in some epochs of a response tend to be locked, while spikes produced earlier or later by the same PNs are not (Laurent, Wehr, and Davidowitz 1996). These epochs of locking are different for different PNs and for different stimuli. Possible reasons for this conditional locking are suggested by modeling experiments (Bazhenov et al. 2001a, 2001b): the strength of locking of any PN spike is correlated with the number of presynaptic LNs that are active in the short period preceding that spike. Consequently, the epochs during which a PN's spikes are locked to the LFP may be determined continuously by the instantaneous state of the network and of the LNs presynaptic to that PN. These observations matter for several reasons. First, they indicate that detecting pair-wise correlations between any two neurons can be very difficult: it is easy to miss those few oscillation cycles when the two examined PNs fire synchronously (Laurent, Wehr, and Davidowitz 1996; Wehr and Laurent 1996). Realizing the transient nature of pair-wise synchronization is crucial to understand the decoding of these signals (see the section, "Decoding," below). Second, they indicate that LFP oscillations arise from a large number of spikes, of which only a fraction are locked at any one cycle. This also is important because, as we will see, the decoding circuits will not react to all spikes equally. The overall PN output during an odor response is thus a complex distributed pattern in which PN spikes can be found at any time but with a bias toward some periodic epochs imposed by collective LN activity. The cells that are active together at different cycles change throughout a response and the spatiotemporal patterns differ for different odors.

Decoding

How is all this decoded? Recent intracellular and tetrode recordings from Kenyon cells (KCs), the intrinsic neurons of the mushroom body (MB), indicate a major transformation of representations between the AL and the MB (Perez-Orive et al. 2002; figure 1-1). PN responses are long lasting, patterned, transiently locked, highly probable when tested over a set of ~20 randomly selected odors, and superimposed on a baseline firing rate of ~4 spikes/s. By contrast, most KC responses are extremely brief (~2 spikes), consequently unpatterned, locked to the LFP, highly improbable over the same odor sets, and superimposed on a baseline-firing rate of 0.005 to 0.025 spike/s (Perez-Orive et al. 2002). Figure 1-1 shows the responses of a typical PN and KC to the same set of sixteen odors. The information content of a KC spike is clearly much higher than for a typical PN. Because KC responses are rare and because the MB contains many more KCs than there are PNs (50,000 vs. 830), odor representations in the MB are sparse (Perez-Orive et al. 2002). The MB thus seems to be sparsening (in space and in time) odor representations.

How is this accomplished? KCs and the circuits surrounding them act as coincidence detectors on the dynamic PN input (Perez-Orive et al. 2002). This results from several cooperating sets of features. First, the olfactory input to KCs is a complex pattern of PN firings distributed in space and in time. Second, anatomy indicates that individual PNs diverge, on average, to ~600 KCs. Third, given this fan-out ratio, the number of PNs (830), and the number of KCs connected to PNs (25,000–50,000), each KC must receive convergent input from 10–20 PNs on average. Fourth, KC responses to PN spikes can be amplified by voltage-dependent nonlinearities that also shorten excitatory postsynaptic potentials (EPSPs) when the input causing them is strong enough (Perez-Orive et al. 2002; Laurent and Naraghi 1994). KCs will thus summate EPSPs preferentially if their timing is synchronized. Fifth, a short feed-forward circuit through inhibitory lateral horn interneurons (LHIs) produces inhibitory postsynaptic potentials (IPSPs) onto KCs that are out of phase with the EPSPs caused by synchronized PNs (Perez-Orive et al. 2002). Because individual LHIs respond to most odors, because LHIs are few (~60), and because they diverge extensively in the MB, they can collectively inhibit all KCs during half of every oscillation cycle caused by any odor. This ensures that KCs can summate PN input only briefly and periodically during the other half of each cycle.

How, then, do KCs respond at all? This is explained by the limited PN convergence onto any KC, by the transient nature of the PN output during odor stimulation, and by a presumably high KC firing threshold. Only when a sufficiently high proportion of the PNs presynaptic to a KC fire synchronously does that KC fire an action potential. The brevity of the KC response could be explained by two observations: first, a large and long-lasting afterhyperpolarization follows each KC spike (Laurent and Naraghi 1994), making EPSP summation less effective and further firing unlikely; second, the evolving nature of the PN output ensures that, within a few cycles, the set of coactive PNs has changed. Sparsening thus results from an asymmetrical influence of periodic excitation and inhibition on each KC: excitation is highly specific while inhibition is not. These results show that neurons can act as coincidence detectors (Abeles 1982; König, Engel, and Singer 1996) and how oscillatory synchronization plays a critical role in this computation.

Significance

Oscillatory synchronization and a set of appropriately tuned ancillary mechanisms can in one step convert a dense, distributed, and redundant stimulus representation into a sparse one. Yet, are oscillations necessary? I would argue that shaping synthetic and specific responses may not be easy, especially in only one step. Synthetic tuning implies the convergence of many inputs onto one neuron. If that neuron must respond only to the coactivation of all (or most) of its converging inputs, it must be able to ignore a pattern in which only a subset of these inputs are vigorously active but respond when all inputs are equally active. In other words, it must be able to select against temporal summation and for spatial summation of (coincident) input: input synchronization and active shortening of the integration window, as found here, is a solution to this problem. It will be interesting to determine whether other solutions—ones that do not use synchrony—are equally efficient.

There are practical considerations. The detection of sparse representations, when they exist, can be difficult, simply because spikes may be extremely rare. If there is no independent reason to suspect the existence of such responses (e.g., intracellular data indicating stimulus-related subthreshold activity), it is easy to miss these few, highly informative action potentials. KC action potentials are highly significant, however, only because they ride on a very low baseline rate. Hence, mechanisms must be invested to secure the contrast between response and no-response. This mode of representation therefore also has a cost that it would be interesting to estimate.

Sparsening has many advantages, especially if it occurs in a structure implicated in learning (such as the MB). As well as reducing overlaps, sparse representations could facilitate storage (fewer synapses need modifying), pattern matching (fewer elements need to be compared), and pattern association: the different attributes of a percept should, in principle, be more easily associated if they require the linking of fewer neuronal elements. Conversely, by combining many converging inputs, specific neurons (KCs in our case) could contribute to the formation of the complex associations that underlie perceptual binding (Gray 1994; Engel, Fries, and Singer 2001; von der Malsberg and Schneider 1986). Sparse, synthetic representations are useful, but they eliminate the detail and segmentability of a representation (gestalt). This is consistent with behavioral and psychophysical observations in olfactory perception (Chandra and Smith 1998; Livermore and Laing 1996; Linster and Smith 1999). Another possible advantage of the phenomena we describe is that they are adaptive. The feed-forward inhibitory loop that sharpens KC tuning could also be seen as a compensatory mechanism for the sloppiness of the oscillatory clock: LFP oscillation frequency usually varies between 15 and 30 Hz from cycle to cycle (Laurent and Davidowitz 1994). Because feed-forward inhibition is locked, cycle by cycle, to each ongoing wave of excitation, delays or advances in the PN output are always compensated for adaptively. The formatting of PN output by the AL can therefore compensate for its own imprecision.

Finally, our results imply that not all spikes are alike: whether a PN spike succeeds in activating its targets will be determined by the timing of that spike relative to the timings of other spikes produced by other neurons around the same time. The relevant information content of a PN spike is thus determined by its temporal correlation with the spikes of other PNs that share the same targets; it cannot be measured meaningfully without the knowledge of these spatiotemporal relationships. Oscillatory synchronization can thus indicate a selective filtering of throughput.

Are Slow Dynamical Patterns Features of a Code?

So far, it seems that decoding of the AL output by KCs makes no explicit use of the dynamical features of PN responses: KCs do not seem to accomplish any kind of sequence decoding across the incoming PN input; rather, each KC selectively assesses the state of a small part of the PN assembly, one fraction of an oscillation cycle at a time, with no apparent memory of activity at previous cycles. For this reason, slow dynamics might appear irrelevant—yet we believe them to be essential. How and why? First, sequence decoding could be accomplished downstream of KCs, for example by extrinsic neurons in the alpha and beta lobes of the MB using unknown spatiotemporal integration mechanisms (Schidberger 1983; MacLeod, Bäcker, and Laurent 1998; Grünewald 1999; figure 1-1c). Alternatively, the relevance of slow dynamics in the AL might be implicit in the KC responses. We proposed that the AL/OB output is a self-organized process whose outcome becomes less ambiguous over time (Friedrich and Laurent 2001; Laurent et al. 2001): I would argue that the evolution of AL/OB patterns may need no decoding per se. In this perspective, dynamics are critical for the optimization of the code but need not be the code itself (i.e., a feature to be decoded); the complicated patterning we observe in AL/OB neurons may simply be part of the process through which the message's format is actively optimized for further processing (learning, association, recall) by downstream areas.

Decoding Temporal Sequences without Explicit Sequence Decoding

Pushing the reasoning further, this strategy might considerably simplify the read-out of spatiotemporal patterns. Consider odor A, represented in the periphery by a physical array of activated glomeruli. This representation overlaps with that of A', a related odorant. By imposing a temporal structure on this representation, the OB/AL unfolds the spatial patterns of A and A' and reduces overlaps. In locust, using the periodic output of the AL, the MB sparsens these representations at each oscillation cycle across a large assembly of KCs. Both transformations decrease the probability of overlap between representations. This implies that a simple postsynaptic integration of KC outputs (sensitive to the identity but not to the order of activation of the responding neurons) might suffice to separate A from A', even if these assemblies contain a few common elements, as a consequence of their relatedness or common root. (This would not be possible with PNs in the AL, especially if many different patterns need to be stored, because PN representations are dense.) Neurons in the alpha and beta lobes, whose dendrites sample the axons of hundreds to thousands of Kenyon cells (Schidberger 1983; MacLeod, Bäcker, and Laurent

1998; Grünewald 1999), could carry out this simple integration. In conclusion, one could imagine that slow temporal patterns, while critical for separation of representations, are never actually decoded as such. More generally, the creation of spatiotemporal representations by circuit dynamics might conceivably be a transient phase in signal processing, used simply to (1) spread out those representations in a larger coding space and (2) facilitate decoding (e.g., sparsening followed by conventional integration).

Circuit Dynamics as a Mechanism for Recall

Forming and storing representations (for odors as for any other feature) is clearly not an end in itself for the brain. Memories are formed because they may be needed later to help decision making and action. In other words, the format of what we call a response probably depends as much on the process that forms a representation as it does on the process that will lead to recall of that representation at a later time: when an animal explores and samples the world, it may not always choose between acquisition and recognition modes. These two operations must therefore be able to occur through the same machinery and the same process. Presumably, evolution applied some selective pressure on brain mechanisms that serve both pattern formation/memorization and recall equally well. Hence, we should consider the possibility that dynamics may be useful not only for representation but also for recognition. For example, reinforcing, through learning, the connections between the neurons that form sequences in a spatiotemporal pattern might facilitate reactivation of the sequence by a corrupted input and thus recognition. This idea (sometimes called *pointer chain*) is implicit in Marr's paper on cerebellum (1969) and developed in Kanerva's study (1988) of sparse distributed memories. Circuit dynamics may also be critical when the animal is actively looking for a particular feature: top-down influences may thus bias and facilitate the recognition of the searched item by more peripheral circuits (Freeman 2000). In brief, circuit dynamics may play roles that we do not understand or even suspect yet because our functional framework, defined by experimental constraints, is often much narrower than that in which brains normally operate.

The Problem with Noise

Much of the processing described exploits mechanisms that should, in principle, be very sensitive to noise. Input decorrelation, for example, requires an operation akin to the amplification of small input differences but not of ones that arise from natural, noisy fluctuations of the stimulus. Also, pattern encoding by KCs relies on rare but highly informative spikes that must not be polluted by spurious ones. How are these problems solved? We do not know, but there are some hints to the possible solutions. The convergence of many (in some cases thousands of) olfactory receptor neurons (ORNs) onto single glomeruli (and thus few output neurons; Axel 1995; Buck 1996; Mombaerts et al. 1996) and the distributed sprinkling of these ORNs on the receptive sheet (Buck 1996; Vassar et al. 1994), limiting the probability of correlated noise, could allow the averaging necessary to increase signal-to-noise ratios. In addition, slow and diffuse communication within individual glomeruli could contribute to averaging or adaptive gain control. This issue clearly needs careful attention. A second potential mechanism for noise reduction is a form of fast learning evidenced in locusts (Stopfer and Laurent 1999). AL circuits seem to undergo stimulus-specific modifications (the nature of which remains unknown) to the extent that successive responses of PNs to the same stimulus rapidly decrease in intensity but become more precise and coordinated with those of other PNs (Stopfer and Laurent 1999). Because olfaction is generally intermittent and because the detection of an odor at one time predicts the detection of the same odor in the very near future (odors rarely disappear suddenly), the AL circuits might operate at low detection threshold at "rest" (hence explaining the high responses in a naive state) but immediately "focus" on a signal once it has been detected (explaining the refined representation after just a few trials). If this form of learning exploits short-term changes in the synapses formed by the activated neurons, only those synapses that are repeatedly activated over successive samplings could be reinforced. In other words, unreliable contamination occurring at some but not all trials (i.e., the noise) would be averaged out. Third, network mechanisms (especially the connectivity matrix of OB/AL circuits) could play a critical role in ensuring stability in the collective output. Recall that the dynamic evolution of the OB/AL output is forced as long as the stimulus lasts (Rabinovich et al. 2001). During that time, the stimulus could define a state space in which the heteroclinic attractor (i.e., the spatiotemporal pattern of PN activity) is unique. There may thus exist particular rules of connectivity that confer stability on their dynamical evolution upon noisy stimulation. This is an area where theory, modeling, neuroanatomy, and physiology will all be necessary. Finally, the mechanisms ensuring that KCs fire only when they detect the right input combination probably need delicate fine-tuning of the balance between excitation and inhibition. Inhibitory feedback from the output of the MB to the dendrites of KCs is known to exist (Grünewald 1999). These pathways might contribute to an adaptive control of KC excitability and thus a more or less constant representation sparseness. In conclusion, the experimental results and proposed principles summarized here require mechanisms that can discriminate noisy from meaningful differences. A few promising candidate mechanisms exist that could ensure or help resistance to noise, but much work is needed in this critical domain.

The Problem with Decorrelation and Perceptual Clusters

While decorrelating input representations is useful in principle, it could also have undesirable effects. For example, the perceptual relatedness between odors (e.g., all citrusy smells) may be lost. Similarly, individual odors at different concentrations usually retain, at least over some range, the same perceptual identity. But if the patterns evoked by different concentrations of the same odor differ even slightly from one another, decorrelation would enhance those differences and possibly preclude their perceptual grouping. How is this potential conflict resolved? Again, we do not know yet but can suggest several possible solutions, each of which needs exploring. The first is that perceptual grouping could be a high-level, learned property. In that scheme, input patterns that end up in the same perceptual group (e.g., several concentrations of jasmine) do not actually evoke related patterns after decorrelation. But because the animal experiences all these concentrations over a same given epoch (e.g., while visiting a given cluster of flowers) and because that period is equally meaningful to all patterns experienced then (e.g., they are all associated with a reward), all patterns evoked by the different concentrations are lumped, downstream of the circuits doing the decorrelation, as "meaning" the same thing (e.g., jasmine). This is a high-level grouping by contingency. A second possibility uses the fact that decorrelation is a temporal process; hence, early phases of a representation (say the first 100 milliseconds) are very similar across related stimuli (Friedrich and Laurent 2001). Provided the brain can hold that fleeting information, it could use it for perceptual grouping or, conversely, ignore it for precise identification using decorrelated patterns. This thus supposes several read-out streams—for example, one for early patterns and one for the entire pattern—and a top-down decision system on what stream to listen to. A third hypothesis is that network dynamics and sparsening never completely orthogonalize representations. In this scheme, the dynamics would be designed such that the spacing between representation clusters increases faster than the spacing between representations within a cluster. Hence, patterns that are similar (e.g., different concentrations or related chemicals) remain more similar to one another than to any random pattern. This type of biased decorrelation might require particular circuit architectures.

General Conclusion

In conclusion, much integrative work is needed to understand the computational organization of olfactory systems. We propose a systems perspective that is based on experimentation with small olfactory brains; exploiting their relative simplicity, we showed that circuit dynamics over multiple timescales and correlation rules play an integral role in optimizing stimulus representations. Dynamical formatting might be a transient phase in the processing of these signals: once representations have been optimized, their apparent initial complexity could be reduced to simple and specific responses carried by few neurons. Circuit dynamics may have other advantages, for example, for memory recall, but experimental support is so far lacking. I also think that knowing better how natural stimuli are distributed within odor space and what the multiple tasks required of olfactory systems are (e.g., learning, recognition, and classification) will help us better understand why olfactory computations are the way they are. I would argue that a traditional, passive "stimulus-response" view of sensory processing hinders our understanding of seemingly complicated modes of operation. In other words, understanding the computations taking place in a circuit can be difficult if we fail to consider individual neurons as parts of a system in action. Our lexicon introduces subtle but real biases in our thinking about sensory processing. Our predisposition as sensory physiologists is to call responses the spike patterns that follow a stimulus; we use these responses to then define "receptive fields." In doing so, we forget that those terms are only meant to be operational. Our thinking about sensory integration seems much too linear and passive: stimulus a → response in area x → response in area y, and so on; in reality, neural circuits are often massively interconnected and reciprocally connected; similarly, our thinking generally ignores the fact that, except for motoneurons, a given neuron never is an end point or its "response" an end product. Hence, how a neuron behaves may be relevant not as a response per se (i.e., something to be analyzed by us to estimate the information it contains about a stimulus, although this is, of course, useful knowledge) but as part of a transformation (possibly extremely complex and distributed) to help further processing (e.g., optimization, storage, recognition, and retrieval) in the area in which the neuron lies (e.g., decorrelation in OB circuits) or in "target" circuits. Thinking about sensory integration in these active terms (i.e., considering "responses" not only as products, but also as ongoing transformations toward some other goal) might be helpful to understand some brain operations. It will be interesting to see whether the principles and mechanisms we propose apply to other, including mammalian, olfactory systems and possibly also to other brain systems involved in the processing of multidimensional signals (e.g., vision; Dong and Atick 1995, Dan, Atick, and Reid 1996; Vinje and Gallant 2000, or action; Bergman and Bar-Gad 2001).

Acknowledgments The work from my laboratory reviewed here was funded by the NSF, NIDCD, and the McKnight, Keck, Sloan, and Sloan-Swartz foundations. I am grateful to Mark Stopfer, Rainer Friedrich, Katrina McLeod, Michael Wehr, Javier Perez-Orive, Ofer Mazor, Stijn Cassenaer, Rachel Wilson, Glenn Turner,

Christophe Pouzat, Vivek Jayaraman, Sarah Farivar, Hanan Davidowitz, Roni Jortner, Alex Holub, Misha Rabinovich, Henry Abarbanel, Ramon Huerta, Thomas Nowotny, Valentin Zighulin, Alejandro Bäcker, Maxim Bazhenov, Pietro Perona, and Erin Schuman for the privilege of working on these problems with them. I thank Peter Cariani for pointing me to Kanerva's book on sparse distributed memories and Karen Heyman for secretarial assistance.

References

Abeles, M. 1982. Role of the cortical neuron, integrator or coincidence detector? *Isr. J. Med. Sci.* 18: 83–92.

Adrian, E. 1942. Olfactory reactions in the brain of the hedgehog. *J. Physiol. (Lond.)* 100: 459–473.

Axel, R. 1995. The molecular logic of smell. *Sci. Am.* 273: 130–137.

Bazhenov, M., M. Stopfer, M. Rabinovich, R. Huerta, H. D. I. Abarbanel, T. J. Sejnowski, and G. Laurent. 2001a. Model of cellular and network mechanisms for temporal patterning in the locust antennal lobe. *Neuron* 30: 569–581.

Bazhenov, M., M. Stopfer, M. Rabinovich, R. Huerta, H. D. I. Abarbanel, T. J. Sejnowski, and G. Laurent. 2001b. Model of transient oscillatory synchronization in the locust antennal lobe. *Neuron* 30: 553–567.

Bergman, H., and I. Bar-Gad. 2001. Stepping out of the box, information processing in the neural networks of the basal ganglia. *Curr. Opin. Neurobiol.* 11: 689–695.

Buck, L. B. 1996. Information coding in the vertebrate olfactory system. *Annu. Rev. Neurosci.* 19: 517–544.

Buonviso, N., M. A. Chaput, and F. Berthommier. 1992. Temporal pattern analysis in pairs of neighboring mitral cells. *J. Neurophysiol.* 68: 417–424.

Burrows, M., J. Boeckh, and J. Esslen. 1982. Physiological and morphological properties of interneurons in the deutocerebrum of male cockroaches with responses to female pheromones. *J. Comp. Physiol. A* 145: 447–457.

Chandra, S., and B. H. Smith. 1998. Analysis of synthetic processing of odour mixtures in the bee (Apis mellifera). *J. Exp. Biol.* 201: 3113–3121.

Dan, Y., J. J. Atick, and R. C. Reid. 1996. Efficient coding of natural scenes in the lateral geniculate nucleus, Experimental test of a computational theory. *J. Neurosci.* 16: 3351–3362.

Dong, D. W., and J. J. Atick. 1995. Temporal decorrelation—a theory of lagged and nonlagged responses in the lateral geniculate-nucleus. *Network-Comp. Neural* 6: 159–178.

Duchamp-Viret, P., and A. Duchamp. 1997. Odor processing in the frog olfactory system. *Prog. Neurobiol.* 53: 561–602.

Engel, A. K., P. Fries, and W. Singer. 2001. Dynamic predictions, oscillations and synchrony in top-down processing. *Nature Reviews Neurosci.* 2: 704–716.

Freeman, W. J. 2000. *Neurodynamics: An Exploration in Mesoscopic Brain Dynamics.* London, Springer-Verlag.

Friedrich, R., and G. Laurent. 2001. Dynamical optimization of odor representations in the olfactory bulb by slow temporal patterning of mitral cell activity. *Science* 291: 889–894.

Friedrich, R. W., and S. I. Korsching. 1997. Combinatorial and chemotopic odorant coding in the zebrafish olfactory bulb visualized by optical imaging. *Neuron* 18: 737–752.

Gelperin, A., and D. W. Tank. 1990. Odour-modulated collective network oscillations of olfactory interneurons in a terrestrial mollusc. *Nature* 345: 437–40.

Gray, C. 1994. Synchronous oscillations in neuronal systems, mechanisms and function. *J. Comput. Neurosci.* 1: 11–38.

Grünewald, B. 1999. Morphology of feedback neurons in the mushroom body of the honeybee, *Apis mellifera. J. Comp. Neurol.* 404: 114–126.

Kanerva, P. 1988. *Sparse Distributed Memory.* Cambridge, Mass.: MIT Press.

Kauer, J. S., and D. Moulton. 1974. Responses of olfactory bulb neurones to odour stimulation of small nasal areas in the salamander. *J. Physiol. (Lond.)* 243: 717–737.

Kauer, J. S., and J. White. 2001. Imaging and coding in the olfactory system. *Annu. Rev. Neurosci.* 24: 963–979.

Knudsen, J. T., L. Tollsten, and L. G. Bergstrom. 1993. Floral scents—a checklist of volatile compounds isolated by headspace techniques. *Phytochemistry* 33: 253–280.

König, P., A. K. Engel, and W. Singer. 1996. Integrator of coincidence detector? The role of the cortical neuron revisited. *Trends Neurosci.* 19: 130–137.

Laing, D. G. 1991. Characteristics of the human sense of smell when processing odor mixtures. In *The Human Sense of Smell*, ed. D. G. Laing, R. L. Doty, and W. Breipohl, 241–259. Berlin: Springer-Verlag.

Laurent, G., and H. Davidowitz. 1994. Encoding of olfactory information with oscillating neural assemblies. *Science* 265: 1872–1875.

Laurent, G., and M. Naraghi. 1994. Odorant-induced oscillations in the mushroom bodies of the locust. *J. Neurosci.* 14: 2993–3004.

Laurent, G., M. Stopfer, R. Friedrich, M. I. Rabinovich, A. Volkovskii, and H. D. I. Abarbanel. 2001. Odor encoding as an active, dynamical process: Experiments, computation and theory. *Annu. Rev. Neurosci.* 24: 263–297.

Laurent, G., M. Wehr, and H. Davidowitz. 1996. Temporal representations of odors in an olfactory network. *J. Neurosci.* 16: 3837–3847.

Linster, C., and B. H. Smith. 1999. Generalization between binary odor mixtures and their components in the rat. *Physiol. Behavior* 66: 701–707.

Livermore, A., and D. G. Laing. 1996. Influence of training and experience on the perception of multicomponent odor mixtures. *J. Exp. Psychol.* 22: 267–277.

MacLeod, K., A. Bäcker, and G. Laurent. 1998. Who reads temporal information contained across synchronized and oscillatory spike trains? *Nature* 395: 693–698.

MacLeod, K., and G. Laurent. 1996. Distinct mechanisms for synchronization and temporal patterning of odor-encoding neural assemblies. *Science* 274: 976–979.

Macrides, F., and S. L. Chorover. 1972. Olfactory bulb units, activity correlated with inhalation cycles and odor quality. *Science* 185: 84–87.

von der Malsburg, C., and W. Schneider. 1986. A neural cocktail-party processor. *Biol. Cybern.* 54: 29–40.

Marr, D. 1969. A theory of cerebellar cortex. *J. Physiol.* 202: 437–470.

Meredith, M. 1986. Patterned response to odor in mammalian olfactory bulb, the influence of intensity. *J. Neurophysiol.* 56: 572–97.

Mombaerts, P., F. Wang, C. Dulac, S. K. Chao, A. Nemes, M. Mendelsohn, J. Edmondson, and R. Axel. 1996. Visualizing an olfactory sensory map. *Cell* 87: 675–686.

Motokizawa, F. 1996. Odor representation and discrimination in mitral tufted cells of the rat olfactory bulb. *Exp. Brain Res.* 112: 24–34.

Perez-Orive, J., O. Mazor, G. Turner, S. Cassenaer, R. Wilson, and G. Laurent. 2002. Oscillations and sparsening of odor representations in the mushroom body. *Science* 297: 359–365.

Rabinovich, M. I., A. Volkovskii, P. Lacanda, R. Huerta, H. D. I. Abarbanel, and G. Laurent. 2001. Dynamical encoding by networks of competing neuron groups, Winnerless competition. *Phys. Rev. Lett.* 87: 68102–1, 4.

Ramon y Cajal, S. 1995. *Histology of the Nervous System.* New York: Oxford University Press.

Spors, H., and A. Grinvald. 2002. Spatio-temporal dynamics of odor representations in the mammalian olfactory bulb. *Neuron* 34: 301–315.

Stopfer, M., S. Bhagavan, B. H. Smith, and G. Laurent. 1997. Impaired odour discrimination on desynchronization of odour-encoding neural assemblies. *Nature* 390: 70–74.

Stopfer, M., and G. Laurent. 1999. Short-term memory in olfactory network dynamics. *Nature* 402: 664–668.

Vassar, R., S. K. Chao, R. Sitcheran, J. M. Nunez, L. B. Vosshall, R. Axel. 1994. Topographic organization of sensory projections to the olfactory bulb. *Cell* 79: 981–991.

Vinje, W. E., and J. L. Gallant. 2000. Sparse coding and decorrelation in primary visual cortex during natural vision. *Science* 287: 1273–1276.

Wehr, M., and G. Laurent. 1996. Odor encoding by temporal sequences of firing in oscillating neural assemblies. *Nature* 384: 162–166.

Wellis, D. P., J. W. Scott, and T. A. Harrison. 1989. Discrimination among odorants by single neurons of the rat olfactory bulb. *J. Neurophysiol.* 61: 1161–1177.

Yokoi, M., K. Mori, and S. Nakanishi. 1995. Refinement of odor molecule tuning by dendrodendritic synaptic inhibition in the olfactory bulb. *Proc. Natl. Acad. Sci. U.S.A.* 92: 3371–3375.

2

Can We Understand the Action of Brains in Natural Environments?

Hermann Wagner and Bernhard Gaese

Introduction

Neuroscience has been very successful in the last 150 years by using largely reductionist approaches. Molecular and cellular approaches are amongst the most fruitful fields in neuroscience at the moment. After the completion of the sequencing of the human genome, many hope that this information will eventually lead to a complete understanding of illnesses and complex behaviors. We argue here that while we are optimistic that this goal can be reached, the route will be long and stony. The reason for our skepticism lies in the complex relations between genotypes and phenotypes. The genetic information provides only more or less limited predispositions for the realization of the phenotype. The "more or less" is part of a sometimes heated debate on "nature or nurture." It is clear, however, that practically all organisms are able to learn and build representations that may obscure the genotype. High-level behavioral functions, especially, are often not directly linked to genetics. Thus, understanding such high-level functions may indeed bring us a new revolution in understanding animal and human behavior, one that goes far beyond what we are able to learn from genetics. This is a much more challenging task than sequencing the genome, because neuroscience data are much more complex than genome or protein databases (Chicurel 2000).

Because organisms are composed of many molecules and cells, we would not understand how these interact to create something like consciousness even if we had worked out the functioning of all channels, cellular propagation mechanisms, and synapses. This insight is reflected by the separate fields of systems and behavioral neuroscience. Nevertheless, even at this "higher" level, neuroscientists are reductionists in that they dissect behavior and study only some aspects of it. Scientists usually work in laboratories. The stimuli are controlled

22

by instruments. The subjects have to attain and remain in a fixed position. The durations of stimuli are as long as the experimenter chooses. Feedback loops are often cut. In the natural environment, on the other hand, stimuli are of varying duration and usually brief; the context in which stimuli occur is constantly changing, subjects are moving, and the subjects have to react very fast. One of our mentors was Werner Reichardt, who studied orientation in flies. These animals perform acrobatic aerial chases. Males are pretty successful in hitting and grasping females in the air. Reaction times are only a few tens of milli-seconds (Wagner 1986). It is intuitively clear that temporal averaging in the range of seconds and the analysis of data averaged over many trials is inap-propriate to understand such behavior. This example brings us to the question we want to tackle.

The Question

One of the major challenges for future research will lie in research trying to reconcile our current reductionist approaches and their results with the real demands of nervous systems: the online operation of brains and organisms in natural environments. In other words, can we understand the action of brains in natural environments?

While the first part of the question is trivial, we want to detail the second half: In which context did brains evolve? What is a natural environment? Maybe for some contemporary humans a lecture hall constitutes a natural environment, maybe some would even accept being fixed while listening to a talk, and then some recordings could be done noninvasively. But this is not the type of environment in which brains evolved and which would be regarded as a natural environment for most brains. What is it then? Natural history films demonstrate this day by day. For example, wolves have to hunt in a coordi-nated way. Otherwise they would have no chance to break, for example, the defense of a male moose. There is a hierarchy in the pack. The alpha male or female is often leading the hunt. The animals bring themselves into the hunting mood by howling, then they trot off. Once they have located possible prey, they have to probe whether it is worthwhile to attack. The predators start to circle around the prey; they watch the prey move; they have to be patient and wait and probe again and again until, finally, they have a chance to attack when the prey makes a mistake. Part of this is certainly due to a genetic program, but lots of movements and actions cannot be preprogrammed because they are not predictable.

Such behaviors and their neural correlates are what we have to explain if we really want to understand brains. We have to explain how one-time events occurring in an ever-changing environment lead to stable behavior. The difficulty is that on the one hand, it is necessary to collect controlled data while on the other hand, we have to cope with the complexity of behavior in the real world. In the following we shall first introduce early concepts by the ethologists to define the problem, then illustrate one case—hunting in the barn owl—in some detail to sketch our philosophy toward a possible way to obtain answers to this question, and finally end with some general remarks.

History of Behavioral Neuroscience

The field of animal behavior has not only attracted filmmakers but also has been scientifically tackled in the twentieth century. Initially the ethologists had developed concepts of how to study behavior. The picture of Lorenz swimming in a pool with his goslings became famous. Lorenz, indeed, was a keen and very precise observer (Lorenz 1967). This may be seen in his early work on mating displays in ducks, which led to his proposal that behavioral elements may serve as a criterion for evolutionary relations. While Lorenz was an excellent observer, Tinbergen did some very basic experiments to demonstrate the inherited and learned components of behavior (Tinbergen 1951). His experiments with bee wolves demonstrated that these animals use landmarks to find their nesting sites. More famous are his experiments with gulls. These experiments led to the notion of the over-optimal stimulus. Although part of these experiments may be criticized, the basic findings have not lost their relevance. Both Lorenz and Tinbergen, and with them many others, described and analyzed many complex behaviors, and they proposed a scheme of classifying these behaviors. They also proposed concepts like those of key stimuli, releasers, and fixed-action patterns. Lorenz even suggested a mechanical model to formalize these findings. These ideas were greeted warmly by many neurobiologists who searched for neural correlates; later, however, neurobiologists were disappointed, because almost all failed to detect such neural principles. What went wrong? In our opinion, the concepts proposed by the ethologists were too far from neurobiological reality. Very few were able to translate these concepts into the language of neurobiologist or theoreticians.

However, even if the concepts of the ethologists were not suitable for neurophysiologists, this did not mean that the observations were obsolete. Indeed, early attempts by von Holst and St. Paul (1960) and Huber (1960) had laid a way to study successfully brains in action through brain stimulation. von Holst and St. Paul (1960) demonstrated that motivational actions of chickens may be influenced by stimulation of defined brain regions. For example, a clucking hen stopped its vocalization when electrically stimulated. Moreover, the reaction depended on the duration of the stimulus. While the hen started to cluck again after brief stimuli, it changed its emotional state after longer stimulation and did not resume vocalization again after such longer stimuli were turned off. Likewise, Huber (1960) could elicit rival calls in crickets by electrically stimulating their mushroom bodies. This indicated that certain behaviors may be initiated by stimulating certain brain regions.

While it was and still is debated how distributed information is coded in brains, it has been clear for a long time that separate areas represent (early) correlates of sensory stimuli. Even more strikingly, visual specialists like primates have huge areas of their brains devoted to visual analysis, while auditory specialists like owls or bats have extended representation of auditory stimuli. Recently it has become clear that the cortex at least is not so specialized as to be able to represent only one modality in one brain region; a rerouting of visual information into the auditory cortex may lead to a functional visual behavior via the temporal lobe (von Melchner, Pallas, and Sur 2000). Apart from the sensory areas, there are associative areas, areas receiving no direct sensory input in invertebrates and vertebrates. In the last forty years, behavioral neuroscientists have tried to identify the role of each area in the representation of stimuli and its role in behavior. This started in work such as that by Hubel and Wiesel (1962), who identified appropriate stimuli for cells in the visual cortex in anesthetized cats by single-cell electrophysiology. In this way, the response properties of these cells could be described. Such work included lesion studies in which portions, often huge ones, of brains were removed in an attempt to demonstrate specific behavioral deficits. It finally led to work with awake, behaving animals in which correlations between behavior and neural responses were possible. The most advanced of these techniques are studies in which electric stimulation caused predictable shifts in behavior (Salzman, Britten, and Newsome 1990).

Despite these successes, many researchers felt uneasy about the information gained in these controlled laboratory situations, with recordings from just one neuron at a time. This led to a variety of new concepts in the last twenty years, including new recording techniques—even without invasive methods like positron emission tomography, functional magnetic resonance imaging, electroencephalography, and electromyography—and allowing for simultaneous recording from many neurons with multielectrode arrays or optical recording methods. In parallel, new concepts on the representation of stimuli arose like synchronization of activity (e.g., Singer and Gray 1995). While all these techniques represent advances in the effort to understand brains, the question still remains whether they are enough to really understand the action of brains in natural environments (see chapters 9 and 19, this volume).

In the rest of this chapter we are approaching answers to this question from the point of view of the neuroethologist. We shall describe our research on the hunting in the barn owl, an animal that provides a good example for our claim (Konishi 1993). Although we have learned quite a lot about the neural mechanisms underlying sound localization in the last thirty years, there is still much more to discover. Thus, we warn the reader in the beginning that we are far from being able to answer the question posed above.

Hunting in the Barn Owl: A Case Study

Trying to understand hunting requires interdisciplinary research. Ethologists can contribute data on the complexity of the behavior, while neurophysiologists study brain functions, and theoreticians might bind all the data together in a mathematical description. Each discipline alone does not suffice to understand the complexities inherent in hunting. The philosophy behind the experiments and data described in this chapter is that by extending the reductionist's approach to more complex but still controlled situations, one might arrive at interpretable data.

Hunting as a Complex Behavior

Hunting is a behavior that may be subdivided into components such as a decision to start hunting, searching for prey, localization of prey, approaching prey, grasping prey, and returning prey to a den or nest. Many of these behaviors can be described by straightforward geometrical relations. Barn owls hunt mainly small rodents such as mice or insectivores such as shrews. Adult animals need about two mice a day for survival. When the nestlings are in the phase of rapid development, each of them demands up to five mice. Since there are often up to seven babies in a nest, the father—who hunts alone when the babies are small and still need the mother's constant attention—has to catch some thirty mice a night. That the birds are indeed very effective hunters of mice has been observed in the wild: during the breeding season the male brings a prey item to the nest every ten to fifteen minutes (Bunn, Warburton, and Wilson 1982). In this time the male has to fly out from the nest, search in an area of several square kilometers, detect, localize, and catch a prey animal, and then fly back to the nest.

Such observations suggest some regularity in hunting and thus seem to emphasize the reflex-like aspects of this behavior. It could also indicate some reliance on memory to relocate places with abundant food, and, indeed, field observations suggest that hunting barn owls follow a favored route time after time (Bunn, Warburton, and Wilson 1982). In the laboratory, capability for memorizing sounds (Quine and Konishi 1974) as well as locations (Knudsen and Knudsen 1996) was demonstrated. However, different hunting strategies have been observed (Brandt and Seebass 1994): barn owls may hunt from a waiting post or by flying slowly over the foraging area. Each individual may adopt either hunting strategy. In addition, different individuals specialize on different prey in the wild. While the main diet of most barn owls is small rodents, there are individuals that prefer shrews. The selection is most probably made by listening to the sounds of the shrews; therefore, the birds must be able to recognize and select shrew sounds from sounds of other animals.

In the laboratory, owls respond with a turn of the head toward novel sounds. This response may be reinforced by feeding the birds if they turn appropriately. The birds then may execute this reflex-like behavior for many trials a day. We have observed, however, that the owls might also stop responding if they fail to associate stimuli with a reward. The owl will then no longer react to the sound, even if it is hungry. In addition, in some cases, even if the owl is well trained and seemingly alert, it does not turn after sound presentation. This shows a high flexibility in the reaction to sounds and suggests that attentional mechanisms might play an important role in sound localization. The importance of attention and neglect is meaningful, because in the wild the owls should be selective and should not make too many errors in hunting. After a failed attempt the owl has to decide whether it pays to try again.

So far we have described hunting from an ethologist's point of view. We now turn to the neurophysiologist's need and formally describe the hunting situation in three-dimensional space (figure 2-1A). Out of the possible three-dimensional coordinate systems that describe the hunting situation, the spherical coordinate system seems to be closest to the natural coordinate system the owl uses (figure 2-1B). A spherical coordinate system consists of two angles, azimuth and elevation, and the distance between the predator and the prey. Azimuth is the horizontal angular deviation of the prey from the midsagittal plane, while elevation is the vertical angular deviation of the prey from the horizontal plane. These two angles can easily be measured in visual or auditory space (figure 2-1C, D). While distance measurement is easy by vision (see the section, "Stereo Vision in Barn Owls," below), not much is known about the third spatial parameter, distance measurement by audition, in the owl. This rather simple description is sufficient in most reductionist approaches. An example where it fails is relative movement between predator and prey.

When trying to understand hunting, one should not forget the adaptations of barn owls to hunting. Owls are one of only a few bird groups that have specialized for life at night. They have developed several specializations for hunting at night or for hunting in environments lacking visibility or with reduced visibility of the prey (the ruff, noise-reducing feathers, frontally oriented eyes, large brain nuclei that process auditory and visual [depth-related] information).

Early Findings on Neural Mechanisms Underlying Sound Localization

The neural mechanisms underlying sound localization have been studied in some detail in the last twenty-five years. This research was initiated by Payne (1971) and continued by Konishi (1973). Subsequently several independent research groups grew out of Konishi's "germ cell." Here only some basic findings can be represented. More detailed overviews may be found in Konishi

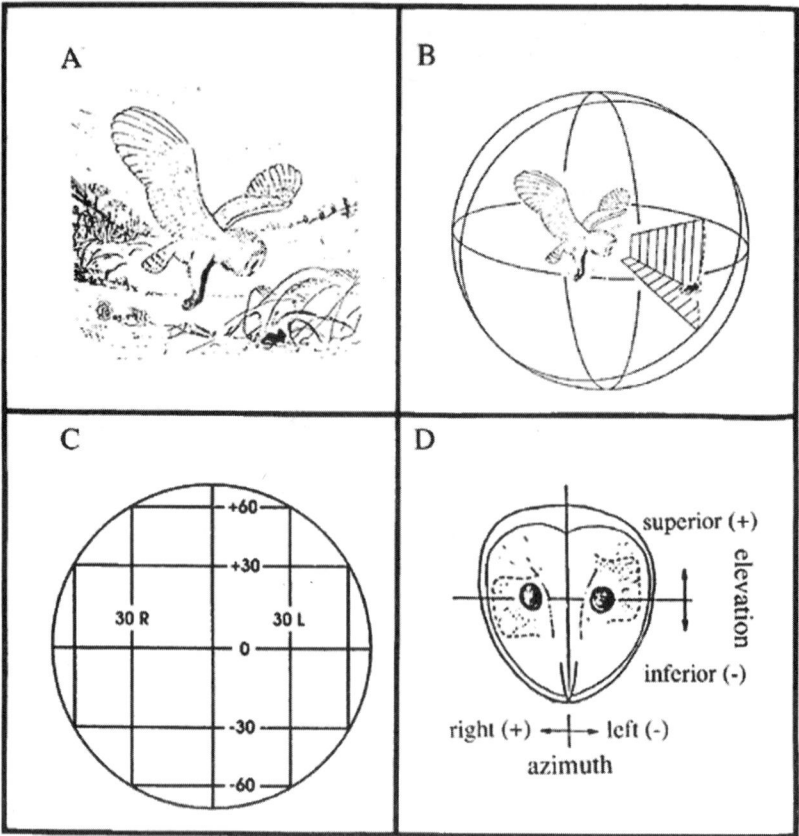

Figure 2-1. The hunting situation. (A) The natural situation. (B) Formalization in a three-dimensional spherical coordinate system with two angles, azimuth and elevation, and the distance between the owl and the mouse. (C) Reduction of the three-dimensional coordinate system to a two-dimensional coordinate system. Only the angles, azimuth and elevation, remain. (D) Centering of the two-dimensional coordinate system on the owl's head.

et al. (1988), Carr (1993), Knudsen (1984, 1999), Takahashi (1989), and Wagner (1999).

Behavioral Correlates

For sound localization, barn owls associate changes in interaural level difference (ILD) with changes in elevation and changes in interaural time difference (ITD) with azimuth (Moiseff 1989). Thus, the ear asymmetry leads to a separation of the coordinates along which ITD and ILD vary and generates a two-dimensional grid, one coordinate of which is formed by ILD, while the other coordinate is formed by ITD. Since both cues are good localization cues, this two-dimensional grid allows the owl to simultaneously determine the location of sounds in azimuth and elevation with high accuracy. The spatial resolution is between one and two degrees in azimuth and elevation (Knudsen 1984 but see Bala, Spitzer, and Takahashi 2003), and thus higher than in any other animal whose hearing has been tested. This precision is improved by the owl's ability to use information about the prey's motion (Payne 1971).

Neural Correlates

Concept of Space-Specific Neuron and Space Map   The initial neurophysiological studies on sound localization in barn owls led to the notion that neurons exist in the auditory system that have restricted spatial receptive fields, the so-called space-specific neurons (Knudsen, Konishi, and Pettigrew 1977; Knudsen and Konishi 1978). One locus in auditory space may be said to be represented by the activity of one (or a few) space-specific neurons. What is important and demonstrates the strength of the approach is the observation that the azimuthal restriction of the space-specific neurons is due to their tuning to ITD, while the elevational restriction is due to their sensitivity to ILD (figure 2-2) (Takahashi, Moiseff, and Konishi 1984). Thus, the behavioral relevance of the neuronal activity is known.

The space-specific neurons may be regarded as representing very specific stimulus features and were taken as an example of a coding scheme resembling the idea of a grandmother neuron. But by itself, the spiking of a space-specific neuron does not tell the organism much. The spiking must be seen in relation to the activity of other auditory neurons representing different aspects of auditory space. Neighborhood relations are often represented in the brain by sensory maps. While in the forebrain these neurons appear to be clustered but not really systematically arranged in a sensory map (Cohen and Knudsen 1995), in the midbrain, specifically the external nucleus of the inferior colliculus (ICx) and the optic tectum (OT), maps of auditory space were identified. The responses of the space-specific neurons represent the behaviorally relevant cues quite well. Since lesions erasing part of the space map (Wagner 1993; Knudsen, Knudsen, and Masino 1993) caused behavioral deficits, it is clear that the space map is important for localization behavior. However, a recovery from the behavioral deficit was observed, demonstrating that the adult auditory system has the capability of plasticity (Wagner 1993).

Initially auditory signals are processed in narrow and largely separated frequency bands. Time is represented by a process termed phase locking: The spikes of a cell occur at a certain phase of the stimulus tone. The information from the two sides converges and creates binaural neurons sensitive to ITDs by processes of delay lines and coincidence detection (Jeffress 1948; Carr and

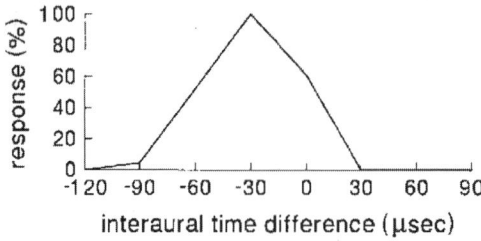

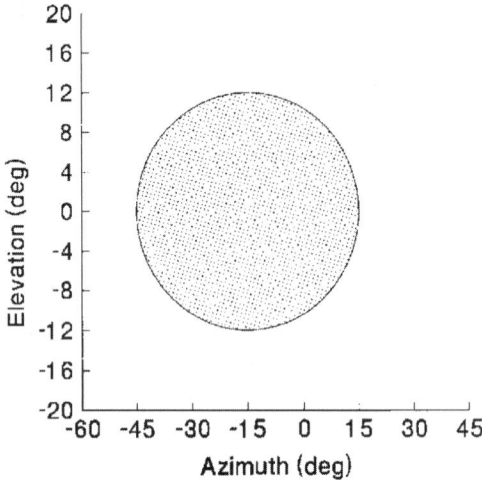

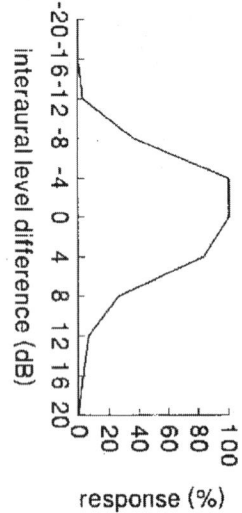

Figure 2-2. Scheme demonstrating the selectivity of a space-specific neuron. *Lower left,* the receptive field of a space-specific neuron in azimuth and elevation is indicated by the dotted circular area. *Top,* an interaural time difference–tuning curve suggesting that the tuning in azimuth of this neuron is due to its selectivity for interaural time difference. *Right,* an interaural level difference (ILD)-tuning curves again, implying that ILD is a very important cue for the elevational selectivity of space-specific neurons.

Konishi 1990). However, the neural responses representing a stimulus after a first coincidence-detection computation are often ambiguous. Thus, the response depends in a periodic manner on ITD. This precludes the representation of one location in space, and thus, false targets or target attributes appear. If a hunting animal wants to survive under evolutionary pressure, it must try to reduce false targets as much as possible. Thus, there should be algorithms after the stage of primary coincidence detection to remove the ambiguity in the response of the neurons performing the initial coincidence detection. These computations take place three synapses after the computation of coincidence detection in the external nucleus of the inferior colliculus. It could be shown that the solution of the false-target problem is achieved by integrating information across frequency. The convergence has to be done in a way to specify one location in space. That the barn owl's brain actually does this was shown in normal owls using ITDs in different frequency channels (Wagner, Takahashi, and Konishi 1987) and recently in owls reared with an acoustic device that caused frequency-dependent shifts in ITD and ILD (Gold and Knudsen 2000). The mechanisms are similar to mechanisms of binding discussed in the visual literature (Singer and Gray 1995) because features of a signal (object) are bound together.

Our Current Approaches

So far we have explained what sound parameter the owl uses to locate prey and how one sound locus is represented in the owl's brain. The picture drawn so far is fairly simple compared with the complexity of the hunting behavior we outlined in the section, "Hunting as a Complex Behavior." We have to explain why owls do not always react to sound stimuli, how owls remember sounds and sound loci, how information about static stimuli is combined with information about the movement of a stimulus, and how all these pieces of information are integrated with other, for example visual and nonsensory, components of the hunting behavior. Currently we are undertaking efforts to understand the neural mechanisms underlying these different components.

Virtual Auditory Worlds for Studying the Contribution of Single Parameters to Sound Localization

As we mentioned in the section, "Behavioral Correlates," previous data suggested that ITD is the main cue for azimuthal sound localization in the barn owl. The exact contributions of ITD could, however, only be tested indirectly because it was not possible to generate a stimulus that contained all relevant spatial information on the one hand and allowed for a clean separation of these parameters on the other hand. While free-field stimuli contain all spatial cues that are available to the auditory system, free-field stimulation does not allow separation of the contribution of the different cues to sound perception. In previous closed-field experiments with barn owls, on the other hand, one or two parameters (ITD or ILD or both) were systematically varied. However, other possible cues that originate from the direction- and frequency-dependent filter characteristics of the external ear were ignored. Such restrictions on stimulus presentation may be overcome with the virtual space technique. If sounds are filtered with the individual transfer functions of the external ears, the so-called head-related transfer functions, before they are played to a subject over headphones, stimuli may be created that contain all relevant spatial cues of a sound. Thus, stimulation with virtual sounds offers the possibility of manipulating monaural or binaural parameters of an acoustic stimulus while leaving other cues as they occur in a natural free-field sound. It was first demonstrated that barn owls responded to virtual auditory stimuli in the same way as they reacted to free-field stimuli (Poganiatz, Nelken, and Wagner 2001). No front-back confusions were observed with virtual stimuli. When the influence of ITD was separated from the influence of all other stimulus parameters by fixing the overall ITD in virtual stimuli to a constant value (+100 μs or −100 μs FixT stimuli) while leaving all other sound characteristics unchanged, the owls always turned to the location specified by the ITD and not by the other sound parameters (figure 2-3). Specifically, ILDs did not have an influence on azimuthal components of head turns. Thus, azimuthal sound localization in the barn owl is only influenced by ITD both in the frontal hemisphere and in large parts of the rear hemisphere. If the ITD of the manipulated virtual stimuli points to different hemispheres in space than the other spatial cues, the sounds contain a binaural combination of spatial information that never occurs under natural conditions. This combination of unusual spatial cues might require a more complex processing of sound location, especially in the forebrain. This is

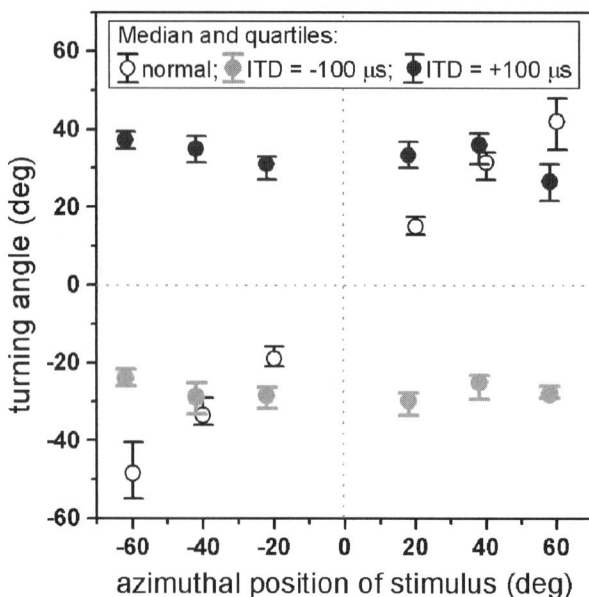

Figure 2-3. Azimuthal components of head turns. Responses to normal and manipulated virtual stimuli (FixT stimuli) of one owl. At each azimuth, about 160 trials with normal and about 20 trials with manipulated stimuli were presented. In the manipulated case, the azimuthal stimulus position on the abscissa means that all spatial cues of the stimulus reproduced a sound at the specific position except for the interaural time difference, which was shifted by +100 μs (FixT$_{100}$) or by −100 μs (FixT$_{-100}$) (normal: open circles; FixT$_{100}$ stimuli: filled circles; FixT$_{-100}$ stimuli: gray filled circles). For clarity, the data points representing FixT$_{100}$ stimuli are shifted slightly to the left, while the data points representing FixT$_{-100}$ stimuli are shifted to the right on the x-axis. Note that the head turn amplitude is independent of the stimulus position for the manipulated stimuli. Elevation was 0° for all stimuli.

indicated by longer response latencies in these conditions. Forebrain pathways have been studied and a contribution to sound localization has been demonstrated (Knudsen, Knudsen, and Masino 1993; Knudsen and Knudsen 1996; Cohen, Miller, and Knudsen 1998). Interestingly, the increase of 60 milliseconds in latency in the opposite-side configuration was very close to what others have seen when invalid cues were presented to barn owls in a cueing paradigm (Johnen, Wagner, and Gaese 2001).

Nonsensory Influences on Sound Localization

Knudsen and Knudsen (1996) have demonstrated an involvement of memory in sound localization. While normal owls could easily make memory-guided head turns toward a remembered location and fly toward that location, owls with unilateral lesions in the arco-pallium refused to fly in many cases when the sound came from the spatial hemifield that had been represented by the lesioned area.

This is, however, not the only nonsensory influence on sound localization. As we have already mentioned, only careful training allows for successful psychophysic experiments in the lab. Owls might otherwise lose interest in the stimulus and might stop reacting to a stimulus at all. This suggests that some kind of attention influences sound localization. To find out more about these influences, two barn owls were tested in a cued discrimination task as introduced by Posner, Snyder, and Davidson (1980) that delivered consistent or inconsistent information about the probable position of an upcoming auditory event to the owl. In the consistent ("valid") condition, an informative visual prestimulus ("cue") in front of the owl pointed to the hemisphere where a subsequent peripheral auditory target would occur after a randomized cue-target delay. In the inconsistent ("invalid") condition, the visual cue pointed to the hemisphere opposite the upcoming auditory target. A clear effect of cue-target configuration was observed in the response latencies. Valid cues led to reduced response latencies compared to response latencies after presentation of invalid cues (figure 2-4). Mean response latencies were significantly reduced by between 14 and 25 percent depending on the owl and target side (Johnen, Wagner, and Gaese 2001). This *validity effect* indicated that the auditory processing speed was facilitated by shifting the attentional focus toward the side of the expected target position.

Interactions of Visual and Auditory Systems

With the across-frequency integration that takes place in the inferior colliculus (IC), the process of representation of static sound sources in the owl's brain is completed and the ICx, the hierarchically highest nucleus in the inferior colliculus, contains a map of auditory space as already explained.

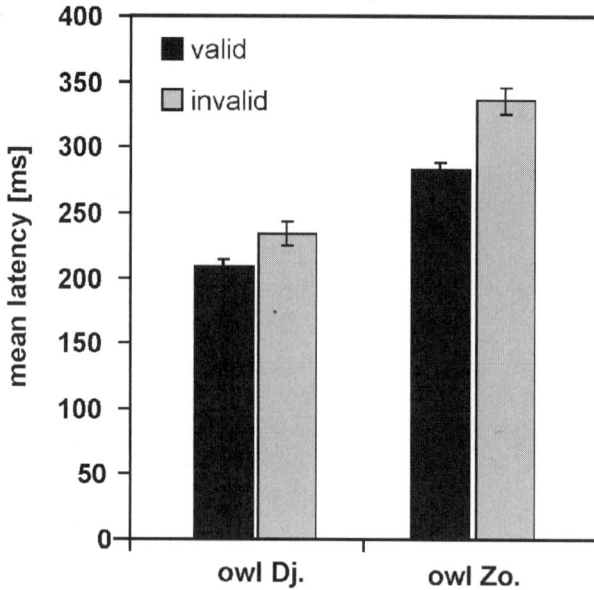

Figure 2-4. Owls can covertly shift attention toward spatial positions where they expect auditory events. Depicted are means and standard error of the mean of response latencies of two owls (Zo. and Dj.) separated for the cue validity (valid or invalid). Depending on owl and target side, mean response latencies were significantly reduced in the valid cue-target condition compared to the invalid condition. In this cross-modal cued discrimination task, an informative visual cue located in front of the owl delivered information about the probable position of an upcoming auditory event. Owls had to maintain fixation after presentation of the cue stimulus (two red LEDs, covering 5° visual angle), but were rewarded for a quick head orienting toward one of four peripheral auditory target stimuli (two in each hemisphere, 30° and 50° from the owls' midline, 0° elevation). Head turnings were measured with a motion tracking system (miniBIRD, Ascension). Response latencies and localization performance of azimuthal head motion trajectories were analyzed.

From the ICx, this information is projected to the OT (Knudsen and Knudsen 1983). In the OT sound localization information is combined with the visual information about a sound source (Knudsen 1982). The OT is a multilayered structure. The OT contains a map of visual space that is evident in most layers. The alignment of the visual and auditory maps is achieved through an instructive signal from the visual system to the auditory pathway (Knudsen and Knudsen 1985). Recently, Luksch, Gauger, and Wagner (2000) filled multipolar neurons in OT with rather restricted dendritic fields intracellularly. These cells projected toward the crossed tecto-bulbar tract via the stratum album centrale and showed arborisations in the ICx and the lateral shell of the central nucleus of the IC. By mass-filling in the stratum griseum centrale this projection could be confirmed in animals as young as embryonic day 18 (E18)

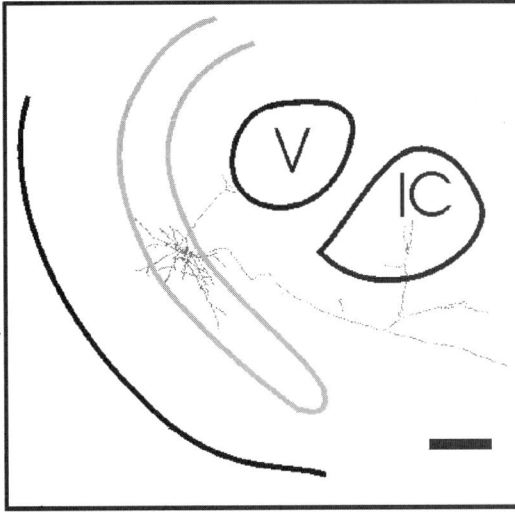

Figure 2-5. Neuron projecting from the optic tectum to the inferior colliculus in an E32 animal. This neuron may transmit information about eye position to the auditory pathway. Labeled structures were reconstructed manually with a camera lucida. The layer stratum griseum centrale of OT is marked in light grey.

(figure 2-5). Since our data span the time from E18 to the time after the eyes open and after the animals start to hear, our results demonstrate that the instructional signal is available early in the sensitive phase of the development of the bimodal map. At about the same time, Hyde and Knudsen (2000) reported a similar projection. Different cells were labeled in the Hyde and Knudsen study as in our study, indicating that several classes of cells might make a connection from the OT to the IC. Luksch, Gauger, and Wagner (2000) proposed that neurons of the owl stratum griseum centrale SGC provide the instructional signal from the OT to the ICx that aligns the auditory and visual maps of space.

Stereo Vision in Barn Owls

Hunting is not only driven by auditory information. Owls use visual cues whenever the target is visible. Since the owl's frontally oriented eyes allow for a large binocular overlap, we tested whether the owl is able to use global stereopsis for distance measurement. Indeed, owls were able to discriminate objects from holes in random-dot stereograms (RDS; van der Willigen, Frost, and Wagner 1998). Moreover, once owls had learned to use this information, they could transfer this information to a different task, motion parallax, without further learning (van der Willigen, Frost, and Wagner 2002).

To understand how stereo information may subserve distance measurement in awake birds, we established a paradigm that allowed us to record neural activity via telemetry in awake fixating owls. First, we demonstrated that owls are able to see illusory contours and that the neural correlate resides within the so-called visual Wulst (Nieder and Wagner 1999). In this brain area there had also been described many disparity-sensitive neurons in anaesthetized preparations (Pettigrew and Konishi 1976; Wagner and Frost 1993, 1994). In awake, fixating animals, these data could be confirmed (Nieder and Wagner 2000). Specifically, the Gabor function provided an appropriate fit to the disparity-tuning curves.

It had been argued that responses in visual area V1 of monkeys represent local but not global stereo information. This argument stems on the one hand from responses of these neurons to anticorrelated RDS (aRDS), in which, in contrast to normal or correlated RDS (cRDS), each black dot is paired with a white dot (and vice versa) in the complementing stereogram. Local detectors reverse their responses profile to stimulation with aRDS, while global detectors do not respond at all. On the other hand, local detectors respond quasi-periodically as a function of disparity, and tuning curves typically exhibit several response peaks, one main peak and other side peaks, while global detectors should suppress the side peaks.

Our data suggested that the visual Wulst contains computations allowing for a representation of global stereo information (Nieder and Wagner 2001). While neurons having a short response latency exhibited many features of local stereo representation, late-responding disparity-sensitive neurons achieved a representation of stereoscopic information postulated for detectors directly relevant for depth perception (figure 2-6). The neurons' response latency was correlated with decreasing responses to false-matched, contrast-inverted images that did not support an owl's depth perception in psychophysical tests. Increasing suppression of response side peaks and enhancement of disparity tuning with response latencies was also observed. Our results suggest a functional hierarchy of disparity processing in the owl's forebrain, leading from spatial filters to more global disparity detectors that may be able to solve the correspondence problem via nonlinear threshold operations.

Similar results were obtained with a hierarchical feedforward network (Lippert, Fleet, and Wagner 2000). These authors designed a three-layered network (input, hidden, and output units) that consisted of physiologically motivated monocular Gabor input filters and created output responses mirroring disparity-selective neurons. In contrast to the responses of most V1 neurons, output neurons of the network were trained to very low baseline activity and, as a result, did not respond to aRDS. This effect was attributed to the nonlinear threshold functions implemented in the model. It is interesting that while output units suppressed responses to false-matched images completely, preceding hidden units still showed substantial responses to disparity in aRDS by profile inversion as well as extensive modulation of the tuning curve and a corresponding higher baseline activity (J. Lippert, personal communication). This shows that hierarchical analysis of disparity information applying nonlinear

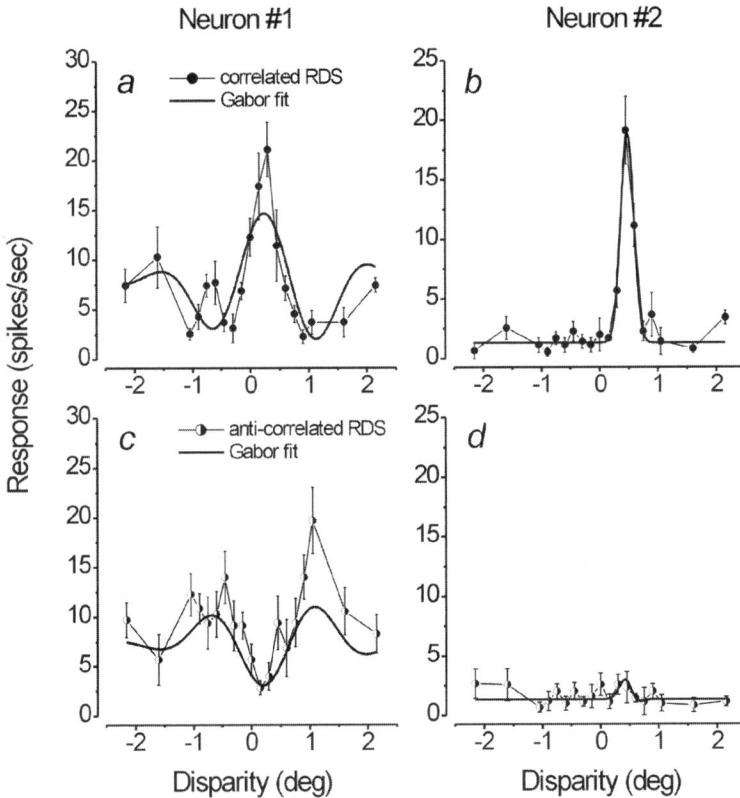

Figure 2-6. Responses of Wulst neurons to random-dot stereograms (RDS). In the correlated condition, a black dot is shown to the left and right eyes, while in the anticorrelated condition a black dot is presented to the left eye, while the right eye sees a white dot at the corresponding position, and vice versa. Responses of two neurons are shown (curves with thin lines and symbols). Gabor functions (thick curves) were fitted simultaneously to both tuning curves of each cell. While neuron 1 showed a pronounced, inverted tuning curve to contrast-inverted images, neuron 2 did respond very weakly to anticorrelated RDS.

threshold function can easily eliminate the major response ambiguities inherent in local detectors along processing stages. Thus, we offered a hierarchical framework to explain a progression from local to global stereo processing.

Planned Approaches

What we have presented in the preceding sections is our current attempt to understand components of hunting in the barn owl under controlled conditions. Each of these aspects has been analyzed separately. These experiments have led to plausible explanations. What has not been done is to integrate the different aspects, that is, to move beyond a reductionist analysis. This is not easy because it requires formulating testable hypotheses on how this integration can happen on a neural basis. On the other hand, by combining these components, new characteristics may arise that are not apparent or even missing in each single component. Similarly, the conditions under which we analyze the behavioral components are still too restricted to reflect the action of the brain in natural environments. Thus, in the future, we have to make progress in several areas. In the following section we shall describe plans for two experiments that are intended to come closer to an answer to our question.

Recordings in Freely Moving Animals

If we want to understand the action of brains in natural environments, we need to work not only with awake but eventually with freely moving animals. Our current work that allows recording neural signals via radio telemetry will serve as the starting point. We plan to implant owls with electrodes and a telemetry setup. After neural activity related to depth vision has been isolated, the owl will be transferred to a perch in a flight chamber. We will attempt to examine disparity-sensitive neurons in unrestricted owls just prior to flight and, if possible, during flight. It is likely that owls rely on stereopsis to make depth analyses prior to striking prey. Because the owls will be able to move freely, they will also be equipped with a head-tracking device as described in van der Willigen, Frost, and Wagner (1998). The head-tracking device will allow us to determine if the owls are relying on motion parallax (through "peering"—side to side movement of the head, which provides depth information) or are relying on stereopsis to make a behavioral response. The behavioral protocol will be similar to that described in Fox, Lehmkuhle, and Bush (1977). The owls will be trained in a room in which barriers will be set up to form a Y-shaped maze. There will be a start perch at the beginning of the maze. Two landing perches and two monitors will be at the end of the two arms of the maze. Initially the owls will be trained to fly from the start perch to both of the landing perches for food reinforcement. Once the owls are acclimated to the maze they will begin behavioral training. The monitors will present two stimuli simultaneously: one monitor will display a cRDS, and the other monitor will display an aRDS. We anticipate that this will be a very difficult experiment. However, if neural activity can be recorded during flight, this experiment will provide extensive insight into the behavioral relevance of disparity sensitive neurons in the Wulst.

Use of Interactive Virtual Auditory Worlds in Sound-Localization Tasks

Motion provides one of the most important cues for survival because it helps to break camouflage of a predator or a prey animal and because it allows predictions about the future path of an object. Indeed, behavioral correlates of acoustic motion have been identified (Grantham 1998) and motion-direction sensitive neurons have been described. While midbrain neurons in mammals seem to reflect more adaptation processes and not "real" motion, in the barn owl, neurons in the midbrain have been show to be sensitive to auditory motion (Wagner and Takahashi 1992, for a review see Wagner, Kautz, and Poganiatz 1997).

So far, we have not been able to obtain a behavioral correlate of acoustic motion sensitivity with moving single sources. Thus, we made efforts to use virtual auditory worlds to create whole-field auditory stimuli. These will be presented virtually via earphones. We are just about to start experiments with interactive stimuli. Since wide-field stimuli do not contain separable single sources, we hope to find out something at the behavioral level about acoustic-motion perception in barn owls.

What Is Necessary in the Future?

What we have presented so far are our approaches to better understanding of hunting behavior in the barn owl. Considerable progress has been made in recent years by using new experimental approaches like virtual-space techniques and telemetric recordings from awake animals. In parallel, we have extended our focus from initially studying the sound localization of static sources and their neural representation to include moving sources, visual and nonsensory information. We are now at a stage where we dare to try to record from completely unrestrained animals. What are the future challenges to us and to others? What are the gains if we can answer the question posed at the beginning of this chapter?

Clearly there are gains for basic neuroscience: we accumulate more knowledge. But the gains go far beyond this. If we understand how brains work, we might be able to use this knowledge to build robots in a similar way, taking advantage of the long evolutionary shaping of brain functions. Indeed, many promising robots are built in close analogy to brain function. Equally important is that there are neurological diseases that might be treated effectively, and prostheses might be constructed if we understand the brain in action. Indeed, cochlear implants are already working. Chapin and colleagues (1999) used recordings from the motor cortex to train rats to use these signals for real-time control of a robot arm. Such work suggests that there may be hope for a high-level, cognitive neuroprosthesis. If we want to be really successful, we need to answer the question posed in the beginning of this chapter. We think that the way to go is to use moving animals, at best freely moving animals, with telemetric signals transmission, to look at natural behaviors. This would preclude the use of modern noninvasive techniques (MRI, PET). This, then may be a new challenge for new model systems in a new neuroethological approach.

Behaviors in which the behavioral relevance is obvious may especially qualify for such an endeavor.

Although we have come a long way, much more is necessary. We close with challenges from a very different direction. There was an exhibition entitled "Adventures into the Mind: The Brain and Thought" (see Lewandowsky and Gruenbein 2000). This exhibition was organized by an artist (Via Lewandowsky) and a lyric poet (Durs Gruenbein). It consisted of seventeen themes housed in seventeen rooms; the last theme was "Will the brain know itself in the future?" This motto was detailed in eighteen questions, most of which are interesting also from a scientific point of view. Here are these questions in our translation: What will happen to the psyche in the age of artificial brains? When will the luxurious brains leave the assembly line? Will the sequencing of the human genome lead to a total knowledge about the brain? Will brain science help us to enjoy our own death as a movie? Will transparent brains soon make governing easier? Will psychology, morality, and religion become neurology in the fifth act? After time and space, will the brain be the last showplace of history? Shall we be able to breed parts of brains like nerve cells and exchange them? Is the brain prepared for the global simultaneity? Will the brain work until the end, when we might soon become old as Methuselah? Will it be necessary to go to school, if we shall have neurosoftware? Will our descent from animals always be present in our brains? How might a perfect brain cope with an imperfect world? Shall we be able to reanimate brains? Will the brain dissolve itself in the media of the future? Will everyone be rich enough to pay for the expensive brain implants? Will it be possible to neuronally load foreign languages before a trip? Will hormone cocktails be able to make humans happy for a whole life?

References

Bala, A. D., M. W. Spitzer, and T. T. Takahashi. 2003. Prediction of auditory spatial activity from neural images on the owl's space map. *Nature* 424: 771–774.

Brandt, T., and C. Seebass. 1994. *Die Schleiereule*. Wiesbaden: Aula-Verlag.

Bunn, D. S., A. B. Warburton, and R. D. S. Wilson. 1982. *The Barn Owl*. London: Poyser.

Carr, C. E. 1933. Processing of temporal information in the brain. *Ann. Rev. Neurosci.* 16: 223–243.

Carr, C. E., and M. Konishi. 1990. A circuit for detection of interaural time differences in the brainstem of the barn owl. *J. Neurosci.* 10: 3227–3246.

Chapin, J. K., K. A. Moxon, R. S. Markowitz, and M. A. Nicolelis. 1999. Real-time control of a robot arm using simultaneously recorded neurons in the motor cortex. *Nat. Neurosci.* 2: 664–670.

Chicurel, M. 2000. Databasing the brain. *Nature* 406: 822–825.

Cohen, Y. E., and E. I. Knudsen. 1995. Binaural tuning of auditory units in the telencephalon archistriatal gaze fields of the barn owl: Local organization but no space map. *J. Neurosci.* 15: 5152–5168.

Cohen, Y. E., G. L. Miller, and E. I. Knudsen. 1998. Forebrain pathway for auditory space processing in the barn owl. *J. Neurophysiol.* 79: 891–902.

Fox, R., S. W. Lehmkuhle, and R. C. Bush. 1977. Stereopsis in the falcon. *Science* 197: 79–81.

Gold, J. I., and E. I. Knudsen. 2000. A site of auditory experience-dependent plasticity in the neural representation of auditory space in the barn owl's inferior colliculus. *J. Neurosci.* 20: 3469–3486.

Grantham, D.W. 1998. Auditory motion aftereffects in the horizontal plane: The effects of spectral region, spatial sector and spatial richness. *Acta Acoustica* 84: 337–347.

von Holst, E., and U. St. Paul. 1960. Vom Wirkungsgefüge der Triebe. *Naturwissenschaften* 47: 409–422.

Hubel, D., and T. Wiesel. 1962. Receptive fields, binocular interaction and functional architecture in the cat's visual cortex. *J. Physiol.* 160: 106–154.

Huber, F. 1960. Untersuchungen über die Funktion des Zentralnervensystems und insbesondere des Gehirns bei der Fortbewegung und der Lauterzeugung der Grillen. *Z. vergl. Physiologie* 44: 60–132.

Hyde, P. S., and E. I. Knudsen. 2000. Topographic projection from the optic tectum to the auditory space map in the inferior colliculus of the barn owl. *J. Comp. Neurol.* 421: 146–160.

Jeffress, L. A. 1948. A place theory of sound localization. *J. Comp. Physiol. Psychol.* 41: 35–39.

Johnen, A., H. Wagner, and B. H. Gaese. 2001. Spatial selective attention modulates localization of sounds in barn owls. *J. Neurophysiol.* 85: 1009–1012.

Knudsen, E. I. 1982. Auditory and visual maps of space in the optic tectum of the owl. *J. Neurosci.* 2: 1177–1194.

Knudsen, E. I. 1984. Synthesis of a neural map of auditory space in the owl. In *Dynamic Aspects of Neocortical Function*, ed. G. M. Edelman, E. W. Gall, and M. W. Cowen, 375–396 (New York: Wiley).

Knudsen, E. I. 1999. Mechanisms of experience-dependent plasticity in the auditory localization pathway of the barn owl. *J. Comp. Physiol.* A185: 305–321.

Knudsen, E. I., and P. F. Knudsen. 1993. Space-mapped auditory projections from the inferior colliculus to the optic tectum in the barn owl. *J. Comp. Neurol.* 218: 187–196.

Knudsen, E. I., and P. F. Knudsen. 1985. Vision guides the adjustment of auditory localization in young barn owls. *Science* 230: 545–548.

Knudsen, E. I., and P. F. Knudsen. 1996. Disruption of auditory spatial working memory by inactivation of the telencephalon archistriatum in barn owls. *Nature* 383: 428–431.

Knudsen, E. I., and M. Konishi. 1978. A neural map of auditory space in the owl. *Science* 200: 795–797.

Knudsen, E. I., P. F. Knudsen, and T. Masino. 1993. Parallel pathways mediating both sound localization and gaze control in the telencephalon and midbrain of the barn owl. *J. Neurosci.* 13: 2837–2852.

Knudsen, E. I., M. Konishi, and J. D. Pettigrew. 1977. Receptive fields of auditory neurons in the owl. *Science* 198: 1278–1280.

Konishi, M. 1973. How the owl tracks its prey. *Am. Sci.* 61: 414–424.

Konishi, M. 1993. Listening with two ears. *Sci. Am.* 268(4): 66–73.

Konishi, M., T. Takahashi, H. Wagner, W. E. Sullivan, and C. E. Carr. 1988. Neurophysiological and anatomical substrates of sound localization in the owl. In: *Auditory Function*, ed. G. M. Edelman, W. E. Gall, and W. M. Cowan, 721–745 (New York: Wiley).

Lewandowsky, V., and D. Gruenbein. 2000. *Gehirn und Denken*. Ostfildern-Ruit: Hatje Cantz.

Lippert, J., D. J. Fleet, and H. Wagner. 2000. Disparity tuning as simulated by a neural net. *Biol. Cybern.* 83: 61–72.

Lorenz, K. 1967. *Uber tierisches und menschliches Verhalten*. Berlin: Deutsche Buch-Gemeinschaft.

Luksch, H., B. Gauger, and H. Wagner. 2000. A candidate pathway for a visual instructional signal to the barn owl's auditory system. *J. Neurosci.* 20: RC70 (1–4).

von Melchner, L., S. L. Pallas, and M. Sur. 2000. Visual behavior mediated by retinal projections directed to the auditory pathway. *Nature* 404: 871–876.

Moiseff, A. 1989. Bi-coordinate sound localization by the barn owl. *J. Comp. Physiol.* A164: 637–644.

Nieder, A. and H. Wagner. 1999. Perception and neuronal coding of subjective contours in the owl. *Nat. Neurosci.* 2: 660–663.

Nieder, A., and H. Wagner. 2000. Horizontal-disparity tuning of neurons in the visual forebrain of the behaving barn owl. *J. Neurophysiol.* 82: 2967–2979.

Nieder, A., and H. Wagner. 2001. Hierarchical processing of horizontal disparity information in the visual forebrain of behaving owls. *J. Neuroscience* 21: 4514–4522.

Payne, R. S. 1971. Acoustic location of prey by barn owls (*Tyto alba*). *J. Exp. Biol.* 54: 535–573.

Pettigrew, J. D., and M. Konishi. 1976. Neurons selective for orientation and binocular disparity in the visual Wulst of the barn owl (*Tyto alba*). *Science* 193: 675–678.

Poganiatz, I., I. Nelken, and H. Wagner. 2001. Sound-localization experiments with barn owls in virtual space: influence of interaural time difference on head-turning behavior. *J. Ass. Res. Otolarnyg.* 2: 1–21.

Posner, M. I., C. R. Snyder, and B. J. Davidson. 1980. Attention and the detection of signals *J. Exp. Psychol.* 2: 160–174.

Quine, D. B., and M. Konishi. 1974. Absolute frequency discrimination in the barn owl. *J. Comp. Physiol.* A 93: 347–360.

Salzman, C. D., K. H. Britten, and W. T. Newsome. 1990. Cortical microstimulation influences perceptual judgments of motion direction. *Nature* 346: 174–177.

Singer, W., and C. M. Gray. 1995. Visual feature integration and the temporal correlation hypothesis. *Ann. Rev. Neurosci.* 18: 555–586.

Takahashi, T. T. 1989. The neural coding of auditory space. *J. Exp. Biol.* 146: 307–322.

Takahashi, T. T., A. Moiseff, and M. Konishi. 1984. Time and intensity cues are processed independently in the auditory system of the owl. *J. Neurosci.* 4: 1781–1786.

Tinbergen, N. 1951. *The Study of Instinct*. Oxford: Oxford University Press.

Wagner, H. 1986. Flight performance and visual control of flight of the free-flying housefly (*Musca domestica* L.). II. Pursuit of Targets. *Phil. Trans. R. Soc. Lond. B* 312: 553–579.

Wagner, H. 1993. Sound-localization deficits induced by lesions in the barn owl's space map. *J. Neurosci.* 13: 371–386.

Wagner, H. 1999. Neural computations in binaural hearing. In: *Psychophysics, physiology and models of hearing*, ed. T. Dau, V. Hohmann, and B. Kollmeier, 169–178 (Singapore: World Scientific).

Wagner, H., and B. Frost. 1993. Disparity sensitive cells in the owl have a characteristic disparity. *Nature* 364: 796–798.

Wagner, H., and B. Frost. 1994. Binocular responses of neurons in the barn owl's visual Wulst. *J. Comp. Physiol.* A174: 661–670.

Wagner, H., and T. T. Takahashi. 1992. Influence of temporal cues on acoustic motion-direction sensitivity of auditory neurons in the owl. *J. Neurophysiol.* 68: 2063–2076.

Wagner, H., T. T. Takahashi, and M. Konishi. 1987. Representation of interaural time difference in the central nucleus of the barn owl's inferior colliculus. *J. Neurosci.* 7: 3105–3116.

Wagner, H., D. Kautz, and I. Poganiatz. 1997. Principles of acoustic motion detection in animals and man. *Trends Neurosci.* 20: 583–588.

van der Willigen, R. F., B. J. Frost, and H. Wagner. 1998. Stereoscopic depth perception in the owl. *Neuroreport* 9: 1233–1237.

van der Willigen, R. F., B. J. Frost, and H. Wagner. 2002. Depth generalization from stereo to motion parallax in the owl. *J. Comp. Physiol.* A 187: 997–1007.

3

Hemisphere Dominance of Brain Function—Which Functions Are Lateralized and Why?

Günter Ehret

Introduction

Common to bilateral animals such as arthropods and vertebrates are central nervous systems with two sides, the left side and the right side. Usually, we believe that bilateral animals are bilaterally symmetric in body structure and function. It is, however, extremely unlikely that even a body consisting of only hundreds or thousands of cells is bilaterally symmetric in all details of cellular number, local cellular specialization, and fine structure down to the molecular level. For simplicity, let us consider organisms having a genetic code without information for any structural and ultrastructural lateral biases. These organisms will not be symmetrical because local influences from the inner and outer environments on developing embryonic or larval tissue and on the metabolism and turnover of adult cells will never, at any given instant of time, be absolutely identical on both sides of the body. Since life implies continuous dynamic interactions with these environments, the argument about unavoidable asymmetry as a consequence of stochastic events impinging on the left and right sides differently leads to the prediction that every bilateral organism—nervous system and sense and effector organs included, of course—is bilaterally asymmetric in its molecular composition and, to various extents, on the level of cells and organs. Hence, it is an illusion to start the discussion of hemisphere dominances of brain functions from the point of a bilaterally symmetric brain. Primates, as we are, with billions of neurons in each brain hemisphere, certainly have to function with a multitude of bilateral asymmetries. Such reasoning runs through the whole book on cerebral lateralization of Geschwind and Galaburda (1987).

On the basis of statistically unavoidable bilateral asymmetries in chemical composition and structure, one may wonder why—on the first view and in many respects—bilateral organisms have a rather symmetrical appearance and behavior. As Levy (1977) pointed out, habitats will require an average symmetrical perception of stimuli and motor output (behavior) if they are without left-right biases acting persistently, or in situations critical for the survival of the animal, on the organisms. From an evolutionary point of view it is clear that an average symmetry of abiotic and biotic factors and structures in the habitat of a given species will favor symmetric behaviors of individual members of this species. I have argued before that bodies of bilateral organisms are asymmetric to some degree, and therefore it is a main task of their nervous systems and brains to produce average symmetric outputs on the basis of initially asymmetric structures. The nervous systems and brains accomplish this task by a strong coupling of the two sides via commissures so that an extensive exchange of information is guaranteed and by various mechanisms of neuronal plasticity which can lead to the necessary adjustments and equalizations of neural activities at the two sides. Furthermore, the central nervous systems of arthropods and other invertebrates (compared to those of vertebrates) work with a rather limited number of neurons, which often can be located as individuals being dedicated to well-defined functions. The reduction of the number of neurons reduces the probability of occurrence of accidental asymmetries between the two sides, however, at the cost of flexibility. The vertebrates, on the other hand, increase the number of neurons in their nervous systems with the advantage of increasing flexibility of behavior and compensate for possibly increasing structural asymmetries between the two sides by adding levels of analysis and control in their brains. For example, the basic neural circuits for the bilateral coordination of walking movements in mammals are located in segments of the spinal cord. Higher control centers are located in the brain stem, opening the possibility of using the legs for various functions besides walking. Control centers in the basal forebrain adjust the occurrence of motor functions to the momentary demands of the individual. Finally, the neocortex can control movements on the basis of experience the animal had made via associative learning (Gallistel 1980).

This example of movement control in mammals clearly demonstrates how difficult, if not impossible, it may be to associate the occurrence of a lateralized behavior, such as handedness, with a certain asymmetry of structure in a well-delimited area of the vertebrate and especially the mammalian brains. Multiple levels of control and the aforementioned neuronal plasticity may restrict and obscure the contribution of a given asymmetrical structure to an observed perceptual or behavioral asymmetry. It seems to me that, at least in mammals, lateralized perceptual and behavioral functions are based more on network than on localized functions in the brain. This is the reason why I propose to start with investigating hemisphere dominances of brain functions rather than structural, cellular, and molecular asymmetries in brains. The latter are certainly neural bases of perceptual and behavioral asymmetries generated in the brain, however, the relationships are largely unclear and most probably complicated in details. Such an approach has both an evolutionary perspective, asking for the common origin and advantages of hemisphere specializations of vertebrate brains, and a perspective of genetic and physiological mechanisms responsible for the realization of hemisphere specializations of certain kinds.

Of Men and Mice

More than a century ago, the lateralization of brain functions was first noted for the human brain when the left-hemisphere dominance of speech functions was discovered (Broca 1861; Dax 1865; Jackson 1874; Wernicke 1874). Since then, a great number of studies using a variety of methods have shown that the left neocortex of most people is specialized for processing and perception of speech, especially with regard to temporal features, semantics and syntax of languages (just a few reviews: Berlin 1977; Kinsbourne 1978; Bradshaw and Nettleton 1981; Mateer 1983; Ojemann 1983; Benson 1986; Nass and Gazzaniga 1987). This left hemisphere specialization is found already in newborns (Witelson and Pallie 1973; Molfese and Molfese 1979; Molfese, Molfese, and Parsons 1983; Woods 1983; Bertoncini et al. 1989) and is present even in deaf people using a sign language (Damasio et al. 1986; Corina, Vaid, and Bellugi 1992; Bavelier et al. 1998). These facts and a left-hemisphere lateralization of speech functions not only in the neocortex but also in the thalamus (Mateer and Ojemann 1983; Hugdahl, Wester, and Asbjørnsen 1990, King et al. 1999) and the automatic discrimination and categorization of speech phonemes in the left hemisphere (Koivisto, 1998; Imaizumi et al. 1998) even without will and without special attention (Kapur et al. 1994; Alho et al. 1998; Rinne et al. 1999), and also for speech played backwards (Kimura and Folb 1968) all indicate that the left-hemisphere dominance is not necessarily bound to an acoustic speech channel, to the understanding of the semantic content of spoken words, or to neocortical functions alone. This suggests that the hemisphere dominance of speech processing may be based on the lateralization of more general mechanisms of handling communication-relevant information in the brain. Cutting (1974) found evidence that there are at least two left-hemisphere dominant mechanisms in speech perception, one as a non–speech-specific mechanism acting in the auditory domain, analyzing complex or time-critical aspects of sounds (Johnsrude et al. 1997), the other as a speech-specific processor analyzing the phonetic structure and thus enabling attribution of a certain meaning to a given sequence of sounds.

If the left-hemisphere dominance in human speech processing reflects the lateralization of general mechanisms in the brain, we can expect an evolutionary history. In fact, a left-hemisphere dominance has been shown from primates down to frogs for (a) detecting phonetic information important to separate the meaning of species-specific calls of macaque monkeys (Petersen et al. 1978, 1984; Beecher et al. 1979), (b) production of social contact calls in marmosets (Hook-Costigan and Rogers 1998), (c) perception of ultrasonic communication calls by mice in a communicative context (Ehret 1987), (d) conditioned responding to a two-tone sequence in rats (Fitch et al. 1993), (e) song control in songbirds (Nottebohm 1970, 1980), and (f) control of vocalizing in a frog (Bauer 1993). Additional evidence from monkeys (Hauser and Andersson, 1994) and mice (Ehret 1987; Koch 1990) demonstrates that the left-hemisphere advantage for processing species-specific vocalizations is not genetically fixed but may be induced through a priming process during social contact with the senders of the vocalizations. For example, mothers discriminate and prefer mouse pup ultrasounds, in figure 3-1 mimicked by 50-kHz tone bursts (Ehret and Haack 1982), against other sounds (e.g., 20-kHz tone bursts) only, if they listen with both ears or the right ear alone (figure 3-1). The right ear is the main and direct source of input to the left auditory mid- and forebrains. Virgin females or males without pup experience behave quite differently. They do not identify pup ultrasounds as relevant releasers of maternal behavior (Koch 1990). If virgin females are conditioned to discriminate 50 kHz from 20 kHz and prefer the 50-kHz tone bursts, they can do it with each ear (each brain hemisphere) as well (figure 3-1). As I have argued before (Ehret 1992), both hemispheres of mice can do sound discrimination, take a decision about what sound is important, and release the appropriate response equally well. The left-hemisphere advantage comes into play only in a communicative situation in which ultrasound can release an instinctive behavior or, in other words, conveys a message that fits into innate releasing mechanisms for the extraction of meaning. In terms of natural selection, the "understanding" of the message in the context of infant-to-mother acoustical communication in the mouse is critical for the survival of the pups if they get out of the nest (e.g., Haack, Markl, and Ehret 1983), and thus, the mother's response is a well-adapted one. These results on the left-hemisphere dominance for semantic processing and species-specific call recognition in mice and monkeys demonstrate no basic differences with the left-hemisphere advantage of human speech perception.

Considering this long evolutionary history, including mammals and possibly dating back to amphibians, of lateralized sound production and perception of certain, often highly communicative vocalizations, we now face the challenge to really demonstrate that these left-hemisphere dominant functions of the brain either are based on the same mechanisms and thus are homologous to each other or have evolved independently several times as special adaptations of a certain species. Let us start with a glimpse of possible mechanisms in the human brain related to the lateralization of speech functions. A general finding from numerous attempts to delimitate the primary and higher-order auditory cortex in humans is that most people without disorders of language functions have a

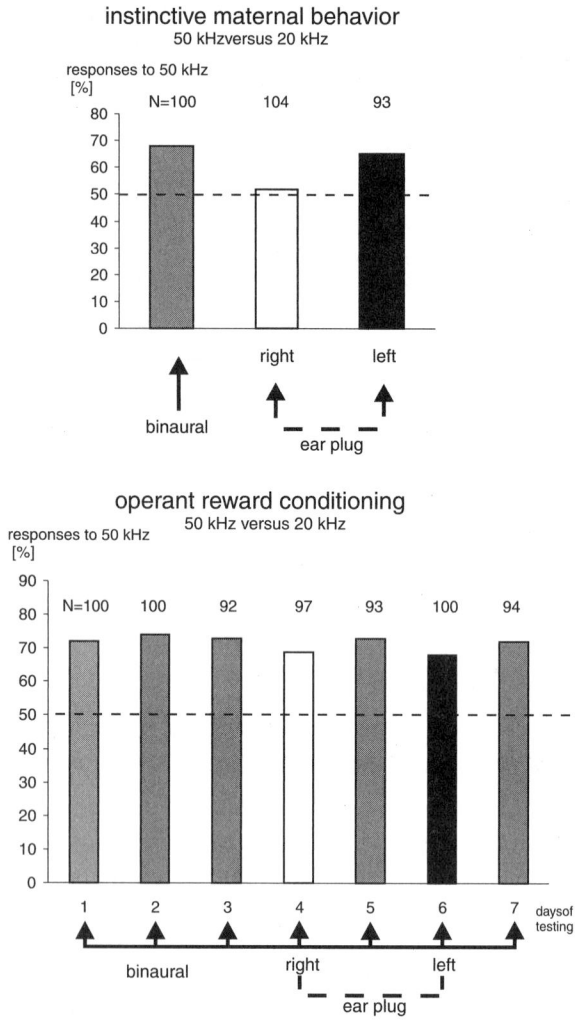

instinctive maternal behavior
50 kHzversus 20 kHz

operant reward conditioning
50 kHz versus 20 kHz

Figure 3-1. *Top*, house mouse mothers had the choice to approach a loudspeaker playing 50-kHz tone bursts that mimicked ultrasonic isolation calls of lost pups or 20-kHz tone bursts of no special meaning. As part of their instinctive maternal behavior, the mothers preferred the 50-kHz tone bursts over the 20-kHz tone bursts with statistical significance, however, only if they listened with both ears or with the right ear (left hemisphere of the brain). If the right ear was plugged, the left brain hemisphere was not sufficiently activated to recognize the ultrasounds. *Bottom*, naive females without experience in pup care preferred neither ultrasonic isolation calls nor 50-kHz tone bursts instinctively but were conditioned to do so. After having reached a preference level in response to 50-kHz similar to that of the mothers (*top*) they were tested binaurally or with the right ear or left ear plugged. These females did not show any ear (brain hemisphere) dominance in this task (modified from Ehret 1987).

larger planum temporale (part of the secondary auditory cortex and Wernicke's area) in the left- compared to the right-side superior temporal gyrus of their neocortical hemispheres (review by Shapleske et al. 1999). This anatomical left-hemisphere dominance in an area of higher auditory processing related to language functions is found already in human newborns (Wada, Clark, and Hamm 1975; Chi, Dooling, and Giles 1977) and chimpanzees (Gannon et al. 1998). In electrophysiological and anatomical mappings of the mouse auditory cortical fields (Stiebler 1987; Stiebler et al. 1997; Geissler and Ehret 2004; figure 3-2), we could show, on the population level, a significantly larger size of the auditory cortex of the left hemisphere, which is based mainly on larger higher-order auditory fields (figure 3-3). In addition, we found an interesting field dorsal

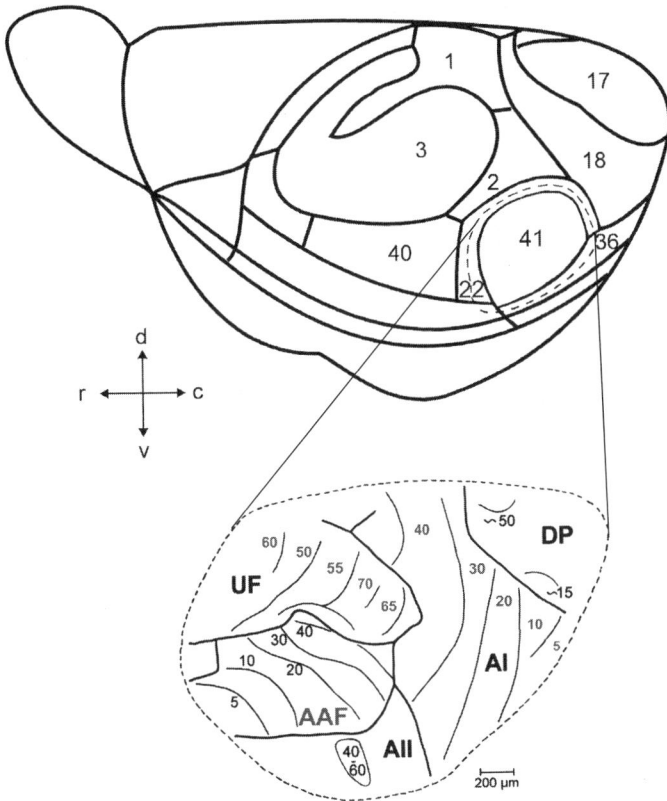

Figure 3-2.   View on the left-side neocortical surface of the mouse with cytologically defined fields indicated (Caviness 1975). The auditory cortical area is enlarged and an example of auditory cortical field arrangement together with isofrequency lines indicated (Stiebler et al. 1997). AI, primary auditory field; AII, secondary auditory field; AAF, anterior auditory field; DP, dorsoposterior field; UF, ultrasonic field; c, caudal; d, dorsal; r, rostral; v, ventral.

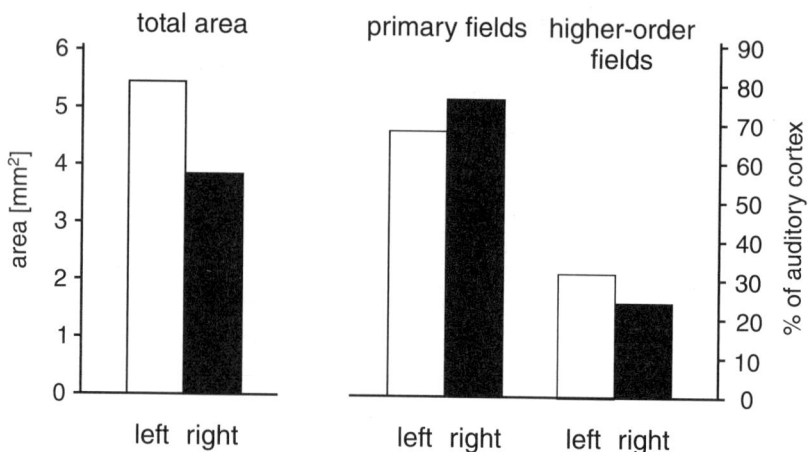

Figure 3-3. Average total area of the mouse auditory cortex of the left and right sides and the percentages of areas covered by primary fields (AI, AAF, UF) or by higher-order fields (AII, DP, and nonspecified areas) on the left and right sides (modified from Stiebler et al. 1997). The left-side auditory cortical area is significantly larger than that on the right side, mainly by a relatively larger size of higher-order fields.

of the auditory cortex having reciprocal connections with the auditory cortex (Hofstetter and Ehret 1992). This dorsal field is highly active in the left hemispheres of mothers when they recognize the ultrasonic calls of their pups (Geissler and Ehret 2004). It is significantly less active in their right hemispheres and in both hemispheres of virgin females without experience with pups who do not recognize the pup calls (Fichtel and Ehret 1999). Functional parallels between ultrasound recognition and the location of the dorsal field (DF in figure 3-4) in mice on the one hand and speech recognition and the location of the Wernicke area in humans on the other suggests that the mouse has a functional area, the dorsal field of the left neocortical hemisphere, that is homologous to the Wernicke area of the human brain.

These comparative aspects provide a first evidence that not only the brain functioning of species-specific call (speech) recognition but also the anatomical basis of this function has a long evolutionary history in the left hemisphere of the brain. The question remains, however, why specializations for communication sound processing are found in the left hemisphere and not in the right hemisphere of the mammalian brain.

Which Functions Are Lateralized?

Besides functions of auditory perception and vocal output described in the previous section, a number of further brain functions have been shown to be

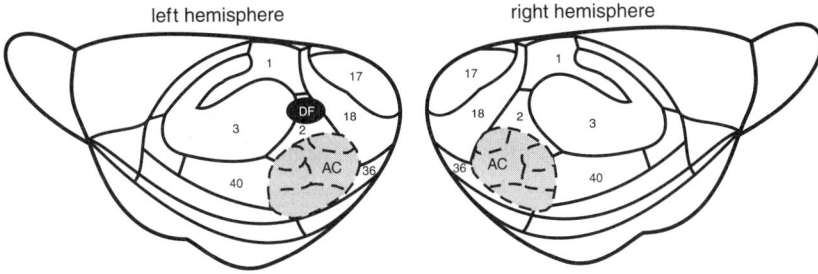

Figure 3-4. Position and average size of the auditory cortex (AC in gray) of the left and right hemispheres of the mouse forebrain. On the left side, we found a dorsal field (DF) having reciprocal connections with the ultrasonic field of the auditory cortex (Hofstetter and Ehret 1992). This DF is selectively activated only on the left side when mothers recognize ultrasonic isolation calls of pups (Fichtel and Ehret 1999; Geissler and Ehret 2004).

lateralized in various vertebrates, including humans. Some general functions listed in table 3-1 will be discussed here. Reviews can be found, for example, in Semmes 1968; Dimond 1979; Walker 1980; Bianki 1983; Geschwind and Galaburda 1987; Silberman and Weingartner 1986; Bisazza, Rogers, and Vallortigara 1998; and Andrew 1999.

A left-hemisphere advantage has been reported (zebrafish, chick, human) for the categorization of stimuli and for forming associations (memory) between stimuli (Andrew 1997, 1999; Koivisto 1998; Miklósi, Andrew, and Savage

Table 3-1   Some Lateralized Brain Functions in Vertebrates

Left-hemisphere advantage	Right-hemisphere advantage
Language (also sign-language), semantic and syntax processing and perception (human)	Perception of pitch, melody and timbre of speech and music (human)
Species-specific call recognition (mouse, monkey)	Visual and acoustic spatial processing (chick, human)
Complex, especially time-critical, sound processing and perception (rat, guinea pig, monkey, human)	Regulation of mood and affect (human)
Vocalization and song control (frog, songbird, monkey, human)	
Categorizations and associations (memory formation) of stimuli (zebrafish, chick, human)	
Perception and expression of positive emotions and display of approach behavior (chick, monkey, human)	Perception and expression of negative emotions and display of avoidance behavior (chick, monkey, human)
Selective (focal) attention to stimuli (zebrafish, chick, monkey, human)	Sustained and global arousal and attention (zebrafish, chick, monkey, human)

1998; McIntosh, Rajah, and Lobaugh 1999; Killgore et al. 2000). The stimuli were perceived by either eye or ear with a right-eye/ear advantage. Another left-hemisphere advantage occurs for the perception and expression of positive emotions (chick, monkey, human) that evoke attractions to certain stimuli and the search for social contact (Tucker 1981; Thompson 1985; Silberman and Weingartner 1986; Hartley et al. 1989; Vallortigara and Andrew 1991; Hook-Costigan and Rogers 1998). The positive emotions can be associated with expectation of a reward in conditioning paradigms, with the perception of friendly or happy faces, and with the sight of, listening to, or expectation to get in contact with social partners. Finally, a left-hemisphere dominance has been noted (zebrafish, chick, monkey, human) in situations in which selective (focal) attention is directed to acoustic or visual stimuli that are actually perceived or whose perception is expected (Hopkins, Morris, and Savage-Runbaugh 1991; Andrew 1997; Grady et al. 1997; Miklósi, Andrew, and Savage 1998; McIntosh, Rajah, and Lobaugh 1999).

In contrast to the left-hemisphere advantages mentioned so far, there are a number of functions in which a right-hemisphere processing dominates (table 3-1). In several respects, the right-hemisphere advantages relate to functions opposite or contrasting with those dominating in the left hemisphere. Although the right hemisphere has considerable language abilities (Gazzaniga and Smylie 1984; Baynes 1990), the left-hemisphere dominance in semantic and syntax processing is contrasted with a right-hemisphere advantage in processing and perception of melody, pitch, and timbre of sounds, including speech and music (Safer and Leventhal 1977; Bradshaw and Nettleton 1981; Sidtis 1984; Riecker et al. 2000). Similarly, advantages in visual and acoustical spatial processing and perception in the right half of the brain (chick, human; Bradshaw and Nettleton 1981; Andrew 1997; Griffiths et al. 1997) contrast with the advantage for complex and time-critical sound processing in the left hemisphere. Mood and affect in monkeys and humans seem generally to be regulated by the right hemisphere, which dominates, at the same time, the perception and expression of negative emotions and avoidance behavior (Tucker 1981; Thompson 1985; Silberman and Weingartner 1986; Hartley et al. 1989; Hook-Costigan and Rogers 1998). Finally, the right hemisphere dominates the generation and maintenance of global arousal and sustained attention in zebrafish, chicks, monkeys, and humans (Dimond 1979; Pardo, Fox, and Raichle 1991; Andrew 1997; Miklósi, Andrew, and Savage 1998; Sturm et al. 1999), while the left hemisphere is leading in selectively directing attention to certain stimuli.

Taking these lateralizations of functions in vertebrates together, we find an important division of general properties and labor between the two hemispheres of the brain. Following, in part, Bianki (1983), the left hemisphere dominates successive, temporal, analytical (selective), and abstract processing and positive emotions, while the right hemisphere dominates simultaneous, spatial, synthetic (general), and concrete processing and negative emotions, all both in perception and response generation. The hypothesis is that lateralizations of brain functions are common to and comparable among vertebrates and express successful evolutionary strategies leading to well-adapted organisms. If we accept this hypothesis, we can postulate that a disturbance or weakness of functional lateralizations in the brain of an individual must be accompanied by certain deficits in perception and behavior or even pathological states. In fact, this is the case. Abnormalities in anatomical, physiological, and chemical lateralizations in the human brain are characteristics of schizophrenic persons (Cowell, Fitch, and Denenberg 1999). These persons show, for example, a reduction or absence of a hemisphere dominance in speech processing and perception, which is a left-hemisphere dominance for most people (Rockstroh et al. 1998). In most autistic children, the right hemisphere is dominant in speech processing. This dominance is likely to shift to the left hemisphere, if these children improve their language abilities (Dawson et al. 1986). Left-handedness, found in about 10 percent of people (see review in Springer and Deutsch 1985), is associated with a general reduction of the degrees of cerebral lateralizations (Sherman and Galaburda 1985) and with higher incidence of immune disease, migraine, and developmental learning disorders, compared to right-handed people (Geschwind and Behan 1982). Thus, left-handedness is not a disease per se, but seems to develop on a background that is responsible for a weakness in laterality of brain functions and for an increased susceptibility to develop several weaknesses of brain and body functions. Fertility in mice (Collins 1985) and humans (Geschwind and Galaburda 1987) is reduced in subjects with reduced or atypical cerebral lateralizations. Since gay men and lesbians also show some atypical functional hemisphere asymmetries (McCormick and Witelson 1994) together with a reduced number of own children, they may be counted in the same category. These examples indicate that evolution favored not just some degree of lateralization of brain functions but a strong and "typical" expression of hemisphere dominances, if such dominances occurred at all.

It is evident that a considerable number of functions have been found to be lateralized in vertebrate brains, with failure of a sufficient strength of lateralization causing various functional deficits. I predict that with modern brain imaging techniques and sophisticated behavioral tests, many more lateralized functions in vertebrate brains will be discovered, especially functions of higher sensory analyses, functions being based on certain emotions or motivations, and functions depending on certain states of attention. The conclusions drawn twenty years ago (Walker 1980) from a review of more than two hundred studies on lateralizations of brain functions—"it is difficult to reject the null hypothesis that the vertebrate nervous system is an entirely symmetrical device, with the possible exception of the brains of humans and canaries"—are certainly obsolete today and are likely to be reversed in the future.

The Unsolved Question of Why

The question about why certain brain functions are lateralized calls for answers from both an evolutionary and a mechanistic point of view. In the previous sections, summarized in table 3-1, we have discussed already several aspects of the evolutionary why. Brain functions such as control of emotions, attention, categorization of stimuli, and coordination of vocalizations are lateralized in comparable ways in the same hemisphere of the brains of fish, amphibians, birds, and mammals. This provides a strong support for the suggestion that these lateralizations have proved to be advantageous for the survival of individual animals and may share common genetic and physiological bases.

There are a number of arguments in the literature about why individuals with strongly lateralized brain functions should be better off compared to those of only weak or no lateralizations. Levy (1977), on the background of research about split-brain patients (for a review, see Springer and Deutsch 1985), believed that a functionally symmetric brain would be a waste, at least in mammals that possess highly developed learning capacity and behavioral plasticity. Since both brain hemispheres can construct internal representations of the world by their own neural mechanisms, it would be a waste of one hemisphere (Levy probably thought of a neocortical hemisphere) if both representations were identical. By having two hemispheres, each with its own specializations mainly on the cognitive level, the intellectual capacity of the brain should be nearly doubled. In the same way Dimond (1979) argued that two complementary halves of the brain could do two different things at the same time—attend to and process different stimuli, prepare different responses—so that the possession of the two independent problem-solving organs could increase the likelihood of flexible and creative solutions to novel problems. Possible conflict between the hemispheres about the leadership in determining the perception and behavior of the individual is solved by giving each hemisphere the ultimate command over a certain set of perceptions and actions. Such a complementary division of labor is visible in table 3-1. Experiments about mouse-killing behavior in rats (Denenberg et al. 1986) and reaction-time studies in humans (Bryden and Bulman-Fleming 1994) support the view that strong lateralizations of brain functions are necessary to generate related behavioral outputs quickly and in a clear-cut way. Thus, strong lateralizations of neural processing in the brain increase the potential, economy, and adaptivity of brain and of the whole behavior.

These approaches to the evolutionary why sound pretty reasonable; however, they are still largely speculative. Even more speculative are approaches to explain the genetic, physiological, and morphological mechanisms leading to lateralizations of certain kinds in the brain. Recent molecular studies of the development of vertebrate embryos have identified more than eighteen genes such as Lefty-1, Lefty-2, and Nodal that are expressed differently in the left and right body sides. They influence the morphogenesis and the asymmetrical placement of inner organs (Tamura, Yonei-Tamura, and Belmonte 1999; Tsukui et al. 1999). Despite this recent progress, it is still unclear whether and how asymmetrical gene expression is involved in producing structural and functional lateralities of the vertebrate brain. A list of more than seventy studies describing various hemisphere asymmetries in the brains of vertebrates can be found in de Lacoste, Horvath, and Woodward (1988). Since (a) different structural asymmetries in the brain can be found in closely related mouse species (Slomianka and West 1987), (b) the types and degrees of asymmetry in a given species often depend on the sex of the individual (McGlone 1977; Robinson et al. 1985; Glick and Shapiro 1985; Diamond 1984; Sherman and Galaburda 1985; Ward, Giguère, and St.-Yves 1985; Collins 1985; Sandhu, Cook, and Diamond 1986; Vallortigara and Andrew 1991), and (c) even identical twins do not necessarily possess the same hemisphere dominances of brain functions (see chapter 5 in Springer and Deutsch 1985 and chapter 13 in Geschwind and Galaburda 1987), it is unlikely that specific genes are directly responsible for the generation of a left- or right-hemisphere dominance of a certain brain function. At present, it seems appropriate to accept the hypothesis that genes do not code for left or right but work on asymmetry gradients, which come about by environmental gradients and random events. Thus, genes determine the degrees but not the directions of asymmetries. This hypothesis has been developed and detailed by Collins (1975, 1977, 1985) on the basis of experiments on the handedness of mice of various strains. Collins's hypothesis allows understanding of functional dominances of the left or right hemisphere of the brain on both the individual and the population levels and of continued influences of internal factors, such as sex hormones, neurotransmitters, and various peptides, and external effects together biasing the development of certain degrees of functional lateralities.

Conclusions

In conclusion, lateralized brain functions are both an expression and a proof of the highly developed plasticity of the vertebrate, especially the mammalian, brain. In general, the right hemisphere's processing seems to represent the default settings of functions that are active and dominating as long as nothing special is required to perceive or do. The left hemisphere's processing, on the other hand, seems to dominate whenever complex and complicated tasks or problems have to be mastered. A beautiful example supporting these general conclusions has been published by Fabbro, Gran, and Gran (1991). They studied the hemisphere dominance of language processing in professional interpreters translating from one into another language simultaneously, for example at European conferences. These persons, contrary to the normal average, show a right-hemisphere advantage for syntax processing of their mother tongues and a left-hemisphere superiority in syntax processing of the languages they translate into their mother tongues. In these persons, syntax processing of their mother tongues is highly automatized and seems to require less attention compared to the syntax processing of the foreign languages they translate. Thus, mother-tongue processing with regard to syntax has become a default state of the brain and is shifted to the right hemisphere, probably to give room for the more demanding syntax processing of the foreign language in the left, the primary language-specialized hemisphere. This example shows that hemisphere dominances of brain functions, even in a single mode such as the speech mode, are plastic and can be adapted in order to reach an optimum of overall functioning of the brain.

Acknowledgments I am grateful to Terry Sejnowski for valuable suggestions on literature. My original work was supported by several grants of the Deutsche Forschungsgemeinschaft.

References

Alho, K., J. F. Connolly, M. Cheour, A. Lehtokoski, M. Huotilainen, J. Virtanen, R. Aulanko, and R. J. Ilmoniemi. 1998. Hemispheric lateralization in preattentive processing of speech sounds. *Neurosci. Lett.* 258: 9–12.

Andrew, R. J. 1997. Left and right hemisphere memory traces: Their formation and fate. Evidence from events during memory formation in the chick. *Laterality* 2: 179–198.

Andrew, R. J. 1999. The differential rolls of right and left sides of the brain in memory formation. *Behav. Brain Res.* 98: 289–295.

Bauer, R. H. 1993. Lateralization of neural control for vocalization by the frog (*Rana pipiens*). *Psychobiology* 21: 243–248.

Bavelier, D., D. Corina, P. Jezzard, V. Clark, A. Karni, A. Lalwani, J. P. Rauschecker, A. Braun, R. Turner, and H. J. Neville. 1998. Hemispheric specialization for English and ASL: left invariance-right variability. *NeuroReport* 9: 1537–1542.

Baynes, K. 1990. Language and reading in the right hemisphere: Highways or byways of the brain? *J. Cogn. Neurosci.* 2: 159–179.

Beecher, M. D., M. R. Petersen, S. R. Zoloth, D. B. Moody, and W. C. Stebbins. 1979. Perception of conspecific vocalizations by Japanese macaques. Evidence for selective attention and neural lateralization. *Brain Behav. Evol.* 16: 443–460.

Benson, D. F. 1986. Aphasia and the lateralization of language. *Cortex* 22: 71–86.

Berlin, C. I. 1977. Hemispheric asymmetry in auditory tasks. In *Lateralization in the Nervous System*, edited by S. Harnard, W. Doty, L. Goldstein, J. Jaynes, and G. Krauthamer, 303–323 (New York: Academic Press).

Bertoncini, J., J. Morais, R. Bijeljac-Babic, S. McAdams, I. Peretz, and J. Mehler. 1989. Dichotic perception and laterality in neonates. *Brain Lang.* 37: 591–605.

Bianki, V. L. 1983. Hemisphere specialization of the animal brain for information processing principles. *Intern. J. Neurosci.* 20: 75–90.

Bisazza, A., L. J. Rogers, and G. Vallortigara. 1998. The origins of cerebral asymmetry: A review of evidence of behavioural and brain lateralization in fishes, reptiles and amphibians. *Neurosci. Biobehav. Revs.* 22: 411–426.

Bradshaw, J., and N. Nettleton. 1981. The nature of hemispheric specialization in man. *Behav. Brain Sci.* 4: 51–91.

Broca, P. 1861. Remarques sur le siège de la faculté du langage articulé, suivie d'une observation d'aphemie. *Bulletin-Société Anatomique de Paris* 2: 330–357.

Bryden, M. P., and M. B. Bulman-Fleming. 1994. Laterality effects in normal subjects: evidence for interhemispheric interactions. *Behav. Brain Res.* 64: 119–129.

Caviness, V. S. 1975. Architectonic map of neocortex of the normal mouse. *J. Comp. Neurol.* 164: 247–264.

Chi, J., E. Dooling, and F. Giles. 1977. Left-right asymmetries of the temporal speech areas of the human fetus. *Arch. Neurol.* 34: 346–348.

Collins, R. L. 1975. When left-handed mice live in right-handed worlds. *Science* 187: 181–184.

Collins, R. L. 1977. Toward an admissible genetic model for the inheritance of the degree and direction of asymmetry. In *Lateralization in the Nervous System,* edited by S. Harnard, R. W. Doty, L. Goldstein, J. Janes, and G. Krauthamer, 137–150 (New York: Academic Press,).

Collins, R. L. 1985. On the inheritance of direction and degree of asymmetry. In *Cerebral Lateralization in Nonhuman Species,* edited by S. D. Glick, 41–71 (New York: Academic Press).

Corina, D. P., J. Vaid, and U. Bellugi. 1992. The linguistic basis of left hemisphere specialization. *Science* 255: 1258–1260.

Cowell, P. E., R. H. Fitch, and V. H. Denenberg. 1999. Laterality in animals: relevance to schizophrenia. *Schizophrenia Bull.* 25: 41–62.

Cutting, J. E. 1974. Two left-hemisphere mechanisms in speech perception. *Percept. Psychophys.* 16: 601–612.

Damasio, A., U. Bellugi, H. Damasio, H. Poizner, and J. van Gilder. 1986. Sign language aphasia during left-hemisphere amytal injection. *Nature* 322: 363–365.

Dawson, G., C. Finley, S. Phillips, and L. Galpert. 1986. Hemispheric specialization and the language abilities of autistic children. *Child Dev.* 57: 1440–1453.

Dax, M. 1865. Lesion de la moitié gauche de l'encéphale coincidant avec l'oublie des signes de la pensée (lu à Montpellier en 1836). *Gazette Hebdomadaire de Médicine et de Chirurgie* 33: 259.

de Lacoste, M. C., D. S. Horvath, and D. J. Woodward. 1988. Prosencephalic asymmetries in Lemuridae. *Brain Behav. Evol.* 31: 296–311.

Denenberg, V. H., J. S. Gall, A. Berrebi, and D.A. Yutzey. 1986. Callosal mediation of cortical inhibition in the lateralized rat brain. *Brain Res.* 397: 327–332.

Diamond, M. 1984. Age, sex, and environmental influences on anatomical asymmetry in rat forebrain. In *Biological Foundations of Cerebral Dominance,* edited by N. Geschwind, A. Galaburda, 134–146 (Cambridge, Mass.: Harvard University Press).

Dimond, S. J. 1979. Symmetry and asymmetry in the vertebrate brain. In *Brain, Behaviour and Evolution,* edited by D. A. Oakley and H. C. Plotkin, 189–218 (London: Methuen).

Ehret, G. 1987. Left-hemisphere advantage in the mouse brain for ultrasound recognition in a communicative context. *Nature* 325: 249–251.

Ehret, G. 1992. Preadaptations in the auditory system of mammals for phoneme perception. In *The Auditory Processing of Speech. From Sounds to Words,* edited by M. E. H. Schouten, 99–112 (Berlin: de Gruyter).

Ehret, G., and B. Haack. 1982. Ultrasound recognition in house mice: key-stimulus configuration and recognition mechanisms. *J. Comp. Physiol. A* 148: 245–251.

Fabbro, F., B. Gran, and L. Gran. 1991. Hemispheric specialization for semantic and syntactic components of language in simultaneous interpreters. *Brain Lang.* 41: 1–42.

Fichtel, I., and G. Ehret. 1999. Perception and recognition discriminated in the mouse auditory cortex by c-Fos labeling. *NeuroReport* 10: 2341–2345.

Fitch, R. H., C. P. Brown, K. O'Connor, and P. Tallal. 1993. Functional lateralization for auditory temporal processing in male and female rats. *Behav. Neurosci.* 107: 844–850.

Gallistel, C. R. 1980. *The Organization of Action: A New Synthesis.* (Hillsdale, N.J.: Lawrence Erlbaum).

Gannon, P. J., R. L. Holloway, D. C. Broadfield, and A. R. Braun. 1998. Asymmetry of chimpanzee planum temporale: Humanlike pattern of Wernicke's brain language area homolog. *Science* 279: 220–222.

Gazzaniga, M. S., and C. S. Smylie. 1984. What does language do for a right hemisphere? In *Handbook of Cognitive Neuroscience,* edited by M. S. Gazzaniga, 199–209 (New York: Plenum Press).

Geissler, D. B., and G. Ehret. 2004 Auditory perception vs. recognition: representation of complex communication sounds in the mouse auditory cortical fields. *Eur. J. Neurosci.* 19: 1027–1040.

Geschwind, N., and P. Behan. 1982. Left-handedness: Association with immune disease, migraine, and developmental learning disorder. *Proc. Natl. Acad. Sci. USA* 79: 5097–5100.

Geschwind, N., and A. M. Galaburda. 1987. *Cerebral Lateralization. Biological Mechanisms, Associations, and Pathology* (Cambridge, Mass.: MIT Press).

Glick, S. D., and R. M. Shapiro. 1985. Functional and neurochemical mechanisms of cerebral lateralization in rats. In *Cerebral Lateralization in Nonhuman Species,* edited by S. D. Glick, 157–183 (New York: Academic Press).

Grady, C. L., J. W. van Meter, J. M. Maisog, P. Pietrini, J. Krasuski, and J. P. Rauschecker. 1997. Attention-related modulation of activity in primary and secondary auditory cortex. *NeuroReport* 8: 2511–2516.

Griffiths, T. D., G. Rees, A. Rees, G. G. R. Green, C. Witton, D. Rowe, C. Büchel, R. Turner, and R. S. J. Frackowiak. 1997. Right parietal cortex is involved in the perception of sound movement in humans. *Nature Neurosci.* 1: 74–79.

Haack, B., H. Markl, and G. Ehret. 1983. Sound communication between parents and offspring. In *The Auditory Psychobiology of the Mouse,* edited by J. F. Willott, 57–97 (Springfield, Ill.: Thomas).

Hartley, L. R., N. Strother, P. K. Arnold, and B. Mulligan. 1989. Lateralization of emotional expression under a neuroleptic drug. *Physiol. Behav.* 45: 917–921.

Hauser, M. D., and K. Andersson. 1994. Left-hemisphere dominance for processing vocalizations in adult, but not infant, rhesus monkeys: Field experiments. *Proc. Natl. Acad. Sci. USA* 91: 3946–3948.

Hofstetter, K. M., and G. Ehret. 1992. The auditory cortex of the mouse: connections of the ultrasonic field. *J. Comp. Neurol.* 323: 370–386.

Hook-Costigan, M. A., and L. J. Rogers. 1998. Lateralized used of the mouth in production of vocalizations by marmosets. *Neuropsychologia* 36: 1265–1273.

Hopkins, W. D., R. D. Morris, and E. S. Savage-Runbaugh. 1991. Evidence for asymmetrical hemispheric priming using known and unknown warning stimuli in two language-trained chimpanzees (*Pan troglodytes*). *J. Exp. Psychol. General* 120: 46–56.

Hugdahl, K., K. Wester, and A. Asbjørnsen. 1990. The role of the left and right thalamus in language asymmetry: Dichotic listening in Parkinson patients undergoing stereotactic thalamotomy. *Brain Lang.* 39: 1–13.

Imaizumi, S., K. Mori, S. Kiritani, H. Hosoi, and M. Tonoike. 1998. Task-dependent laterality for cue decoding during spoken language processing. *NeuroReport* 9: 899–903.

Jackson, J. H. 1874. On the duality of the brain. In *Selected Writings of John Hughlings Jackson Vol. II*, edited by J. Taylor (London: Hodder and Stoughton).

Johnsrude, I. S., R. J. Zatorre, B. A. Milner, and A. C. Evans. 1997. Left-hemisphere specialization for the processing of acoustic transients. *NeuroReport* 8: 1761–1765.

Kapur, S., R. Rose, P. F. Liddle, R. B. Zipursky, G. M. Brown, D. Stuss, S. Houle, and E. Tulving. 1994. The role of the left prefrontal cortex in verbal processing: semantic processing or willed action? *NeuroReport* 5: 2193–2196.

Killgore, W. D. S., D. J. Casasanto, D. A. Yurgelun-Todd, J. A. Maldjian, and J. A. Detre. 2000. Functional activation of the left amygdala and hippocampus during associative encoding. *NeuroReport* 11: 2259–2263.

Kimura, D., and F. Folb. 1968. Neural processing of backwards-speech sounds. *Science* 161: 395–396.

King, C., T. Nicol, T. McGee, and M. Kraus. 1999. Thalamic asymmetry is related to acoustic signal complexity. *Neurosci. Lett.* 267: 89–92.

Kinsbourne, M., ed. 1978. *Asymmetrical Function of the Brain* (Cambridge: Cambridge University Press).

Koch, M. 1990. *Neuroendokrinologische Aspekte des Jungeneintrageverhaltens bei der Hausmaus (Mus musculus)*, *Konstanzer Dissertationen* vol. 265 (Konstanz: Hartung-Gorre, Konstanz).

Koivisto, M. 1998. Categorical priming in the cerebral hemispheres: automatic in the left hemisphere, postlexical in the right hemisphere? *Neuropsychologia* 36: 661–668.

Levy, J. 1977. The mammalian brain and the adaptive advantage of cerebral asymmetry. *Ann. NY Acad. Sci.* 299: 264–272.

Mateer, C. A. 1983. Motor and perceptual functions of the left hemisphere and their interaction. In *Language Functions and Brain Organization*, edited by S. J. Segalowitz, 145–170 (New York: Academic Press).

Mateer, C. A., and G. A. Ojemann. 1983. Thalamic mechanisms in language and memory. In *Language Functions and Brain Organization*, edited by S. F. Segalowitz, 171–191 (New York: Academic Press).

McCormick, C. M., and S. F. Witelson. 1994. Functional cerebral asymmetry and sexual orientation in men and women. *Behav. Neurosci.* 108: 525–531.

McGlone, J. 1977. Sex differences in the cerebral organization of verbal functions in patients with unilateral brain lesions. *Brain* 100: 775–793.

McIntosh, A. R., M. N. Rajah, and N. J. Lobaugh. 1999. Interactions of prefrontal cortex in relation to awareness in sensory learning. *Science* 284: 1531–1533.

Miklósi, A., J. Andrew, and H. Savage. 1998. Behavioural lateralisation of the tetrapod type in the zebrafish (*Brachydanio rerio*). *Physiol. Behav.* 63: 127–135.

Molfese, D. L., and V. J. Molfese. 1979. Hemisphere and stimulus differences as reflected in the cortical responses of newborn infants to speech stimuli. *Dev. Psychol.* 15: 505–511.

Molfese, V. J., D. L. Molfese, and C. Parsons. 1983. Hemisphere processing of phonological information. In *Language Functions and Brain Organization*, edited by S. J. Sagalowitz, 29–49 (New York: Academic Press).

Nass, R. D., and M. S. Gazzaniga. 1987. Lateralization and specialization in human central nervous system. In *Handbook of Physiology*, section 1, vol. 5, parts 1 and 2, *Higher Functions of the Brain*, edited by S. Plum, 701–761 (Bethesda, Md.: American Physiological Society).

Nottebohm, F. 1970. Ontogeny of bird song. *Science* 167: 950–956.

Nottebohm, F. 1980. Brain pathways for vocal learning in birds: a review of the first 10 years. *Prog. Psychobiol. Physiol. Psychol.* 9: 85–124.

Ojemann, G. A. 1983. Brain organization for language from the perspective of electrical stimulation mapping. *Behav. Brain Sci.* 2: 189–230.

Pardo, J. V., P. T. Fox, and M. E. Raichle. 1991. Localization of a human system for sustained attention by positron emission tomography. *Nature* 349: 61–64.

Petersen, M., M. Beecher, S. Zoloth, D. Moody, and W. Stebbins. 1978. Neural lateralization of species-specific vocalizations by Japanese macaques (*Macaca fuscata*). *Science* 202: 324–327.

Petersen, M. R., M. D. Beecher, S. R. Zoloth, S. Green, P. R. Marler, D. B. Moody, and W. C. Stebbins. 1984. Neural lateralization of vocalizations by Japanese macaques: communicative significance is more important than acoustic structure. *Behav. Neurosci.* 98: 779–790.

Riecker, A., H. Ackermann, D. Wildgruber, G. Dogil, and W. Grodd. 2000. Opposite hemispheric lateralization effects during speaking and singing at motor cortex, insula and cerebellum. *NeuroReport* 11: 1997–2000.

Rinne, T., K. Alho, P. Alku, M. Holi, J. Sinkkonen, J. Virtanen, O. Bertrand, and R. Näätänen. 1999. Analysis of speech sounds is left-hemisphere predominant at 100–150 ms after sound onset. *NeuroReport* 10: 1113–1117.

Robinson, T. E., J. B. Becker, D. M. Cant, and A. Mansour. 1985. Variation in the pattern of behavioral and brain asymmetries due to sex differences. In *Cerebral Lateralization in Nonhuman Species*, edited by S.D. Glick, 185–231 (New York: Academic Press).

Rockstroh, D., B. A. Clementz, C. Pantev, L. D. Blumenfeld, A. Sterr, and T. Elbert. 1998. Failure of dominant left-hemispheric activation to right-ear stimulation in schizophrenia. *NeuroReport* 9: 3819–3822.

Safer, M. A., and H. Leventhal. 1977. Ear differences in evaluating emotional tones of voice and verbal content. *J. Exp. Psychol., Human Percept. Perform.* 3: 75–82.

Sandhu, S., P. Cook, and M. C. Diamond. 1986. Rat cerebral cortical estrogen receptors: Male-female, right-left. *Exp. Neurol.* 92: 186–196.

Semmes, J. 1968. Hemispheric specialization: A possible clue to mechanism. *Neuropsychologia* 6: 11–26

Shapleske, J., S. L. Rossell, P. W. R. Woodruff, and A. S. David. 1999. The planum temporale: a systematic, quantitative review of its structural, functional and clinical significance. *Brain Res. Revs.* 29: 26–49.

Sherman, G. F., and A. M. Galaburda. 1985. Asymmetries in anatomy and pathology in the rodent brain. In *Cerebral Lateralization in Nonhuman Species*, edited by S. D. Glick, 89–107 (New York: Academic Press).

Sidtis, J. J. 1984. Music, pitch perception, and the mechanisms of cortical hearing. In *Handbook of Cognitive Neuroscience*, edited by M. J. Gazzaniga, 91–114 (New York: Plenum Press).

Silberman, E. K., and H. Weingartner. 1986. Hemispheric lateralization of functions related to emotion. *Brain Cognit.* 5: 322–353.

Slomianka, L., and M. J. West. 1987. Asymmetry in the hippocampal region specific for one of two closely related species of wild mice. *Brain Res.* 436: 69–75.

Springer, S. P., and G. Deutsch. 1985. *Left Brain, Right Brain* (New York: Freeman).

Stiebler, I. 1987. A distinct ultrasound-processing area in the auditory cortex of the mouse. *Naturwissenschaften* 74: 96–97.

Stiebler, I., R. Neulist, I. Fichtel, and G. Ehret. 1997. The auditory cortex of the house mouse: left-right differences, tonotopic organization and quantitative analysis of frequency representation. *J. Comp. Physiol. A* 181: 559–571.

Sturm, W., A. de Simone, B. J. Krause, K. Specht, V. Hesselmann, I. Radermacher, H. Herzog, L. Tellmann, H. W. Müller-Gärtner, and K. Willmes. 1999. Functional anatomy of intrinsic alertness: Evidence for a frontal-parietal-thalamic-brainstem network in the right hemisphere. *Neuropsychologia* 37: 797–805.

Tamura, K., S. Yonei-Tamura, and J. C. I. Belmonte. 1999. Molecular basis of left-right asymmetry. *Dev. Growth Differ.* 41: 645–656.

Thompson, J. K. 1985. Right brain, left brain; left phase, right phase: Hemisphericity and the expression of facial emotion. *Cortex* 21: 281–299.

Tsukui, T., J. Capdevila, K. Tamura, P. Ruiz-Lozano, C. Rodriguez-Estaban, S. Yonei-Tamura, J. Magallón, R. A. S. Chandraratna, K. Chien, B. Blumberg, R. M. Evans, and J. C. I. Belmonte. 1999. Multiple left-right asymmetry defects in Shh^{-II-} mutant mice unveil a convergence of the Shh and retinoic acid pathways in the control of *Lefty-1*. *Proc. Natl. Acad. Sci. USA* 96: 11376–11381.

Tucker, D. M. 1981. Lateral brain function, emotion, and conceptualization. *Psychol. Bull.* 89: 19–46.

Vallortigara, G., and R. J. Andrew. 1991. Lateralization of response by chicks to change in a model partner. *Anim. Behav.* 41: 187–194.

Wada, J. A., R. Clark, and A. Hamm. 1975. Cerebral hemispheric asymmetry in humans. *Arch. Neurol.* 32: 239–246.

Walker, S. F. 1980. Lateralization of functions in the vertebrate brain: A review. *Brit. J. Psychol.* 71: 329–367.

Ward, R., L. Giguère, and M. St.-Yves. 1985. Some behavioral differences between strongly and weakly lateralized mice. *Behav. Genet.* 16: 575–584.

Wernicke, E. C. 1874. *Der aphasische Symptomenkomplex* (Bresslau: Cohn & Weigart).

Witelson, S. F., and W. Pallie. 1973. Left hemisphere specialization for language in the newborn: neuroanatomical evidence of asymmetry. *Brain* 96: 641–647.

Woods, B. T. 1983. Is the left hemisphere specialized for language at birth? *Trends Neurosci.* 6 (1983): 115–117.

PART II

HOW IS THE CEREBRAL CORTEX ORGANIZED?

4

What Is the Function of the Thalamus?

S. Murray Sherman

What is the function of the thalamus? This may seem an odd question to pose because we know that virtually all information reaching the cortex, and thus conscious perception, passes through the thalamus. Thus an easy answer to the question is that the thalamus serves as a relay for the flow of information to the cortex. But if thalamus is a simple relay, as has been suggested on the basis of receptive field studies, which indicate that the receptive fields of relay cells projecting to the cortex are little different from those of their sensory afferents (reviewed in Sherman and Guillery 1996, 2001), why do we even have one? That is, why does the retina not project directly to the cortex instead of relaying through the lateral geniculate nucleus? The same question could be posed for the primary somatosensory and auditory relays. While I cannot answer any of these posed questions in detail or with any confidence, I can suggest insights that indicate that the relay role of the thalamus is complex and dynamically changing in ways that influence the nature of information reaching the cortex. Furthermore, based on a new interpretation of a variety of data, I can suggest that the thalamus does not merely relay information up to the cortex but also serves a vital role in cortico-cortical communication. To arrive at these conclusions, I shall review some of the intrinsic properties of thalamic relay cells as well as some of the functional circuitry of the thalamus and thalamocortical interactions. Although most of our knowledge of thalamic function derives from study of the lateral geniculate nucleus, the main principles described below appear to apply to all of thalamus.

The Low-Threshold Ca^{2+} Spike

Like any respectable neuron, the thalamic relay cell is endowed with a rich variety of gated membrane conductances, mostly gated by voltage (for details, see

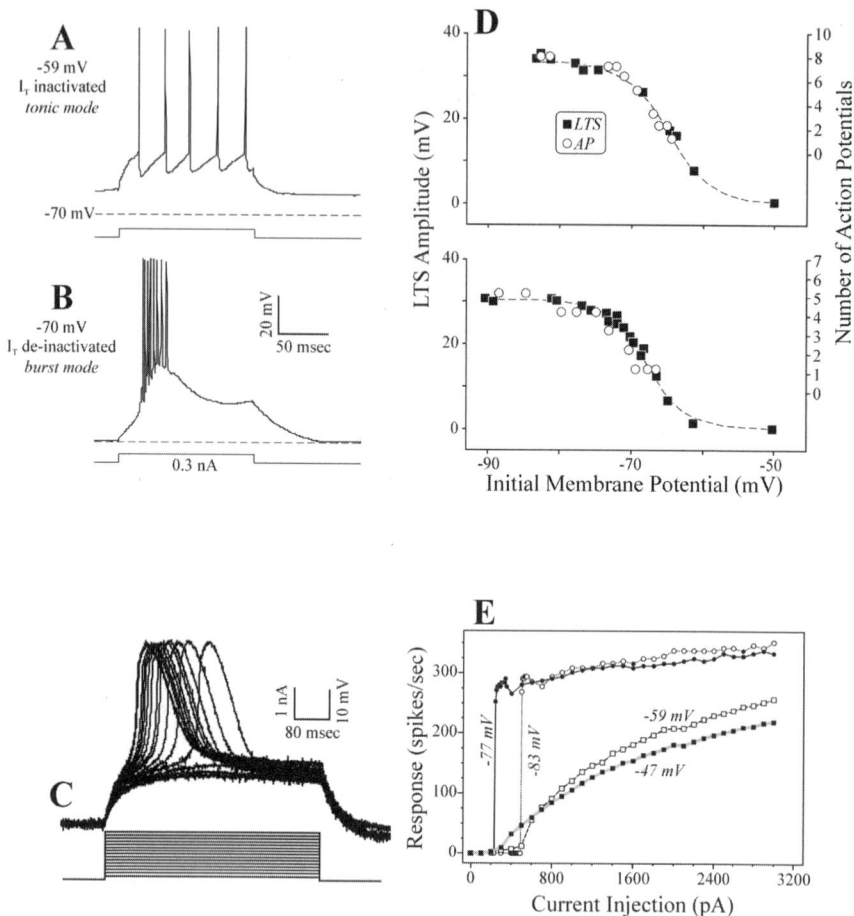

Figure 4-1. Voltage dependent properties of I_T and the low-threshold spike (LTS) shown in relay cells of the cat's lateral geniculate nucleus recorded intracellularly in an in vitro slice preparation. (A, B) Voltage dependency of the LTS. Responses are shown to the same depolarizing current pulse delivered intracellularly from two different initial holding potentials. When the cell is relatively depolarized (A), I_T is inactivated, and the cell responds to the suprathreshold stimulus with a stream of unitary action potentials. This is the tonic mode of firing. When the cell is sufficiently hyperpolarized to de-inactivate I_T (B), the current pulse activates an LTS with eight action potentials riding its crest. This is the burst mode of firing. (C) The all-or-none nature of LTSs activated from hyperpolarized cells in the presence of TTX. Current pulses were injected starting at 200 pA in amplitude and incremented in 10-pA steps. Smaller (subthreshold) pulses led to pure resistive-capacitative responses, but all larger (suprathreshold) pulses led to an LTS. Much like conventional action potentials, the LTSs are all the same amplitude regardless of the amplitude of suprathreshold depolarizing pulses, although there is latency variability for smaller suprathreshold pulses. Redrawn from Zhan et al. (1999). (D) Voltage dependency of amplitude of LTS and burst response. Examples for two cells are shown. The more hyperpolarized the cell before being activated (initial membrane potential), the larger the LTS and the more action potentials (AP) in the burst. The number of action potentials were measured first and then TTX was applied to isolate the LTS

McCormick and Huguenard 1992; Sherman and Guillery 1996, 2001). However, one that seems particularly important to relay functions and that has received considerable attention is the voltage-gated Ca^{2+} conductance underlying the *low-threshold spike*. This involves T-type Ca^{2+} channels that can exist in three states: *inactivated*, when the membrane has been relatively depolarized (beyond about -60 to -65 mV) for >50–100 msec[1]; *de-inactivated*, when the membrane has been relatively hyperpolarized (beyond about -60 to -65 mV) for >50–100 msec; and *activated*, which occurs when the channels are de-inactivated and then the membrane is sufficiently depolarized (e.g., via an EPSP). Activation of the channels opens them and leads to an inward current, I_T, carried by influx of Ca^{2+}, and this, in turn, leads to a largely all-or-none depolarization, which is the low-threshold spike. This spike is usually large enough to evoke a burst of conventional action potentials that ride its crest. This property of the low-threshold spike is ubiquitous for the thalamus: every relay cell of every thalamic nucleus of every mammalian species so far studied displays this property.

The properties of the T channel are qualitatively very similar to the more familiar Na^+ channel associated with the action potential, since this Na^+ channel has the same three states and is both voltage- and time-gated. However, there are three quantitative differences: (1) The Ca^{2+} channel operates in a regime more hyperpolarized by roughly 20 mV. (2) The time constants for inactivation and de-inactivation are almost two orders of magnitude slower for the Ca^{2+} channel. An interesting aspect of this is that once the membranes repolarize after an action potential, the Na^+ channel takes roughly 1 msec to de-inactivate, and this determines the absolute refractory period. After the membranes repolarize following a low-threshold spike, the Ca^{2+} channel takes roughly 100 msec to de-inactivate, and thus the limit to the frequency of low threshold spiking is about 10 Hz (Mukherjee and Kaplan 1995; Smith et al. 2000). (3) The Na^+ channels are found all along the axon, permitting propagation of the action potential down the axon to cortex, whereas the Ca^{2+} channels are found in appreciable numbers only in the soma and dendrites. Thus the only message reaching the cortex from relay cells is via action potentials. However, although the low-threshold spike does not propagate down the axon, it does propagate to the axon hillock and thus affects the firing of action potentials.

Figure 4-1 shows some of the features of the low-threshold spike from in vitro recording of cells of the cat's lateral geniculate nucleus. When a

for measurement. Redrawn from Zhan, Cox, and Sherman (2000). (E) Input-output relationship for one cell. The input variable is the amplitude of the depolarizing current pulse, and the output is the firing frequency of the cell. To compare burst and tonic firing, the firing frequency was determined by the first six action potentials of the response, since this cell usually exhibited six action potentials per burst in this experiment. The initial holding potentials are shown, and -47 mV and -59 mV reflects tonic mode, whereas -77 mV and -83 mV reflects burst mode. Redrawn from Zhan et al. (2000).

thalamic relay cell receives a depolarizing pulse (or EPSP), its response depends on its initial level of membrane polarization since this determines the state of I_T. If the cell is relatively depolarized with I_T inactivated (figure 4-1A), the response is a tonic stream of unitary action potentials; this represents the *tonic mode* of firing. If relatively hyperpolarized with I_T de-inactivated (figure 4-1B), I_T becomes activated, producing a low-threshold spike and burst of action potentials; this represents the *burst mode* of firing.

As noted, the low-threshold spike is evoked in an all-or-none manner, and this means that from any level of I_T de-inactivation, an evoked low-threshold spike is relatively invariant in size, meaning that larger suprathreshold depolarizations (or EPSPs) do not evoke larger low-threshold spikes (figure 4-1C). However, more initial hyperpolarization, on average, de-inactivates more Ca^{2+} channels (and thus more I_T), producing a larger low-threshold spike (figure 4-1D). This, in turn, produces a larger burst of action potentials relayed to the cortex (figure 4-1D).

Two obvious differences between firing modes can be deduced so far. First, note that tonic firing persists as long as the stimulus applied is suprathreshold, while the burst evoked would be the same for a wide range of stimulus durations. This has to do partly with the long refractory period for the low-threshold spike noted above. Second, during tonic firing, one would expect a larger EPSP to produce a higher firing rate, and it does (figure 4-1E). However, a larger EPSP does not produce a larger low-threshold spike or burst of action potentials, so there is a very nonlinear relationship between input and output during burst firing (figure 4-1E).

Role of Firing Modes in Thalamic Relays

Insights into the significance of these two relay modes come largely from studies of visual response properties of neurons in the lateral geniculate nucleus of lightly anesthetized and awake, behaving cats and monkeys. Studies of behaving animals make clear that both response modes are useful in relaying information to cortex, although in the fully alert animal, tonic firing is more common (McCarley, Benoit, and Barrionuevo 1983; Guido and Weyand 1995; Ramcharan, Gnadt, and Sherman 2000).[2] Both firing modes have also been described during thalamic recording in alert humans (Lenz et al. 1998; Radhakrishnan et al. 1999; Magnin, Morel, and Jeanmonod 2000). Studies of lightly anesthetized animals provide a more quantitative appreciation of the information relayed during both modes.

An analysis of raw information content relayed suggests that the amount is roughly the same during tonic and burst firing (Reinagel et al. 1999). However, the nature of the information differs (Sherman 1996), and that can be appreciated from responses to sinusoidal gratings drifting through the receptive field (figure 4-2A, B, *lower sections*). Here, during intracellular recording in vivo, response mode can be controlled by current injection that biases the membrane polarization to more depolarized to create tonic firing or more hyperpolarized for burst firing. During tonic firing, the response profile is sinusoidal, mirroring the input (as predicted from figure 4-1E) and reflecting a high degree of linear summation. During burst firing to the same stimulus, the response profile deviates from a sinusoidal shape, reflecting nonlinear distortion. Figure 4-2C shows for a population of geniculate cells in the cat that there is a dramatic difference in linearity between firing modes: tonic firing always results in better linearity.

There seems to be an obvious advantage for improved linearity with tonic firing: if the cortex is to reconstruct the outside world accurately, it requires information relayed through the thalamus to have minimal nonlinear distortion. What, then, might be an offsetting advantage for burst firing? Again, the neural responses provide a clue. Note that spontaneous activity is considerably higher during tonic firing (figure 4-2A, B, *upper sections*). The higher spontaneous activity actually helps sustain linearity by preventing inhibitory visual stimuli from bottoming out the response, leading to a nonlinearity via rectification of the response. It is more interesting that we can think of the spontaneous activity as background noise against which the signal—the visual response—must be detected. A glance at figure 4-2A, B suggests that the signal-to-noise ratio is higher during burst firing, which, in turn, suggests better signal detectability during burst firing. Detectability was assessed by the construction of receiver operating characteristic curves (for details, see Green and Swets 1966; Swets 1973; Macmillan and Creelman 1991), and the result (figure 4-2D) shows a dramatic advantage for burst firing on this measure (Guido et al. 1995).

Other, more subtle differences between burst and tonic firing exist (see, e.g., Mukherjee and Kaplan 1995; Sherman 1996; Smith et al. 2000), and more dramatic ones may emerge, but at present the most salient differences for relay function are these differences in linearity and detectability: tonic firing is better for linearity, and burst firing is better for signal detection. A speculative hypothesis that incorporates these differences goes as follows. If a group of geniculate cells has receptive fields in an unattended part of visual field—unattended because the animal is attending elsewhere, is using another sensory modality, is drowsy, and so on—these relay cells might be held in burst mode so they can more efficiently signal the presence of a novel, potentially interesting or threatening stimulus. Once the stimulus is detected, the response mode may be shifted to tonic to enable a more accurate analysis of the newly detected stimulus. For this speculative hypothesis to be plausible, there must be efficient ways for the brain to control the response mode of thalamic relay cells according to behavioral state.

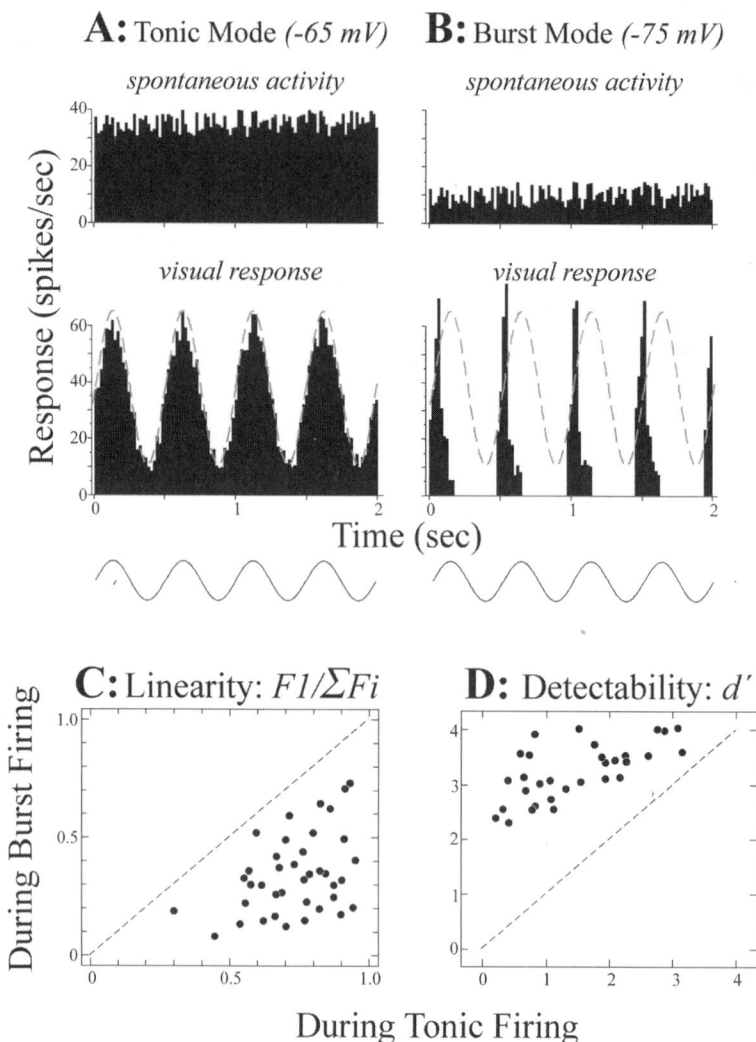

A: Tonic Mode *(-65 mV)*   **B:** Burst Mode *(-75 mV)*

spontaneous activity   *spontaneous activity*

visual response   *visual response*

Time (sec)

C: Linearity: $F1/\Sigma Fi$ **D:** Detectability: d'

During Burst Firing

During Tonic Firing

Figure 4-2. Tonic and burst responses of relay cells from the cat's lateral geniculate nucleus to visual stimulation during in vivo recording. (A, B) Average response histograms of responses of one cell to four cycles of drifting sinusoidal grating (bottom) and during spontaneous activity (top). The sinusoidal contrast changes resulting from the drifting grating are shown below the histograms. The cell was recorded intracellularly, and current injected through the recording electrode was used to bias membrane potential to more depolarized (−65 mV), producing tonic firing (A), or more hyperpolarized (−75 mV), producing burst firing (B). (C, D) Response linearity (C) and signal detectability (D) during tonic and burst firing (for details of how many of the data points were derived, see Guido, Lu, and Sherman 1992; Guido et al. 1995). Each point in the scatter plots reflects data from one relay cell of the cat's lateral geniculate nucleus recorded in vivo during visual stimulation, and the plots compare the response during tonic firing on the abscissa versus burst firing on the ordinate. The dashed

Control of Response Mode

The obvious place to look for ways to control response mode is in thalamic circuitry, since synaptic inputs to relay cells can alter membrane potential and thereby control the state of I_T. These inputs affect relay cells by releasing neurotransmitters that act on the postsynaptic cell via various postsynaptic receptors. These receptors come in two basic classes: ionotropic and metabotropic. Examples of the former receptors in thalamic functioning are AMPA and NMDA (for glutamate), $GABA_A$, and nicotinic (for acetylcholine); examples of the latter receptors are metabotropic glutamate, $GABA_B$, and muscarinic (for acetylcholine). The differences between ionotropic and metabotropic receptors are critical in their ability to control response mode.

Ionotropic and Metabotropic Receptors

Details of differences between these receptor types can be found elsewhere (Nicoll, Malenka, and Kauer 1990; Mott and Lewis 1994; Pin and Bockaert 1995; Pin and Duvoisin 1995; Recasens and Vignes 1995; Brown et al. 1997; Isaac et al. 1997), and only some are briefly outlined here. Transmission via ionotropic receptors is simpler and, as a result, much faster. The receptor itself is a complex, transmembrane protein that usually has an ion channel embedded within it. Binding of the neurotransmitter leads to a conformational change in the receptor that exposes the ion channel, thereby allowing flow of charged ions into or out of the cell. This leads to a postsynaptic potential (PSP) that is very fast, typically with a latency <1 msec and a peak duration of 10–20 msec or less. Metabotropic receptors have a rather indirect link, usually via G-proteins and second messenger pathways, to ion channels. For these, binding of the neurotransmitter unleashes a cascade of biochemical reactions that eventually leads to opening or closing of an ion channel, which in the case of thalamic relay cells is usually a K^+ channel. Opening the channel increases the flow of K^+ out of the cell, producing an IPSP, while closing the channel does the opposite, producing an EPSP. But the PSPs are slow, typically with a latency ≥ 10 msec and a duration of hundreds of msec or longer.

As noted in the section "The Low-Threshold Ca^{2+} Spike" above, to change the firing mode of a thalamic relay cell requires sustaining for $\geq \sim 100$ msec a

---

line of slope 1 is also shown in each plot. Linearity (C) was determined from Fourier analysis, and a linearity index was created by dividing the first Fourier component, which is linear, by the sum of the higher order distortion components. Thus the higher the index, the more linear the response. Note that every single cell shows more linearity during tonic firing. For detectability (D), d' values were determined from receiver operating characteristic curve analysis (for details, see Green and Swets 1966; Swets 1973; Macmillan and Creelman 1991). Note that every single cell shows better detectability during burst firing.

hyperpolarization (to switch to burst mode) or depolarization (to switch to tonic mode). Fast PSPs via ionotropic receptors are ill suited to do this, although it is possible with temporal summation for activation of ionotropic receptors to manage the job. However, activation of metabotropic receptors seems ideally suited to control response mode because the PSPs evoked are sustained enough to do the job.

Thalamic Circuitry and Receptor Types

Figure 4-3 summarizes for the lateral geniculate nucleus most of the inputs to relay cells and the receptor types they activate (reviewed in Sherman and

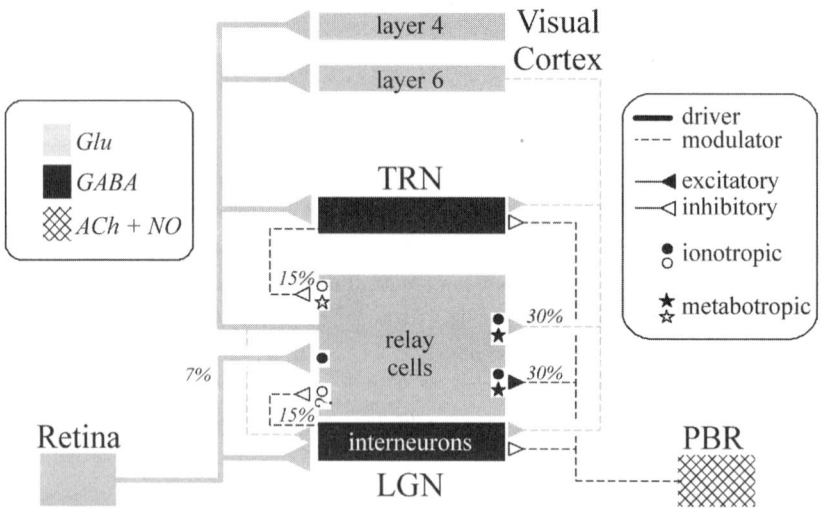

Figure 4-3. Circuitry of the lateral geniculate nucleus showing inputs to relay cells plus neurotransmitters and related receptors. Other thalamic nuclei seem to be organized along the same pattern (Sherman and Guillery 1996, 2001). The retinal input activates only ionotropic receptors, whereas all nonretinal inputs activate metabotropic receptors and often ionotropic receptors as well. The question mark related to interneuronal input indicates uncertainty whether metabotropic receptors are involved. The percentages indicate approximate contribution of each of the inputs to relay cells in terms of the number of actual synapses provided. Not shown for simplicity are other small modulatory inputs to geniculate relay cells; these include noradrenergic inputs from the parabrachial region, serotonergic inputs from the dorsal raphé nucleus, and histaminergic inputs from the tuberomamillary nucleus of the hypothalamus (for details, see Sherman and Guillery 1996, 2001). The relative number of synapses onto relay cells from various inputs is also shown. *Abbreviations:* ACh, acetylcholine; GABA, γ-aminobutyric acid; Glu, glutamate; LGN, lateral geniculate nucleus; NO, nitric oxide; PBR, parabrachial region of the brainstem; TRN, thalamic reticular nucleus.

Guillery 1996, 2001). Other thalamic nuclei have similar patterns of circuitry. What is striking about the pattern shown in figure 4-3 is that retinal input, which is the primary sensory input to be relayed to the cortex, activates only ionotropic receptors, whereas all of the nonretinal afferent pathways activate metabotropic receptors (and often ionotropic as well). The retinal input is glutamatergic, and the nonretinal inputs include a glutamatergic feedback input from layer six of the cortex, a cholinergic input from the brainstem parabrachial region, and a GABAergic input from intrinsic interneurons and cells of the adjacent thalamic reticular nucleus. Omitted for simplicity are much smaller inputs that also activate metabotropic receptors: a noradrenergic input, also from the parabrachial region, a serotonergic input from the dorsal raphé nucleus, and a histaminergic input from the tuberomamillary nucleus of the hypothalamus (reviewed in Sherman and Guillery 1996, 2001).

This pattern has several implications. The fact that retinal inputs activate only ionotropic receptors means that the EPSPs used in retinogeniculate transmission are relatively fast. Thus retinal input itself is less likely to change response mode, and this makes sense because the bulk of control should be the responsibility of pathways that reflect the dynamic needs of the system, not the primary sensory input to be relayed. Also, the slow EPSPs related to metabotropic receptors cannot reflect fast changes in the afferent input pattern. In fact, these slow PSPs act like a low pass temporal filter, filtering out fast input signals. Thus the association of only ionotropic receptors with retinal input helps preserve fast-changing temporal events in the signals relayed to the cortex, and this seems ideal for sensory transmission.

The concentration of metabotropic receptors with nonretinal inputs means that these pathways are well designed to control response mode. The direct inputs to relay cells from layer six of the cortex and from the parabrachial region are able to produce sustained EPSPs via activation of metabotropic receptors, and this should effectively switch firing mode from burst to tonic. Some evidence for this exists. Activation of metabotropic glutamate receptors from firing of corticogeniculate axons does indeed promote tonic firing (McCormick and von Krosigk 1992; Godwin, Vaughan, and Sherman 1996) in geniculate relay cells, as does activation of inputs from the parabrachial region (Lu, Guido, and Sherman 1993). The opposite—a switch from tonic to burst firing—is possible from activation of $GABA_B$ receptors, which can be achieved from reticular cells and possibly from interneurons. While experimental evidence for this has not yet been gathered, it seems a plausible scenario. Note that external control of firing mode by this scenario rests with cortical and brainstem input (see figure 4-3): directly, they depolarize relay cells to effect tonic firing; indirectly, via reticular (and possibly also interneuronal) input, they hyperpolarize to effect burst firing.

Drivers and Modulators

Afferent inputs to thalamic relay cells are not all the same but can be divided into at least two functionally distinct groups: drivers and modulators (Sherman and Guillery 1998). Drivers are the inputs that convey the basic information to be relayed to the cortex. In the lateral geniculate nucleus, this is the retinal input, and in other primary sensory relay nuclei, drivers can also be clearly defined as the medial lemniscal afferents for the ventral posterior nucleus (the primary somatosensory relay) and the inferior collicular afferents for the ventral part of the medial geniculate nucleus (the primary auditory relay). Drivers can also be clearly recognized for some other thalamic relays, but in many, the identity of the drivers is not yet obvious. By definition, any afferents that are not drivers are modulators. Their job is to provide modulation of the thalamic relay. An example of modulation is the control of response mode as described in the previous paragraph. A more detailed description of the distinction between drivers and modulators can be found elsewhere (Sherman and Guillery 1998), but a brief summary of there differences is as follows (see figure 4-3; see also Sherman and Guillery 1996, 1998, 2001):

- The driver (retinal) afferents innervate relay cells and interneurons but fail to innervate the thalamic reticular nucleus, whereas the modulator (nonretinal) afferents innervate the thalamic reticular nucleus as well.
- Driver inputs activate only ionotropic receptors, whereas modulator inputs activate metabotropic receptors as well and sometimes do so exclusively.
- Driver axons exhibit a morphology known as type 2 (Guillery 1966), having thick axons with richly branched, flowery, dense terminal arbors. Most modulatory inputs have type-1 axons (Guillery 1966), which involve thin axons with few preterminal branches and terminals *en passant* or on short side branches.
- Driver synaptic terminals are larger than any others in the geniculate neuropil, and they contact proximal dendrites, often in glomeruli and exhibiting triadic arrangements with terminals from interneurons.
- Driver axons often (perhaps always, as is the case for retinal axons) branch to innervate extrathalamic structures; at least some modulatory afferents (i.e., from cortex and the thalamic reticular nucleus) innervate only thalamic structures; whether brainstem modulatory afferents branch to innervate extrathalamic targets is not presently known.

The numbers of synapses provided by drivers versus modulators is also interesting (see figure 4-3). In the lateral geniculate nucleus, the driver (retinal) input provides only 5 to 10 percent of synaptic input to relay cells, and limited evidence from other sensory thalamic relays also indicates that the driver inputs provide a small minority of synapses there as well. It has been argued elsewhere (Sherman and Guillery 1998, 2001) that this makes sense because bringing the basic information to a thalamic nucleus does not require as much synaptic investment as would be required for fine modulation of the relay.

A further interesting speculation is that this idea of driver inputs to an area being small in number may not be limited to the thalamus. For example, it now seems clear that the main information is brought to layer-four cells in the visual cortex by geniculocortical axons (Reid and Alonso 1995, 1996; Ferster, Chung, and Wheat 1996; Chung and Ferster 1998), yet these provide only about 5 to 10 percent of the synaptic inputs to these cells (Ahmed et al. 1994; Latawiec, Martin, and Meskanaite 2000). Perhaps the similarity of these numbers with those of retinal inputs to relay cells is a coincidence, but maybe not, and perhaps it signifies a general property of drivers in the central nervous system. For example, an analysis of synaptic counts on spinal motoneurons indicates that Ia afferents, which constitute a major driver input, provide <5 percent of the synaptic terminals to these cells (reviewed on p. 462 of Henneman and Mendell 1981). It will be interesting to see how general beyond the sensory thalamic relays this finding is that driver inputs constitute a small minority of synapses.

An important point that is reiterated below is that in many areas of the brain there is a tendency to equate functional importance with the size of the input to an area. This, of course, completely ignores the fact that different inputs may be functionally quite different and thus cannot be compared anatomically. If that strategy were applied to the lateral geniculate nucleus, one would come to the silly conclusion that retinal input is of minor importance to a thalamic nucleus that relays brainstem parabrachial information to the cortex!

First- and Higher-Order Relays

Layer-Six versus Layer-Five Corticothalamic Inputs

Guillery (1995) first refocused attention on a long-known curiosity about thalamic relays: that while all seem to receive a generally reciprocal, modulatory feedback from layer six of the cortex, some in addition receive an input from layer five. These latter thalamic relays do not have an obvious subcortical driver input, and the suggestion has been offered that their driver inputs come instead from layer five of certain cortical areas (Guillery 1995; Sherman and Guillery 1998, 2001). Examples of nuclei that receive layer-five afferents are the pulvinar,[3] the posterior medial nucleus,[4] the magnocellular division of the medial geniculate nucleus (as opposed to the ventral division, which receives inferior collicular input), the medial dorsal nucleus, and others (for details, see Guillery 1995; Sherman and Guillery 1996, 2001). It is not clear if the layer-five innervation supplies the entirety of each of these nuclei or only certain as-yet-undefined segments, and there may be sparse layer-five input to certain primary sensory nuclei like the ventral posterior nucleus. It is clear that the layer-five afferents are quite unlike layer-six afferents but bear a striking anatomical resemblance to driver afferents as described for the main sensory relays (compare with the driver/modulator bulleted list above; for details, see Sherman and Guillery 1996, 1998, 2001; Vidnyánszky et al. 1996):

- They innervate dorsal thalamic nuclei but fail to innervate the thalamic reticular nucleus with collateral branches even though they pass through this region *en route* to their dorsal thalamic target.
- Where studied, layer-five inputs activate only ionotropic receptors on relay cells, whereas layer-six inputs activate metabotropic receptors as well.
- Their axons are thick, with type-2 morphology and terminal fields, while layer-six afferents have type-1 morphology.
- Their synaptic terminals are quite large and seem to innervate proximal dendrites, often in glomeruli and exhibiting triadic arrangements with terminals from interneurons.
- Many, if not all, branch to innervate extrathalamic targets.

The above list matches point-for-point the earlier list that distinguishes drivers from modulators. It thus seems reasonable to regard these layer-five afferents as drivers in the same sense that we consider retinal afferents as the drivers for the lateral geniculate nucleus. If so, this may offer an important insight into the function of these relays with layer-five inputs, because no obvious subcortical driver inputs had been previously suggested.

Two Types of Thalamic Relay

Thus some thalamic relays receive their driver input from subcortical sources, like the retina, brainstem, and so on, and relay this information to the cortex, whereas others receive their main driver input from the cortex itself and relay this information to another cortical area. Some relay nuclei, like perhaps pulvinar, may receive both subcortical and layer-five driver inputs, but unless individual relay cells receive both types—a genuine possibility, but one that is ignored here due to our ignorance on the subject—functionally, the subcortical and layer-five relays can be considered as distinct types. We can consider the former type of relay as *first order* because it represents a first pass of the relevant information into the cortex and the latter as *higher order*, because it represents a relay of information that has already reached cortex but from one cortical area to another (see figure 4-4 and Guillery 1995).

It is interesting that when thought of this way, each of the main sensory systems has both types of relay: for vision, the lateral geniculate nucleus is the first-order relay, and most of the pulvinar is the higher-order relay; for somesthesis, the ventral posterior nucleus is the first-order relay, and the medial portion of the posterior complex is the higher-order relay; and for audition, the ventral part of the medial geniculate nucleus is the first-order

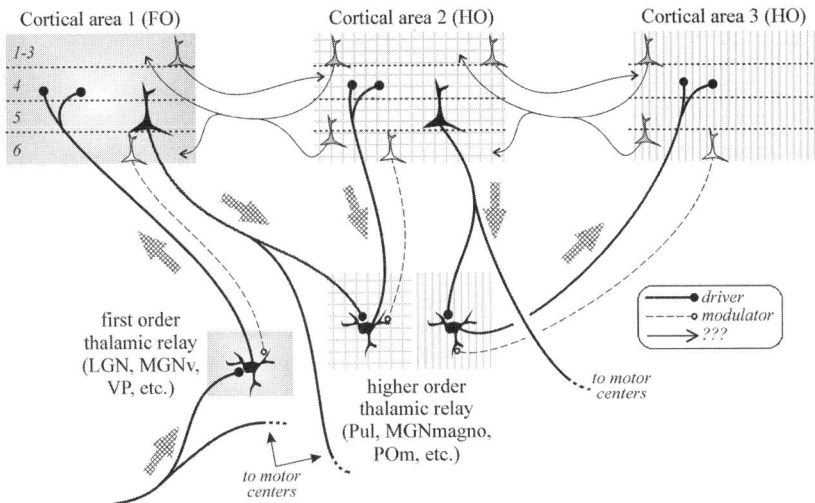

Figure 4-4. Schema of the hypothesis that corticocortical information flow involves a higher-order thalamic relay. The first-order relay (e.g., lateral geniculate nucleus) relays a driver (e.g., retinal) input to primary cortex (e.g., V1). From here, information is relayed among cortical areas via corticothalamocortical paths involving different regions of a higher-order thalamic nucleus (e.g., the pulvinar) routes. Thick, dark pathways represent the drivers, and thin, lighter pathways with dashed lines represent the modulators. The nature of direct corticocortical projections (thin solid lines) is ambiguous as to identity as driver or modulator. *Abbreviations:* FO, first order; HO, higher order; LGN, lateral geniculate nucleus; MGNmagno, magnocellular region of medial geniculate nucleus; MGNv, ventral region of medial geniculate nucleus; POm, posterior medial nucleus; Pul, pulvinar; VP, ventral posterior nucleus.

relay, and the magnocellular part is the higher-order relay. Olfactory information reaches the cortex in an unusual way that makes it difficult to fit into this duality, but much olfactory information is ultimately relayed to the cortex via the medial dorsal nucleus. It is interesting that much of the medial dorsal nucleus, a very large thalamic relay that innervates the frontal cortex, is a higher-order relay. How far this neat duality can be applied for other types of information relayed through the thalamus remains to be determined.

Implications for Corticocortical Communication

The visual world in carnivores and primates is analyzed by many different areas of cortex (more than thirty in monkeys) in the occipital, parietal, and temporal lobes (for reviews, see Van Essen 1985; Felleman and Van Essen 1991; Van Essen, Anderson, and Felleman 1992). Attempts to understand how these areas communicate with one another in visual analysis has to date focused almost entirely on direct corticocortical connections, which are rich and often reciprocal (see figure 2 of Van Essen, Anderson, and Felleman 1992 for details). Strategies exist to distinguish feedforward from feedback pathways among these connections (Felleman and Van Essen, 1991). The basic notion here, which is challenged later in this chapter, is that visual information enters the striate cortex from the lateral geniculate nucleus, and in a more or less hierarchical set of feedforward connections, the information is passed from the striate cortex to higher and higher areas, with many feedback connections, the function of which remains obscure. Note that, according to this view, once the information reaches the cortex from the lateral geniculate nucleus, it stays within the cortex, being routed effectively only amongst cortical areas. Among other drawbacks, this view of visual processing has little regard for the pulvinar, which is a much larger thalamic structure than is the lateral geniculate nucleus and which seems to innervate all of the extrastriate visual areas.

Apparently, the main and perhaps sole reason this view is so widely held is due to the very massive nature of direct corticocortical connections. Indeed, each cortical area receives the vast majority of its extrinsic afferents from other cortical areas and rather little from subcortical structures, like the thalamus. But this linking of functional importance of a pathway with its size is the very thinking that, as I suggested at the end of this section's chapter on drivers and modulators, would lead one to conclude that retinal input to the lateral geniculate nucleus is functionally of little consequence.

An alternative view of corticocortical communication is offered here to contrast with the traditional one. For cortical afferents, just as for those of the thalamus, it may be that drivers and modulators exist. The drivers carry the main information, and identifying them among afferents to a cortical area becomes supremely important. Perhaps the driver inputs are a small minority, as in the thalamus. Then a blind concentration on large pathways, which describes most direct corticocortical connections, may be misleading. That is, perhaps only a small minority of these direct pathways are drivers, with the rest being modulators.

The most extreme view, which I offer here for clarity, is that *none* of the direct corticocortical projections is a driver, and instead they are all modulators. The drivers, then, are the thalamocortical afferents. By this extreme version, the information route for corticocortical communication travels from layer five of one area down to a higher order thalamic relay (i.e., pulvinar for visual cortical communication) and then back up to the target cortical area (see figure 4-4). Just as retinal information passes through a thalamic relay before reaching the cortex, a more general rule may be that *any* new information coming into a cortical area, whether originating subcortically or in another cortical area, benefits from a thalamic relay.

A less extreme and perhaps more plausible hypothesis is that one important route for corticocortical communication via drivers involves a relay through higher-order thalamic nuclei, but that another route involves some of the direct corticocortical connections, presumably a minority, the rest being modulatory. If so, there remains an important difference between corticocortical drivers and those involving corticothalamocortical routes. Information carried by the former stays strictly within the cortex, but that carried by the latter pathway also informs other parts of the subcortical central nervous system. Whatever the ultimate accuracy of this hypothesis, it does draw attention to the need to avoid treating all connections among cortical areas as functionally equivalent.

Conclusions

We are far from a definitive answer to the question posed: What is the function of the thalamus? However, recent research offers several glimpses of partial answers. This is probably just the tip of the iceberg, meaning that as we learn more about thalamic relays, we are likely to see more and more key functions attributed to the thalamus. The suggestions here take two forms.

First, the complex cell and circuit properties of thalamic nuclei leave little doubt that the relay of information to the cortex is an active and mutable process. Clearly, these thalamic relays can affect the nature of information arriving in the cortex. How these different relay properties are controlled is a related issue of great importance. Specific suggestions have been offered here about how circuit properties control a voltage-dependent conductance, I_T, in relay cells to control responsiveness, and how this could affect the nature of information relayed to the cortex. However, this control of tonic and burst response modes is likely to be just one of many mechanisms by which thalamic relays can control the flow of information to cortex.

Second, not only do thalamic relays play an active role in relaying information to the cortex, they may also play a key role in corticocortical communication. The discovery that many thalamic regions seem to receive their driving input from layer five of cortex itself leads to the suggestion that much of corticocortical communication involves a route through the thalamus, with the same advantages of having a thalamic relay for this route as exists for relaying, say, retinal information to the cortex. As a corollary, the direct connections among cortical areas need to be reconsidered with regard to the nature of these pathways and the possibility that many, and perhaps all, are modulatory in nature. Thus the full impact of the thalamus may be much more than simply controlling the flow of information to the cortex: it may remain an active partner in all cortical computations.

Notes

1. Actually, the state of the T channel is a complex function of voltage and time, so stronger polarizations take less time to affect the change in the channel's inactivation state.

2. We are dealing here only with the function of thalamic relays in the behaving animal. During sleep and in certain pathological conditions, relay cells tend to burst rhythmically at frequencies varying up to about 10 Hz, and large assemblies of these cells manage to synchronize their firing. This is very different from the bursting seen in lightly anesthetized and behaving animals, which is nonrhythmic (Steriade and McCarley 1990; Steriade, McCormick, and Sejnowski 1993; Ramcharan, Gnadt, and Sherman 2000).

3. For simplicity, the term *pulvinar* includes the lateral posterior nucleus in carnivores.

4. Terminology across species can often be confusing. The primate equivalent to the medial portion of the posterior complex in rodents and carnivores is the anterior or "oral" part of pulvinar. The nonprimate terminology is used here, and so the pulvinar (which includes what is sometimes called the lateral posterior nucleus) is a structure associated essentially only with vision.

References

Ahmed B, Anderson JC, Douglas RJ, Martin KAC, Nelson JC (1994) Polyneuronal innervation of spiny stellate neurons in cat visual cortex. *J Comp Neurol* 341: 39–49.

Brown DA, Abogadie FC, Allen TG, Buckley NJ, Caulfield MP, Delmas P, Haley JE, Lamas JA, Selyanko AA (1997) Muscarinic mechanisms in nerve cells. *Life Sciences* 60: 1137–1144.

Chung S, Ferster D (1998) Strength and orientation tuning of the thalamic input to simple cells revealed by electrically evoked cortical suppression. *Neuron* 20: 1177–1189.

Felleman DJ, Van Essen DC (1991) Distributed hierarchical processing in the primate cerebral cortex. *Cerebral Cortex* 1: 1–47.

Ferster D, Chung S, Wheat H (1996) Orientation selectivity of thalamic input to simple cells of cat visual cortex. *Nature* 380: 249–252.

Godwin DW, Vaughan JW, Sherman SM (1996) Metabotropic glutamate receptors switch visual response mode of lateral geniculate nucleus cells from burst to tonic. *J Neurophysiol* 76: 1800–1816.

Green DM, Swets JA (1966) *Signal Detection Theory and Psychophysics.* New York: Wiley.

Guido W, Lu S-M, Sherman SM (1992) Relative contributions of burst and tonic responses to the receptive field properties of lateral geniculate neurons in the cat. *J Neurophysiol* 68: 2199–2211.

Guido W, Lu S-M, Vaughan JW, Godwin DW, Sherman SM (1995) Receiver operating characteristic (ROC) analysis of neurons in the cat's lateral geniculate nucleus during tonic and burst response mode. *Visual Neurosci* 12: 723–741.

Guido W, Weyand T (1995) Burst responses in thalamic relay cells of the awake behaving cat. *J Neurophysiol* 74: 1782–1786.

Guillery RW (1966) A study of Golgi preparations from the dorsal lateral geniculate nucleus of the adult cat. *J Comp Neurol* 128: 21–50.

Guillery RW (1995) Anatomical evidence concerning the role of the thalamus in corticocortical communication: A brief review. *J Anat* 187: 583–592.

Henneman E, Mendell LM (1981) Functional organization of motoneuron pool and its inputs. In *Handbook of Physiology*, section 1, vol II, part 1, edited by Brooks VB, 423–507. (Bethesda, Md.: American Physiological Society).

Isaac JTR, Crair MC, Nicoll RA, Malenka RC (1997) Silent synapses during development of thalamocortical inputs. *Neuron* 18: 269–280.

Latawiec D, Martin KAC, Meskenaite V (2000) Termination of the geniculocortical projection in the striate cortex of macaque monkey: A quantitative immunoelectron microscopic study. *J Comp Neurol* 419: 306–319.

Lenz FA, Garonzik IM, Zirh TA, Dougherty PM (1998) Neuronal activity in the region of the thalamic principal sensory nucleus (ventralis caudalis) in patients with pain following amputations. *Neurosci* 86: 1065–1081.

Lu S-M, Guido W, Sherman SM (1993) The brainstem parabrachial region controls mode of response to visual stimulation of neurons in the cat's lateral geniculate nucleus. *Visual Neurosci* 10: 631–642.

Macmillan NA, Creelman CD (1991) *Detection Theory: A User's Guide.* Cambridge: Cambridge University Press.

Magnin M, Morel A, Jeanmonod D (2000) Single-unit analysis of the pallidum, thalamus and subthalamic nucleus in Parkinsonian patients. *Neurosci* 96: 549–564.

McCarley RW, Benoit O, Barrionuevo G (1983) Lateral geniculate nucleus unitary discharge in sleep and waking: state- and rate-specific aspects. *J Neurophysiol* 50: 798–818.

McCormick DA, Huguenard JR (1992) A model of the electrophysiological properties of thalamocortical relay neurons. *J Neurophysiol* 68: 1384–1400.

McCormick DA, von Krosigk M (1992) Corticothalamic activation modulates thalamic firing through glutamate "metabotropic" receptors. *Proc Natl Acad Sci USA* 89: 2774–2778.

Mott DD, Lewis DV (1994) The pharmacology and function of central GABA$_B$ receptors. *Int Rev of Neurobiol* 36: 97–223.

Mukherjee P, Kaplan E (1995) Dynamics of neurons in the cat lateral geniculate nucleus: in vivo electrophysiology and computational modeling. *J Neurophysiol* 74: 1222–1243.

Nicoll RA, Malenka RC, Kauer JA (1990) Functional comparison of neurotransmitter receptor subtypes in mammalian central nervous system. *Physiol Rev* 70: 513–565.

Pin JP, Bockaert J (1995) Get receptive to metabotropic glutamate receptors. *Curr Opin Neurobiol* 5: 342–349.

Pin JP, Duvoisin R (1995) The metabotropic glutamate receptors: structure and functions. *Neuropharmacol* 34: 1–26.

Radhakrishnan V, Tsoukatos J, Davis KD, Tasker RR, Lozano AM, Dostrovsky JO (1999) A comparison of the burst activity of lateral thalamic neurons in chronic pain and non-pain patients. *Pain* 80: 567–575.

Ramcharan EJ, Gnadt JW, Sherman SM (2000) Burst and tonic firing in thalamic cells of unanesthetized, behaving monkeys. *Visual Neurosci* 17: 55–62.

Recasens M, Vignes M (1995) Excitatory amino acid metabotropic receptor subtypes and calcium regulation. *Ann N Y Acad Sci* 757: 418–429.

Reid RC, Alonso JM (1995) Specificity of monosynaptic connections from thalamus to visual cortex. *Nature* 378: 281–284.

Reid RC, Alonso JM (1996) The processing and encoding of information in the visual cortex. *Curr Opin Neurobiol* 6: 475–480.

Reinagel P, Godwin DW, Sherman SM, Koch C (1999) Encoding of visual information by LGN bursts. *J Neurophysiol* 81: 2558–2569.

Sherman SM (1996) Dual response modes in lateral geniculate neurons: mechanisms and functions. *Visual Neurosci* 13: 205–213.

Sherman SM, Guillery RW (1996) The functional organization of thalamocortical relays. *J Neurophysiol* 76: 1367–1395.

Sherman SM, Guillery RW (1998) On the actions that one nerve cell can have on another: Distinguishing "drivers" from "modulators." *Proc Natl Acad Sci USA* 95: 7121–7126.

Sherman SM, Guillery RW (2001) *Exploring the Thalamus.* San Diego: Academic Press.

Smith GD, Cox CL, Sherman SM, Rinzel J (2000) Fourier analysis of sinusoidally-driven thalamocortical relay neurons and a minimal integrate-and-fire-or-burst model. *J Neurophysiol* 83: 588–610.

Steriade M, McCarley RW (1990) *Brainstem Control of Wakefulness and Sleep.* New York: Plenum Press.

Steriade M, McCormick DA, Sejnowski TJ (1993) Thalamocortical oscillations in the sleeping and aroused brain. *Science* 262: 679–685.

Swets JA (1973) The relative operating characteristic in psychology. *Science* 182: 990–1000.

Van Essen DC (1985) Functional organization of primate visual cortex. In *Cerebral Cortex*, vol. 3, edited by Peters A, Jones EG, 259–329 (New York: Plenum).

Van Essen DC, Anderson CH, Felleman DJ (1992) Information processing in the primate visual system: an integrated systems perspective. *Science* 255: 419–423.

Vidnyánszky Z, Görcs TJ, Négyessy L, Borostyánkoi Z, Kuhn R, Knöpfel T, Hámori J (1996) Immunocytochemical visualization of the mGluR1a metabotropic glutamate receptor at synapses of corticothalamic terminals originating from area 17 of the rat. *Eur J Neurosci* 8: 1061–1071.

Zhan XJ, Cox CL, Rinzel J, Sherman SM (1999) Current clamp and modeling studies of low threshold calcium spikes in cells of the cat's lateral geniculate nucleus. *J Neurophysiol* 81: 2360–2373.

Zhan XJ, Cox CL, Sherman SM (2000) Dendritic depolarization efficiently attenuates low threshold calcium spikes in thalamic relay cells. *J Neurosci* 20: 3909–3914.

5

a neuronal representation of the external mind. (handwritten annotation)

What Is a Neuronal Map, How Does It Arise, and What Is It Good For?

J. Leo van Hemmen

Introduction

Defining a "map" to be a neuronal representation of the outside world, we are facing three closely related problems: what does a map as a neuronal representation mean, how does it arise, and what is it good for? The solution to this circle of problems is fundamental to understanding how animals (and humans) relate their own position to that of a stimulus in the world surrounding them. We concentrate on elementary temporal maps, such as the interaural time difference (ITD) map in the barn owl, since they seem to contain the essentials of map formation. In so doing, we focus on the auditory system and its functional analogs. A key to our understanding of how a temporal-feature, and maybe any, map arises is either genetic coding or analyzing the occurrence of pre- and postsynaptic events in a narrow learning window. This, then, leads to the question: Do pre- and postsynaptic spike events "add" or "subtract"? We are thereby led to a closely related problem: what are the localization universals of the auditory system and, more generally, of any sensory system exploiting phase locking?

Is there a mapping problem? This seems to be the first question arising when one ponders the origin and meaning of a neuronal map. Let us therefore start by trying to define what we think a neuronal map is: a representation of the outside world. The trouble, however, is in the word *representation*. What does that mean? Apparently, then, there *is* a mapping problem, and we first embark on delineating what it may look like. We do so by studying three examples, the sand scorpion, the barn owl, and the paddle fish. These animals are inhabiting three totally different media—sand, air, and water—and using three totally different techniques to implement what we think to be a map to localize their prey in the world surrounding them.

I realize some authors need not agree with my point of view that temporal maps are at the roots of the mapping problem. In a sense it is a claim that may or may not be fulfilled in the time to come. I will give strong evidence, however, underlining the key role of temporal maps. For a slightly different but very interesting point of view, the reader may consult Knudsen, du Lac, and Esterly (1987). These authors coined the phrase *computational map*, and here we will see what some of these computations may look like. For an excellent review of data relevant to the present context, one may consult Buonomano and Merzenich (1998).

The sand scorpion (*Paruroctonus mesaensis*), a night hunter, provides us with what I think to be one of the simplest maps known to exist at present (Stürzl, Kempter, and van Hemmen 2000; Brownell and van Hemmen 2001) in that eight coding neurons determine the prey direction, an angle in the range 0°–360°. They do so through an interplay of excitation, opening a *time window*, and inhibition, closing it. The larger the time window, the bigger the probability that a coding neuron can fire. As we will see, the interplay is made possible by the substrate the animal is living on: sand. Furthermore, sand waves give rise to some kind of phase locking of the mechanoreceptors receiving prey-generated input and, hence, of the coding neurons connected to them. In other words, *phase locking* means that neuronal firing is positively correlated with amplitudes of waves in a medium, here sand.

The barn owl (*Tyto alba*), also hunting at night, is able to perform a very accurate localization of its prey, say, mice. We will focus on its azimuthal localization, which hinges on determining ITDs and reaches a spatial accuracy of 1°–2° corresponding to a temporal resolution better than 10 milliseconds. Determining ITDs means that the barn owl combines inputs from both ears (Konishi 1993). The first stage in the ascending auditory pathway where this happens is the laminar nucleus, the real object of our study. Each of the two laminar nuclei contains about twelve thousand neurons and obtains input from the magnocellular nuclei, one on the right and one on the left, receiving input signals from the cochlear nuclei, which are directly behind the ears.

The cochlea itself performs a frequency decomposition so that a tonotopic organization shows up in the structures appearing thereafter. Hence there is no harm in assuming that specific laminar neurons obtain input spikes with a specific frequency, once again exhibiting a phase locking. How, then, does a barn owl map a sound source, say a squeaking mouse (its favorite food), into the laminar nucleus, and is there any relation between the source "location" at different frequencies? Moreover, laminar neurons function as coincidence detectors as they combine signals from left and right ear. That is to say, instead of *subtraction*, as in the sand scorpion, they exploit *addition* as their key operation.

Our third example is the paddlefish (*Polyodon spathula*) living in the muddy Mississippi river, where its eyes are no good (Wilkens et al. 1997). The juvenile fish feeds on zooplankton, in particular, water fleas (*Daphnia*). Since a water flea has a small but permanent dipole moment, it is detected by electroreceptors on the long flattened "rostrum" in front of the paddlefish's mouth. The localization mechanism, however, is completely different from the previous two and is meant to serve as a contrast.

The above three animals either do not have or do not use vision to detect their prey. In so doing they generate a neuronal representation, a map, of their outside world. Once we have understood what a map is and how it works, the next question is how does it arise? The typical biologist's answer is genetic coding, but, for instance, a barn owl would need too much of it for fixing its axonal wiring. Furthermore, in this way the young bird cannot adapt to sudden changes in its outside world. And finally, after three weeks the young bird's head is full grown, but sound localization does not function yet.

In the last decade of the twentieth century the idea came up that synaptic efficacies evolve under the influence of external stimuli and internal interactions among the neurons. This idea has led to considerable insight in, for instance, the evolution of the visual cortex, but the underlying algorithms do not work for the time domain, as in the auditory system. For that we need new ideas. Here I will present evidence both for synaptic learning through a *learning window*, a kind of "map formation" at a single neuron, and for map formation in the true sense of the phrase as a consequence of interaction between different developing synapses at different neurons. In passing I note that these ideas also lead to orientation and direction preference in the spatiotemporal context of the visual cortex, so it is worthwhile to study them in the simplest possible context, the time domain as such. The final section of this article combines the data provided by various examples and presents the question that is in the title together with its underlying rationale.

Computational neuroscience is meanwhile able to provide the experimental neurobiologist with *quantitative* predictions that underline the importance of mathematical modeling. Both reference and reverence to Hilbert's twenty-three problems proposed at the beginning of the twentieth century therefore seems more than justified. Hilbert too (Hilbert, 1902) was motivated by examples to see the general context. Our own first example is truly exemplary.

The Sand Scorpion's Exquisite Maps

Arachnids, a large class of eight-legged arthropods such as scorpions and spiders, have exceptional ability to localize their prey using vibrational cues from the substrate (Barth 1985, 1998). Here we analyze the neurobiological mechanism of extracting the spatial direction of a stimulus from a temporally encoded signal at the sense organs of the sand scorpion *P. mesaensis*. Figure 5-1 shows a sand scorpion in the dark under ultraviolet radiation. It is fluorescent in this condition; a satisfactory explanation of this fluorescence is not available yet. It is a nocturnal animal that lives in the deserts of Southern California and

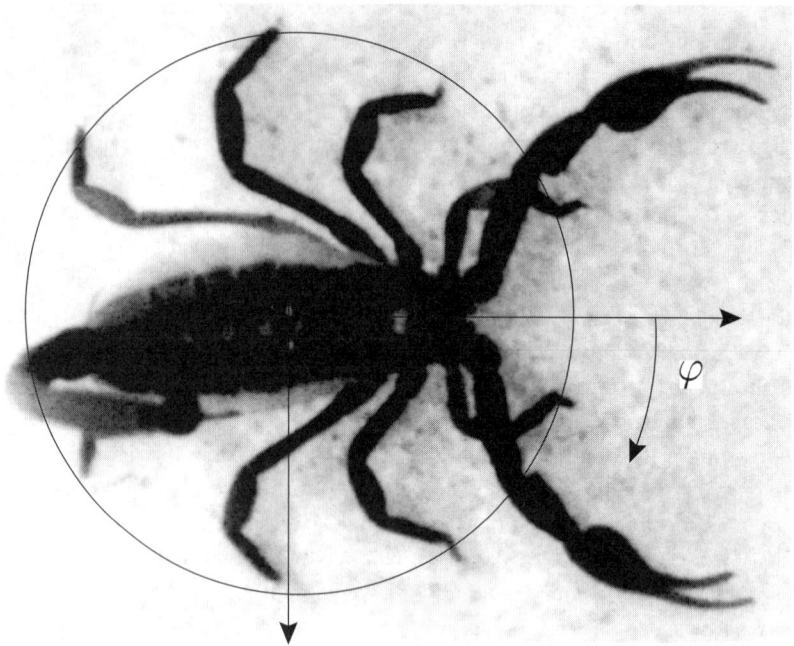

Figure 5-1. Desert scorpion *Paruroctonus mesaensis* as seen from above. It is in a defense position with its eight tarsi (feet) on a circle with radius $r = 2.5$ cm and its huge pedipalps in front. Tail and venom gland are ready for attack. The picture is a negative of a scorpion that is fluorescent in the dark under the influence of ultraviolet light. The stimulus angle is $\varphi = \varphi_s$. The circular arrangement of the mechanoreceptors at the joint of tarsus and basitarsus has been indicated. Reprinted with permission from Stürzl, Kempter, and van Hemmen (2000).

feeds mainly on small insects and other scorpions. Its eyes are rudimentary. During daytime it stays in a burrow about thirty centimeters below the surface to escape surface temperatures of over $60°$C, but after sunset it appears and, standing on the sand, waits for its prey.

Surface (Rayleigh) waves are essential to the scorpion's facility of determining the prey's direction through sense organs, the basitarsal compound slit sensilla (BCSS), mechanoreceptors that are located just above the joint of tarsus (foot) and basitarsus of each of its eight legs (Brownell 1984; Brownell and Polis 2001; Brownell and van Hemmen 2001). The BCSS are on a circle (Brownell and Farley, 1979a, 1979b) of radius $r \approx 2.5$ cm at angles $\gamma_i = \pm 18°, \pm 54°, \pm 90°, \pm 140°$, where $0°$ is ahead; see figure 5-1.

The mechanoreceptors, though very sensitive $(0.1 \, \text{nm} = 10^{-10} \, \text{m})$, may but need not fire so that all a theory can, and does (Stürzl, Kempter, and van Hemmen 2000), predict is probabilities. The neuronal setup is shown in figure 5-2. Prey, perhaps a moth, is moving and in so doing generates surface

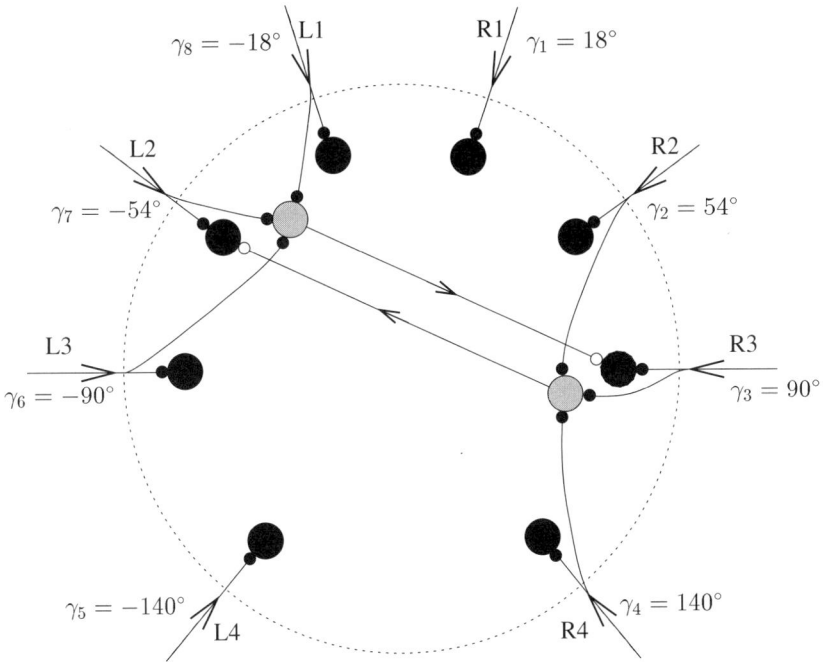

Figure 5-2. Diagram of the sand scorpion's eight command neurons (*black, inner circle*) corresponding to the mechanoreceptors (*outer circle*) at the end of the leg whose direction they are encoding. For two of them, $i = 3$ and $i = 7$ corresponding to R3 and L2, respectively, the inhibitory partner neurons (gray) are shown as well. The inhibitory triad of R3 consists of L1, L2, and L3; that of L2 is made up of R2, R3, and R4.

waves whose spectral decomposition shows a dominant peak at 300 Hz. These Rayleigh waves hit, say, the BCSS of leg R3 first. In the scorpion's brain there are eight command neurons, one corresponding to each leg and "coding" its direction. Command neuron R3 is now excited first as well as a neighboring inhibitory neuron that receives input from the *triad* R2–R4 and inhibits command neuron L2 opposite to it. Command neurons L2 and L3 at the other side receive their excitatory input about one millisecond later, so that it is highly probable that they have received inhibition from "their" triad before and stay practically silent while the excitation of the neurons in between interpolates between these two extremes.

Through the *interplay* between excitation and inhibition (Stürzl, Kempter, and van Hemmen 2000; Brownell and van Hemmen, 2001) the command neurons constitute a kind of committee that decides about the direction the animal will adopt by means of a *vector code* (Georgopoulos, Schwartz, and Kettner, 1986; Salinas and Abbot, 1994; Lewis, 1999). To see what this

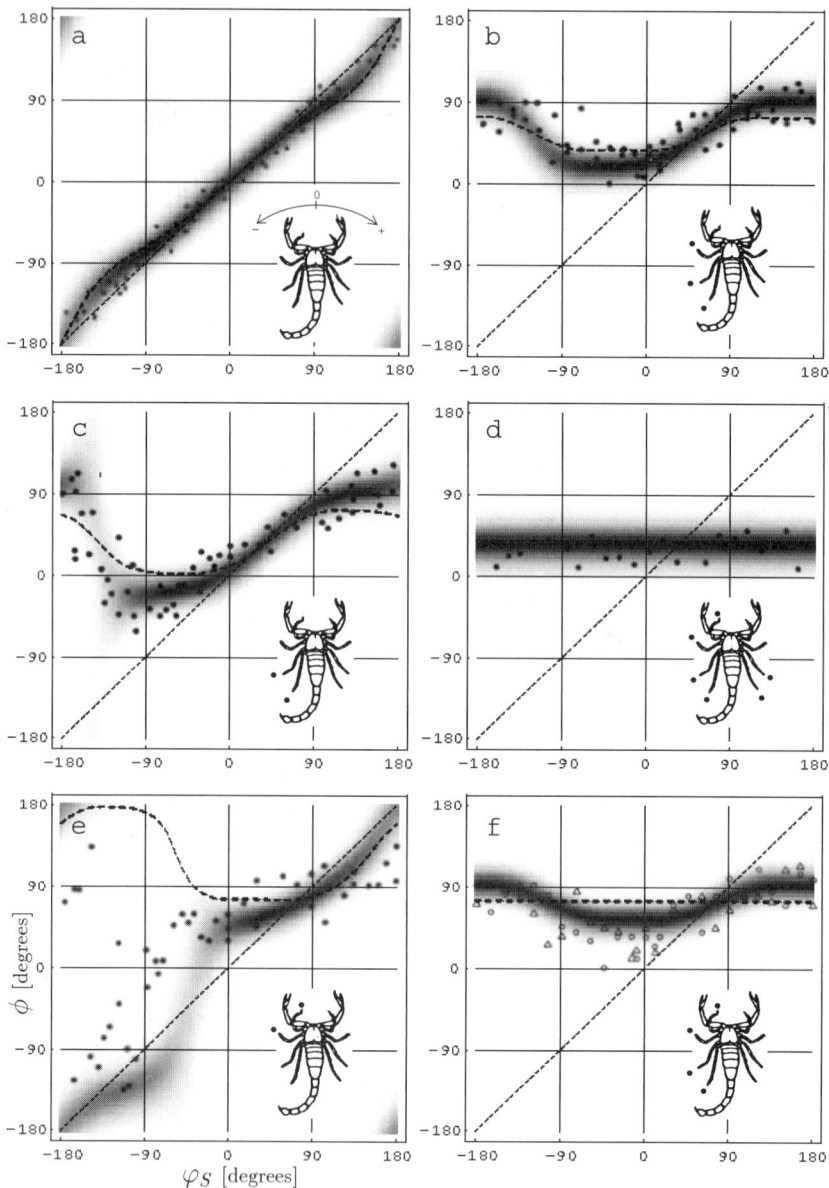

Figure 5-3. Response angle ϕ of a scorpion (vertical axis) as a function of the stimulus angle φ_s (horizontal axis). (a) Systematic deviation of the response of an intact animal that hardly ever manages the complete turn φ_s. If it were perfect, the performance would have been a straight line, the dashed diagonal. (b)–(f) Ablated mechanoreceptors (BCSS) are indicated by dots at the end of the tarsi. Both the theoretically computed probability density $P(\varphi)$ (dark shadings) and experimental points (dots) are indicated. Experimental data have been taken from Brownell and Farley (1979a, 1979b) and Brownell (1984). For model details, see Stürzl, Kempter, and van Hemmen (2000). If the inhibitory triad is replaced by

means, let Δt be a short time interval, say, 0.5 seconds, during which the prey is moving, and let v_i be the number of spikes produced by command neuron i as a consequence of the stimuli generated by the prey. Then the direction to go is given by the weighted vector sum

$$\mathbf{v} = \sum_{i=1}^{8} v_i \mathbf{e}_i \tag{5.1}$$

where $\mathbf{e}_i$ is a unit vector in the direction γ_i of leg i; compare figure 5-2. This is a vector code generated by the sand scorpion's map $\{v_i; 1 \leq i \leq 8\}$. It describes a part of the surrounding world, all directions between $-180°$ and $180°$. In its simplicity the map is truly exquisite. The anatomical rationale behind the vector code of the command neurons is that the legs' motor neurons are in their direct neighborhood (Gronenberg 1989, 1990).

As figure 5-3 shows, Nature has taken care of being sure about the inhibition: A single inhibitory input instead of a triad cannot explain the experimental data. Otherwise the coding would be modified strongly, whereas a single BCSS missing is one out of eight constituting a committee. As yet it is not clear whether a sand scorpion acquires its localization abilities through genetic coding or synaptic training. For the barn owl the latter seems practically unavoidable. We therefore turn to synaptic learning rules first.

Synaptic Dynamics inside a Learning Window

Let us focus on excitatory synapse n out of a collection of $1 \leq n \leq N$ synapses on the dendritic tree of neuron i. Spikes arrive at synapse n at times t^n and the postsynaptic neuron fires at times t_i. The notion of "Hebbian" learning dates back to Donald Hebb's 1949 classic *The Organization of Behavior—A Neurophysiological Theory* (Orbach 1998). On page 62 of Hebb's book one can find the now famous "neurophysiological postulate": "When an axon of cell A is near enough to excite a cell B and *repeatedly* or *persistently* takes part in firing it, some growth process or metabolic change takes place in one or both cells such that A's efficiency, as one of the cells firing B, is increased."

Hebb pointed out that he expected learning to take place at the synapses but carefully avoided any specification. The present implementation (Herz et al. 1989; van Hemmen et al. 1990; Gerstner et al. 1996; van Hemmen 2001) through a learning window W (see figure 5-4) specifies[1] how the synaptic efficacy is to increase or decrease according to the arrival and departure times t^n and t_i of pre- and postsynaptic spikes, respectively,

---

a *single* inhibitory neuron, we find the dashed line as the mean response; the agreement with experiment is in general less good.

$$\Delta J_i(t)|_{\text{Hebb}} = \eta \left[\sum_{t-T_l \le t_i, t^n < t} W(t_i - t^n) \right]. \tag{5.2}$$

We have added here all contributions during a learning session of duration T_l. Each individual contribution is small since η is small, and equation 5.2 is therefore called "infinitesimal" learning. For an excitatory synapse it is natural that $W(t_i - t^n) \ge 0$ for $t_i - t^n < 0$, meaning that the postsynaptic spike appears *after* the presynaptic one (as it should), whereas $W(t_i - t^n) \le 0$, if the presynaptic spike is "too late" which is no good. For inhibitory synapses W's definition is to be taken just the other way around. The temporal width of W is restricted, and rightly so since an action potential's t_i is noticed rather rapidly at any synapse through backpropagation (Stuart et al. 1997).

In addition, a synapse may change by the mere existence of pre- and postsynaptic spikes, so that altogether we obtain the following *local* learning rule

$$\Delta J_i(t) = \eta \left[\sum_{t-T_l \le t_i^f < t} w^{\text{in}} + \sum_{t-T_l \le t^n < t} w^{\text{out}} + \sum_{t-T_l \le t_i^f, t^n < t} W(t_i^f - t^n) \right]. \tag{5.3}$$

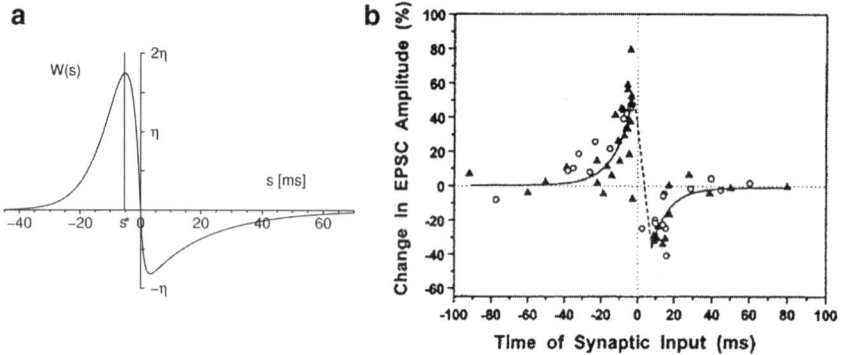

Figure 5-4. (a) A learning window W in units of the learning parameter η as a function of the delay $s = t_i^f - t^n$ between presynaptic spike arrival at synapse i at time t_i^f and postsynaptic firing at time t^n. If $W(s)$ is positive (negative) for some s, then the synaptic efficacy J_i is increased (decreased). The increase of J_i is most efficient if a presynaptic spike arrives a few milliseconds before ($s < 0$) the postsynaptic neuron starts firing ($s = 0$). For $|s| \to \infty$ we have $W(s) \to 0$. The form of the learning window and parameter values are as described in equation 5.5 (see Note). Taken from Kempter, Gerstner, and van Hemmen (1999). (b) Experimentally obtained learning window of a cell in rat hippocampus. Reprinted by permission from MacMillan Publishers Ltd: Zhang et al. 1998. A critical window for cooperation and competition among developing retinotectal synapses. *Nature* 395: 41, Fig. 5, copyright 1998. The similarity with part (a) is evident. It is important to realize that the width of the learning window is to be in agreement with other neuronal time constants. In the auditory system, for instance, these are nearly two orders of magnitude smaller so that the learning window's width is to scale accordingly.

It is local both in space and in time. The notion of learning window has been proposed on theoretical grounds (Gerstner et al. 1996) as have been the other terms, but there is meanwhile ample evidence for both the presynaptic w^{in} (Brown and Chattarji 1994) and postsynaptic w^{out} (Brown and Chattarji 1994; Buonomano and Merzenich 1998; Linden 1999) and the Hebbian contributions given by the learning window W (Linden 1999; Paulsen and Sejnowski 2000; Bi and Poo 2001), to mention just a few review papers.

These processes are called "homosynaptic" since they all happen at the very same synapse. We now turn to a "heterosynaptic" process, where changes of one synapse are transported through the axon it is connected to, with other synapses receiving input from the same axon (Bonhoeffer, Staiger, and Aertsen 1989; Tao et al. 2000). More precisely, a synapse-specific weight change $\left(\frac{d}{dt}J_{mn}\right)_{local}$ due to pre- and postsynaptic spikes at the very same synapse also has a small effect on all other ($m' \neq m$) synaptic weights $J_{m'n}$ connected to the *same* axon n. The total change dJ_{mn}/dt of a specific synaptic strength is thus given by

$$\frac{d}{dt}J_{mn} = \sum_{m' \in \text{axon } n} (\delta_{mm'} + \rho)\left(\frac{d}{dt}J_{m'n}\right)_{local} \qquad (5.4)$$

where ρ is small and positive. There is experimental evidence (Tao et al. 2000) that the heterosynaptic contributions ($m' \neq m$) may be very weak, say of the order $\rho \approx 1/100$. As we will see, in the long run their effect is large.

The Barn Owl's Temporal Map

An azimuthal direction is mapped one to one onto a time difference between two ears, the ITD. This suggests a one-dimensional map in the brain. The way in which it functions was conceived more than half a century ago by Jeffress (1948); see figure 5-5. With hindsight it is fair to say it was pure theory, but we will see in what sense he was right. The key ingredients are axons as delay lines and neurons as coincidence detectors. They fire at a maximal rate when signals from both ears arrive simultaneously, or, in other words, when ipsi- and contralateral delays compensate the ITD.

In the barn owl the first binaural interaction occurs in the tonotopically organized laminar nucleus. Information is conveyed through phase-locked spikes in a frequency range between 1 and 8 kHz. Of course, a single neuron cannot fire at such a high rate, but phase locking means that, say, if the neuron fires, then it does so only if the sound amplitude is maximal. The period of a 5-kHz signal is 200 μs. The initial scatter of the axonal delays (one millisecond) exceeds it by about an order of magnitude, whereas the barn

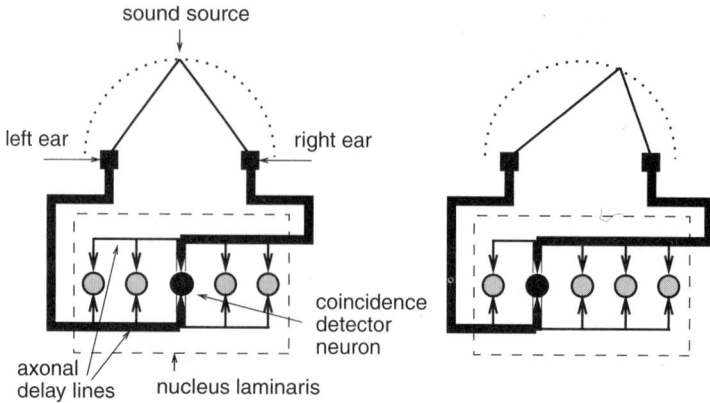

Figure 5-5. Jeffress suggested in 1948 a neural circuit in the barn owl's laminar nucleus for processing interaural time differences (ITDs). By means of such a circuit an ITD is transformed into a characteristic spatial activity pattern of the neurons (horizontal row of gray/black disks). Starting from the sound source, the sound reaches both ears with an azimuth-dependent delay. In the ears the sound is transformed into action potentials. These travel along axons that serve as delay lines to the laminar nucleus. There axons from the left and right ears converge at neurons which act as coincidence detectors. If spikes arrive at a coincidence detector neuron "simultaneously," the neuron shows up high activity (dark disk). Temporally displaced inputs cause a weaker response (lighter disks). A comparison of both halves of the figure shows that coincident input requires a compensation of a shorter delay from the sound source to one ear by a longer axonal delay from the other ear. Depending on the azimuthal position of the sound source, there is a characteristic spatial pattern of neuronal activity, which is processed by "higher" brain areas. Though the Jeffress scheme is a nice idea, it has—as most cybernetic schemes do—drawbacks: (1) in reality there is not a *single* delay line, but hundreds (altogether many thousands) of delay lines, and (2) one still has to show that the hardware does what the scheme wants it to do.

owl's accuracy in sound localization (<10 μs) is about two orders of magnitude smaller, that is, better. So we face a true problem: how does the barn owl manage to fine-tune its wetware that precisely?

Tuning Synapses at a Single Neuron

The learning rule (5.3) has been used to resolve the Konishi paradox (Konishi 1993), which focuses on individual laminar neurons. They reach a temporal precision that is better by at least an order of magnitude than their individual membrane time constants. The explanation (Gerstner et al. 1996) is based on three observations:

    (i) the rise of the postsynaptic potential is steep, which is de facto a rewording of the fact that laminar neurons have (practically) no dendritic tree,

(ii) synaptic activity leads to time constants as small as $0.1\,\text{ms} = 100\,\mu s$ because of outward rectifying currents, and

(iii) a *collective* selection process singles out specific, "well-timed," synapses. It is based on the learning window. The total synaptic input determines when the postsynaptic neuron fires, but, once it does so, only synapses of spikes that arrive "in time" are strengthened, whereas the others are weakened. In this way we end up with an *indirect interaction* among the several hundred synapses located at a specific laminar neuron. At the end only those with the "right timing" survive. Resembling evolutionary selection, it is map formation in a nutshell and, as such, due to interaction among its constituents, a key element of any map formation. Furthermore, it is "self-organized," meaning that there is no outside source deciding what to do.

The upshot here is a tuning of the synapses and hence a sharp phase locking with a temporal precision of $20\,\mu s$ (Gerstner et al. 1996) so that presynaptic spikes from a certain cochlear frequency domain arrive *coherently* at the receiving neuron. This is a necessary condition for map formation because it guarantees a local precision. There being several laminar neurons doing the very same job in a certain frequency range, a combination of them, that is, a population code, leads to a $5\,\mu s$ precision (Gerstner et al. 1996).

Though necessary, the above mechanism is not sufficient for map formation as a neuronal representation of the outside world. In the barn owl's case of azimuthal sound localization, different directions should be mapped onto different locations in the laminar nucleus (Carr and Konishi 1990; Carr 1993; Konishi 1993) but we are still at the level of a single neuron.

Temporal Map Formation

In other words, we have not understood yet how a temporal map along a laminar axon (compare figure 5-6) and through the whole tonotopic range of the barn owl's laminar nucleus arises. As in the single-neuron case, we need an interaction among synapses, but now at *different* neurons. In the visual cortex, there is the intracortical interaction in addition to the input from the lateral geniculate nucleus; only a combination of both (Bartsch and van Hemmen 2001) gives a satisfying explanation of the experimental data. In the laminar nucleus, however, neither excitatory nor inhibitory "intralaminar" interaction has been found. The available inhibition stems mainly from the superior olivary nucleus so as to vary the neuronal threshold of discharge *with intensity* since otherwise monaural input might take over (Peña et al. 1996).

Where, then, does the interaction between synapses at different neurons come from? The answer (Kempter et al. 2001) may be quite simple. Axons enter the laminar nucleus from two sides, corresponding to both ears; see figures 5-5 and 5-6a, the latter being quite schematic—hence clear. Their "best

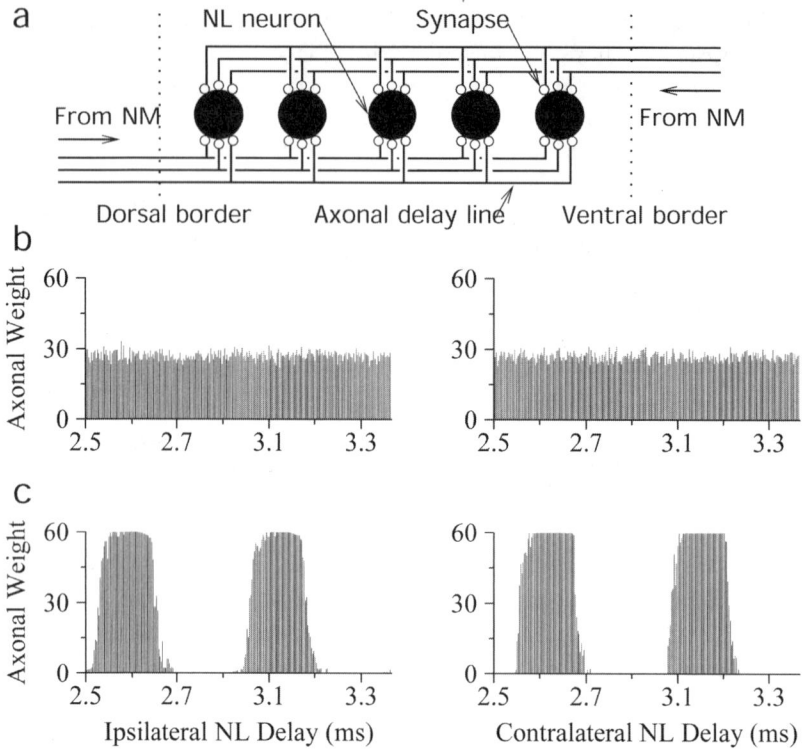

Figure 5-6.   Axonal delay selection in a single-frequency layer of the barn owl's laminar nucleus (NL). (a) Contra- and ipsilateral afferents (full lines) from the magnocellular nucleus (NM) enter NL ventrally and dorsally, respectively, at the laminar borders (vertical dotted lines). They have synaptic contacts (small open circles) with all laminar neurons (large filled circles). (b) In young owls, the distribution of NL delays (from one papilla to the border of NL) is broad, that is, one millisecond, with respect to the best frequency (here $f = 3$ kHz) of the considered NL layer. The "axonal weight" is the sum of all synaptic weights of an axon. (c) After applying the learning algorithm presented by Kempter et al. (2001), laminar delays roughly differ by multiples of the period $T = f^{-1}$; compare Carr and Konishi (1990, figure 9C). In other words, the learning rule selects axons with suitable delays.

frequencies" are in a narrow range, a so-called isofrequency lamina. The axons contact different neurons, about thirty in a slice as shown in figure 6a, and we have just seen that only well-timed synapses and, hence, axons survive since synapses connected to the *same* axon influence each other through equation 5-4.

Reality, however, also dictates that axons bifurcate and, in so doing, generate a frequency spread $\Delta f \approx 1/3$ octave. This, then, leads to a map through the whole tonotopic range (Leibold et al. 2001). That is to say, along the tonotopic axis, which is orthogonal to the dorso-ventral one. In this way neighboring neurons are tuned to the same ITD and, hence, to the same spatial direction, the notion of neighborhood being extended continuously through all frequencies beyond 1 kHz.

Thus a coherent map arises consisting of parallel "stripes" to the tonotopic axis. Furthermore, going from left to right in figure 5-6, we get a continuous coding of spatial directions; see also figure 3 of Knudsen, du Lac, and Esterly (1987). In addition, an interaural level (= intensity) difference map codes elevation. The two merge higher up in the auditory stream, in the external nucleus of the inferior colliculus; see Knudsen, du Lac, and Esterly (1987) and Cohen and Knudsen (1999).

Summary

For the barn owl, apparently two different processes are relevant to generating a temporal map. First, homosynaptic tuning of synapses at the same neuron gives rise to accuracy in time. Second, heterosynaptic tuning of synapses that are at different neurons but connected to the same axon then leads to a map in the true sense, that is, a neuronal representation of (aspects of) the outside world. In this case the map must extend over *many* frequencies to lift the degeneracy that, for a given period T belonging to a frequency f, cannot discern directions whose ITD differ by nT for some integer n; compare figure 5-6c. Phrased differently, the former is related to the map's functioning, the latter to the way in which it comes about. Figure 5-7 shows that a combination of the two works well. Coincidence detection, addition of input from left and right ear, is the driving mechanism.

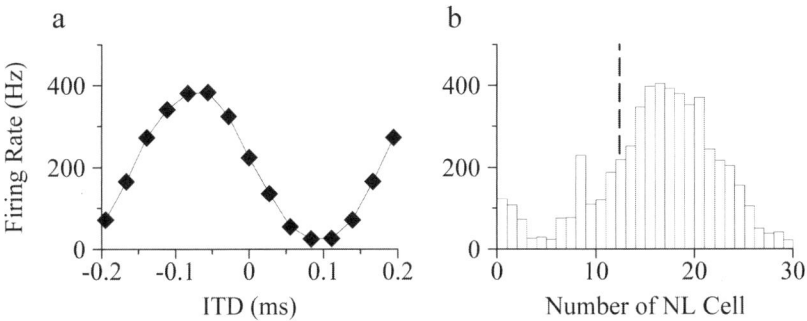

Figure 5-7. Firing rates of laminar neurons after learning. (a) ITD tuning curve of a single neuron. (b) ITD map given by the firing rate as a function of the position of the neuron along the array for fixed ITD = 0. It is a population code of the direction corresponding to the given ITD. The dashed line indicates the cell shown in (a).

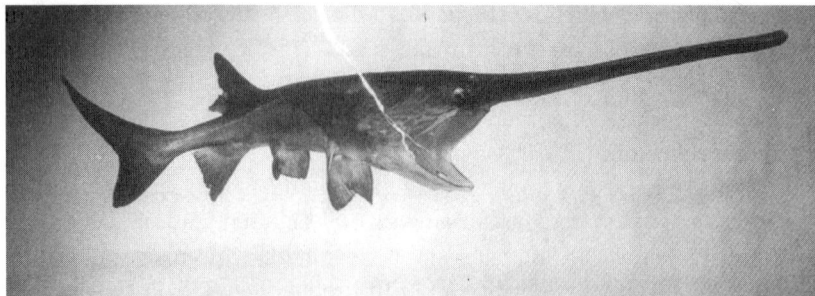

Figure 5-8. Juvenile paddlefish snapping at an artificial dipole at the end of the white coated wires (*center*). The huge rostrum in front of its mouth—which is huge, too—is covered by electroreceptors, which are similar to the ampullae of Lorenzini of sharks and rays. The adult animal simply opens its mouth to filter the water but apparently does not use the electroreceptors anymore. Photograph courtesy of Professor Lon A. Wilkens (University of Missouri at St. Louis).

The Paddlefish

Figure 5-8 shows a juvenile paddlefish. It feeds on zooplankton, especially water fleas (*Daphnia*), which through their small but permanent dipole moment stimulate passive electroreceptors on the "rostrum" in front of the mouth, provided the distance is less than 1.5–2 cm; *Daphnia* are caught individually. The two sides of the rostrum detect prey independently of each other. Since the average speed of the fish with respect to muddy river water surrounding it is about 10–20 cm/s, the time of stimulation is of the order of two hundred milliseconds.

How, then, does the fish locate its prey? Though the final proof has not been given yet, I expect that a somatosensory map, here a neuronal representation of the surface of the rostrum, does the job. The fish "follows" the signal on its rostrum and behaves accordingly in that it simply opens its mouth if a water flea is directly below the rostrum or else turns around first. No precise timing is needed, and all that is required is a straightforward map from the two sides of the rostrum into, presumably, the dorsal octavolateral nucleus.

Conclusion: What Is the Problem?

This chapter started with the questions, what is a neuronal map, how does it arise, and what is it good for? To see what all this might mean, we analyze our three examples from a higher point of view, through a bird's eye, so to speak.

Sand Scorpion

The sand scorpion has eight command neurons standing for eight directions γ_i as given by figure 5-1, corresponding to its eight legs, and determining the direction to go via the vector code given by equation 5.1. We have a *distribution* of activities that depend on the prey direction and, hence, on the width of the command neurons' time window. A time window is opened by the excitation coming from the BCSS whose direction the command neuron is encoding and closed by its inhibitory triad, opposite to it. If the prey is nearest to leg $R3$, then command neuron $R3$'s rate is maximal, opposite to it the rate is minimal, and the other command neurons in general interpolate between these two extremes—a true map. It is a neuronal representation of $(-180°, 180°)$ through the firing rates $\{v_i; 1 \le i \le 8\}$ and, in fact, it is one of the simplest known in neurobiology. What it is good for is plain.

The command neurons' functioning is clear. How it comes about is less clear. It may well be that the young scorpion has a different diet (P. H. Brownell, personal communication). Since the vector code is generated by a trade-off of excitation and inhibition and is also relatively simple, one might speculate that genetic coding gives rise to the whole setup. It is tempting to speculate a bit more. Not only does a sand scorpion determine the prey's direction, it also "knows" its distance as it stops at the "right" position, if it is near enough (≤ 15 cm). This suggests that the animal also has a one-dimensional map of the distance. Together the two maps would provide a neuronal representation in terms of two-dimensional *polar coordinates*.

There is another remarkable aspect of the sand scorpion's map in that its functioning is based on a trade-off of excitation and inhibition. We may compare this with the arithmetical operation "subtraction" since inhibition annihilates excitation through a spatial synaptic subtraction.

Barn Owl

In the barn owl, elementary maps of both azimuth and elevation exist. A distance map as such has not been found yet. The two directions are both contained in but strictly smaller than $(-180°, 180°)$. That is to say, both are one-dimensional, generated by ITD and interaural level difference (ILD), respectively, and together they give rise to a neuronal representation of three-dimensional *spherical coordinates* ϕ and θ. For the ITD map a coincidence detection à la Jeffress (1948) as given in figure 5-5 is meanwhile widely accepted (Carr 1993; Konishi 1993; Carr 1995). There is a huge difference, however, between Jeffress's original idea and what is known at present.

While figure 5-5 with two delay lines, one coming from the right and one coming from the left, is a caricature of Jeffress's 1948 idea, there are in reality thousands of delay lines, that is, axons and about twelve thousand laminar neurons that process the information coming from both ears. Though meanwhile both the Konishi paradox (Gerstner et al. 1996) and the map formation problem in the laminar nucleus (Kempter et al. 2001; Leibold et al. 2001) have been solved, the problem of what this map is really good for is, despite a clever ansatz (Cohen and Knudsen, 1999), open.

One might even argue there need not be any map at all. According to Kalmijn (1997), sharks and rays *have no map* of their electrosensory surroundings but locate their (stationary) prey through a clever algorithm: the fish's axis is to be kept at a more or less constant angle $|\theta| < 90°$ with respect to stimulus-generated lines of the electric field. So what we are seeing here in the sand scorpion, barn owl, and paddlefish is a *fata morgana*. We just have to look better. Nevertheless maps are quite real for the barn owl (Carr and Konishi 1990; Peña and Konishi 2001).

The key mechanism in coincidence detection is the neuronal equivalent of the arithmetical operation "addition" where double excitation is a simple spatial synaptic summation. Because a neuron is a threshold element looking through a time window, it can discern single from double input by putting the threshold right. Elementary maps such as the ITD one use addition. It is quite surprising that, once the ITD and ILD maps are available, they are put together at the external nucleus of the inferior colliculus through "multiplication" (Peña and Konishi 2001). The two coordinates ϕ and θ are orthogonal so that, with hindsight, it makes an awful lot of sense, if in a two-dimensional map a specific direction (ϕ_0, θ_0) is singled out by $1 \times 1 = 1$ whereas the "other" directions $(\phi, \theta) \neq (\phi_0, \theta_0)$ give $1 \times 0 = 0$.

Multiplication and separating 1 from 0 looks indeed much smarter than discerning $1 + 1 = 2$ from $1 + 0 = 1$, though in both cases thresholding must do the job. In generating the ITD map the owl converts a very precise time code into a much more robust rate code. Such a procedure generates the 1s we were just analyzing. Once they exist, putting them together is a different story. Finally, it may be useful to realize that, one stage further, acoustic and optic maps are put on top of and compared with each other in the optic tectum.

Paddlefish

What the rostral map is good for is evident: localizing the prey with respect to the rostrum so that the fish's mouth is opened at the right spot. How it works is also more or less evident. How, then, does it arise? For the visual cortex as a neuronal representation of retinal images detailed theories have been developed—see, for example, Bartsch and van Hemmen (2001) and references quoted therein. Accordingly there is ample reason to assume that, on a genetic substrate, a similar mechanism as in the vertebrate visual system governs the development of a rostral map in the the dorsal octavolateral, or subsequent, nucleus of the paddlefish. In accordance with its visual analog, it may even be sensitive to direction. What the fish then does with this map is a different, highly nontrivial, but nonfundamental problem.

Universals?

In a sense, asking "What is a neuronal map, how does it arise, and what is it good for?" is looking for an answer to a deeper question: are there universals, that is, universal properties, mechanisms, and origins of a neuronal map? Is there a universal procedure generating a temporal map on a genetic substrate? It looks as if a learning window with its temporal asymmetry is one of them. We are thereby led to a closely related problem: what are the localization universals of the auditory system and, more generally, of any sensory system exploiting phase locking? Are "addition" and "subtraction" of pre- and postsynaptic spikes the only two? The future has to decide whether universals exist and, if so, what they look like. If not, can we define a neuronal map according to its generating mechanism, its functioning, or both?

Note

1. A mathematical expression for the learning window is for instance (Kempter et al. 2001)

$$W(s) = \begin{cases} \exp[-(s-\hat{s})/\tau_1]\left[2\left(1+(s-\hat{s})\frac{\tau_1+\tau_2}{\tau_1\,\tau_2}\right)-\left(1+(s-\hat{s})\frac{\tau_0+\tau_1}{\tau_0\,\tau_1}\right)\right] & \text{for } s \geq \hat{s} \\ 2\exp[(s-\hat{s})/\tau_2]-\exp[(s-\hat{s})/\tau_0] & \text{for } s < \hat{s} \end{cases}$$

(5.5)

The parameter values in figure 5-4 are $\tau_0 = 0.025\,\text{ms}$, $\tau_1 = 0.15\,\text{ms}$, $\tau_2 = 0.25\,\text{ms}$, and $\hat{s} = -0.005\,\text{ms}$.

Acknowledgments   It is a great pleasure to the author to thank Catherine Carr for supplying precise barn-owl data, Armin Bartsch, Christian Leibold, and Wolfgang Stürzl for providing the figures in a most enjoyable collaboration, and Christian Leibold for his help in preparing the final version.

References

Bartsch, A. and van Hemmen, J. L. 2001. Combined Hebbian development of geniculo-cortical and lateral connectivity in a model of primary visual cortex. *Biol. Cybern.* 84: 41–55.

Barth, F. G., ed. 1985. *Neurobiology of Arachnids.* Berlin: Springer.

Barth, F. G. 1998. The vibrational sense of spiders. In *Comparative Hearing: Insects,* ed. R. R. Hoy, A. N. Popper, and R. R. Fay, 230–278. New York: Springer.

Bi, G. and Poo, M. 2001. Synaptic modification by correlated activity: Hebb's postulate revisited. *Annu. Rev. Neurosci.,* 24: 139–166.

Bonhoeffer, T., Staiger, V., and Aertsen, A. 1989. *Proc. Natl. Acad. Sci. USA* 86: 8112–8116.

Brown, T. H., and Chattarji, S. 1994. Hebbian synaptic plasticity: Evolution of the contemporary concept. In *Models of Neural Networks II*, ed. E. Domany, J. L. van Hemmen, and K. Schulten, 287–314. New York: Springer.

Brownell, P. H. 1984. Prey detection by the sand scorpion. *Sci. Am.* 251(6): 94–105.

Brownell, P. H., and Farley, R. D. 1979a. Detection of vibrations in sand by tarsal sense organs of the nocturnal scorpion, *Paruroctonus mesaensis. J. Comp. Physiol.* 131: 23–30.

Brownell, P. H., and Farley, R. D. 1979b. Orientation to vibrations in sand by the nocturnal scorpion *Paruroctonus Mesaensis*: Mechanisms of target localization. *J. Comp. Physiol.* 131: 31–38.

Brownell, P. H., and Polis, G. 2001. *Scorpion Biology and Research*. Oxford: Oxford University Press.

Brownell, P. H., and van Hemmen, J. L. 2001. Vibrational sensitivity and a computational theory for prey-localizing behavior in sand scorpions. *Am. Zool.* 41(5): 1229–1240.

Buonomano, D. V., and Merzenich, M. M. 1998. Cortical plasticity: from synapses to maps. *Annu. Rev. Neurosci.* 21: 149–186.

Carr, C. E. 1993. Processing of temporal information in the brain. *Annu. Rev. Neurosci.* 16: 223–243.

Carr, C. E., 1995. The development of nucleus laminaris in the barn owl. In *Advances in Hearing Research*, ed. G. A. Manley, G. M. Klump, C. Köppl, H. Fastl, and H. Oeckinghaus, 24–30. Singapore: World Scientific.

Carr, C. E., and Konishi, M. 1990. A circuit for detection of interaural time differences in the brain stem of the barn owl. *J. Neurosci.* 10: 3227–3246.

Cohen, Y. E., and Knudsen, E. I. 1999. Maps versus clusters: Different representations of auditory space in the midbrain and forebrain. *Trends Neurosci.* 22: 128–135.

Georgopoulos, A., Schwartz, A. B., and Kettner, R. E. 1986. *Science* 233: 1416–1419.

Gerstner, W., Kempter, R., van Hemmen, J. L., and Wagner, H. 1996. A neuronal learning rule for sub-millisecond temporal coding. *Nature* 383: 76–78.

Gronenberg, W. 1989. Anatomical and physiological observations on the organization of mechanoreceptors and local interneurons in the central nervous system of the wandering spider *Cupiennius salei. Cell Tissue Res.* 258: 163–175.

Gronenberg, W. 1990. The organization of plurisegmental mechanosensitive interneurons in the central nervous system of the wandering spider *Cupiennius salei. Cell Tissue Res.* 260: 49–61. We hypothesize that in the wandering spider *Cupiennius salei* the homologous organ is SLT3.

Hebb, D. O. 1949. *The Organization of Behavior—A Neuropsychological Theory*. New York: Wiley.

van Hemmen, J. L. 2001. Theory of synaptic plasticity. In *Handbook of Biological Physics, Vol. 4: Neuro-Informatics, Neural Modelling*, ed. Moss, F., and Gielen, S., 771–823. Amsterdam: Elsevier.

van Hemmen, J. L., Gerstner, W., Herz, A. V. M., Kühn, R., and Vaas, M. 1990. Encoding and decoding of patterns which are correlated in space and time. In

Konnektionismus in Artificial Intelligence und Kognitionsforschung, ed. G. Dorffner, 153–162. Berlin: Springer.

Herz, A. V. M, Sulzer, B., Kühn, R., and van Hemmen, J. L. 1989. Hebbian learning reconsidered: Representation of static and dynamic objects in associative neural nets. *Biol. Cybern.* 60: 457–467.

Hilbert, D. 1902. Mathematical problems. *Bull. Amer. Math. Soc.* 8: 437–479, an English translation of the German original "Mathematische Probleme" in *Nachr. Kgl. Ges. d. Wiss. zu Göttingen, math.-phys. Klasse* (1900)(3): 253–297.

Jeffress, L. A. 1948. A place theory of sound localization. *J. Comp. Physiol. Psychol.* 41: 35–39.

Kalmijn, A. J. 1997. Electric and near-field acoustic detection, a comparative study. *Acta Physiol. Scand.*, Suppl. 638: 25–38.

Kempter, R., Gerstner, W., and van Hemmen, J. L. 1999. Hebbian learning and spiking neurons. *Phys. Rev. E* 59: 4498–4514.

Kempter, R., Leibold, C., Wagner, H., and van Hemmen, J. L. 2001. Formation of temporal feature maps by axonal propagation of synaptic learning. *Proc. Natl. Acad. Sci. USA* 98: 4166–4171.

Knudsen, E. I., du Lac, S., and Esterly S. D. 1987. Computational maps in the brain. *Annu. Rev. Neurosci.* 10: 41–65.

Koch, C. 1999. *Biophysics of Computation: Information Processing in Single Neurons.* New York: Oxford University Press.

Konishi, M. 1993. Listening with two ears. *Sci. Am.* 268(4): 34–41.

Lamperti, J. 1996. *Probability.* 2nd ed., New York: Wiley. (Orig. pub. 1966.)

Leibold, C., Raach, A. W., Kempter, R., and van Hemmen, J. L. 2001. Across-frequency synchronization of interaural time-difference maps. In *Göttingen Neurobiology Report 2001*, ed. Elsner, N., and Kreutzberg, G. W., 1037 Stuttgart: Thieme.

Lewis, J. E. 1999. Sensory processing and the network mechanisms for reading neuronal population codes. *J. Comp. Physiol. A* 185: 373–378.

Linden, D. J. 1999. The return of the spike: Postsynaptic action potentials and the induction of LTP and LTD. *Neuron* 22: 661–666.

Orbach, J. 1998. *The Neuropsychological Theories of Lashley and Hebb.* Lanham, Md.: University Press of America.

Paulsen, O., and Sejnowski, T. J. 2000. Natural patterns of activity and long-term synaptic plasticity. *Current Opinion in Neurobiology,* 10(2): 172–179.

Peña, J. L., Viete, S., Albeck, Y., and Konishi, M. 1996. Tolerance to sound intensity of binaural coincidence detection in the nucleus laminaris of the owl. *J Neurosci.* 16: 7046–7054.

Peña, J. L., and Konishi, M. 2001. Auditory spatial receptive fields created by multiplication. *Science* 292: 249–252.

Salinas, E., and Abbott, L. F. 1994. Vector reconstruction from firing rates. *J. Comput. Neurosci.* 1: 89–107.

Stuart, G., Spruston, N., Sakmann, B., and Häusser, M. 1997. Action potential initiation and backpropagation in neurons of the mammalian CNS. *Trends Neurosci.* 20: 125–131.

Stürzl, W., Kempter, R., and van Hemmen, J. L. 2000. Theory of arachnid prey localization. *Phys. Rev. Lett.* 84: 5668–5671.

Tao, H.-z. W., Zhang, L. I., Bi, G., and Poo, M. 2000. Selective presynaptic propagation of long-term potentiation in defined neural networks. *J. Neurosci.* 20: 3233–3243.

Wilkens, L. A., Russell, D. F., Pei, X., and Gurgens, C. 1997. The paddlefish rostrum functions as an electrosensory antenna in plankton feeding. *Proc. R. Soc. Lond. B* 264: 1723–1729.

Zhang, L. I., Tao, H. W., Holt, C. E., Harris, W. A., and Poo, M. 1998. A critical window for cooperation and competition among developing retinotectal synapses. *Nature* 395: 37–44.

6

What Is Fed Back?

Jean Bullier

Introduction

As we enter the twenty-first century, we witness a shift in the types of models used to explain the processing of sensory information by the cerebral cortex. Until now, these models have overwhelmingly emphasized the *feedforward* direction for the transfer of sensory information, and it is only recently that a more balanced view of the cortical network has emerged. The following chapter summarizes this evolution and focuses on the question of the role of *feedback* connections in the processing of sensory information. I will use the visual system as a model since most of what is known concerning the processing of sensory inputs by the brain comes from studying that system.

A Feedforward Model

The architecture of the primate visual system is characterized by a succession of processing stages: the retina, the lateral geniculate nucleus (LGN), area V1, and a large number of functional cortical areas (figure 6-1A; for more precise and recent diagrams see Bullier 2003). It is known that in primates, 90 percent of the retinal input is relayed to area V1 and that the network of interconnected cortical areas accomplishes the major part of the processing, giving us a conscious impression of the world and enabling us to act on this world under visual control. Cortical computation involves interactions between thousands of neurons. Some interactions occur between neighboring neurons within a given cortical area through the very dense network of *intrinsic* connections (also called *horizontal* connections). In terms of density of connections, the local

A

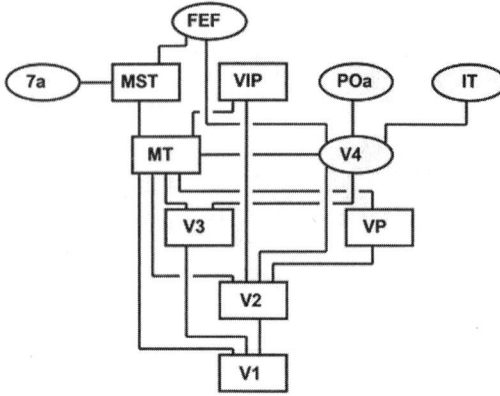

B

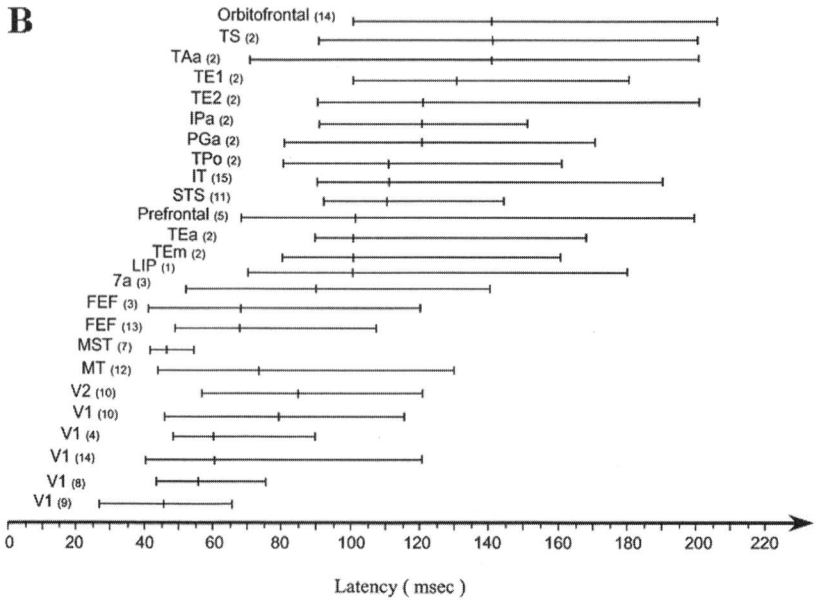

Latency (msec)

Figure 6-1.   (A) Hierarchical organization of cortical areas. This model presents the
different cortical areas of the primate visual system staged at different levels
according to a simple rule: areas of low-order stages send feedforward connections
to the upper levels, whereas high-order areas send feedback connections to areas at a
lower level (modified from Van Essen and Maunsell 1983). (B) Latencies of visual
responses of neurons in different cortical areas. For each area, the central tick marks
the median latency, and the extreme ticks the tenth and ninetieth percentiles.
Numbers in parentheses refer to bibliographic references given in Bullier (2001).
Note that the shortest latencies do not always correspond to the lowest stages of the
hierarchy. In particular, areas MT, MST, and FEF have very short latencies despite
their being placed at the highest levels in the earliest version of the hierarchy (A).

network is the densest, being estimated at 60 percent of the connections (Kennedy, Barone, and Falchier 2000).

In addition to this local network, there are connections that link together the cortical areas, as shown for areas V1, V2, and MT in figure 6-2. Among these interarea connections, it is usual to distinguish between interhemispheric connections that link together areas of opposite cerebral hemispheres and two main types of intrahemispheric connections, the feedforward and the feedback connections. These connections differ in terms of morphology: feedforward connections arise mostly from supragranular layers (layers 2 and 3) and terminate in and around layer 4. Feedback connections arise preferentially from deep layers and terminate outside layer 4 (Felleman and Van Essen 1991; Salin

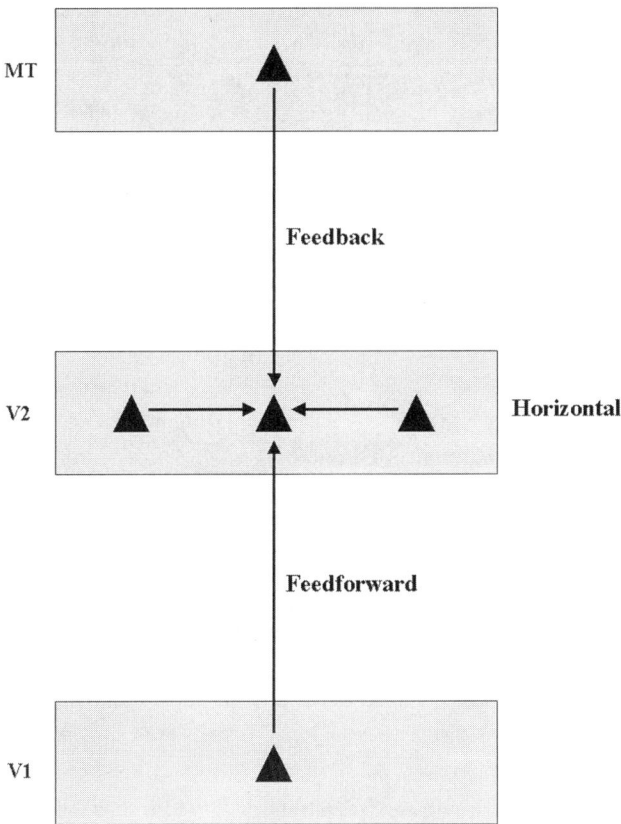

Figure 6-2. Schematic representation of three different cortical areas, V1, V2, and MT, and their interconnections. A cortical neuron in area V2 receives converging information from horizontal (also called intrinsic) connections within V2, feedforward connections from V1, and feedback connections from MT (and other areas). Much of the recent research deals with understanding the differences between the roles of these three different types of connections.

and Bullier 1995). Feedforward connections carry information from lower-order areas (like V1 or V2) to higher-order areas (like V4 and MT), whereas feedback connections connect areas in the reverse direction. In most cases of connections between the thalamus and cortex and between cortical areas, these connections are bidirectional: if area A sends feedforward connections to area B, area B sends feedback connections to area A. The relative density of feedforward and feedback connections differs for different areas: an area like V1 receives a small contingent of feedforward connections from the LGN and most of its external input from feedback connections (mainly from V2), whereas area V4 receives a similar number of feedforward and feedback connections (Kennedy, Barone, and Falchier 2000).

Until now, most functional models of information processing by the visual cortex have emphasized the feedforward connections and largely ignored the local intrinsic and feedback connections. According to such feedforward models, vision is understood in terms of a series of processing stages with more and more elaborate filtering properties for the receptive fields of neurons at higher levels. Areas that play an essential role in pattern recognition and visuomotor interactions are placed at the highest levels of the hierarchy of cortical areas, with V1 at the bottom and hippocampal formation at the top (figure 6-1A; for more recent and detailed versions see Bullier 2003). The models assume that the selectivity of a high-level neuron is achieved by a proper combination of the different feedforward inputs from lower-order areas. The best example is provided by the mechanisms underlying orientation selectivity of neurons in area V1. Although the question has been studied for more than forty years, it is still a matter of debate whether orientation selectivity is simply the result of a proper selection of thalamic inputs with receptive fields aligned in the optimal orientation (Ferster, Chung, and Wheat 1996; Hubel 1996) or whether intrinsic excitatory (Douglas and Martin 1991) and/or inhibitory inputs (Borg-Graham, Monier, and Fregnac 1998) are necessary for shaping the selectivity of these neurons.

Beyond V1, several attempts have been made to explain the receptive field selectivities of neurons in terms of their feedforward inputs. In area LS of the cat (presumed homologue of monkey area MT), the patchy organization of axon terminals that are grouped according to their optimal direction is in keeping with idea that direction selectivity is the result of a proper arrangement of direction selective inputs from areas seventeen and eighteen (Sherk 1990). Similarly, Movshon, and Newsome (1996) have argued that the direction selectivity of neurons in area MT is due to the appropriate combination of inputs from V1 that show homogeneous receptive field properties and the proper optimal direction.

The models of cortical plasticity that explain the progressive refinement of receptive field properties of cortical neurons during development are also based on a combination of feedforward inputs. For example, it is known that monocular deprivation and surgery-induced strabismus induces strong changes in ocular dominance of neurons in area V1. The explanation of such changes has been based on the idea that feedforward thalamic inputs from the two eyes compete at the level of their input in the cortex (layer 4 of V1), the most active input wins over the less active one, and the observed shifts in ocular dominance therefore simply reflect the results of this competition. The higher stages of the hierarchy are supposed to simply follow the reorganization that has taken place at a lower level.

The Turning Point

Several recent lines of evidence have led to a reexamination of the feedforward model. The first evidence comes from studying the timing of information transfer at the different stages of the cortical hierarchy (figure 6-1). A feedforward model based on a series of processing stages from V1 to the highest stages of the hierarchy requires that a certain amount of processing time is spent at each stage. Thus, higher levels of the hierarchy should show longer latencies to visual stimulation. In fact, as shown by Nowak and Bullier (1997; figure 6-1B) and Schmolesky and his collaborators (Schmolesky et al. 1998), there are many examples of areas that have similar latency distributions and are located at very different stages of the hierarchy. For example, consider areas V2, MT, and FEF. Although they belong to different hierarchical levels (figure 6-1A), they have similar latencies to visual stimulation (figure 6-1B). What is shown in figure 6-1B and even more clearly demonstrated by Schmolesky et al. (1998) is that areas of the dorsal stream (areas V3d , MT, MST, LIP, and FEF) are activated almost simultaneously with V1. This is probably related to the role that neurons in these areas play in visuomotor interactions that require the maximal processing speed (Rossetti, Pisella, and Pélisson 2000) and to the fact that these areas are mainly driven by the magnocellular component of the LGN input that reaches area V1 some 20 milliseconds earlier than the parvocellular inputs in monkeys (Nowak and Bullier 1997), as well as in humans (Baseler and Sutter 1997). Areas of the ventral stream such as V4, TEO, and TE show progressively longer latencies to visual stimulation that are consistent with a cascade of processing stages, each stage taking on average 10 milliseconds (Nowak and Bullier, 1997).

The second set of evidence comes from reports showing that the feedforward connections are not the sole determinant of the response strength and selectivity of visual cortical neurons. For example, a series of reports by the group of Eysel clearly demonstrate a major role for the network of local horizontal connections in shaping the orientation selectivity of neurons in area V1 (Crook, Kisvarday, and Eysel 1997). An example of their results (figure 6-3) shows the orientation tuning curves of two neurons recorded at the site RS in area V1 and

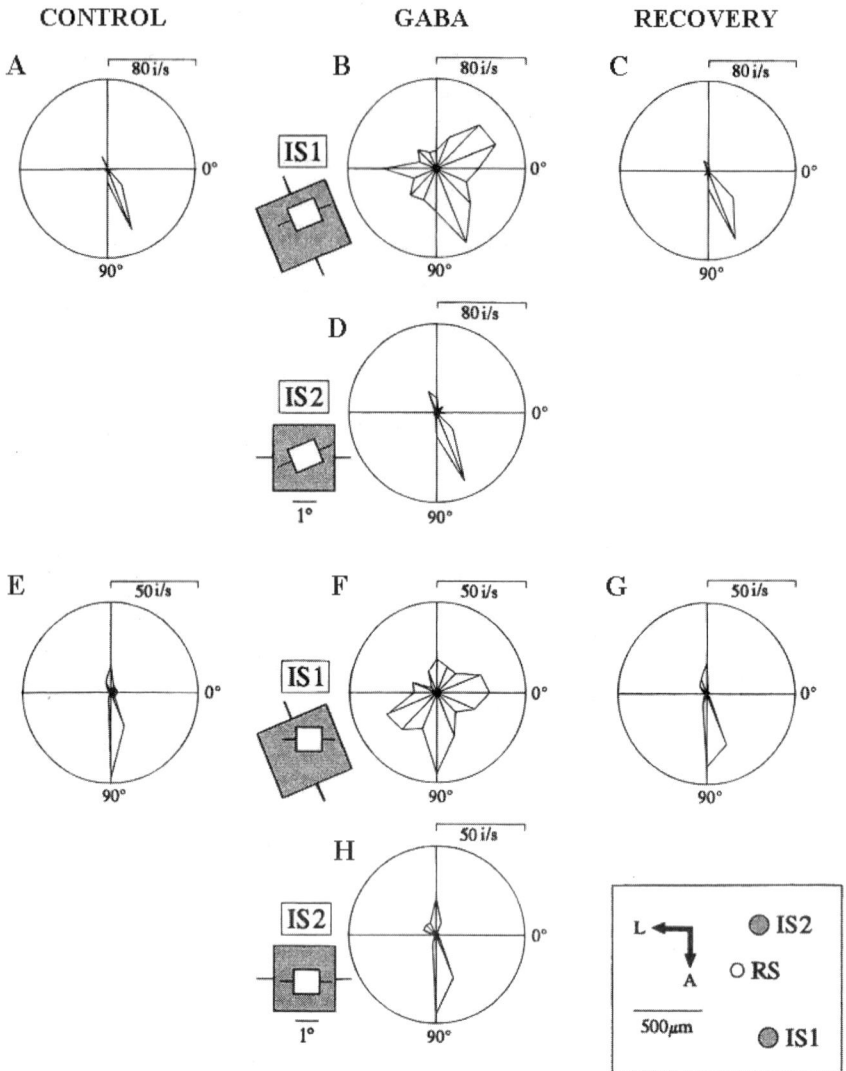

CONTROL   GABA   RECOVERY

Figure 6-3. Effects of inactivation by GABA iontophoresis of the local network of horizontal connections on orientation tuning in area V1 (modified from Crook, Kisvarday, and Eysel, 1997). (A) Polar plot of the orientation tuning of a V1 neuron recorded at the site RS (sites illustrated in the lower right insert). (B) Orientation tuning of the neuron during GABA inactivation at the site IS1 with optimal orientation orthogonal to that of the recorded neuron. There is a substantial broadening of the tuning curve. (C) Recovery run: orientation tuning curve of the recorded neuron after the effects of GABA at site IS1 have vanished. (D) Effect of inactivating the site IS2 (with optimal orientation close to that of the recorded neuron): no effect is observed on the response strength or the orientation tuning curve of the recorded neuron. (E) Orientation tuning curve of another neuron recorded at the same site RS. (F) Effect of inactivating the site IS1 with optimal orientation close to orthogonal to that of the recorded neuron: there is an

108

the effects on these tuning curves of reversible inactivation of the sites IS1 and IS2 by GABA iontophoresis. The first neuron (figure 6-3A–D) has an optimal orientation of 30 degrees with respect to the horizontal and is strongly direction selective (figure 6-3A). The cortical site IS1 with a preferred orientation orthogonal to that of the recorded neuron is inactivated with GABA. This leads to a considerable broadening of the orientation selectivity of the recorded neuron (figure 6-3B) that totally recovers after the effects of GABA have vanished (figure 6-3C). When the site IS2 with optimal orientation parallel to that of the recorded neuron is inactivated, practically no effect is observed (figure 6-3D). A similar specific effect is observed for another neuron also recorded at the site RS (figure 6-3E–H). Again, the inactivation of the site with orthogonal optimal orientation leads to an almost complete loss of orientation selectivity (figure 6-3F), whereas very little effect is observed for inactivation of the iso-oriented site IS2 with (figure 6-3H). Since GABA silences reversibly cortical neurons within a few hundred microns from the iontophoresis micropipette and does not affect axons, these results demonstrate that a proper combination of afferent thalamic axons is not sufficient to achieve orientation selectivity in a cortical neuron but that a major role is devoted to local excitatory and inhibitory connections. Given that inactivation of the IS1 and IS2 sites broadens the tuning curves but does not change the optimal orientation, it is likely that the optimal orientation of a cortical neuron depends on the selection of thalamocortical axons that drive it. However, it is equally obvious that the sharp orientation selectivity displayed by the cortical neurons illustrated in figure 6-3 is mainly the result of processing by the local network of horizontal connections. Note that the effects of inactivation of local connections are specific to regions with optimal orientation orthogonal to that of the recorded neuron (figure 6-3B, D, F, and H) and that the broadening of the tuning curve is achieved by responses increases at nonoptimal orientations. This strongly suggests that the orientation tuning curve is mainly shaped by inhibitory mechanisms for nonoptimal orientations, in keeping with the early conclusions of Sillito (Sillito et al. 1980) and the recent results of Borg-Graham and Frégnac (Borg-Graham, Monier, and Fregnac 1998).

The importance of the local connectivity network in shaping the responses of neurons in the visual cortex is further demonstrated by a recent report from the group of Grinvald that shows that a cortical neuron emits a spike (during spontaneous or driven firing) preferentially for a given distribution of activity on the cortical surface surrounding the recorded neuron (Tsodyks et al. 1999). The spatiotemporal state of cortical activity most often associated with the

---

almost complete loss of selectivity to orientation. (G) Recovery run after the effects of inactivating site IS1 have vanished. (H) Effect of inactivating the site IS2 with same optimal orientation as the recorded neuron: again, no effect is observed on the orientation tuning curve of the recorded neuron.

appearance of a spike is the same whether the spike is emitted in response to a visual stimulus at the optimal orientation or whether the spike is emitted during spontaneous firing. This suggests that the spatiotemporal pattern of local cortical activity acts as a gate in such a way that a strong cortical response to a given thalamic input will be produced only when the local cortical network is in a permissive state. Thus, the local network of intrinsic connections plays a major role in modulating the excitability of a cortical neuron and the feedforward input may be more properly compared to a trigger than to an input governing the receptive field selectivity. Such a trigger role is consistent with the small number of feedforward thalamic inputs to a given neuron of layer 4 in V1 compared to the very rich array of connections from local neighboring neurons (Douglas et al. 1995; Latawiec, Martin, and Meskenaite 2000; Peters, Payne, and Budd 1994).

Responses of cortical neurons also depend on the activity of feedback connections. This has been shown in experiments in which the responses of neurons in low-order cortical areas were modified by inactivation of higher-order areas by cooling (Bullier et al. 2001; Hupé, James, Girard, and Bullier 2001; Hupé, James, Girard, Payne, and Bullier 2001; Hupé et al. 1998). An example is shown in figure 6-4A: the response of a neuron recorded in area V3 to a moving bar activating the receptive field center is substantially decreased when area MT is reversibly inactivated by cooling (histograms labeled "Center"). Such an effect was observed on about 40 percent of the sampled neurons in areas V1, V2, and V3. Figure 6-4B shows that the mean response decrease for the affected neurons was stronger for stimuli of low salience (contrast of the moving bar to the stationary textured background). The set of histograms in figure 6-4A labeled "Center + background" presents the responses of the neuron to the same bar moving with the low-contrast textured background (see figure 6-4B). Because of the inhibitory influence of the RF surround on the center response, the response to the bar and background is much weaker than that to the bar alone (compare the control responses to those of the center and center and background stimuli). During inactivation of area MT, the response to "Center + background" was not decreased like the response to the bar alone, but was slightly increased (compare the histograms in figure 6-4A during cooling). This results from a decrease of the inhibitory influence of the RF surround upon the RF center mechanism. As shown for the population data (figure 6-4C), this decrease was particularly strong in the case of low salience stimuli for which the background suppression is almost abolished in the absence of feedback from area MT. Thus, feedback connections appear to potentiate the center response and the center-surround inhibitory interactions for low salience stimuli.

Thus, the stimulus selectivities and center-surround interactions of neurons in the visual cortex are not solely determined by the proper arrangement of their feedforward input but also depend on the activity of neurons in the

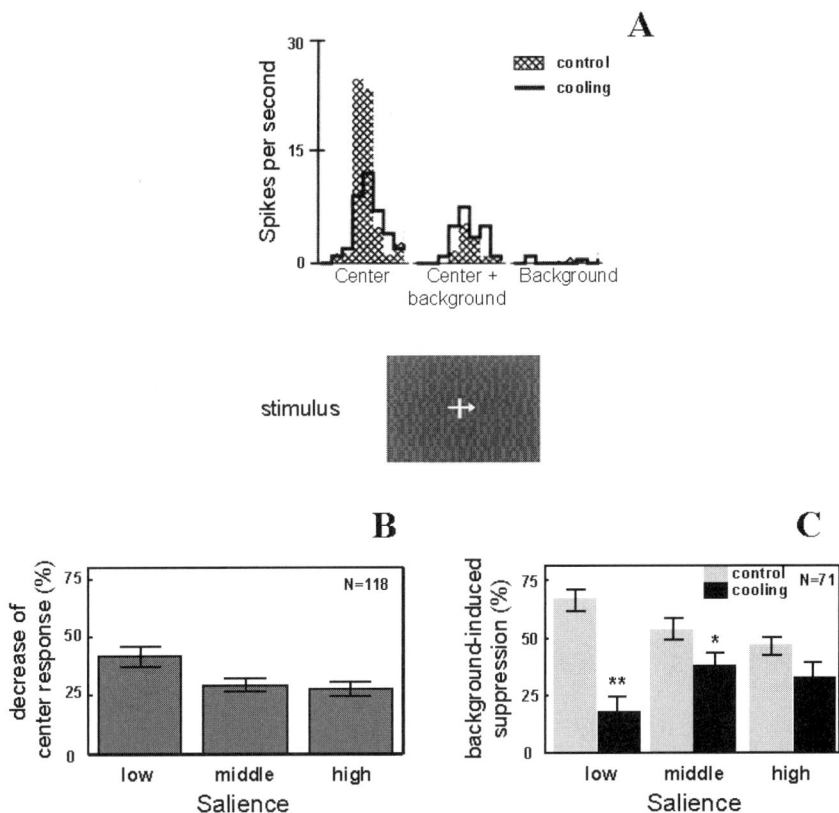

Figure 6-4. Effect of cooling inactivation of area MT on the responses of neurons in lower-order areas. (A) Responses of a V3 neuron to a light bar moving across the receptive field center during control and cooling of area MT (modified from Hupé et al. 1998). The set of histograms labeled "Center" corresponds to the responses to the bar moving across the static textured background. The response of the neuron is strongly diminished by MT inactivation. The set of histograms labeled "Center + background" corresponds to the responses recorded when both the central bar and the textured background are moving. The response during control is much weaker than the response to the center stimulus due to the strong inhibitory effect of the surround mechanism. On the contrary, during cooling of MT, the response to "Center + background" is only marginally weaker than to the center stimulus. Thus, the inhibitory action of the surround is diminished by inactivation of MT. The "Background" set of histograms shows that there is no response to the movement of the background stimulus alone. (B) Average percent decrease of neuron responses in areas V1, V2, and V3 to center stimulus during MT inactivation for different amount of salience of the center stimulus compared to the background (the salience corresponds to the ratio of the contrasts of the center bar and the background; see Hupé et al. 1998 for definition; modified from Bullier, Hupé, James, and Girard, 2001). This shows that at low salience about 40 percent of the response to the center bar is under the control of feedback connections from MT. (C) Percentage of suppression of the center response of neurons in areas V1, V2, and V3 by the moving textured background (modified from Bullier, Hupé, James, and Girard, 2001). Note that at low salience there is an almost complete disappearance of the suppression when MT is inactivated. This means that at low salience, the major part of the surround suppression is under the control of the feedback connections.

local network of horizontal connections and of feedback connections from higher-order areas. The importance of the local cortical network in shaping the neuronal selectivity has also been emphasized recently by reports showing that the mechanisms of developmental plasticity do not simply involve the competition between two sets of feedforward inputs. Although it has been known for some time that following monocular deprivation, the shift in ocular dominance of neurons in V1 is more rapid in supragranular layers than in layer 4, the presumed site of the competition between the inputs coming from the two eyes, the consequences for models of synaptic competition were not fully understood. In a recent study, Stryker and his collaborators targeted specifically the regions of area V1 containing neurons that shift their ocular dominance following short periods of monocular deprivation. They showed that in these regions, neurons in the supragranular and infragranular layers rapidly shift their ocular dominance after the deprivation, whereas the ocular dominance of layer-4 neurons remains unchanged for long periods afterwards (Trachtenberg, Trepel, and Stryker 2000). This suggests that the supragranular and infragranular layer neurons do not simply shift their ocular dominance as a result of the competition between afferent inputs at the lower level of layer 4. Instead, ocular dominance shifts occur rapidly at a higher processing level (the supra- and infragranular layers) and the switch in eye dominance at the level of thalamic inputs is likely to follow as a result from information being fed back from upper and lower layers. It remains to be seen whether similar mechanisms are at play in other cases of plastic reorganization, such as the consequences of rearing animals in stroboscopic light, which have also been interpreted as resulting from interactions between competing inputs in layer 4 (Humphrey, Saul, and Feidler 1998).

What Is Fed Back?

Thus, in models of the visual cortex, the emphasis is presently shifting from feedforward models to models that include horizontal as well as feedback connections to explain the receptive field properties of cortical neurons. This shift also corresponds to a larger importance given to higher levels of the brain in cognitive issues: no longer considered mainly as processing sensory inputs for generating motor outputs, the brain is viewed as a self-sustained machine processing mainly internal information and sampling occasionally the external world. The recent renewal of interest in brain rhythmic activity is also a consequence of this change in perspective that leads the community of brain scientists to abandon a view of the brain as a simple input-output machine.

The processing of incoming sensory input depends heavily on the state of the brain at a given moment. This has been particularly evident in the recent development of studies of attention on cortical processing of visual information. The internal processes of the brain are also an essential ingredient in ensuring perceptual constancy, recognition, and classification of objects. It is generally assumed that effects of the internal states of the brain on the processing of sensory information are transferred by top-down or feedback connections. The role of top-down influences in higher-level vision remains somewhat of a mystery and will constitute one of the most important questions that we will have to answer in the future. In the following sections, I briefly review what is known about a number of top-down influences that have been attributed to the feedback connections in the visual system.

Are Feedback Connections Driving or Modulating?

A couple of recent papers have formalized the old idea that some connections act as drivers, other as modulators (Crick and Koch 1998; Sherman and Guillery 1998). Driver connections are supposed to be small in number and have strong, focused, and rapid effects on their target neurons; modulators have symmetrical properties: they are numerous and have weak, slow, and diffuse actions. It is likely that such a dichotomy is artificial and that there is a continuum of connection types from the strongest and most rapid ones to the very weak and slow. However, as argued by Crick and Koch (1998), it is probably important to distinguish between strong and weak connections for understanding the function of a network and solving a number of questions such as that of the stability of the network. Typical examples of drivers and modulators are given in both papers by comparing the thalamocortical and corticothalamic connections for relay neurons of the thalamus. For example, LGN inputs into layer 4 concern a small number of synaptic contacts compared to the cortical connections (Latawiec, Martin, and Meskenaite 2000; Peters, Payne, and Budd 1994), but they appear to have a strong driving force, as evidenced by the effects of reversible inactivation of the LGN (Malpeli 1983; Nealey and Maunsell 1994) and the large sizes of the synaptic potentials (Stratford et al. 1996). On the other hand, corticothalamic axons are extremely numerous (Sherman and Koch 1986) but do not appear to have a very strong effect, as evidenced by the variable and weak results obtained by reversible inactivation of cortex on thalamic firing (see review in Steriade, Jones, and McCormick 1997). The effects also appear to be slow because most thalamocortical axons are approximately ten times slower than corticothalamic axons (review in Nowak and Bullier 1997) and appear to activate metabotropic receptors that give slow responses (Godwin, Vaughan, and Sherman 1996; McCormick and von Krosigk 1992).

The distinction between drivers and modulators that appears so clear between thalamocortical and corticothalamic connections has been extended to the feedforward and feedback corticocortical connections by Crick and Koch (Crick and Koch 1998). Feedforward connections, like thalamocortical connections, terminate in layer 4, whereas feedback connections terminate outside layer 4, in layers which may be less essential for driving a cortical area. There is indeed evidence to support the idea that feedforward connections are of the driver type. Inactivation of area V1 leads to a complete cessation of activity in area V2, which receives a strong feedforward connection from V1. Since V2 is an important feedforward input to area V4, the loss of activity in V4 during inactivation of V1 also suggests that the feedforward connections from V2 to V4 are of a driver type. The lack of activity in inferotemporal cortex following lesion of area V1 is also in keeping with the idea that feedforward connections are drivers (see review in Bullier, Girard, and Salin 1994).

It may be, however, that this equation between feedforward and driver is an oversimplification due to the small number of feedforward connections that have been tested. In particular, the fact that data from primates come only from inactivation of area V1 may bias the conclusions because this action stops or strongly decreases firing in many areas. Thus, what is interpreted as the result of inactivating the feedforward connection from V1 to area A may in fact reflect the influence of inactivating V1 on the whole network to which area A belongs. Given the strong reciprocal connections between areas V2 and V4, it would be interesting to test whether area V4 stops responding when V2 instead of V1 is inactivated. Furthermore, the conclusion that feedforward connections carry the major drive to a cortical area is questioned by the residual activity observed in MT after inactivation or lesion of V1 in macaque monkeys (see review in Bullier, Girard, and Salin, 1994): is the input from V1 a driver or a modulator input? Consideration of results in other species also calls for caution because inactivation of feedforward connections often does not lead to much changes in the target area (like in the case of area seventeen to area eighteen connections in the cat; see review in Bullier, Girard, and Salin 1994).

In conclusion, the idea that feedforward connections have a strong driver-type effect on the target area may be true only in such special cases as the thalamic inputs from relay nuclei into layer 4 of primary cortical areas, or in the case of the projection from V1 to V2 which is the main target of V1. Results from the V1 to MT connections in macaque monkeys and from the feedforward connections in the cat suggest that for other types of corticocortical connections, the situation may be less clear-cut.

Less is known concerning the action of feedback connections. As mentioned earlier in this section, the published accounts of effects of inactivating the corticothalamic pathway suggest that the role of this input is not to drive the thalamic neurons but possibly to modulate the activity driven by the retinal inputs. It is not certain that the cortical feedback connections necessarily have the same functional role as the corticothalamic connections. For example, there is a clear difference in timing between corticothalamic and corticocortical connections. Most corticothalamic connections are much slower than the thalamocortical connections (Nowak and Bullier 1997; Steriade, Jones and McCormick, 1997), whereas the conduction speed of corticocortical feedback connections is as fast as that of feedforward connections (Girard, Hupé, and Bullier 2001; Nowak and Bullier 1997). This suggests that the role of feedback connections is not limited to that of a slow and diffuse modulating action on the activity of neurons in lower-order areas. Our finding of a small number of cells in area V2 that completely stopped responding to visual stimulation during inactivation of area MT (Hupé et al. 1998) also suggests that feedback connections can have a very strong effect on their target neurons. The results of Miyashita and his colleagues (Tomita et al. 1999) demonstrating neural drive in memory recall by feedback connections from the frontal cortex also illustrate the possibility for feedback connections to act as drivers under certain conditions.

However, results of our experiments with inactivation of cortical areas are generally consistent with a driver/modulator difference between feedforward and feedback connections. Consider what is observed in area V2 when V1 is inactivated: despite the presence of a strong residual activity in area MT and feedback connections from MT to V2, there is no response of neurons in V2 (Girard and Bullier 1989), even though we have also shown that inactivating area MT changes the responses of a number of neurons in V2 (Bullier et al. 2001; Hupé et al. 1998) when V1 is active. This suggests that feedback connections cannot drive V2 neurons in the absence of a feedforward input from V1 but can modulate the responses of neurons in V2 when they are driven by their feedforward inputs from V1. Consideration of the responses to MT inactivation in areas V1, V2, and V3 also points to the same conclusion: when MT is inactivated, the effects are observed only on the response of the neuron to stimulation of the RF center mechanism, which presumably reflect the feedforward input to the neuron (Hupé, James, Girard, Payne, and Bullier 2001). This suggests that feedback connections act as a gain control of the response of the neuron to the feedforward input. Thus these results enable us to define more precisely the differences between drivers and modulators: a driver-type connection can activate a neuron, even without the concomitant action of other inputs. Modulator-type connections are not potent enough to raise the membrane potential up to firing threshold in the absence of other inputs. They act in a nonlinear fashion by changing the gain of the neuron response to a given stimulus.

Such a difference suggests that feedforward and feedback connections use different types of receptors, that their axon terminals contact different parts of the neuron, or both. Evidence has been presented that feedforward and feedback connections have different synaptic mechanisms (Nowak, James, and Bullier 1997; Shao and Burkhalter 1996); however, the validity of these results is limited by the use of electrical stimulation that does not distinguish between a direct orthodromic drive and an antidromic activation of recurrent collaterals (Nowak, James, and Bullier 1997). It will be necessary to target specific feedforward and feedback circuits with other activation techniques than electrical stimulation before we understand the differences between feedforward and feedback connections at the synaptic level.

It has been known for a long time that axons of different functional groups of neurons conduct at very different speeds. Although this has been a major theme in the study of the retinogeniculate connections (Stone 1983), it is only recently that it has been considered an important aspect of corticocortical connectivity. As mentioned in the section of this chapter headed "The Turning Point" (figure 6-1B), neurons in different cortical areas of the macaque monkey visual system have similar latencies to visual stimulation. This raises the question of the timing of activation of a cortical neuron by a given afferent connection. It is interesting to note that the thickest axons that are known to conduct faster also have very large terminal boutons, suggesting that their action on the target neuron is very potent in terms of synaptic activation. This has been observed in thalamocortical interactions (Rockland 1998; Steriade, Jones and McCormick, 1997) and in corticocortical connections (Rockland 1989; Rockland 1995). Thus, the driver-type connections may correspond to the thick axons with heavy terminal boutons that can drive a target neuron even when its membrane potential is far from the spike threshold. Thick axons and heavy terminal boutons are indeed found in the terminals of thalamo-cortical axons in layer 4 of V1 (Freund et al. 1989; Humphrey et al. 1985) in keeping with their classification as drivers. On the other hand, most corticothalamic axons have thin axons and small boutons *en passant* (Rockland 1998; Steriade, Jones, and McCormick 1997), which is in keeping with their being classified as modulators.

If thick axons always correspond to driver connections, the presence of a few thick corticopulvinar axons originating in layer 5 with large axons and heavy terminals (Rockland 1998; Steriade, Jones, and McCormick 1997) suggest that not all corticothalamic connections should be regarded as modulators. Indeed, as argued by Sherman and Guillery (1998), such corticothalamic drivers with thick axons may be an important component of the feedforward transfer of information between visual cortical areas. A similar case may be made for the fastest feedback connections that may correspond to drivers. For example, it is known that area FEF is activated very early (figure 6-1B), much earlier than some of the inferotemporal cortex regions to which it is connected (Bullier, Schall, and Morel 1996). It would be interesting to determine whether some of its input into these areas involves rapid axons and is of the driver type.

In conclusion, although feedback connections in general appear to produce a nonlinear effect on their targets by acting on the gain of the response to the feedforward inputs, and thus act as modulators, it remains to be seen whether this is true also for the thickest axons. Feedback connections are very diverse in their morphology, their topographical organization on the cortical surface, and the laminar distribution of their terminals (Bullier 2003; Salin and Bullier 1995). It is likely that one cannot speak of the functional role of feedback connections in general; it will be necessary to distinguish between different sets of feedback connections that play different roles in the cortical processing of visual information. This will require isolating specific circuits and activating and inactivating different inputs to a recorded neuron.

What Are the Roles of Horizontal and Feedback Connections in Global-to-Local Interactions?

Pattern recognition and classification is possible with a purely feedforward model when a single stimulus is present in the visual field, as is the case in most laboratory situations. However, vision in a cluttered environment requires computations that combine global information concerning how the local details fit in the large picture with local information with a high degree of precision (Mumford 1993). It is usually assumed that, at the single-neuron level, such global-to-local interactions correspond to the center-surround interactions that are observed in the receptive fields of most cortical neurons, as illustrated in the example presented in figure 6-4A. As shown originally in area V1 by Bishop and Henry (Bishop, Coombs, and Henry 1973) and more recently by Freeman and his colleagues (Walker, Ohzawa, and Freeman 1999) and in area V5 by Orban's group (Xiao et al. 1997), the receptive field surround, unlike that of retinal ganglion cells, is usually not circular. Most often, the surround corresponds to a single region that produces antagonistic inhibitory interactions with the activity generated by stimulation of the receptive field center. Such center-surround interactions could be produced by local horizontal connections and/or by feedback connections.

We saw earlier that horizontal connections play a role in shaping the orientation selectivity of cortical neurons in V1 (figure 6-3). The question is whether center-surround interactions are also mediated by horizontal connections or whether feedback connections are involved. We tested the effect of center-surround interactions in two preparations: (1) inactivation of area MT by cooling and recording in areas V1, V2, and V3 and (2) GABA inactivation of area V2 and recording in V1. In the case of MT inactivation, we did find modifications of the suppression of the center response by surround stimulation (figure 6-4A). However, as shown in figure 6-4C, the effects were observed mostly when the stimulus was of low salience with respect to the background. In other words, the center-surround interactions in areas V1, V2, and V3 depend on the feedback connections from MT when the stimulus is masked by a noisy background of similar contrast. When the stimulus is clearly distinguishable from the background by its high contrast, feedback connections appear to play a limited role in setting up the center-surround interactions. This corresponds to the small effect of MT inactivation on background suppression at middle and high salience in areas V1, V2, and V3 (figure 6-4C; Bullier et al. 2001; Hupé et al. 1998). A similar conclusion was reached in the study of V2 inactivation on the center-surround interactions in V1: here the stimuli activating the center and surround mechanisms were of high contrast, and no effect of feedback inactivation was seen on the center-surround interactions, at least for stimuli limited to a few degrees away from the RF center (Hupé, James, Girard, and Bullier 2001). Thus, contrary to the prediction of a recent theoretical paper (Rao and Ballard 1999), the proximal center-surround interactions that are observed with high-contrast stimuli do not appear to depend on feedback connections. It is therefore likely that the local horizontal connections are the major factor underlying short-range center-surround interactions in cortical neurons.

One argument frequently used to distinguish between effects mediated by local and feedback connections is that of the response timing. The idea is that feedback connections being driven by higher-order areas must be delayed because of the supposedly longer latencies in these areas, whereas the local connections must have a rapid effect because of their short lengths. It is ironic that we found exactly the opposite to be true: feedback effects from MT and V2 to V1 neurons are not delayed with respect to the response onset by more than 10 milliseconds and do not appear to target neurons with late latencies (Hupé, James, Girard, and Bullier 2001; Hupé, James, Girard, Payne, and Bullier 2001). This finding is in keeping with our demonstration that feedback connections are fast conducting (Girard, Hupé, and Bullier 2001) and that latencies of neurons in higher-order areas are not substantially longer than those of low-order areas (figure 6-2B). On the contrary, local horizontal connections appear to be slow-conducting (Girard, Hupé, and Bullier 2001) and their effects to be delayed by several tens of milliseconds (Bringuier et al. 1999). Thus, although some delayed influences may be mediated by feedback connections as argued by Lamme and his colleagues (Lamme, Super, and Spekreijse 1998), it is clear that feedback connections can also act on the early part of the response, so it is not possible to differentiate local versus feedback connections by the latency of the effect. For example, the orientation-specific center-surround interaction in V1 neurons is delayed by 40 milliseconds with respect to the onset of the center response (Hupé, James, Girard, and Bullier 2001). Despite this substantial delay, this effect does not appear to depend on the feedback from area V2 (Hupé, James, Girard, and Bullier 2001) that provides the bulk of the feedback connections to V1 (Kennedy, Barone, and Falchier 2000). It is likely that this delayed effect is mediated by local horizontal connections.

What, then, is the role of the feedback connections in global-to-local interactions? We saw that in the case of low-contrast stimuli masked by the background, feedback connections help in differentiating figure from background by potentiating the inhibitory center-surround interactions. Another likely role for the feedback connection is that of reorganization of the local connectivity. As demonstrated by experiments in awake monkeys (Gilbert et al. 2000; Kapadia, Westheimer, and Gilbert 1999), the receptive field organization of V1 neurons depends on the general context in which the stimulus is embedded, and the changes resulting from modifications of the environment cannot be predicted by simple center-surround interactions as described earlier in this section. It is likely that such reorganization is under the control of the feedback connections that would therefore modify the receptive fields of low-order areas to optimize their processing abilities. Another example of such a reorganization was provided by recent results showing that feedback connections play a role in shaping the responses of neurons in low-order areas to plaid-type stimuli (Schmidt et al. 2000).

One of the interesting questions concerning this modification of receptive field properties by feedback connections concerns the level at which the modifications are made. It is known, for example, that dorsal and ventral visual cortical areas exchange relatively few connections at levels higher than areas V4 and MT (Baizer, Ungerleider, and Desimone 1991; Morel and Bullier 1990). Since the dorsal stream is activated earlier than the ventral stream (Nowak and Bullier 1997), during processing of visual information it is likely that feedback information will mostly flow from dorsal areas to ventral areas. Information from dorsal cortical areas does not appear to be sent directly to high-order areas of the ventral stream, such areas Tep and Tea, but instead to be sent back to low processing levels such as areas V1, V2, and V4 to be used for computations in the ventral stream. Direct evidence for this hypothesis has been provided by results from Orban's laboratory (Sary, Vogels, and Orban 1993). Neurons in inferotemporal cortex are selective to the orientation of kinetic boundaries (boundaries between streams of random dots moving in different directions or different velocities; Sary, Vogels, and Orban 1993). Although lesions of area MT appear to disrupt the ability of monkeys to distinguish objects defined by kinetic boundaries (Marcar and Cowey, 1992), no neurons selective to the orientation of kinetic boundaries were found in area MT (Marcar et al. 1995), whereas they have been found in small numbers in area V2 (Marcar et al. 2000). It is likely that such properties are retroinjected by combinations of feedback inputs from area MT or MST to area V2 and that this information can be used by neurons in the ventral stream to identify shapes based on the differential speed of the random dots. In other words, feedback connections would act on the building blocks of cortical processing more than on the highest stages of processing of sensory information in the ventral stream.

This idea of retroinjection of information from the dorsal stream to V1 and V2 neurons processing information for the ventral stream corresponds to the proposition made by Mumford of considering areas V1 and V2 as "active blackboards" for posting the results of computations so they can be used by other streams of computations (Mumford 1992). Viewed in this way, feedback connections would play an essential role in integrating the processing done by the neurons of the ventral and dorsal streams. Thus, it will be particularly interesting to study the interactions between the dorsal and ventral streams in the different cytochrome oxidase bands of area V2 that correspond to the main inputs to the dorsal and ventral streams.

Is Attention Mediated by Feedback Influences?

With the development of techniques of functional brain imaging, there has been recently a major increase in the number of studies devoted to the neural mechanisms of attention. It is usually assumed that attention effects observed in lower-order areas are under the control of higher-order areas through feedback connections. I will show below that although there are indications that this might be the case, there is little published evidence for such a role of feedback connections, and much remains to be done before we understand the mechanisms underlying the attention switches that characterize our processing of visual information.

Attention is a mechanism used by the nervous system to reduce the load of processing sensory signals by selecting relevant inputs and ignoring the others (Tsotsos 1990). For example, when a subject is searching for a target among a population of distractors, the identification is rapid and independent of the number of distractors if the target differs from the distractors by a single feature (like color or shape). In this situation of "pop-out," the attention is distributed over the array of target and distractors. When there is more than one parameter and the target differs from the distractors by a conjunction of features, in most cases the processing load is too important for the target to pop out, and the subject uses a strategy of focusing its attention successively on each of the items in the array to detect the target (Treisman and Gelade 1980).

The simplest way to test attention effects on visual processing is to ask the subject to pay attention to one part of the visual field (for example one hemifield) and to compare the responses of neurons when their receptive fields are in the attended region with the responses measured when the attention is shifted to another part of the visual field. It has been one of the main experimental paradigms used recently to investigate the effect of attention on the processing of visual information in different structures of the visual system. This type of attention is called spatial or covert attention, and it has been compared to a searchlight that is directed to different regions of the visual field to optimize processing (Crick 1984). Spatial attention has been shown to influence the gain of single neurons in several extrastriate cortical areas of the primate visual system (Reynolds, Pasternak, and Desimone 2000; Treue and Maunsell 1999).

Concerning area V1, however, the role of attention in modulating the neuronal responses is subject to debate. Because V1 receives only feedback connections from other cortical areas, it is a particularly interesting region for testing the role of feedback connections in attention. Most reports from electrophysiological studies in monkeys have stressed that attention effects are null or weaker in V1 than in extrastriate cortical areas (Haenny and Schiller 1988; Luck et al. 1997; McAdams and Maunsell 1999). On the other hand, several single-unit studies have reported clear effects of attention at the single-cell level in V1 (Gilbert et al. 2000; Motter 1993; Roelfsema, Lamme, and Spekreijse 1998). In humans, it has been argued from visual evoked potential (VEP) data that spatial attention potentiates neuronal responses only in the extrastriate cortex and that such effects are not observed in V1 (Martinez et al. 1999). In contrast, recent imaging studies with fMRI have documented very strong effects of spatial attention on blood flow in V1 (Brefczynski and DeYoe 1999; Kastner et al. 1999; Somers et al. 1999). The effects demonstrated by these studies concern mainly the resting blood flow that is substantially increased by attention (Kastner et al. 1999), in a manner predictive of the performance of the subjects in a visual detection task (Ress, Backus, and Heeger 2000). The discrepancy between the magnitude of the effects observed on single-unit responses and blood flow may simply reflect the fact that to increase by a small amount the responses of neurons corresponding to a given region of the visual field involves activating a very large number of neurons because of the small receptive fields of neurons in V1. Another reason may be related to the difference in experimental paradigms. For example, it has been shown that the effects of attention on single-unit responses are much stronger for low-contrast than for high-contrast stimuli (Reynolds, Pasternak, and Desimone 2000). Since VEP studies and most single-unit experiments have been done with high-contrast stimuli, they may have missed significant effects of attention on neuronal responses in V1.

The observation that the effects of attention on single-unit responses are much stronger for low-contrast stimuli (Reynolds, Pasternak, and Desimone 2000) suggests that attention regulates the contrast gain of the neuron responses in V1. Such a nonlinear aspect of attention is reminiscent of the nonlinear effect of feedback connections on figure-ground discrimination, which had a stronger effect for low-salience stimuli, as mentioned in the sections "The Turning Point" and "What Are the Roles of Horizontal and Feedback Connections in Global-to-Local Interactions" above. This suggests, therefore, that response enhancements evoked by attention and, in the case of figure-ground discrimination, in low salience conditions may use a similar mechanism of nonlinear gain control through feedback connections.

The parietofrontal network appears to be strongly activated during many different tasks requiring attention (Kanwisher and Wojciulik 2000). Feedback inputs from the parietal cortex probably play a role in directing spatial attention, since it is known that parietal lesions produce attention deficits that are usually interpreted as a failure of the system to disengage attention to a given part of the visual field (Desimone and Duncan 1995). Consistent with this hypothesis, recent experiments showed that when the function of the parietal cortex is disturbed with transcranial magnetic stimulation, the serial search strategy in a conjunction task is substantially slowed down, whereas reaction times in a pop-out out situation are not modified (Ashbridge, Walsh, and Cowey 1997). Whether the influence from the parietal cortex is transferred through feedback connections or through other connection systems such as the corticopulvinocortical loops (Desimone and Duncan 1995) remains to be directly tested. If corticocortical connections are involved in the control of attention in the ventral stream by the frontoparietal network, it is likely that this involves feedback projections to mid-level areas such as V2 or V4 because of the small number of direct connections between parietal and inferotemporal cortex (Baizer, Ungerleider, and Desimone 1991; Morel and Bullier 1990).

Attention is usually assumed to potentiate the neuronal responses, as evidenced by response increases observed in most cortical neurons when attention is directed to their receptive fields in absence of other distracting stimuli. However, measurement of attention effects when two competing stimuli are presented within the receptive field center reveals substantial inhibitory influences on the responses to the unattended stimulus (Moran and Desimone 1985; Reynolds, Chelazzi, and Desimone 1999; Treue and Maunsell 1996). Thus, this effect of attention differs from that of the searchlight hypothesis that assumes a potentiating effect of attention on the neurons responding to the proper location in visual space or feature space (Koch and Ullman 1985; Treisman and Gelade 1980). Here, attention shrinks down the RF center onto the attended stimulus. In such a biased competition model of attention (Desimone and Duncan 1995), higher-order areas are supposed to inhibit the responses of neurons at lower stages when the nonpreferred stimulus is chosen as the focus of attention. To discover whether feedback connections or other types of mechanisms are involved in this push-pull action is clearly on the agenda for the future (Tsotsos et al. 1995). The mechanisms used may be similar to those demonstrated for the modifications of center-surround interactions in case of low-salience stimuli (figure 6-4).

Another paradigm that has been used to study attention is that of selecting a given object or set of objects defined by their characteristics, such as color, form, orientation, or general shape. At the single-cell level, this has been mainly studied in the inferotemporal cortex by Desimone and his collaborators, using a delayed-match-to-sample paradigm for a set of two complex images (Chelazzi, Miller, Duncan, and Desimone 1993). These studies showed that when the animal is cued to the optimal stimulus, the responses of IT neurons remain at a higher level until the animal generates a behavioral response, whereas the response to the nonpreferred stimulus is strongly decreased after the initial visually evoked response. Similar results have been found in area V4 when the animal had to choose between two sets of stimuli depending on their color (Motter 1994). Although these results are usually interpreted in terms of attention, they could as well be interpreted in terms of selection of an object as a target for eye movements, as is done for studies in frontal cortex that produce similar results (Schall 1997).

A similar paradigm of object selection according to features has been used with imaging techniques. In one of the earliest imaging studies, it was shown that blood flow in different cortical areas was increased when attention was paid to the color or movement direction of a complex stimulus (Corbetta et al. 1991). Similar methods have been used to differentiate between regions of the fusiform gyrus responding to objects or faces (Kanwisher and Wojciulik 2000) or between cortical regions responding to the position in the visual field or the type of faces (Haxby et al. 1994). More recently, Watanabe and his colleagues used a composite stimulus made of a superposition of translating and expanding random dots (Watanabe et al. 1998). They showed that attending to the set of expanding dots increased the blood flow in area V5 and the surrounding areas but not in V1, whereas attending to the translating dots increased the blood flow in V1 as well as in V5. Their finding was interpreted in terms of attention being specifically mediated to V1 by feedback connections for translating dots, whereas attention to expanding dots would not activate feedback connections. It is interesting to remark that area V1 is organized in a columnar fashion for orientation and direction of movement in visual space, whereas there is no columnar organization for expanding and contracting movements in visual space. When attention is shifted to a parameter organized in columns, the observed effects could be mediated through feedback connections that are organized, at least in some cases, in columnar fashion (see review in Bullier 2003). The interpretation of Watanabe and colleagues of their findings in terms of feedback connections is consistent with our results showing that feedback connections play an important role in differentiating figure from background (Bullier et al. 2001; Hupé et al. 1998). Thus, it is likely that feedback connections are involved in the role of selective attention for extracting specific features from noisy or ambiguous figures. This is likely to be particularly easy to observe when the features to disambiguate are organized in a columnar fashion in the target area.

So far we have been interested in top-down studies of attention that are characterized by the fact that the shift in attention precedes the visual stimulus. In many cases, it is the visual stimulus that produces an attention shift and an eye movement to foveate the attention-grabbing stimulus. In fact, there is a great deal of similarity in the mechanisms and cortical areas involved in directing gaze and attention (Corbetta et al. 1998). Obviously, it is of strong survival value to orient rapidly toward a potentially menacing stimulus. This may be the reason why fast attention switching appears to involve mainly the dorsal stream and the M channel that responds rapidly to visual stimuli (figure 6-1B; Steinman, Steinman, and Lehmkuhle 1997) and why attention appears to accelerate the processing of visual stimuli by speeding up specifically the responses of the M channel (Di Russo and Spinelli 1999). It is not clear whether these mechanisms necessarily involve feedback connections or other loops through the pulvinar. However, the observation that feedback effects can be extremely rapid (Girard, Hupé and Bullier, 2001; Hupé, James, Girard, Payne, and Bullier 2001) is in keeping with their potential role in rapid attention shifts.

In conclusion, although several lines of evidence suggest that feedback connections may be involved in generating the effects of attention in low-order visual areas, there has not been any direct test of this hypothesis. At the moment, it appears that the most interesting avenue is to explore the possibility that feedback connections are used to focus attention on a given parameter (color, movement speed, or shape) in a multiparameter stimulus. Such a role in selective attention is in keeping with that of reconfiguration of network mentioned in the section on global/local interactions.

Role of Feedback Connections in Matching Internal Models to Sensory Data

In the first two sections of this chapter we saw the limits of purely feedforward models of visual processing by the brain. Although these models remain dominant for neurobiologists, they cannot properly account for much of higher-level processing, such as pattern recognition, object classification, and mental imagery (Ullman 1996). A number of theoreticians have proposed that the visual cortex is not a feedforward classifier but that it contains a number of generative models of the world (Hinton 2000). Such generative models work by generating virtual sensory data that are compared with the incoming stream of sensory information. Models such as those proposed by Mumford (1993), Ullman (1995) and Rao and Ballard (1999) belong to the class of generative models. In these models, feedback connections play a very different role from that assumed in feedforward models: instead of adjusting the processing of sensory data depending on the incoming volley of information and computations made in higher order areas, they transmit virtual low-level data generated by the model, and the processing consists in

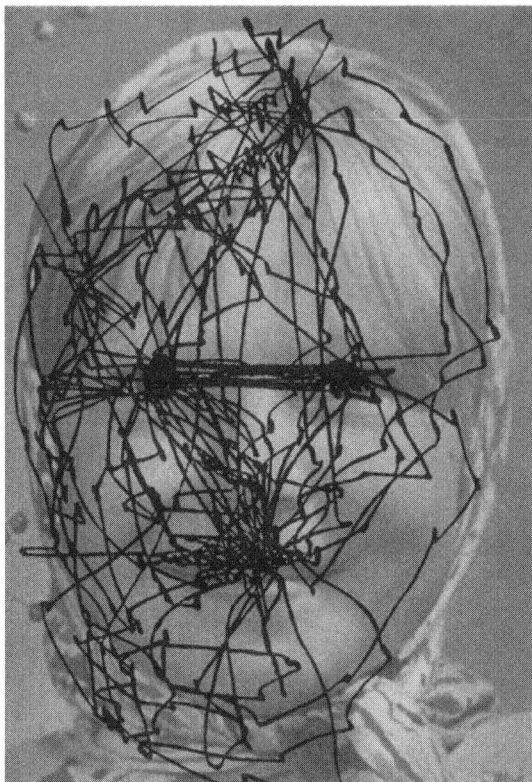

Figure 6-5. Pattern of the eye movements of a subject observing the picture of the face of a girl (after Yarbus 1967).

calculating the differences between these virtual data and the data sampled by the senses. Perception is no longer an operation of classification but of matching the reality to the state of the generative model. Because of its similarity with Helmholz's unconscious inference theory of perception, this type of model is called a Helmholtz machine (Hinton et al. 1995).

Clearly, there is first a conceptual revolution to be made in the world of neurobiologists before the Helmholtz machines replace the feedforward models. Then it will be necessary to answer several questions: what are the neural correlates of generative models, of the parameters of internal representations? The recent explosion of imaging studies during mental imagery (Goebel et al. 1998; Kosslyn et al. 1995; Le Bihan et al. 1993), visual illusions (Mendola et al. 1999; Seghier et al. 2000) or hallucinations (Dierks et al. 1999; Ffytche et al. 1998) has given a few clues as to where and how interactions between the generative models and the sensory stream could take place, but much remains to be done before we understand how the internal models interact with the samples of the reality transmitted by our sensory organs.

One of the lines of attack that could be productive is to study the mechanisms of prediction by the brain. If the brain uses a generative model of the visual world, it must use it to make predictions. So far, very little effort has been devoted to the temporal aspects of visual processing from one gaze fixation to the next. Most research groups assume that there is no information before the onset of a visual stimulus, and they simply measure the neuronal responses after stimulus onset. A few studies have tested the effects of changing the information before the onset of a stimulus or the change in fixation on the responses of neurons (Assad and Maunsell 1995; Duhamel, Colby, and Goldberg 1992) and they showed that, indeed, prior information influences the responses of neurons, and there is a certain amount of prediction that is made by the visual cortex. As usual, the development of such studies may only come when an experimental block has been superseded, that of studying neurons in awake monkeys with fixed heads and fixed gazes. Unlike the experimental monkey, we are not perceiving the world with a fixed gaze, we are exploring it. How does the generative internal model and the incoming stream of sensory information interact during visual exploration of our visual world? Consider the famous picture by Yarbus (1967) of the eye movements of a subject looking at a girl's face (figure 6-5): Which neuronal responses differ between subsequent fixations on the same part of the picture? Which parameters of the generative model are updated when the gaze returns for the fifth time on the right eye of the girl? Will we know the answer at the end of this century?

References

Ashbridge, E., Walsh, V., and Cowey, A. (1997). Temporal aspects of visual search studied by transcranial magnetic stimulation. *Neuropsychologia 35*, 1121–1131.

Assad, J.A., and Maunsell, J.H.R. (1995). Neuronal correlates of inferred motion in primate posterior parietal carter. *Nature 373*, 518–521.

Baizer, J.S., Ungerleider, L.G., and Desimone, R. (1991). Organization of visual inputs to the inferior temporal and posterior parietal cortex in macaques. *J. Neurosci. 11*, 168–190.

Baseler, H.A., and Sutter, E.E. (1997). M and P components of the VEP and their visual field distribution. *Vision Res., 37*, 675–90.

Bishop, P.O., Coombs, J.S., and Henry, G.H. (1973). Receptive fields of simple cells in the cat striate cortex. *J. Physiol., 231*, 31–60.

Borg-Graham, L.J., Monier, C., and Fregnac, Y. (1998). Visual input evokes transient and strong shunting inhibition in visual cortical neurons. *Nature, 393*, 369–373.

Brefczynski, J.A., and DeYoe, E.A. (1999). A physiological correlate of the "spotlight" of visual attention. *Nat. Neurosci., 4*, 370–374.

Bringuier, V., Chavane, F., Glaeser, L., and Fregnac, Y. (1999). Horizontal propagation of visual activity in the synaptic integration field of area 17 neurons. *Science, 283*, 695–699.

Bullier, J. (2003). Cortical connections and functional interactions between visual cortical areas. In *The Neuropsychology of Vision*, edited by M. Fahle and M. Greenlee, 23–67. Oxford: Oxford University Press.

Bullier, J., Girard, P., and Salin, P.A. (1994). The role of area 17 in the transfer of information to extrastriate visual cortex. In *Primary visual cortex in primates*, edited by A. Peters and K. S. Rockland, 301–330. New York: Plenum Press.

Bullier, J., Hupé, J.-M., James, A., and Girard, P. (2001). The role of feedback connections in shaping the responses of visual cortical neurons. *Progr. Brain Res.,134*, 193–204.

Bullier, J., Schall, J.D., and Morel, A. (1996). Functional streams in occipito-frontal connections in the monkey. *Behavioural Brain Res., 76*, 89–97.

Chelazzi, L., Miller, E.K., Duncan, J., and Desimone, R. (1993). A neural basis for visual search in inferior temporal cortex. *Nature, 363*, 345–347.

Corbetta, M., Akbudak, E., Conturo, T.E., Snyder, A.Z., Ollinger, J.M., Drury, H.A., Linenweber, M.R., Petersen, S.E., Raichle, M.E., Van Essen, D.C., and Shulman, G.L. (1998). A common network of functional areas for attention and eye movements. *Neuron, 21*, 761–773.

Corbetta, M., Miezin, F.M., Dobmeyer, S., Shulman, G.L., and Petersen, S.E. (1991). Selective and divided attention during visual discriminations of shape, color, and speed: Functional anatomy by positron emission tomography. *J Neurosci., 11*, 2383–2402.

Crick, F. (1984). The function of the thalamic reticular complex: The searchlight hypothesis. *Proc. Natl. Acad. Sci. USA, 81*, 4586–4590.

Crick, F., and Koch, C. (1998). Constraints on cortical and thalamic projections: The no-strong loops hypothesis. *Nature, 391*, 245–250.

Crook, J.M., Kisvarday, Z.F., and Eysel, U.T. (1997). GABA-induced inactivation of functionally characterized sites in cat striate cortex: effects on orientation tuning and direction selectivity. *Vis. Neurosci., 14*, 141–158.

Desimone, R., and Duncan, J. (1995). Neural mechanisms of selective visual attention. *Annu. Rev. Neurosci., 18*, 193–222.

Dierks, T., Linden, D.E., Jandl, M., Formisano, E., Goebel, R., Lanfermann, H., and Singer, W. (1999). Activation of Heschl's gyrus during auditory hallucinations. *Neuron, 22*, 615–621.

Di Russo, F., and Spinelli, D. (1999). Spatial attention has different effects on the magno- and parvocellular pathways. *Neuroreport, 10*, 2755–2762.

Douglas, R.J., Koch, C., Mahowald, M., Martin, K.A., and Suarez, H.H. (1995). Recurrent excitation in neocortical circuits. *Science, 269*, 981–985.

Douglas, R.J., and Martin, K.A. (1991). A functional microcircuit for cat visual cortex. *J. Physiol. (Lond), 440*, 735–769.

Duhamel, J.-R., Colby, C.L., and Goldberg, M.E. (1992). The updating of the representation of visual space in parietal cortex by intended eye movements. *Science, 255*, 90–92.

Felleman, D.J., and Van Essen, D.C. (1991). Distributed hierarchical processing in the primate cerebral cortex. *Cereb. Cortex, 1*, 1–47.

Ferster, D., Chung, S., and Wheat, H. (1996). Orientation selectivity of thalamic input to simple cells of cat visual cortex. *Nature, 380*, 249–52.

Ffytche, D.H., Howard, R.J., Brammer, M.J., David, A., Woodruff, P., and Williams, S. (1998). The anatomy of conscious vision: an fMRI study of visual hallucinations. *Nat. Neurosci., 1*, 738–742.

Freund, T.F., Martin, K.A.C., Soltesz, I., Somogyi, P., and Whitteridge, D. (1989). Arborisation pattern and postsynaptic targets of physiologically identified thalamocortical afferents in striate cortex of the macaque monkey. *J. Comp. Neurol., 289,* 315–336.

Gilbert, C., Ito, M., Kapadia, M., and Westheimer, G. (2000). Interactions between attention, context and learning in primary visual cortex. *Vision Res., 40,* 1217–26.

Girard, P., and Bullier, J. (1989). Visual activity in area V2 during reversible inactivation of area 17 in the macaque monkey. *J. Neurophysiol., 62,* 1287–1302.

Girard, P., Hupé, J.M., and Bullier, J. (2001). Feedforward and feedback connections between areas V1 and V2 of the monkey have similar rapid conduction velocities. *J. Neurophysiol., 85,* 1328–1331.

Godwin, D.W., Vaughan, J.W., and Sherman, S.M. (1996). Metabotropic glutamate receptors switch visual response mode of lateral geniculate nucleus cells from burst to tonic. *J. Neurophysiol., 76,* 1800–1816.

Goebel, R., Khorram-Sefat, D., Muckli, L., Hacker, H., and Singer, W. (1998). The constructive nature of vision: Direct evidence from functional magnetic resonance imaging studies of apparent motion and motion imagery. *Eur. J. Neurosci., 10,* 1563–1573.

Haenny, P.E., and Schiller, P.H. (1988). State dependent activity in monkey visual cortex. I. Single cell activity in V1 and V4 on visual tasks. *Exp. Brain Res., 69,* 225–244.

Haxby, J.V., Horwitz, B., Ungerleider, L.G., Ma. Maisog, J., Pietrini, P., and Grady, C.L. (1994). The functional organization of human extrastriate cortex: A PET-rCBF study of selective attention to faces and locations. *J. Neurosci., 14,* 6336–6353.

Hinton, G.E. (2000). Computation by neural networks. *Nature Neuroscience, 3,* 1170.

Hinton, G.E., Dayan, P., Frey, B.J., and Neal, R.M. (1995). The "wake-sleep" algorithm for unsupervised neural networks. *Science, 268,* 1158–1161.

Hubel, D. (1996). A big step along the visual pathway. *Nature, 380,* 197–198.

Humphrey, A.L., Saul, A.B., and Feidler, J.C. (1998). Strobe rearing prevents the convergence of inputs with different response timing onto area 17 simple cells. *J. Neurophysiol., 80,* 3005–3020.

Humphrey, A.L., Sur, M., Uhlrich, D.J., and Sherman., S.M. (1985). Projection patterns of individual X- and Y-cell axons from the lateral geniculate nucleus to cortical area 17 in the cat. *J. Comp. Neurol., 233,* 159–189.

Hupé, J.M., James, A.C., Girard, P., and Bullier, J. (2001). Response modulation by static texture surround in area V1 of the macaque monkey do not depend on feedback connections from V2. *J. Neurophysiol., 85,* 146–163.

Hupé, J.M., James, A.C., Girard, P., S., L., Payne, B., and Bullier, J. (2001). Feedback connections act on the early part of the responses in monkey visual cortex. *J. Neurophysiol., 85,* 134–145.

Hupé, J.M., James, A.C., Payne, B.R., Lomber, S.G., Girard, P., and Bullier, J. (1998). Cortical feedback improves discrimination between figure and background by V1, V2 and V3 neurons. *Nature, 394,* 784–787.

Kanwisher, N., and Wojciulik, E. (2000). Visual attention: Insights from brain imaging. *Nature Reviews Neuroscience, 1,* 91–100.

Kapadia, M.K., Westheimer, G., and Gilbert, C.D. (1999). Dynamics of spatial summation in primary visual cortex of alert monkeys. *Proc. Natl. Acad. Sci. USA, 96,* 12073–12078.

Kastner, S., Pinsk, M.A., De Weerd, P., Desimone, R., and Ungerleider, L.G. (1999). Increased activity in human visual cortex during directed attention in the absence of visual stimulation. *Neuron, 22,* 751–761.

Kennedy, H., Barone, P., and Falchier, A. (2000). Relative contributions of feedforward and feedback inputs to individual areas. *Eur. J. Neurosci., 12, Supp.* S489–489.

Koch, C., and Ullman, S. (1985). Shifts in selective visual attention: Towards the underlying neural circuitry. *Human Neurobiol., 4,* 219–227.

Kosslyn, S.M., Thompson, W.L., Klim, I.J., and Alpert, N.M. (1995). Topographical representations of mental images in primary visual cortex. *Nature, 378,* 496–498.

Lamme, V.A., Super, H., and Spekreijse, H. (1998). Feedforward, horizontal, and feedback processing in the visual cortex. *Curr. Opin. Neurobiol., 8,* 529–535.

Latawiec, D., Martin, K.A., and Meskenaite, V. (2000). Termination of the geniculocortical projection in the striate cortex of macaque monkey: A quantitative immunoelectron microscopic study. *J. Comp. Neurol., 419,* 306–319.

Le Bihan, D., Turner, R., Zeffiro, T.A., Cuénod, C.A., Jezzard, P., and Bonnerot, V. (1993). Activation of human primary visual cortex during visual recall: A magnetic resonnance imaging study. *Proc. Natl. Acad. Sci. USA, 90,* 11802–11805.

Luck, S.J., Chelazzi, L., Hillyard, S.A., and Desimone, R. (1997). Neural mechanisms of spatial selective attention in areas V1, V2 and V4 of macaque visual cortex. *J. Neurophysiol., 77,* 24–42.

Malpeli, J.G. (1983). Activity of cells in area 17 of the cat in absence of input from layer A of lateral geniculate nucleus. *J. Neurophysiol., 49,* 595–610.

Marcar, V.L., and Cowey, A. (1992). The effect of removing superior cortical motion areas in macaque monkey. II. Motion discrimination using random dot displays. *Eur. J. Neurosci., 4,* 1228–1237.

Marcar, V.L., Raiguel, S.E., Xiao, D., and Orban, G.A. (2000). Processing of kinetically defined boundaries in areas V1 and V2 of the macaque monkey. *J. Neurophysiol., 84,* 2786–2798.

Marcar, V.L., Xiao, D.-K., Raiguel, S.E., Maes, H., and Orban, G.A. (1995). Processing of kinetically defined boundaries in the cortical motion area MT of the macaque monkey. *J. Neurophysiol., 74,* 1238–1268.

Martinez, A., Anllo-Vento, L., Sereno, M.I., Frank, L.R., Buxton, R.B., Dubowitz, D.J., Wong, E.C., Hinrichs, H., Heinze, H.J., and Hillyard, S.A. (1999). Involvement of striate and extrastriate visual cortical areas in spatial attention. *Nat. Neurosci., 2,* 364–369.

McAdams, C.J., and Maunsell, J.H.R. (1999). Effects of attention on orientation-tuning functions of single neurons in macaque cortical area V4. *J. Neurosci., 19,* 431–441.

McCormick, D.A., and von Krosigk, M. (1992). Corticothalamic activation modulates thalamic firing through glutamate "metabotropic" receptors. *Proc. Natl. Acad. Sci. USA, 89,* 2774–2778.

Mendola, J., Dale, A.M., Fischl, B., Liu, A.K., and Tootell, R.B.H. (1999). The representation of illusory and real contours in human cortical visual areas revealed by functional magnetic resonance imaging. *J. Neurosci., 19,* 8560–72.

Moran, J., and Desimone, R. (1985). Selective attention gates visual processing in the extrastriate cortex. *Science, 229,* 782–784.

Morel, A., and Bullier, J. (1990). Anatomical segregation of two cortical visual pathways in the macaque monkey. *Visual Neurosci., 4,* 555–578.

Motter, B. (1994). Neural correlates of attentive selection for color or luminance in extrastriate area V4. *J. Neurosci., 14,* 2178–2189.

Motter, B.C. (1993). Focal attention produces spatially selective processing in visual cortical areas V1, V2 and V4 in the presence of competing stimuli. *J. Neurophysiol., 70,* 909–919.

Movshon, J.A., and Newsome, W.T. (1996). Visual response properties of striate cortical neurons projecting to area MT in macaque monkeys. *J. Neurosci., 16,* 7733–41.

Mumford, D. (1992). On the computational architecture of the neocortex. II. The role of cortico-cortical connections. *Biol. Cybernetics, 66,* 241–251.

Mumford, D. (1993). Neuronal architecture for pattern-theoretic problems. In *Large scale neuronal theories of the brain,* edited by C. Koch and J.L. Davis, 125–152. Cambridge, Mass.: MIT Press.

Nealey, T.A., and Maunsell, J.H. (1994). Magnocellular and parvocellular contributions to the responses of neurons in macaque striate cortex. *J. Neurosci., 14,* 2069–2079.

Nowak, L.G., and Bullier, J. (1997). The timing of information transfer in the visual system. In *Extrastriate visual cortex in primates,* edited by J.H. Kaas, K.L. Rockland, and A.L. Peters, 205–241. New York: Plenum Press.

Nowak, L.G., James, A.C., and Bullier, J. (1997). Corticocortical connections between visual areas 17 and 18a of the rat studied in vitro: Spatial and temporal organisation of functional synaptic responses. *Exp. Brain Res., 117,* 283–305.

Peters, A., Payne, B.R., and Budd, J. (1994). A numerical analysis of the geniculocortical input to striate cortex in the monkey. *Cereb. Cortex, 4,* 215–29.

Rao, R.P., and Ballard, D.H. (1999). Predictive coding in the visual cortex: a functional interpretation of some extra-classical receptive-field effects. *Nat. Neurosci., 2,* 79–87.

Ress, D., Backus, B., and Heeger, D. (2000). Activity in primary visual cortex predicts performance in a visual detection task. *Nat. Neurosci., 3,* 940–945.

Reynolds, J.H., Chelazzi, L., and Desimone, R. (1999). Competitive mechanisms subserve attention in macaque areas V2 and V4. *J. Neurosci., 19,* 1736–53.

Reynolds, J.H., Pasternak, T., and Desimone, R. (2000). Attention increases sensitivity of V4 neurons. *Neuron, 26,* 703–14.

Rockland, K.S. (1989). Bistratified distribution of terminal arbors of individual axons projecting from area V1 to middle temporal area (MT) in the macaque monkey. *Visual Neurosci., 3,* 155–170.

Rockland, K.S. (1995). Morphology of individual axons projecting from area V2 to Mt in the macaque. *J. Comp. Neurol., 355,* 15–26.

Rockland, K.S. (1998). Convergence and branching patterns of round, type 2 corticopulvinar axons. *J. Comp. Neurol., 390,* 515–36.

Roelfsema, P.R., Lamme, V.A., and Spekreijse, H. (1998). Object-based attention in the primary visual cortex of the macaque monkey. *Nature, 395,* 376–81.

Rossetti, Y., Pisella, L., and Pélisson, D. (2000). Eye blindness and hand sight: Temporal aspects of visuo-motor processing. *Visual Cognition, 7,* 785–809.

Salin, P.-A., and Bullier, J. (1995). Corticocortical connections in the visual system: Structure and function. *Physiol. Reviews, 75,* 107–154.

Sary, G., Vogels, R., and Orban, G.A. (1993). Cue invariant shape selectivity of macaque inferior temporal neurons. *Science, 260,* 995–997.

Schall, J.D. (1997). Visuomotor areas of the frontal lobe. In *Cerebral cortex*, vol. 12, edited by J.H. Kaas, K.L. Rockland, and A.L. Peters, 527–638. New York: Plenum Press.

Schmidt, K.E., Goebel, R., Castelo-Branco, M., Lomber, S., Payne, B.R., and Galuske, R.A.W. (2000). Global motion representation in primary visual cortex of the cat and the influence of feedback. *Eur. J. Neurosci., 12*, 75.

Schmolesky, M.T., Wang, Y., Hanes, D.P., Thompson, K.G., Leutgeb, S., Schall, J.D., and Leventhal, A.G. (1998). Signal timing across the macaque visual system. *J. Neurophysiol., 79*, 3272–3278.

Seghier, M., Dojat, M., Delon-Martin, C., Rubin, C., Warnking, J., Segebarth, C., and Bullier, J. (2000). Moving illusory contours activate primary visual cortex: An FRMI study. *Cereb. Cortex, 10*, 663–670.

Shao, Z.W., and Burkhalter, A. (1996). Different balance of excitation and inhibition in forward and feedback circuits of rat visual cortex. *J. Neurosci., 16*, 7353–7365.

Sherk, H. (1990). Functional organization of input from areas 17 and 18 to an extrastriate area in the cat. *J. Neurosci., 10*, 2780–2790.

Sherman, S.M., and Guillery, R.W. (1998). On the actions that one nerve cell can have on another: distinguishing "drivers" from "modulators." *Proc. Natl. Acad. Sci. USA, 95*, 7121–7126.

Sherman, S.M., and Koch, C. (1986). The control of retinogeniculate transmission in the mammalian lateral geniculate nucleus. *Exp. Brain Res., 63*, 1–20.

Sillito, A.M., Kemp, J.A., Milson, J.A., and Berardi., N. (1980). A re-evaluation of the mechanisms underlying simple cell orientation selectivity. *Brain Res., 194*, 517–520.

Somers, D.C., Dale, A.M., Seiffert, A.E., and Tootell, R.B. (1999). Functional MRI reveals spatially specific attentional modulation in human primary visual cortex. *Proc. Natl. Acad. Sci. USA, 96*, 1663–1668.

Steinman, B.A., Steinman, S.B., and Lehmkuhle, S. (1997). Transient visual attention is dominated by the magnocellular stream. *Vision Res., 37*, 17–23.

Steriade, M., Jones, E.G., and McCormick, D.A. (1997) *Thalamus*. Volume 1, Organisation and Function. Amsterdam: Elsevier.

Stone, J. (1983) *Parallel Processing in the Visual System: The Classification of Retinal Ganglion Cells and Its Impact on the Neurobiology of Vision*. New York: Plenum Press.

Stratford, K.J., Tarczy-Hornoch, K., Martin, K.A.C., Bannister, N.J., and Jack, J.J.B. (1996). Excitatory synaptic inputs to spiny stellate cells in cat visual cortex. *Nature, 382*, 258–261.

Tomita, H., Ohbayashi, M., Nakahara, K., Hasegawa, I., and Miyashita, Y. (1999). Top-down signal from prefrontal cortex in executive control of memory retrieval. *Nature, 401*, 699–703.

Trachtenberg, J.T., Trepel, C., and Stryker, M.P. (2000). Rapid extragranular plasticity in the absence of thalamocortical plasticity in the developing primary visual cortex. *Science, 287*, 2029–2032.

Treisman, A., and Gelade, G. (1980). A feature-integration theory of attention. *Cognit. Psychol., 12*, 97–136.

Treue, S., and Maunsell, J.H. (1999). Effects of attention on the processing of motion in macaque middle temporal and medial superior temporal visual cortical areas. *J. Neurosci., 19*, 7591–7602.

Treue, S., and Maunsell, J.H.R. (1996). Attentional modulation of visual motion processing in cortical areas MT and MST. *Nature, 382*, 539–541.

Tsodyks, M., Kenet, T., Grinvald, A., and Arieli, A. (1999). Linking spontaneous activity of single cortical neurons and the underlying functional architecture. *Science, 286*, 1943–1946.

Tsotsos, J.K. (1990). Analyzing vision at the complexity level. *Behav. Brain Sci., 13*, 423–469.

Tsotsos, J.K., Culhane, S.M., Wai, W.Y.K., Lai, Y., Davis, N., and Nuflo, F. (1995). Modeling visual attention via selective tuning. *Artificial Intelligence, 78*, 507–545.

Ullman, S. (1995). Sequence seeking and counter streams—A computational model for bidirectional information-flow in the visual-cortex. *Cereb. Cortex, 5*, 1–11.

Ullman, S. (1996) *High Level Vision.* Cambridge, Mass.: MIT Press.

Van Essen, D.C., and Maunsell, J.H.R. (1983). Hierarchical organization and functional streams in the visual cortex. *Trends Neurosci., 6*, 370–375.

Walker, G.A., Ohzawa, I., and Freeman, R.D. (1999). Asymmetric suppression outside the classical receptive field of the visual cortex. *J. Neurosci., 19*, 10536–10553.

Watanabe, T., Harner, A.M., Miyauchi, S., Sasaki, Y., Nielsen, M., Palomo, D., and Mukai, I. (1998). Task-dependent influences of attention on the activation of human primary visual cortex. *Proc. Natl. Acad. Sci. USA, 95*, 11489–11492.

Xiao, D.K., Raiguel, S., Marcar, V., and Orban, G.A. (1997). The spatial distribution of the antagonistic surround of MT/V5 neurons. *Cereb. Cortex, 7*, 662–677.

Yarbus, A.L. (1967) *Eye Movements and Vision.* New York: Plenum Press.

PART III

HOW DO NEURONS INTERACT?

7

How Can the Brain Be So Fast?

Wulfram Gerstner

Introduction

It is often thought that neurons are comparatively slow elements. How, then, is it possible that the human brain as a whole can solve complex tasks such as pattern recognition within a few hundred milliseconds—much more rapidly than any current computer vision system? In this chapter we critically review some of the arguments that have suggested that neurons are slow, and discuss a reaction time experiment that shows that the brain as a whole is nevertheless fast. In the final part of the chapter, we indicate some directions of research towards solving the apparent conflict; in particular we speculate on the role of noise.

In What Sense Are Neurons Slow?

What are the time scales that might limit neuronal information processing? First, signal processing speed may be limited by the *internal time constants* of neuronal dynamics such as the membrane time constant, synaptic time constant, or dendritic integration time. Second, signal processing speed may also be limited by the *coding scheme* used to transmit information. In this section, we review both the time scale set by neuronal time constants and the timing constraints induced by rate coding.

Neuronal Time Constants

If a neuron is stimulated by a short intracellular current pulse, the cell responds by a change of membrane potential. After the end of the pulse, the membrane potential decays slowly back to its resting value. The time constant of the decay defines the (passive) membrane time constant τ_m of the neuron. Typical values of the membrane time constant of cortical neurons are in the range of 20–50 milliseconds. Since the membrane potential responds to a short current pulse of, say, one millisecond with a voltage excursion that extends over tens of milliseconds, the exact information about the timing of the pulse appears "smeared out" over time. If we consider an input that changes each millisecond (i.e., an input that is modulated at a rate of 1 kHz), the membrane potential modulation is hardly perceptible if the neuronal time constant is in the range of 20 milliseconds. More generally, any temporal information that is significantly faster than τ_m is strongly suppressed, if not lost. This conclusion is, in fact, correct for any passive filter with a time constant τ_m. We say that a filter with time constant τ_m has a "cut-off frequency" of $1/\tau_m$, that is, input with a modulation frequency $f > 1/\tau_m$ is strongly suppressed; see figure 7-1.

In a more realistic scenario, a neuron is not stimulated by an artificially induced current pulse but by the arrival of a presynaptic action potential. The result of synaptic transmission is a conductance change of the postsynaptic neuron, and hence a postsynaptic current pulse. The duration of the current pulse is characterized by a time constant τ_s. Depending on the type of synapse, this time constant can range from less than one millisecond for non–N-methyl-D-aspartate (NMDA) to over 100 milliseconds for NMDA channels.

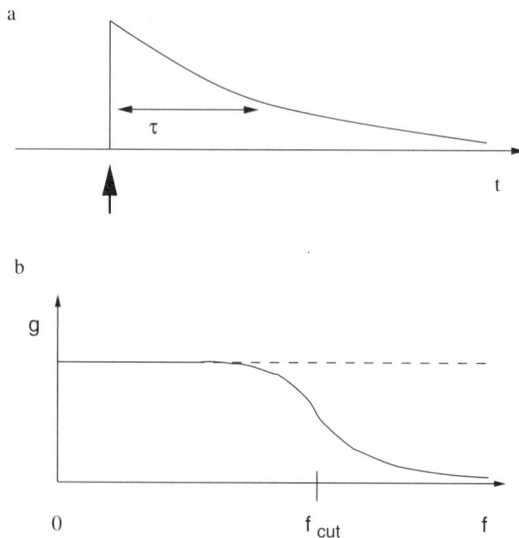

Figure 7-1. (a) A low-pass filter stimulated by an input pulse (arrow) responds with a broad pulse characterized by a time constant τ. (b) The "gain" g of signal transmission has a cut-off frequency at $f_{cut} = 1/\tau$. The dashed line indicates the identity filter without cut-off frequency (schematic figure).

Synaptic filtering induces therefore a cut-off frequency of $1/\tau_s$. A further filtering of synaptic input is due to the spatial extension of the neuron. In particular, synaptic input at a distal point of a (passive) dendrite has a longer-lasting but less pronounced effect at the soma than input into the soma itself. The time scale of dendritic integration is again controlled by the membrane time constant τ_m.

Coding

Apart from intrinsic time constants, neuronal processing may also be limited by the coding scheme. We recall that neurons communicate by short electrical pulses, the so-called action potentials or spikes. A straightforward interpretation of the early studies of Adrian (Adrian 1926) as well of typical receptive-field studies (Mountcastle 1957; Hubel and Wiesel 1959) seems to suggest that the external input is represented in the brain by neuronal firing rates. In a standard single-neuron experiment, the *firing rate* v can be defined by counting the number of action potentials $n_{AP}(t; t + T_W)$ that occur in a interval T_W, divided by T_W, that is, $v = n_{AP}(t; t + T_W)/T_W$. The length of the time window T_W depends on the type of experiment. Since interspike intervals vary from one spike to the next, we need to average over several spikes in order to get a reasonable estimate of the firing rate. In many experiments, the duration T_W of the time window is therefore taken in the range of 100 milliseconds or longer; see figure 7-2.

If temporal averaging is not just an experimental convenience but the actual code used by the neurons to transmit information, then a neuron further down in the processing stream would have to average its input spike trains over at least 100 milliseconds in order to "decode" the message of the presynaptic neuron. Because of temporal averaging, transmission of input that is modulated at 10 Hz or faster would then be difficult.

In summary, a naive rate coding picture with a time window of $T_W = 100$ milliseconds imposes a cut-off frequency of 10 Hz. If we neglect coding constraints and focus on neuronal time constants, a membrane time constant of 20–30 millisecond seems to imply a cut-off frequency of signal transmission in the range of 30–50 Hz.

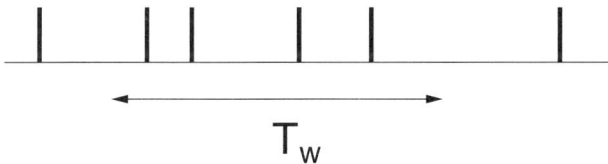

Figure 7-2. Definition of a firing rate by a spike count in a time window of duration T_W.

In What Sense Are Brains Fast?

Reaction time experiments offer a direct way to measure the performance of a system as a whole. We focus on one example (Thorpe, Fize, and Marlot 1996) of such an experiment which stands in the tradition of earlier reaction time studies. A human subject looks at images of real-world scenes such as landscapes, houses, flowers, or animals in a natural setting. The images are flashed for a short time. Whenever the subject sees an animal, he is supposed to release a button that he holds down otherwise. The reaction time defined as the interval between the onset of the image and the response is measured. Typical reaction times are in the range of 400 milliseconds, but some correct responses occur already after 200 milliseconds; see figure 7-3. Moreover, an electroencephalograph (EEG) signal recorded during the experiment exhibits already after 150 milliseconds significant differences between images with and those without animals. The EEG results suggest that the recognition and classification of an image is essentially completed after 150 milliseconds. Typically more than 90 percent of the responses are correct. Many of the remaining mistakes must be attributed to insufficient control of the finger rather than to incorrect classification.

From the field of computer vision, it is known that recognition and classification of real-world images is a hard task. None of the existing algorithms is capable of solving a complex task such as the recognition of animals in a natural setting which comprises situations as different as grasshoppers, flying birds, a flock of sheep, the head of a giraffe, or a green frog on a green leaf. Moreover, current algorithms are slow and are far from reaching recognition in the range of a few hundred milliseconds—despite the fact that computer hardware is extremely rapid.

Typical computational approaches involve several preprocessing steps including filtering and segmentation between foreground and background before the final classification is made. The brain uses probably a similar sequence of processing steps starting at the retina leading over the visual cortex to association areas before a final decision is made and the picture classified as "animal" or "nonanimal." If we estimate the number of processing steps between six and eight (the actual number might in fact be even higher), then

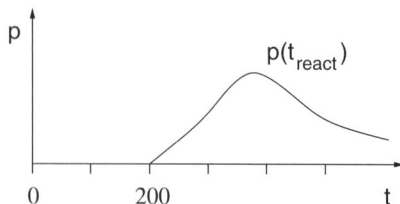

Figure 7-3. Distribution $p(t_{react})$ in a reaction time experiment (schematic).

the processing time per step is between 20 and 25 milliseconds. During twenty milliseconds, most neurons will at most emit a single spike—how then is information processing possible?

Elements of an Answer

In the two preceding sections I have outlined what appears to be a paradox: The brain as a whole is fast despite the fact that neurons as the individual processing elements seem to be slow. In this section I would like to point to some elements that might help to solve the apparent paradox. I start by re-considering the coding problem. I then return to the problem of time constants which leads us to questions about the role of noise in the brain.

Constraints Imposed by Neuronal Coding

Rate coding defined as a spike count in a time window of $T_W = 100$ milliseconds or longer is inconsistent with reaction time experiments. The time window T_W for estimating the rate imposes a minimal processing time per "processing step." To simplify the picture, we think of each processing step as a layer in a feedforward architecture; see figure 7-4.

If neurons in layer $n + 1$ have to wait a time T_W before they can estimate the firing rate of neurons in layer n, then a recognition time of 150 milliseconds would allow at most two layers with $T_W = 75$ milliseconds, or at most three layers with $T_W = 50$ milliseconds. A maximum of two or three processing steps would require an extremely "flat" or "parallel" brain architecture. Note that information flow via lateral or feedback connections has to be counted as normal processing steps that add a standard decoding time T_W.

With a more realistic (but still conservative) estimate of about six to eight processing steps, the processing time per step is in the range of 20 to 25 milliseconds. Indeed, a detailed analysis of neural activity in the cortex has shown that a large amount of information about the stimulus is contained in the

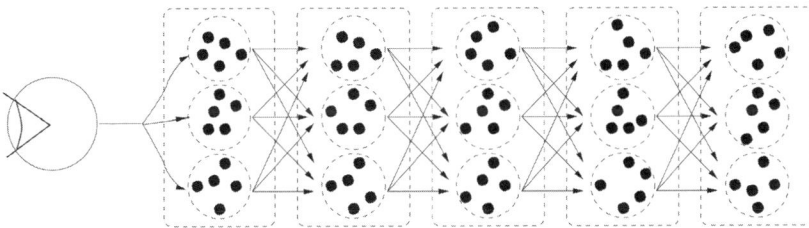

Figure 7-4. A simple feedforward processing scheme in five steps. Visual input is transmitted from left to right starting from the eye on the left-hand side.

first 20 to 30 milliseconds after stimulus onset (Optican and Richmond 1987). Moreover, the initial transient after stimulus onset is often a fairly reliable well-timed event with little spike jitter (Marsalek, Koch, and Maunsell 1997). Spike-triggered averaging ("reverse-correlation") experiments show that visual neurons can resolve temporal structure on the time scale of twenty milliseconds, for example (Eckhorn, Krause, and Nelson 1993).

Since the majority of neurons will emit at most a single spike within 20 milliseconds, the notion of a rate code defined as a *spike count* in $T_W = 20$ milliseconds is questionable, but other coding schemes and other definitions of a "rate" are possible. In simple systems it has been possible to "read the code" of single neurons and reconstruct the stimulus from the spike train (Bialek et al. 1991; Rieke et al. 1996). Linear stimulus reconstruction can be seen as a "rate" defined by averaging over a short and smooth time window whose form is adapted to the stimulus ensemble (Theunissen and Miller 1995). Given the large number of neurons in the cortex, another potential coding scheme seems to be a "rate" that is defined by a population average rather than a temporal average. From a pure coding point of view, signal processing by population activity could be very fast and only limited by the rise time of an action potential (i.e., approximately one millisecond). While the exact neuronal code is still unknown, reaction time studies and other experimental evidence impose an important timing constraint: we know that the code must be fast so as to allow "decoding" in less than 30 milliseconds.

Constraints Imposed by Neuronal Time Constants and Noise

Let us now simply assume that neurons use a rapid coding scheme. Since the temporal performance of the system can still be hampered by the insufficient speed of neuronal dynamics, we return to our discussion of the membrane time constant. A membrane time constant of $\tau_m = 25$ milliseconds would appear to imply a cut-off frequency of about 40 Hz which would just be enough for about six processing steps in 150 milliseconds. We may wonder, however, whether the passive membrane time constant τ_m really is a limiting factor or whether it is somehow possible to increase the cut-off frequency beyond 40 Hz.

First, the passive membrane time constant characterizes the neuronal dynamics at the resting potential. Cortical neurons in vivo are, however, subject to a continuous bombardment with presynaptic spikes that arrive due to spontaneous cortical activity. The background activity shifts the mean membrane potential close to threshold, induces voltage fluctuations, and increases the conductance. The *effective* membrane time constant τ_m^{eff} seems to be shorter by a factor of two to ten and can be below ten milliseconds. (Bernander et al. 1991; Destexhe and Pare 1999). A shorter membrane time constant implies an increased cut-off frequency.

Second, action potentials are triggered if the membrane potential reaches a critical "threshold" value. At least in simple neuron models of the integrate-and-fire type, the threshold process counterbalances the integration with the membrane time constant. In fact, in a population of noise-free spiking neurons that are firing asynchronously the signal transmission properties are *not* limited by the membrane time constant. In other words, there is no cut-off frequency (Knight 1972; Gerstner 2000).

In the presence of noise, the picture is somewhat more complicated. The cut-off frequency depends on the type and amount of noise. For a large amount of white noise, the signal transmission properties are indeed limited by the (effective) membrane time constant τ_m^{eff} (Gerstner 2000). On the other hand, for correlated noise (Brunel et al. 2001) or slow noise in the parameters (Knight 1972; Gerstner 2000) such as slow fluctuations in the firing threshold, signal transmission is not limited by the membrane time constant. In this case, the cut-off frequency is only limited by the *synaptic* time constant τ_s—which can be shorter than one millisecond for non-NMDA synapses. A similar result holds if the neuronal code is not the population activity itself but rather the *variance* of the activity (Silbergeld et al. 2004); in that case the signal consists of the fluctuations that are usually attributed to noise.

These theoretical insights hold not only for feedforward but also for lateral and feedback connections. Thus parallel processing in highly connected neuronal architectures can be rapid (Treves, Rolls, and Simmen 1997). A system with "slow" noise can therefore be extremely fast. The time per processing step is only limited by the pure axonal transmission and synaptic time.

A necessary condition for this and similar fast signal transmission schemes is that prior to stimulus onset neurons are in a state of spontaneous activity, that is, different neurons fire asynchronously at a low rate so that at least some of the neurons are always close to threshold. Such a scheme would not require specialized cells with fast ion currents as they are, for example found in barn owl auditory system (Carr 1993; Reyes, Rubel, and Spain 1994). Moreover, such a scheme indicates a functional role for spontaneous cortical activity—currently spontaneous activity is barely understood. At the moment these are speculations. Experiments and analysis in the next decades will probably shed new light on the information processing speed of neural system.

References

E. D. Adrian. 1926. The impulses produced by sensory nerve endings. *J. Physiol. (London)* 61: 49–72.

Ö. Bernander, R. J. Douglas, K. A. C. Martin, and C. Koch, Synaptic background activity influences spatiotemporal integration in single pyramidal cells. *Proc. Natl. Acad. Sci. USA* 88 (1991): 11569–11573.

W. Bialek, F. Rieke, R. R. de Ruyter van Stevenick, and D. Warland, Reading a neural code. *Science* 252 (1991): 1854–1857.

N. Brunel, F. S. Chance, N. Fourcaud, and L. F. Abbott, Effects of synaptic noise and filtering on the frequency response of spiking neurons. *Phys. Rev. Lett.* 86 (2001): 2186–2189.

C. E. Carr, Processing of temporal information in the brain. *Annual Rev. Neurosci.* 16 (1993): 223–43.

A. Destexhe and D. Pare, Impact of network activity on the integrative properties of neocortical pyramidal neurons in vivo. *J. Neurophysiol.* 81 (1999): 1531–1547.

R. Eckhorn, F. Krause, and J. L. Nelson, The rf-cinematogram: a cross-correlation technique for mapping several visual fields at once. *Biol. Cybern.* 69 (1993): 37–55.

W. Gerstner, Population dynamics of spiking neurons: fast transients, asynchronous states and locking. *Neural Computation* 12 (2000): 43–89.

D. H. Hubel and T. N. Wiesel, Receptive fields of single neurons in the cat's striate cortex. *J. Physiol.* 148 (1959): 574–591.

B. W. Knight, Dynamics of encoding in a population of neurons. *J. Gen. Physiology* 59 (1972): 734–766.

P. Marsalek, C. Koch, and J. Maunsell, On the relationship between synaptic input and spike output jitter in individual neurons. *Proc. Natl. Acad. Sci. USA* 94 (1997): 735–740.

V. B. Mountcastle, Modality and topographic properties of single neurons of cat's somatosensory cortex. *J. Neurophysiol.* 20 (1957): 408–434.

L. M. Optican and B. J. Richmond, Temporal encoding of two-dimensional patterns by single units in primate inferior temporal cortex. 3. Information theoretic analysis. *J. Neurophysiol.* 57 (1987): 162–178.

E. D. Reyes, E. W. Rubel, and W. J. Spain, Membrane properties underlying the firing of neurons in the avian cochlear nucleus. *J. Neurosci.* 14 (1994): 5352–5364.

F. Rieke, D. Warland, R. de Ruyter van Steveninck, and W. Bialek, *Spikes— Exploring the neural code.* (MIT Press, Cambridge, MA, 1996)

G. Silbergeld, M. Bethge, H. Markram, K. Pawelzik, and M. Tsodyks, Dynamics of population rate codes in ensembles of neocortical neurons. *J. Neurophysiol.* 91 (2004): 704–709.

F. Theunissen, and J. Miller, Temporal encoding in nervous systems: a rigorous definition. *J. Comput. Neurosci.*, 2 (1995): 149–162.

S. Thorpe, D. Fize, and C. Marlot, Speed of processing in the human visual system. *Nature* 381 (1996): 520–522.

A. Treves, E. T. Rolls, M. Simmen. 1997. Time for retrieval in recurrent associative memories. *Physica D* 107: 392–400.

8

What Is the Neural Code?

C. van Vreeswijk

Brains, even those of simple animals, are enormously complex structures, and a detailed understanding of how the brain as a whole works is still far beyond us. Luckily brains are organized in such a way that the specific tasks they perform are largely constrained to different subregions. For example, the different sensory areas of the cortex, visual, auditory, or somatosensory, specialize in analyzing the sensory inputs of different modalities, and the motor-cortex formulates the commands that, through the spinal cord, drive the muscles in voluntary movement. These regions can be further subdivided in areas that that perform subtasks, such as extracting movements or color from the visual input.

It is this property of the central nervous system allows us to study the central nervous system scientifically. To do meaningful systems neuroscience it is not necessary to be a hubristic holist, only satisfied by understanding the whole system at once. A more modest reductionist approach is possible. Because of this organization in areas specializing in (sub)tasks, useful information can be obtained by studying how these different areas subserve their putative function. Once this is better understood, one can go back, "put the brain together again," hypothesize how the system as whole works, and test these predictions.

To properly understand how the anatomy and physiology of a particular area allows it to perform its function, it is of prime importance to know how the information it receives and the results it sends out are coded. At a very basic level this is well known. The neurons in the area receive spike trains from the cells that project to it, and the area's cells that project to other areas likewise send out spike trains. However, that says relatively little. For comparison, this book is written using the Roman alphabet. Knowing this is useful, but to get anything out of this book, the reader has to realize that it is written in English and be able to read that language. Likewise, the neuronal network has be able to extract the information it needs to perform its function from the incoming spike trains and put the outcome of its computation in the outgoing spike trains in such a way that other parts of the central or peripheral nervous system have access to this result. Thus determining which code is used in information transmission between brain areas is a sine qua non for our ability to relate mechanisms operating in neuronal networks to the functions these areas perform.

Candidates for the Neural Code

In general it is not easy to determine the code, but for early sensory areas one is able to say something about it. This is because in these areas one knows what information the neurons should be conveying. Thus one can test a candidate code by presenting many stimuli and determining whether this code conveys any information about these stimuli. Unfortunately, even if one has determined that using this code one obtains information about the stimulus, this does not guarantee that this is the code that is actually used. There may be a different code that also conveys information about the stimulus. Nevertheless, if one considers several candidate codes, one can say which is the code that is most likely to be used by determining which code gives the most information about the stimulus.

Since the seminal work of Adrian (1926) it has been recognized that firing rates of sensory neurons change in a consistent manner with the sensory input. This has given rise to the idea that neurons code their information in the neuronal firing rate. Much subsequent work has tacitly assumed such a rate model in trying to understand how the different sensory cortices represent the stimuli. A clear example of this is the work of Hubel and Wiesel (1962). In their study of neurons in the primary visual cortex they investigated which stimulus induced the neurons to fire the most action potentials, in other words, to which stimuli the neurons respond with the highest firing rate.

However, Adrian's work already showed that the individual spikes arrive in a highly irregular manner (Burns and Webb 1976). In fact, the total number of emitted spikes varies substantially from trial to trial, when the same stimulus is shown repeatedly (Richmond, Optican, and Spitzer 1990). This has prompted neuroscientists to formulate very different hypotheses about the nature of the neural code. At one extreme, it has been argued that neurons encode their information in relatively slowly changing rates, and the irregularity in the spike trains reflect noise in the system (Georgopolous, Schwartz, and Kettner 1986), while in the other extreme this irregularity *is* the code, the precise timing of every spike carries additional information about the input (Abeles 1991).

The attraction of the latter view is that in it every action in the brain has a purpose. This agrees with the idea that evolutionary pressure will have allowed only the survival of those species in which the brain works close to optimally. Conversely, a rate code suffers from the assumption that there is a lot of noise in the system, and this noise will impact detrimentally on the brain's function. So it would seem that, if a rate code were used, brains would have evolved to a state in which the noise is minimal, that is the neuronal activity would be regular.

Unfortunately very few testable models exist that take the extreme view that the timing of each spike is important. The model that comes closest to this ideal is the model of the syn-fire chain proposed by Abeles (1991). In this model, an input sequentially elicits synchronized activity in different pools of neurons, each of which is connected to a relatively large number of cells in the next pool. Modeling of such an architecture shows that, due to the large number of connections from one pool to the next, synchronization between cells in each pool builds up rapidly, and, because neurons participate in different pools, which will be activated with different delays, the activity of individual cells is very irregular. A prediction of this model, which is testable with current techniques, is that there are unitary event which occur much more often than chance. Namely, if neuron 1 belongs pool A, neuron 2 to pool B, and neurons 3 to pool C, where in a syn-fire chain pool B is activated a time t_1 after pool A and pool C with a delay t_2, then if neuron 1 fires at time t, there is an increased probability that neurons 2 and 3 fire at times $t + t_1$ and $t + t_2$ respectively. Thus, in multicell recordings, one can look for delay triples that occur much more often than chance. However, for this one needs a adequate null hypothesis which defines the chance level. In their original study, Abeles and colleagues (1993), calculated the probability of triplets occurring under the assumption that the neurons fired Poissonian with a constant rate, and they found that there were triplets that occurred much more frequently that could be explained by this model. As was pointed out later (Oram et al. 1999), this procedure underestimates the probability some triplets occurring very often if the underlying rates vary in a correlated manner. A better test of the syn-fire chain hypothesis is given by taking the original data, and for each spike train randomly shuffling the spikes within windows of, for example, 10 milliseconds. The syn-fire hypothesis predicts that this would decrease the count for the triplets that occur most often. When this test was performed, the counts for the shuffled data were statistically hardly different from those in the original data (Oram et al. 1999). Thus, evidence for the syn-fire architecture in the cortex is weak at best.

An intermediate model, which has gained much popularity of the last decade, is a model in which not only the rates of the neurons but also the correlations in their activity carry information. Such a model has been proposed by von der Malsburg to solve the binding problem (von der Malsburg 1985), the problem of determining which property analyzed in one cortical area, goes with which property determined in another area. A pure rate code has difficulty with binding. If multiple objects are present and elevated rates in different populations of neurons indicate features of the object, how does the cortex decide which features belong to the same object? A correlation based mechanism would have no problem with this. It could accommodate multiple objects by having correlations between activities representative of the same object in different areas, with no correlation between activities due to different objects.

Experimental work on the primary visual cortex initiated in the lab of Singer (Gray et al. 1999; Gray and Singer 1989) has shown that there is an increased level of synchronization in the activity of simple cells in V1 with different receptive fields, if these cells are activated by a stimulus consisting of a single long bar which overlaps both cells' receptive fields, when compared to stimulation by two short bars, each overlapping the receptive field of one of the cells. Thus the level of synchronization seems to encode whether the stimulation is due to one or more objects. Apart from these functionally relevant correlations found in the primary visual cortex, the only other area in which correlations have been shown to be reliably relate to stimuli is the auditory cortex (deCharms and Merzenich 1996). Furthermore, it should be noted that by no means all experimentalists looking for correlations related to function in the striate cortex observe them. Therefore, only in a very small number of the cortical areas have indications of correlation based coding been observed. Furthermore, these correlations are evidently so small that competent neuroscientists can disagree about whether they do exist at all. If they are so small one may also wonder how the rest of the central nervous system could possibly use them reliably to extract information about the animals environment.

Thus, at present, the only candidate for a neural code for which there has been reliable experimental evidence in all cortical, subcortical, and peripheral areas that have been investigated is the rate code. While other putative codes have not been excluded by the data, evidence for them is still weak at best.

Nevertheless, the assumption of a rate code, with its implication that the irregularity of the neuronal activity is due to noise, seems to be extremely inefficient. So despite the overwhelming evidence that the neuronal firing rates vary consistently with the function the network is performing, the efficiency assumption would seem to argue against assuming a rate code. However, the arguments against the efficiency of a rate code in which the irregularity is assumed to be no more than noise are to a large degree heuristic and make (unstated) assumptions which may not be applicable to what is going on in the central nervous system. Furthermore, such arguments only aim to show that irregularly active neurons are less efficient all other things being equal. It should be evident that all other things are not equal. The fact that the neurons fire with highly variable interspike intervals has important effects on the dynamics of neural networks.

In the rest of this chapter I will look at these effects and describe more quantitatively the impact of neuronal spike variability on the information transmission capabilities of neuronal networks. This will be done by looking at simplified mathematical models in which these questions can be studied analytically. This approach has the advantage that a detailed understanding of the issues involved is possible, allowing for extrapolation to more realistic model networks. A further advantage of studying these questions in mathematical models is that this approach forces one to make explicit the assumptions made about the system. As I will show this can lead to nonintuitive results. But first I will show how irregular activity could possibly arise and show that a mechanism that leads to a representation in which the input is coded by the mean rate of a population of irregularly active neurons has other properties which are highly desirable.

Creating Irregular Activity

The overwhelming majority of numerical simulations of models of large networks of neurons show systems that settle in an activity pattern in which the neurons fire in a very regular pattern. The reason for this is easy to understand: as in the real brain, the neurons in these networks receive inputs from many other cells. Assuming that the network does not synchronize significantly, this means that each of the neurons receives a large number of uncorrelated inputs. The central limit theorem implies that the sum of all these inputs should be roughly constant with only small fluctuations around the mean. If the model's neurons replicate the behavior of real neurons, a nearly constant input should lead to an output that is highly regular unless the mean input is very close to the threshold input. If the latter is the case, the neurons will fire irregularly, but their rates will be very low, because even during positive fluctuations, the total input will be just above threshold.

On the other hand, when there is a substantial synchronization in the network, the fluctuations in the total input can be quite large. However, due to the extensive feedback in such systems, synchronized activity will usually force the whole network to modulate its activity in lockstep. Such globally synchronized activity usually evolves to a state in which the network modulations are periodic. For networks in which the individual neurons have a sufficiently rich dynamics, synchronized activity can evolve to a chaotic state (Bush and Douglas 1991, Hansel and Sompolinsky 1996) with irregularly active neurons. But all current models that show such synchronized chaos imply a level of synchrony between the cells that is much higher than electrophysiological measurements indicate for biological systems.

What is needed is a system in which the input into the cells has temporal fluctuations that are substantial and are, to prevent massive synchronization, at best very weakly correlated between cells. Let us consider in some more detail the input into a cell which receives K uncorrelated inputs, where K is large. For simplicity assume that these input can be either "on" or "off," that is, the state of input i is, at time t, described by a variable $\sigma_i(t)$ that is either 0 or 1. Each of the synaptic inputs is of equal strength, J. Thus the total input I into the cell is

$$I(t) = J \sum_{i=1}^{K} \sigma_i(t) \tag{8.1}$$

If the probability of an input being "on" is v, the mean and variance of the total input are given by $\langle I \rangle = JKv$ and $\text{Var}(I) = J^2Kv(1-v)$ respectively. For such a cell to have a biologically plausible output rate, the total input should be of the same order as the threshold input, so that, in units of the threshold input, $Jv \propto 1/K$. Thus, either J is small or v is small (or both). However, v, the probability that an input is active, being means that rate of the cells projecting to the target cell is small, or that the synapses have a very high failure rate (Tsodyks and Sejnowski 1995). The latter does not appear to be the case, and the former would imply an extremely high output rate for input neurons with plausible rates. Therefore one has to assume that the synaptic strength J is of the order $1/K$. This implies that the input variance is also of order $1/K$. Thus, because K is large, the fluctuations in the input are small, and hence the cells fire very regularly.

This argument neglects the fact that neurons receive both excitatory and inhibitory inputs. To also take inhibition, one has to consider a cell that receives K excitatory and K inhibitory inputs, denoted by σ_i^E and σ_i^I, with synaptic strengths J_E and $-J_I$ respectively. If the respective probabilities of being active are given by v_E and v_I, the mean input satisfies $\langle I \rangle = K(J_Ev_E - J_Iv_I)$, while its variance is given by $\text{Var}(I) = K[J_E^2v_E(1-v_E) + J_I^2v_I(1-v_I)]$.

To get a highly variable output of the neuron, the standard deviation in the input has to be at least of the same order as the threshold input. For this to be the case, the synaptic strengths can be no smaller than of order $1/\sqrt{K}$. This means, however, that the mean input, $\langle I \rangle$, is $\langle I \rangle = \sqrt{K}(j_Ev_E - j_Iv_I)$, where I have used $J_A = j_A/\sqrt{K}$, with $A = E, I$. The average input is potentially extremely high, of order $\sqrt{K}$, but if the total mean excitatory and inhibitory inputs nearly cancel each other, that is, $j_Ev_E - j_Iv_I \propto 1/\sqrt{K}$, both the mean and standard deviation of the input are of the same order as the threshold current, and hence the neuron fires irregularly.

This scenario explains how the neurons in the cortex could have highly irregular activity, even though they receive input through a very large number of synapses. Problematic seems to be that this mechanism requires a precise tuning of the parameters. The probabilities v_E and v_I have to be chosen carefully to assure that $j_E \nu_E - j_I \nu_I \propto 1/\sqrt{K}$. Surprisingly, this fine-tuning problem disappears if one considers networks of interconnected neurons that receives external input (van Vreeswijk and Sompolinsky 1996, 1998): Consider a network of N excitatory cells, N inhibitory cells, with a probability K/N for a synaptic connection from any cell to any other cells. The strength of the synapses depends on both the projecting and target neurons, with the strengths being denoted as $j_{EE}/\sqrt{K}$, for connections from excitatory to excitatory cells, $-j_{EI}/\sqrt{K}$, for those from inhibitory to excitatory cells, and so on. The cells also receive excitatory input from outside the network. Assume that this input is constant and has the same order of magnitude as the total excitatory and inhibitory inputs. These external inputs are denoted by $\sqrt{K} j_{0A} \nu_0$, where $A = E(I)$ for the excitatory (inhibitory) neurons, and ν_0 denotes the external rate. Both the total number of cells per population, N, and the average number of synapses per cell from one of these populations, K, are large, but $N \gg K$. This should be compared to a cortical column, in which $N \approx 10^5$, and $K \approx 10^3$. Because K/N is very small, one can assume that the inputs into each cell are uncorrelated. This leads means and variances of the inputs which satisfy

$$\langle I_A \rangle = \sqrt{K}[j_{A0}\nu_0 + j_{AE}\nu_E - j_{AI}\nu_I] \tag{8.2}$$

and

$$\mathrm{Var}(I_A) = j_{AE}^2 \nu_E + j_{AI}^2 \nu_I. \tag{8.3}$$

for the excitatory ($A = E$) and inhibitory ($A = I$) populations. (Here we obtain $j_{AB}^2 \nu_B$ rather than $j_{AB}^2 \nu_B (1 - \nu_B)$ for the contribution of population B to $\mathrm{Var}(I_A)$ because the number of inputs from B is not fixed at K but varies between $K - \sqrt{K}$ and $K + \sqrt{K}$.)

To give the network dynamics, we assume that the neurons are randomly updated, and their output set to 1 is the total input exceeds their threshold and 0 otherwise. Thus the network evolution satisfies

$$\tau_A \frac{d}{dt} \nu_A = -\nu_A + H\left(-\frac{\langle I_A \rangle}{\sqrt{\mathrm{Var}(I_A)}}\right). \tag{8.4}$$

Here τ_E (τ_I) is the average time between updates of a cell in the excitatory (inhibitory) population. H is the complementary error function, which has the property that it goes smoothly from $H(x) = 1$ for $x \to -\infty$, to $H(x) = 0$ for $x \to \infty$. Under some mild restrictions, namely that the inhibitory feedback strength j_{EI} is sufficient and the inhibitory population responds rapidly enough (τ_I is not to large), it can be shown that the network always evolves to a balanced state, in which mean inputs $\langle I_A \rangle$ are of order 1 (van Vreeswijk and Sompolinsky 1996). Thus, a precise tuning of the parameters is not necessary. The network feedback acts in such a way that the network automatically finds an operating point in which the neurons receive an input which is balanced.

An important advantage that such model networks have, over more classical models, is that they can track changes in their input much more rapidly that traditional model networks. The reason for this is that even a small change, δv_0, in the rate v_0 of the cells that project to the network, leads to a large change in the total input of the cells, $\delta I_A = \sqrt{K} j_{A0} \delta v_0$, and hence a large drive to change the rates of the network neurons. But the change in activity δv_A of these cells needs to be only small to restore the balance, $\delta v_A \approx \delta v_0$. This results in the network very rapidly adjusting its rate with the input.

Figure 8-1 shows this effect. It shows the activity of the excitatory population in a network that receives an input which is initially constant and then linearly increases for a while, after which it is kept constant again. The graph shows three traces. One shows the response the network would have if it could react infinitely fast to the change in input. The second is the response of the balanced network, while the third shows how a traditional network of threshold linear neurons tracks the input. To make the comparison between them fair, the units in the last two models have the same neuronal time constants. Clearly the balanced network, with its strong input, is able to follow changes in input much more rapidly.

This suggests the following design principle: It is important to have a system that can respond to input changes rapidly. To get this the network needs to

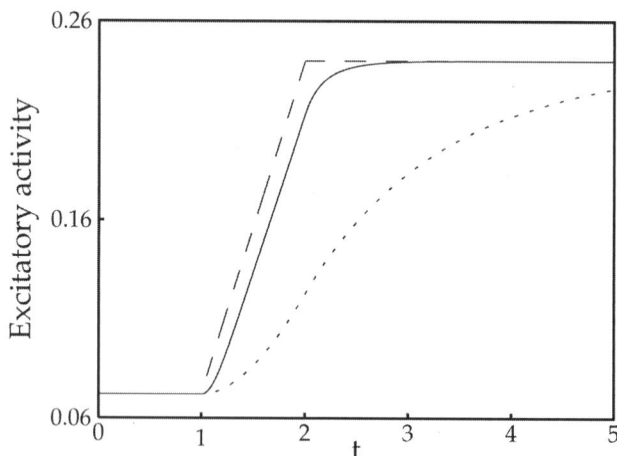

Figure 8-1. The response of a balanced network with strong synapses as the input is changed. The figure shows the average activity of the excitatory population in a balanced network (solid line), a tradition network with threshold linear neurons (dotted line), and a network that reacts instantaneous to the input (dashed line). Figure adapted from van Vreeswijk and Sompolinsky (1996, 1998).

receive an external input which is strongly magnified, together with a strong inhibitory network feedback. For the network to be useful, there should also be tuned excitatory feedback that can enhance modes of the feedforward input which are of relevance. These kinds of systems, however, tend to be highly unstable. In the balanced network this instability is suppressed by ensuring that the coupling is very sparse. The resultant network has the desirable properties of (1) enhancement of input modes of interest, (2) rapid response to changes in input, and (3) a stable response. The fact that the output of the network neurons is also highly irregular is in this view merely an epiphenomenon.

Functional Implications

These design arguments may or may not be plausible, but they are concerned with the mean firing rates. Neurons higher up in the hierarchy do not have access to these mean firing rates, but these have to be inferred from the spike output of the neuronal population. It is probably this inference which leads to most resistance to a rate-based neural code. It is generally assumed that noisy neurons encode the firing rate much less efficiently than neurons that fire more regularly. Let us consider the reason for this and assess whether a better understanding of the issues involved could not lead to a different conclusion.

The balanced network model suggest that the individual neurons function more or less as renewal processes in which the input is encoded in the mean output rate of the neuron. Consider, therefore, a neuron in which the inter-spike interval distribution, $p_v(t)$, varies with the rate v as

$$p_v(t) = vP(vt). \qquad (8.5)$$

If the rate, v, is constant over a time window, $0 < t < T$, the rate can be estimated by dividing the total number of spikes fired in this window by the time T. It is well known that if the duration of the measurement period, T, is long, the estimate of the rate, v_{est}, will have a larger accuracy if the coefficient of variation, C_V, is smaller. Indeed for large T, one can show (Smith 1951),

$$v_{est} = v \pm C_V \sqrt{\frac{v}{T}}. \qquad (8.6)$$

Thus it would seem that a neuron with a small C_V would be better able to convey information about its firing rate than a neuron with a high C_V. Given that cortical neurons fire almost in a Poissonian manner, this would imply that if the cortex indeed uses a rate code, it is extremely inefficient.

This argument, however, overlooks some important facts. The animal has to operate in a world that is not static, in which it has to respond rapidly to stimuli. The brain does not have the time to average over a long time window to estimate the firing rate. Stimuli are also coded by many neurons in parallel. Therefore, the estimate is based on the output of many cells observed over a short time window. It has been shown experimentally that visual stimuli can be processed extremely rapidly (Thorpe, Fize, and Marlot 1996), on a time scale in which not more one spike per cell is available. To capture this, consider a situation in which the rate v is coded in a large number, N, of cells, and the output of all these cells is observed over a very short time T. If T is small compared to the average inter-spike interval, each cell will either fire one spike or no spike at all. The probability of firing one spike is, for each cell, $p = vt$, and the rate can be estimated by dividing the total number of spikes by NT. Thus the estimated rate will satisfy

$$v_{est} = v \pm \sqrt{\frac{v}{NT}}. \tag{8.7}$$

Notice that this is the same for processes with a high C_V and processes with a low C_V. This is due to the fact that measuring period, T, is so short that the probability of more than one spike is negligible. As a result, the shape of the interspike interval distribution is irrelevant for the estimate.

Thus, when considering information transmission over a very short time window using many neurons, one reaches the conclusion that the irregularity of the spike train is of no importance. In contrast, when one neuron or a few neurons are used and the time window is long, the coefficient of variation affects the accuracy of the estimate strongly. To see how this difference changes with the length of the time window, T, consider two processes that transmit the rate v. One is a Poisson process, for which the ISI distribution satisfies $p_v(t) = v \exp(-vt)$. The other is a fourth-order gamma process with $p_v(t) = (4v)^4 t^3 \exp(-4vt)/3!$. Both these processes have the same mean firing rate v, but the first process has a C_V of 1, while for the second $C_V = 1/2$. Figure 8-2 shows the ratio between the standard error of the fourth-order gamma process, σ_4, and the error for the Poisson process, σ_1, as a function of T. For $vT \approx 0$, the ratio is 1. For $vT \approx 1$, that is, the time at which there is a reasonable chance of more than one spike, the measurement errors start to deviate, and only for large T does the ratio of the two errors approach 0.5.

Optimal Coding

We have seen that if the signal is coded in a massively parallel way, the error in the reconstruction does depends only weakly on the irregularity of the firing of the neurons provided that the activity of the cells is uncorrelated. To further my claim that the brain uses a rate code, I will now show that if we assume that the neurons fire irregularly, the code used in the cortex is optimal under certain assumptions.

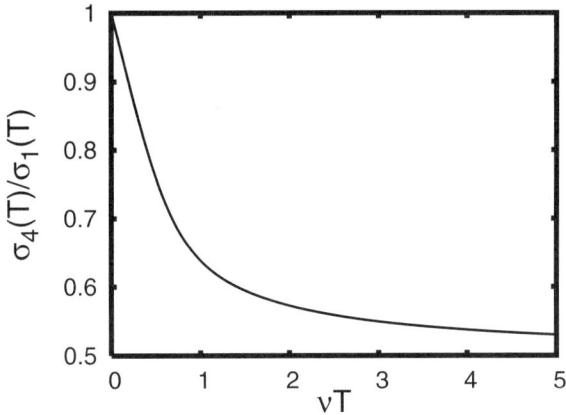

Figure 8-2. The ratio of the estimation error for a fourth-order gamma process, σ_4, and the estimation error for a Poisson process, $\sigma_1(T)$, plotted against vT, the measurement time relative to the average interspike interval.

Optimal coding has recently been the subject of vigorous research. For example, in (Olshausen and Field 1996, 1997) optimal encoding of natural visual stimuli was investigated by studying how the pixel intensities of photographs could be optimally transmitted in noisy channel. A set of K pictures was taken, each consisting of M pixels. For each picture, κ, the pixel intensity, $I_i(\kappa)$, of pixel i was determined. In natural scenes not all spatial frequencies have, on average, the same amplitude. This makes the encoding by the pixel intensities redundant. To overcome this, a whitening filter was applied, which got rid of the second-order redundancy. These whitened intensities, $z_i(\kappa)$, had to be transmitted. The signal that was transmitted consisted of N firing rates, $v_i(\kappa)$, which were linearly dependent on the whitened intensities

$$v_i(\kappa) = \sum_{j=1}^{M} W_{ij} z_j(\kappa) \quad \text{for } i = 1, 2, \ldots, N \qquad (8.8)$$

The outputs, $R_i(\kappa)$ of the system is equal to $v_i(\kappa)$ plot some Gaussian noise

$$R_i(\kappa) = v_i(\kappa) + \sigma \xi_i(\kappa), \qquad (8.9)$$

where the variables $\xi_i(\kappa)$ are independently drawn from a unitary Gaussian distribution. The task that Olshausen and Field studied was the following: A picture is presented, and we receive the outputs R_i from the system. Based on these outputs, we make the best guess which picture was presented. How good our guess is will depend on which matrix W_{ij} we used. The goal of this exercise was to find the matrix W that leads to the best guess. A trivial way to increase the accuracy of the guess is multiplying the matrix by a large number. This increases the rates v_i, and the noise will become small compared to the signal. But if one does this, the rates one has to transmit become large, and this will cost more energy. To keep the energy consumption acceptable, the weights W were therefore constrained to keep the average of $\sum_{i=1}^{N} R_i^2$ below the maximum value NR_{MAX}^2.

If one makes this assumption, there are still many matrices W that lead to equally good guesses. The reason for this is that both the constraint, $\sum_i R_i^2 \leq NR_{MAX}^2$, and the noise, $\sigma\xi_i$, are invariant under rotation. If one makes the rotation

$$
\begin{aligned}
\tilde{v}_1 &= v_1 \cos(\theta) - v_2 \sin(\theta) \\
\tilde{v}_2 &= v_1 \sin(\theta) + v_2 \cos(\theta)
\end{aligned}
\tag{8.10}
$$

and

$$
\begin{aligned}
\tilde{R}_1 &= R_1 \cos(\theta) - R_2 \sin(\theta) \\
\tilde{R}_2 &= R_1 \sin(\theta) + R_2 \cos(\theta)
\end{aligned}
\tag{8.11}
$$

one will have that $\tilde{R}_1^2 + \tilde{R}_2^2 = R_1^2 + R_2^2$ and

$$
\begin{aligned}
\tilde{R}_1 &= \tilde{v}_1 + \sigma\tilde{\xi}_1 \\
\tilde{R}_2 &= \tilde{v}_2 + \sigma\tilde{\xi}_2
\end{aligned}
\tag{8.12}
$$

where $\tilde{\xi}_1 = \xi_1 \cos(\theta) - \xi_2 \sin(\theta)$ and $\tilde{\xi}_1 = \xi_1 \sin(\theta) + \xi_2 \cos(\theta)$ are still uncorrelated unitary Gaussians. Thus, this rotation leads to a valid output which is invariant under the constraint and gives an equally good estimate. Clearly this is true for a rotation for any pair of neurons, i and j, and even for many rotations in sequence. Therefore all transformations under the rotation group in the M dimensional output space lead to an equally good guess of the identity of the picture.

To reduce the space of optimal solutions, Olshausen and Field added an extra constraint, namely that the response of the neurons should be as sparse as possible. This means that the output, $R_i(n)$, of neuron i is small for most pictures, n, while it is large for a small subset of pictures. This extra constraint leads to an optimal solution for the weights W which is unique, up to a permutation. If we now investigate characteristics of the neurons' responses, we see that the neurons respond to light spots at different positions very similarly to the simple in the primary visual cortex.

The sparseness of the representation can be seen as the result of a strategy in which the neurons represent independent components of the stimuli (Bell and Sejnowski 1995). Thus the fact that one has to impose the extra constraint of sparseness to obtain response properties of the neurons similar to those of simple cells in the primary visual cortex has lead people to infer to that the primary visual cortex is organized in such a way that it represents visual stimuli with a minimal loss of information *and* so that the neurons code for independent components.

It should be noted, however, that the output neurons in this model have a few unrealistic features. First of all the rates v_i can be both either positive or negative, and secondly the trial-to-trial variability is here modeled by adding a Gaussian noise term to the input. The amplitude of the noise is independent of the firing rate. Obviously, for real neurons the firing rate has to be non-negative, and as has been shown (Dean 1981), for repeated presentation of the same stimulation the variance in the number of spikes a neuron in the sensory area emits is approximately proportional to the average number of spikes. The optimal representation of the stimulus, the representation in which the effects of the noise is minimal, depends on the kind of noise that is present in the system.

In particular, we saw that in the model described above choosing a representation that maximizes the probability of a correct guess leads to a degeneracy because both the constraint and the noise distribution are invariant under rotation. Having a different constraint and/or noise model will abolish this degeneracy, so in a different model this degeneracy may not show up, and therefore one may not have to add an extra constraint. But whether in such a model the representation will be sparse will depend on the details of the model.

Let us therefore consider a model that is simple enough so that we can understand what is going on but yet hopefully is sufficiently close to biology that its results are relevant for neuroscience. This means that we want a model in which the neurons emit spikes, and the variance in the spike count is linear with the average. Modeling the neurons as renewal processes will accomplish this, and we have seen in the previous section that if we only consider very short time windows all renewal processes behave as Poisson processes, so we might just as well consider these. Consequently we will consider a system where the neurons are modeled by Poisson processes whose rates v_i are a function of the stimulus. The noise is in the number of spikes the neurons emit. In a time interval of length T neuron i emits n_i spikes with probability $(v_i T)^{n_i} \exp(-v_i T)/n_i!$. In such a model a natural constraint on the overall rate is one in which the total number of spikes is, on average, constant.

Let us first investigate such a system in a simple setting. Consider a two-dimensional input space (s_1, s_2). All stimuli fall along the x or y axis of this space, and the probability density, P, of the stimuli is given by

$$P(s_1, s_2) = \frac{1}{4} \left[e^{-|s_1|} \delta(s_2) + e^{-|s_2|} \delta(s_1) \right]. \tag{8.13}$$

The stimulus is encoded by four neurons whose rates v_n are given by

$$v_n(s_1, s_2) = A \left(s_1 \cos \left[\theta + n\frac{\pi}{2} \right] + s_2 \sin \left[\theta + n\frac{\pi}{2} \right] \right)_+. \tag{8.14}$$

Here I have used $(\cdots)_+$ to denote the half rectifying function, $(x)_+ = 0$ for $x < 0$ and $(x)_+ = x$ for $x \geq 0$. The stimulus is projected on the output axes,

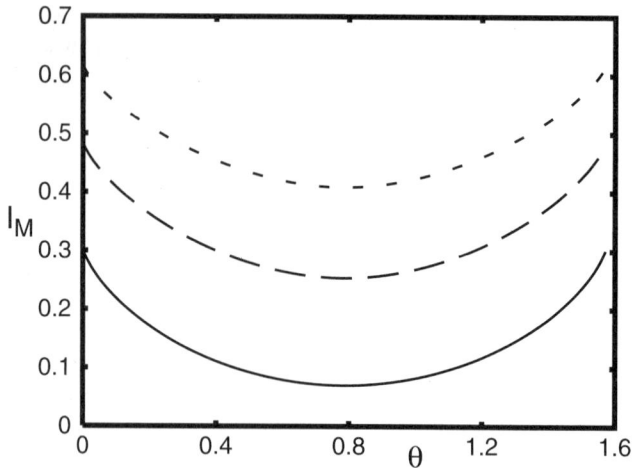

Figure 8-3. The mutual information, I_M, between the spike counts and the stimulus as a function of the angle θ between the input and output axes for different measurement times T. Figure adapted from van Vreeswijk (2001).

which make an angle θ with the input axes, and each of the output axes is encoded by two neurons, one encoding for the positive direction, the other for the negative one. The scaling variable A is chosen such that the average of v_n is 1, $A = (|cos(\theta)| + |sin(\theta)|)^{-1}$. For this model one can calculate how much information the output (n_1, n_2, n_3, n_4) gives about the stimulus, (s_1, s_2), by determining the mutual information between the spike counts and the stimulus (Cover and Thomas 1991).

The mutual information between the spike count and the stimulus is plotted against the angle θ for different length of the measurement time, $T = 1$ (solid line), $T = 2$ (dashed line), and $T = 3$ (dotted line). For all measurement times most information is conveyed if the angle is 0 modulo $\pi/2$, and the least information for $\pi/4$ modulo $\pi/2$. The difference between minimum and maximum value is large for short measurement intervals, and decreases as T becomes longer. Indeed, as T becomes very large the dependence on the angle disappears.

One can also show that the response of the neurons has the maximal sparseness for $\theta = 0 \mod(\pi/2)$, and the sparseness is minimal for $\theta = \pi/4 \mod(\pi/2)$. So for this model requiring that the output is maximally informative about the stimulus leads to a sparse representation even though sparseness was not imposed as an extra constraint.

This suggests that in a model of simple cells in the primary visual cortex that uses Poisson neurons, one can require only that the cells convey the maximal amount of information about the stimulus, and that this is enough get a sparse representation of the input. To test this hypothesis we consider as inputs natural scenes. As before, the pixel intensities are whitened and the firing rates v_i depend on the whitened pixel intensities z_i, as

$$v_i(\kappa) = \left(\sum_{j=1}^{K} W_{ij} z_j(\kappa) \right)_+ , \qquad (8.15)$$

and, for a measurement period T, draw the number of spikes for neuron i according a Poisson process with a mean $v_i T$. Next the weights W_{ij} that maximize the mutual information between the spike counts $\{n_i\}$ and κ are determined. Once this is done, one finds indeed that the representation is sparse and the response of the neurons in this model mimic the response of simple cells in the primary visual cortex. Note that for the Poisson model these results are obtained without having to add an extra constraint of maximal sparseness.

Figure 8-4 shows the receptive fields of forty neurons in this model. These receptive fields show the amplitude of the response to a light spot as on gray scale as function of the position of the spot for a model trained on stimuli of ten by ten pixels seen through a circular aperture. The receptive fields of these cells are approximately Gabor filters, similar to those of simple cells in the primary visual cortex.

If we model the neurons as Poisson processes, we automatically obtain a representation of the stimulus that is sparse if we require that this representation conveys the maximal amount of information about about the stimulus. Since Poisson processes mimic the activity of real neurons more closely that Gaussian channels, this would argue that from the fact the simple cells in the primary visual cortex have a sparse representation of natural visual stimuli, we cannot infer that the visual cortex is set up to yield a sparse representation or to extract independent components of the stimulus. The sparseness of the representation could simply be a byproduct of minimizing the effect of the noise.

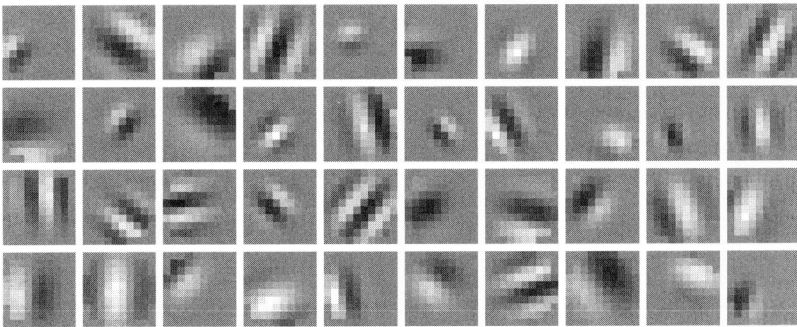

Figure 8-4. The receptive fields of forty neurons in the Poisson model of primary visual cortex. Figure adapted from van Vreeswijk (2001).

Discussion

Concentrating on information transmission in the sensory part of the cortex, we have looked at the question, what is the code which areas in the cortex use to exchange information? Experimentally there is overwhelming evidence that the firing rate of neurons carry information about the sensory input. Other candidates for the neural code have much weaker experimental evidence in their favor. One argument against a rate code is that a rate code is very inefficient if the neurons fire as irregularly as those in the brain. I have presented a model of a cortical network that is capable to react rapidly to changes in the stimuli and shown that such a model has as a byproduct highly irregular activity of the individual cells. The inefficiency of irregularly active neurons is also rather modest if one considers parallel transmission using many cells with an averaging over very short time periods.

I have applied the assumption that the neurons transmit a firing rate using highly irregular activity to the most extensively studied cortical area, the primary visual cortex. The assumption that the cells represent the natural visual stimuli optimally, using irregularly spike activity leads, to response properties that are similar to those of simple cells in the primary visual cortex.

All of this suggests that neurons in the brain do indeed communicate through a rate code. While this may be the rule, there are undoubtedly exceptions. It seems quite clear for example that the primary auditory cortex does not use a rate code (deCharms and Merzenich 1996). This may be because a noisy rate code is not well suited to encode naturally occurring auditory signals. The fact is that we have very little theoretical understanding about when such a code does well and when it does not. Such theoretical insight would be very valuable to infer which areas of the brain are likely to use a rate code and will also help us to identify in which areas other interesting codes may be used.

References

Abeles, M. (1991) *Corticonics: Neural Circuitry of the Cerebral Cortex.* Cambridge: Cambridge University Press.

Abeles, M., Bergman, H., Margalit, E., and Vaadia, E. (1993) Spatiotemporal firing patterns in the frontal cortex of behaving monkeys. (1993) *J. Neurophysiol.* 70: 1629–1638.

Adrian, E. D. (1926) The impulses produced by sensory nerve endings: Part I. *J. Physiol. (Lond.)* 61: 49–72.

Bell, A. J., and Sejnowski, T. J. (1995) An information maximization approach to blind seperation and blind deconvolution. *Neural Comput.* 7: 1129–1159.

Burns, D. B., and Webb, A. C. (1976) The spontaneous activity of neurons in the cat's visual cortex. *Proc. R. Soc. Lond. B* 194: 211–223.

Bush, P. C., and Douglas, R. J. (1991) Synchronization of bursting action potentials in discharges in a model network of neocortical neurons. *Neural Comput.* 3: 19–30.

Cover, T. M., and Thomas, J. A. (1991) *Information Theory* (New York: Wiley and Sons).

Dean, A. F. (1981) The variability of discharge of simple cells in cat striate cortex. *Exp. Brain. Res.* 44: 437–440.

deCharms, R. C., and Merzenich, M. M. (1996) Primary cortical representation of sounds by the coordination of action-potential timing. *Nature* 381: 610–613.

Georgopoulos, A. P., Schwartz, A. B., and Kettner, R. E. (1986) Neuronal population coding of movement direction. *Science* 233: 1416–1419.

Gray, C. M., Konig, P., Engel, A. K., and Singer, W. (1989) Synchronization of oscillatory neuronal responses in cat striate cortex: temporal properties. *Vis. Neurosci.* 8: 337–347.

Gray, C. M., and Singer, W. (1989) Oscillatory responses in cat visual cortex exhibits inter-columnar synchronization which reflects global stimulus properties. *Proc. Natl. Acad. Sci. USA* 86: 1698–1702.

Hansel, D., and Sompolinsky, H. (1996) Chaos and synchrony in a model of a hypercolumn in visual cortex. *J. Comput. Neurosci.* 3: 7–34.

Hubel, D. H., and Wiesel, T. N. (1962) Receptive fields, binocular interaction and functional architecture in the cat's visual cortex. *J. Physiol. (Lond.)* 160: 106–154.

von der Malsburg, C. (1985) Nervous structures with dynamical links. *Ber. Bunsenges. Phys. Chem.* 89: 703–710.

Olshausen, B. A., and Field, D. J. (1996) Emergence of simple-cell receptive field properties by learning a sparse code for natural images. *Nature* 381: 607–609.

Olshausen, B. A., and Field, D. J. (1997) Sparse coding with an over-complete basis set: A strategy employed by V1? *Vision Research* 37: 3311–3325.

Oram, M. W., Wiener, M. C., Lestienne, R., and Richmond, B. J. (1999) Stochastic nature of precisely timed spike patterns in visual system neuronal response. *J. Neurophysiol.* 81: 3021–3033.

Richmond, B. J., Optican, L. M., and Spitzer, H. (1990) Two dimensional patterns are represented in temporally modulated activity of striate cortex cells. *J. Neurophysiol.* 64: 351–369.

Smith, W. L. (1951) On the variance of the spike count of renewal processes. *Biometrica* 46: 1–34.

Thorpe, S. J., Fize, D., and Marlot, C. (1996) Speed of processing in the human visual system. *Nature* 381: 520–522.

Tsodyks, M., and Sejnowski, T. (1995) Rapid state switching in balanced cortical network models. *Network* 6: 111–124.

van Vreeswijk, C. (2001) Whence sparseness? In *Advances in Neural Information Processing Systems 13*, ed. Leen, T. K., Dietterich, T. G., and Tresp, V., 189–195 (Cambridge, Mass.: MIT Press).

van Vreeswijk, C., and Sompolinsky, H. (1996) Chaos in neuronal networks with balanced excitatory and inhibitory activity. *Science* 274: 1724–1726.

van Vreeswijk, C., and Sompolinsky, H. (1998) Chaotic balanced state in a model of cortical circuits. *Neural Comput.* 10: 1321–1371.

9

Are Single Cortical Neurons Soloists or Are They Obedient Members of a Huge Orchestra?

Tal Kenet, Amos Arieli, Misha Tsodyks, and Amiram Grinvald

Spontaneous cortical activity of single neurons is often either dismissed as noise or is regarded as carrying no functional significance, and hence it is ignored. Previous reports as well as our recent research suggest that such concepts should be revised. Particularly important is exploring the collective coherent activity of interconnected neurons, namely coherent population activity of neuronal assemblies. Recent advances in real-time optical imaging based on voltage sensitive dyes (VSDI) have facilitated exploration of population activity and its intimate relationship to the activity of individual cortical neurons. It has been shown by in vivo intracellular recordings that the dye signal measures the sum of the membrane potential changes in all the neuronal elements in the imaged area, emphasizing subthreshold synaptic potentials in neuronal arborizations originating from neurons in all cortical layers whose dendrites reach the superficial cortical layers. Thus, in addition to being suitable for measurements of population activity, the VSDI has allowed us to image the rather elusive activity in neuronal dendrites that cannot be readily explored by single-unit recordings. Surprisingly, we found that at the population level the amplitude of this type of ongoing activity is of the same order of magnitude as evoked activity. Since often the ongoing activity is very large, it may reach the threshold for neuronal action potentials without any external input. Therefore we expected it to play a major role in cortical function and decided to explore it further. We also found that this ongoing activity exhibited high synchronization over many millimeters of cortex. We then investigated the influence of ongoing activity on the evoked response and showed that the two interact very strongly. Taken together these findings indicate that ongoing activity cannot be dismissed as mere noise. Next, we demonstrated that indeed ongoing activity carries a specific functional significance. We found that cortical states that were previously associated only with evoked activity can actually be observed also in the absence of stimulation; for example, the cortical representation of a given orientation may appear without any visual input. This demonstration suggests that ongoing activity may also play a major role in other cortical functions by providing a neuronal substrate for the dependence of sensory information processing on context, behavior, memory, and other aspects of cognitive function. Our results suggest that an attempt to decipher the neural code, namely the processing strategies used by the mammalian CNS, may remain incomplete if subthreshold dendritic activity is not considered and the role of spontaneous population activity is ignored.

Listening to the Orchestra

The music of an orchestra is the harmonic sum of the music played by all its individual members playing in a coherent fashion. Each instrument may follow a different individual tune, but in such a coherent way that the sum of the individual notes, the tunes played by all instruments, is music. Carrying this crude analogy to neuronal function, we can liken groups of neurons to an orchestra. We feel we know many of the tunes the orchestra usually plays, because we know what each player is playing (e.g., single-unit responses to a specific stimulus presentation). In well-explored primary cortical areas, we may even say we are familiar with a lot of the repertoire the orchestra plays during a performance (functional maps of orientation selectivity, ocular dominance, etc.). To understand ongoing activity, we face the challenge of trying to understand what the orchestra is playing when we don't have a program in front of us, we don't hear all the players, and we don't even know when the music begins and ends.

Over the years, many groups have been trying to do just that. It was often found that the spontaneous firing of neocortical neurons, that is, the individual players of our orchestra, is a noisy, stochastic process (Softky and Koch 1993; Shadlen and Newsome 1994; Van Vreeswijk and Sompolinsky 1996; Amit and Brunel 1997). In 1949, Hebb introduced the concept of cell assembly, according to which cortical neurons organize dynamically into functional groups by the temporal structure of their spike activity. Such neuronal ensembles, or groups of instruments playing similar notes in the orchestra, were difficult to study; a technique was needed that would provide both high temporal and high spatial resolutions. So for many years since the concept was first introduced, little was known about the spatiotemporal organization of this coherent activity, its interactions with stimulus-evoked activity, and its significance to cortical information processing and decoding. Optical imaging of voltage sensitive dyes, described in the next section, allowed us for the first time to directly study these neuronal assemblies and thus attempt to answer the question, can spontaneous activity, either at the single neuron or at the population level, be dismissed as mere noise? Specifically, we asked, how large is the spontaneous activity, how is it spread in space and time, whether or not it influences the evoked responses, and if so how, and finally, whether spontaneous activity can exhibit a spatial structure that can be linked to cortical functionality.

The Method

Real-time optical imaging based on voltage-sensitive dyes (Grinvald 1984; Orbach, Cohen, and Grinvald 1985; Grinvald et al. 1999) is a useful tool for imaging the membrane potential changes of large neuronal populations extending across a few millimeters in the cortex. Hence, this technique (figure 9-1) provides an accurate real-time view of neuronal activity spread across several neocortical hypercolumns, whose task in primary cortices is to process sensory input. Currently optical imaging of voltage-sensitive dyes is the only technique that has both the temporal and spatial resolution for approaching the question posed above.

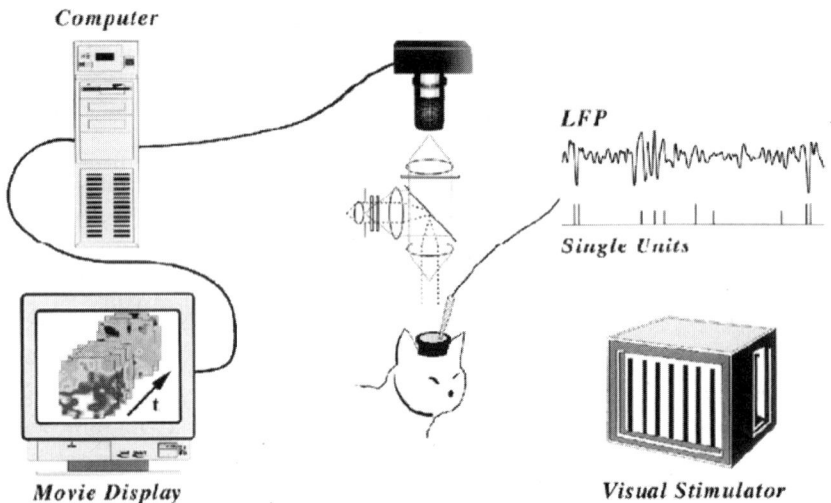

Figure 9-1. The setup for real-time optical imaging. The exposed cortex was stained for two hours by topical application of a voltage-sensitive dye. An image of a 1- to 7-mm square area of primary visual cortex was projected onto a fast camera with the aid of a macroscope. Computer-controlled visual stimuli were presented on a video monitor. Images of the cortical fluorescence were taken at 100 to 1000 Hz. The output of the fast camera was displayed in slow motion on an RGB monitor using either grayscale, color-coded, or surface plot images. Local field potentials, multiunit or single-unit activity, or intracellular recording were performed simultaneously. Modified from Arieli et al. (1995) and Tsodyks et al. (1999).

What Does the Cortical Dye Signal Represent?

In simple preparations, where single cells are distinctly visible by a single pixel, the dye signal looks just like an intracellular electrical recording (Salzberg, Davila, and Cohen 1973; Salzburg et al. 1977; Grinvald, Salzberg, and Cohen 1977; Grinvald et al. 1981; Grinvald, Manker, and Segal, 1982). However, it is important to note that the dye signals recorded or imaged from the neocortex are different from those recorded from single cells or their individual processes in simpler nervous system. In optical recordings from cortical tissue, the optical signal does not have single-cell resolution. Rather, it represents the sum of membrane potential changes in both pre- and post-synaptic neuronal elements, as well as a possible contribution from the depolarization of neighboring glial cells (Konnerth and Orkand 1986; Lev-Ram and Grinvald 1986). Since the optical signals measure the sum of the membrane potential changes over membrane area, optical recording can easily detect slow subthreshold synaptic potentials in the extensive dendritic arborization. Thus, optical signals, when properly analyzed, can provide information about aspects of cortical processing by neuronal populations, which usually cannot be obtained from single-unit or intracellular recordings.

Visualization of Coherent Assemblies

A neuronal assembly may be defined as a group of neurons that cooperate to perform a specific computation required for a specific task. The activity of cells in an assembly is time locked (coherent). However, the cells that comprise an assembly may be spatially intermixed with cells in other neuronal assemblies that are performing different computational tasks. Therefore, techniques that can visualize only the average population activity in a given cortical region are not adequate for the study of neuronal assemblies. What is needed is a method to discriminate between the operations of several colocalized assemblies. This problem was tackled precisely by making use of the fact that activity of the neurons in an assembly is time locked (Arieli et al. 1995). The firing of a single neuron served as a time reference to selectively visualize only activity that was synchronized with it, that is, only the activity in the assembly it belonged to. To study the spatiotemporal organization of neuronal assemblies we combined single-unit recordings and subsequent spike-triggered averaging of the optical recordings. With sufficient averaging, the activity of neuronal assemblies not time-locked to the reference neuron was averaged out (figure 9-2), enabling the selective visualization of the reference neuron's assembly.

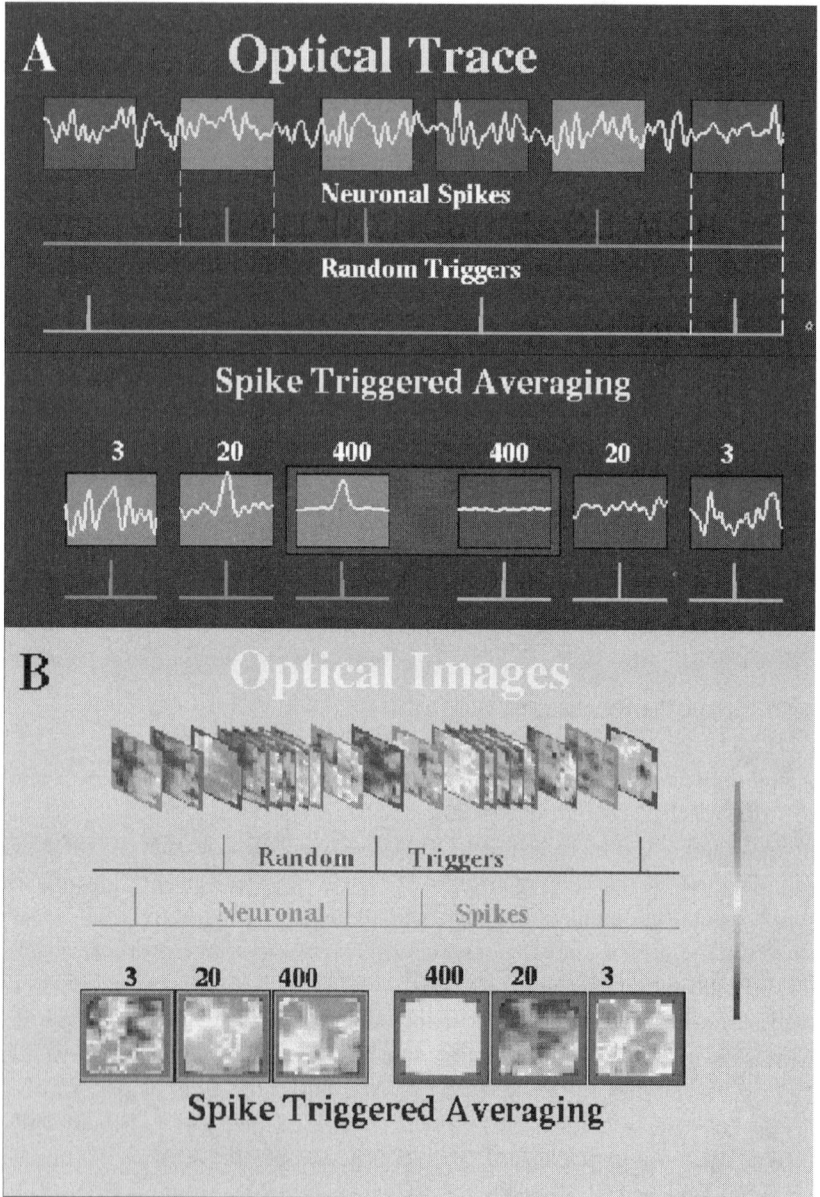

Figure 9-2. The procedure for selective visualization of the dynamics of coherent neuronal assemblies. See color insert.

Experimental Procedure

Obviously, if the coherent activity results only from a small population of scattered neurons it is going to be difficult to detect experimentally. Therefore, in practice, for studying the relationship of the activity of a single cortical neuron and that of the coherent population it is important that the size of the population will be relatively large. In other words, it is important to select situations when a large enough fraction of cortical neurons will exhibit coherent activity with that of a single cell in the imaged area. We chose to explore this question in areas seventeen and eighteen of the anesthetized cat. Orientation tuning was chosen as the specific functional property because the majority of neurons in the striate cortex of the anesthetized cat are tuned for the orientation of bars or gratings (Hubel and Wiesel 1962). Furthermore, the functional architecture and intracortical circuitry underlying orientation tuning is well known (Hubel, Weisel, and Stryker 1977; Bonhoeffer and Grinvald 1991).

Simultaneous recordings of single-unit activity and real-time optical imaging from a surrounding region were performed in areas seventeen and eighteen of the visual cortex of anesthetized cats (for a detailed description of the technique see Arieli et al. 1995, 1996; Tsodyks et al. 1999; Shoham et al. 1999). For the study of evoked activity, a drifting grating of either an optimal or a nonoptimal orientation was presented, while for the study of spontaneous activity the room was left dark and the eyes closed. In order to assure that we had enough spikes so that we could perform statistical analyses safely, we selected only neurons exhibiting a relatively high spontaneous spike rate when the room was left dark (at least two spikes per second, which is a high rate for an anesthetized animal). To assure that these neurons have a well-defined function associated with their activity, we next characterized the orientation tuning properties of these neurons and further selected only the neurons with sharp tuning preference and robust response.

Noise or Music?

Is ongoing activity randomly distributed in space and time, or is its distribution somehow structured? Does it interact with evoked responses, and if so how? Is it noise, or is it an expression of an intrinsic cortical mechanism, which is useful for cortical processing? The answers to these questions are critical if one wishes to understand the nature of spontaneous activity. Such answers emerged from two recent studies from our lab, dealing with these questions precisely.

What Is the Relative Amplitude of Ongoing Activity?

We used the approach described in the previous section to image coherent activity during ongoing and sensory evoked activity (Arieli et al. 1995). Surprisingly, we found that the amplitude of this coherent ongoing activity, recorded optically, was often almost as large as the activity evoked by optimal visual stimulation (Arieli et al. 1996b; Kenet et al. 2003; figure 9-3). Inspecting

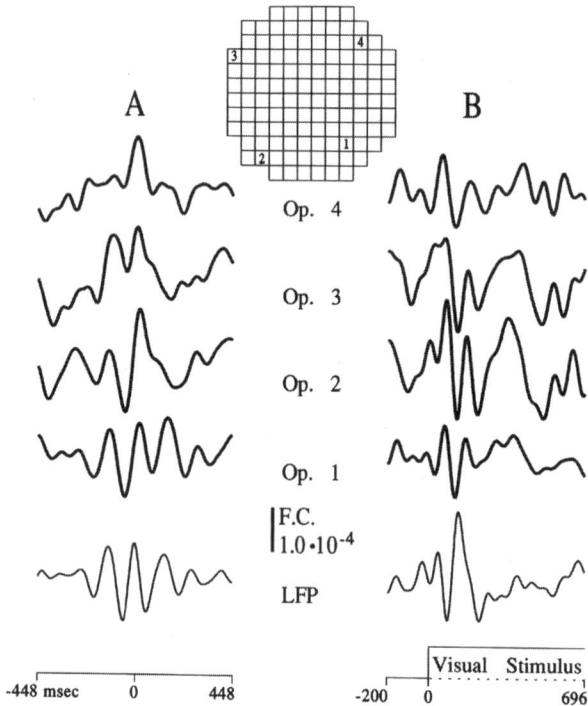

Figure 9-3.   The relative amplitude of ongoing activity. (A) The STA of spontaneous optical and electrical signals: thirty-five spikes of a single neuron were used for the STA. The first optical trace above the LFP (Op. 1) shows a similar wave as the LFP but with a phase shift. The three other traces (Op. 2–4) from different cortical loci had a different time course with peaks that coincided with various peaks of the LFP. This comparison suggests that different components of the LFP originated at different sites across the cortex. (B) The visual evoked response (VER) of the LFP and the optical signal were obtained in a subsequent recording session from exactly the same area. The signals were averaged on the onset of thirty-five grating stimuli delivered in the preferred orientation of the unit activity. The averaged LFP and optical signal (Op. 1–4) show a significant evoked response. Note that the amplitudes of the STA for spontaneous activity and of the VER were similar. The scale at the bottom shows the timing of the visual stimuli. Raw data filtered 2–14 Hz, $\sigma = 7$ msec in (A) and (B). Modified from Arieli et al. (1995).

our entire data set, we found that, on average, the amplitude of the ongoing activity that was directly and reproducibly related to the spontaneous spikes of a single neuron was as high as 54 percent of the amplitude of the visually evoked response by optimal sensory stimulation, recorded optically.

How Is Ongoing Activity Distributed in Space and Time?

Using the technique of spike triggered averaging explained above, we found that in 88 percent of the neurons recorded during spontaneous activity (eyes closed), a significant correlation was found between the occurrence of a spike and the optical signal recorded in a large cortical region surrounding the recording site. This result indicates that spontaneous activity of single neurons is not an independent process, but is time-locked to the firing or to the synaptic inputs from numerous neurons, all activated in a coherent fashion even without a sensory input. Furthermore, coherent activity was detected even at distant cortical sites up to 6 mm apart.

The spontaneous activity of two adjacent neurons, isolated by the same electrode and sharing the same orientation preference, was often correlated with two different spatiotemporal patterns of coherent activity, suggesting that adjacent neurons in the same orientation column can belong to different neuronal assemblies (Arieli et al. 1995).

Does Ongoing Activity Affect Cortical Processing and Behavior?

In the mammalian visual cortex, evoked responses to repeated presentation of the same stimulus exhibit a large variability. It has been found that this variability in the spatiotemporal patterns of the evoked activity results from the ongoing activity, reflecting the dynamic state of the cortical network (Arieli et al. 1996b). In spite of the large variability, the evoked responses in single trials can be predicted by taking account of the preceding ongoing activity (figure 9-4). On average, the ongoing dynamics are not affected by the response, and therefore the prediction is valid as long as the ongoing activity pattern, which may continue to change during the evoked response, is still similar to the initial state (e.g., 50–100 milliseconds). In a follow-up study in collaboration with Dr. Eilon Vaadia's group, we found that also in the motor cortex of behaving monkeys the large trial-by-trial variability of evoked activity (as measured by LFP) results from the ongoing activity dynamics. In addition, we found direct evidence for the behavioral significance of this effect: ongoing activity, prior to the presentation of the visual cue, affects the actual motor behavior of the monkey as well (Arieli et al. 1996b).

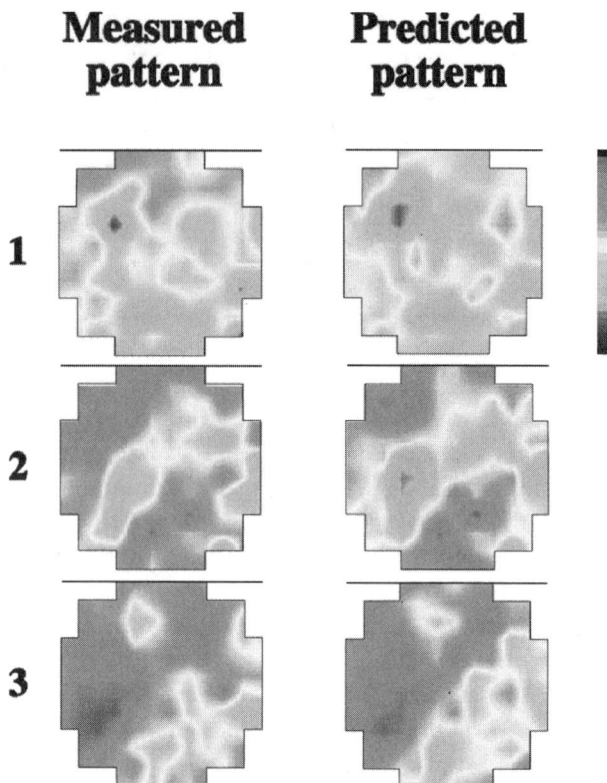

Figure 9-4. Predicting cortical evoked responses in spite of their large variability. See color insert.

Are Single Neurons Soloists or Well-Tuned, Obedient Players?

All the above results indicate that ongoing activity plays a major role in affecting cortical response and is not noisy, but rather exhibits a large degree of temporal coherence. Yet temporal coherence could result from many various sources. We therefore focused on the spatiotemporal characteristics and the functional significance associated with the ongoing activity. Knowing that ongoing activity over a large area is highly correlated with the spontaneous action potentials of a single neuron, we chose to approach this question precisely by looking at the relationship between times of spontaneous action potentials and the spatial patterns of the coherent activity. More specifically, we wanted to know if there is a link between the ongoing patterns of cortical population activity that occur simultaneously with the action potentials of a single neuron (hereafter the *spontaneous regime*) and the reproducible pattern of activity evoked by the presentation of a well-defined stimulus, optimal for that neuron (hereafter the *evoked regime*). In cat areas seventeen and eighteen such reproducible patterns for each stimulus orientation are referred to as the orientations maps. We calculated the relationship between the probability for a neuron to fire an action potential, and the pattern of the population activity in that region of the cortex, both in the presence and in the absence of a visual input. We began by looking at the spatiotemporal relationship during the evoked regime so that we can later compare these results to the results obtained for the spontaneous regime. In other words, we first study the role the neuron plays in the orchestra when we let the orchestra play a familiar tune by presenting it with a stimulus. We can then look at whether the neuron played the same role in the orchestra when the tune was no longer imposed, that is, in the absence of stimulation.

As already mentioned the vast majority of single neurons in the striate cortex have a well-defined tuning curve. This means they are selective for stimuli of specific orientations (a much smaller fraction is selective for direction and spatial frequency; Shoham et al. 1999; Shmuel and Grinvald 1996); hence one can easily choose the preferred orientation stimulus for a given neuron to maximize its firing. Similarly, it is possible to simultaneously map out the response of a large population of neurons, a whole orchestra of them in fact, to this same stimulus using optical imaging based on voltage sensitive dyes. The pattern of evoked cortical activity for a particular stimulus attribute is defined as the functional map related to this stimulus attribute (i.e., orientation maps, direction maps, spatial frequency maps, etc.). It can be obtained with high precision if signal averaging is used to remove the ever-changing ongoing activity from the individual response. The orientation maps in early visual cortex are known to be correlated with the underlying anatomical structure of patches hundreds of micron in diameter interconnected by long-range horizontal connections (Gilbert 1993; Malach et al. 1993; Fitzpatrick 1996; Bosking et al. 1997). Here we define these spatial patterns of coherently active neurons as a cortical state, a state that is related to the intracortical connectivity, and once it emerges it implies that a given response property is now represented in the cortex. For example, the spatial pattern of coherently active neurons evoked by a vertical grating is a cortical state and a representation of this stimulus in that cortical area.

Single Neurons and Population Activity in the Evoked Regime

In the evoked regime, the relationship between the single neuron response and the response of the entire population is trivial and can easily be quantified. To determine the pattern of the evoked cortical state for which a given neuron has a maximal firing rate, we averaged over all patterns observed at the times corresponding to action potentials that were evoked by the optimal stimulus (Tsodyks et al. 1999). The spatial pattern thus observed is illustrated in figure 9-5A. We refer to the spatial pattern observed when the optimal stimulus for a given neuron was used to evoke population activity as the neuron's preferred cortical state (PCS). To minimize the possibility that a few distinct states were mixed in the PCS spatial pattern when this procedure was applied, we limited our choice to well-defined neurons with a sharp tuning preference and a robust response yet without direction selectivity. We then calculated the single condition orientation map from the same data by averaging the spatial patterns of activity, at the time of stimulus onset (figure 9-5B). Note that a single condition map is calculated by looking directly at the average response to repeated presentation of that same stimulus. As is expected, the single condition map (figure 9-5B) obtained by triggering on the stimulus onset is almost identical to the preferred cortical state of the neuron (figure 9-5A) obtained by triggering on the neuron's action potentials. This is trivial because by definition, the neuron fired most vigorously when the

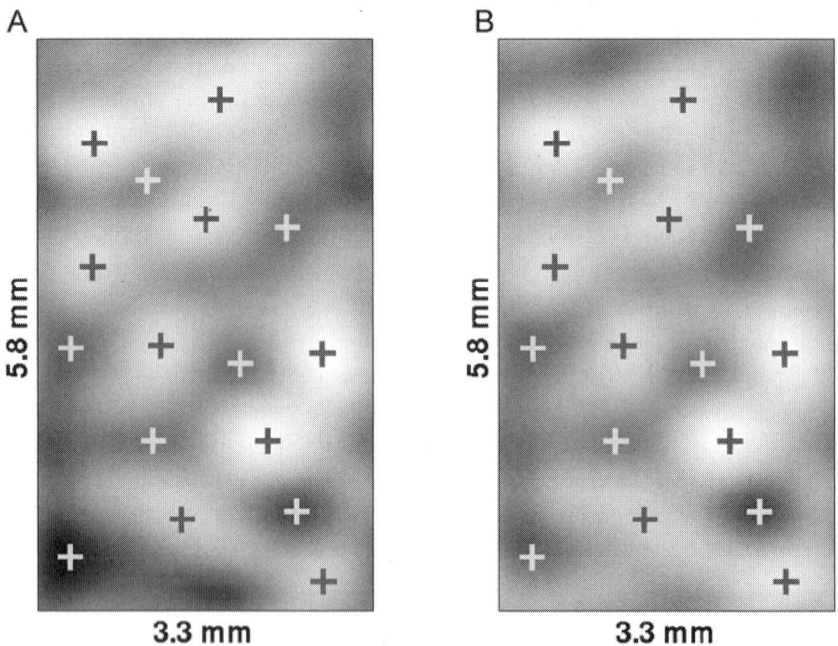

Figure 9-5.   The preferred cortical state of a neuron versus the functional map obtained using its optimal stimulus. (A) The preferred cortical state of the neuron, computed by averaging all time frames that coincided with spikes during the evoked session where the neuron's optimal stimulus was used. (B) The single-condition functional orientation map obtained from the same recording session by triggering on stimulus onset and disregarding spike times. As expected, the two are very similar. Modified from Tsodyks et al. (1999).

stimulus was on, and this is also the period of time from which the map was computed.

Hence it is not at all surprising that during the presentation of a full-field stimulus optimal for the neuron, the neuron is an obedient member of a large orchestra; it is not a soloist in the sense that its spike train lacks an individual imprint. This fact was verified and quantified by using the population response to predict the observed spike train of the neuron. We did this by simply looking at the similarity between the instantaneous pattern of population activity, that is, each and every frame in the recording session, and the neuron's preferred cortical state. As a measure of the similarity between the two patterns, we used their correlation coefficient. The time course of this similarity index is shown in figure 9-6A. We were also able to compare the observed spike train plotted in figure 9-6C with the time course of the correlation coefficients (figure 9-6A). Each presentation of the optimal stimulus evoked increased spiking activity (bursts) and, as expected, the pattern of population activity became more similar to the neuron's PCS, as reflected by higher values of correlation coefficients. The meaning behind the values of the correlation coefficients merits a small digression: The correlation is done between an average pattern (the PCS) and the single frames. To assess the significance of the correlation values we obtained, we compared them to the correlation between the map (which is again an averaged pattern triggered on stimulus and not on the evoked spikes) and the single frames that were used to obtain the map during stimulation. The maximal values of correlation obtained in each of the cases are, in fact, similar. Since we know that the correlation between the map and one of its components is significant, this means that the high correlation between the PCS and single frames is also significant. The fact that the highest values of the correlation coefficients do not reach values as high as those observed between similar maps obtained from different trials, for instance, is simply explained by the single trial variability (see Kenet et al. 2003 for more details). The residual noise in the recording of instantaneous activity patterns of the single trial variability results from the ongoing activity preceding the stimulus (Arieli et al. 1996a) or the potential mixing of several cortical states in one PCS during the averaging.

To quantify the distribution of correlation coefficients over an entire recording session one can simply look at the corresponding histogram (figure 9-6B). This histogram has a small tail in the high positive values, as expected, since the stimulus "pushed" the network to spend more time in the neuron's preferred cortical state.

What is more interesting, though, is what happens at times coincident with action potentials, as plotted in figure 9-6C. The subset of the histogram that includes only spikes times is shown in figure 9-6D. We see, as expected, that spikes occurred mostly when the instantaneous pattern of population activity was similar to the PCS. Another way of assessing the resemblance

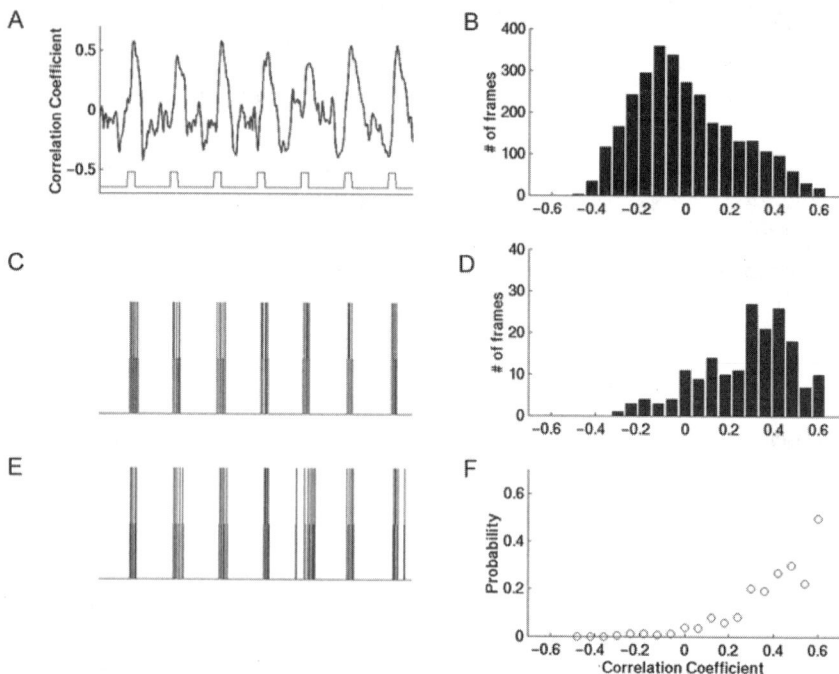

Figure 9-6. Predicting evoked activity of a single neuron from population activity. (A) The time course of the correlation coefficients in a time window from the recording session. (B) The histogram of distribution of correlation coefficients over the entire recording session. (C) Spikes that occurred in the same window as plotted in (A). (D) The subset of the histogram plotted in (B), which includes only spike times. We see, as expected, that spikes occurred mostly when the instantaneous pattern of population activity was similar to the PCS. We compared the observed spike train plotted in (C) with the time course of the correlation coefficients (A). Each presentation of the optimal stimulus evoked increased spiking activity (bursts) and, as expected, the pattern of population activity became more similar to the neuron's PCS, as reflected by higher values of correlation coefficients. (E) A prediction of the observed spike train of the neuron, from the similarity of the ongoing pattern to the PCs. (F) We compute the probability for a neuron to fire an action potential at any given value of the correlation coefficient, by dividing the bottom histogram (D) by the top one (B). This yields a steadily increasing function past a certain threshold. At low values of correlation coefficients, the firing rate is practically zero, and above a certain value it increases monotonically. This graph implies that a neuron's instantaneous firing rate is higher the more similar the instantaneous population activity pattern is to its functional architecture. Modified from Tsodyks et al. (1999).

between the discrete spike train (figure 9-6C) and the continuous correlation trace (figure 9-6A) is to reconstruct a spike train from the correlation trace. We do this by computing the probability for a neuron to fire an action potential at any given value of the correlation coefficient, obtained in turn by dividing the bottom histogram (figure 9-6D) by the top one (figure 9-6B). To obtain the predicted instantaneous firing rate of a neuron, we further divided this probability by the duration of the time frame. This yields a steadily increasing function past a certain threshold (figure 9-6F). At low values of correlation coefficients the firing rate is practically zero, and above a certain value, it increases monotonically. This graph implies that a neuron's instantaneous firing rate is higher the more similar the instantaneous population activity pattern is to its functional architecture. We can use these probabilities as instantaneous rate in a Poisson process to predict a spike train of the neuron from the similarity of the ongoing pattern to the PCS (figure 9-6E). Evidently, the observed and predicted spike trains are very similar in shape, meaning the instantaneous state of the population activity can indeed predict how likely a neuron is to fire at any given moment. Note, however, that this procedure predicted well the occurrence of bursts in the observed spike train rather than the exact timing of single action potentials.

The next intriguing question we asked was, what happens if we use a stimulus that is not optimal to the neuron? Does the neuron still prefer to play the same tune? Namely, does the neuron still fire when the instantaneous pattern of population activity is more similar to the optimal stimuli (even though we presented nonoptimal stimuli)? We repeated the exact procedure as with the optimal stimulus for a recording session where the spike activity of the same neuron was recorded, but the stimulus was of an orientation orthogonal to the one preferred by the neuron. As expected, since now the PCS does not coincide with the state evoked by the stimulus, the histogram of the distribution of correlation coefficients has no tail in the positive domain (figure 9-7A). In fact, the histogram is indeed slightly shifted toward the negative domain. However, spikes were still most likely to occur when the correlation coefficients were high, as is obvious by looking only at the subset of this histogram that corresponds to the spike times (figure 9-7B).

Since the stimulus was not optimal, fewer spikes occurred, and more averaging was required to see the similarity. Therefore, for this case of the orthogonal to optimal stimulus we calculated here the peristimulus time histogram (PSTH) rather than the spike train itself. Thus, we can again predict the probability of spike occurrence per correlation coefficient (figure 9-7C) and use it to predict the PSTH of the neuron. Comparing the observed PSTH of the neuron (9-7D) and the predicted PSTH (figure 9-7E), we found that the two were indeed quite similar. We see once again that the PCS of the neuron (calculated using optimal stimuli) together with the population activity that stemmed from a stimulus not preferred by the neuron can nonetheless predict this neuron's response. Figure 9-7C shows that as in the case of an optimal stimulus, the more similar the instantaneous population is to the neuron's PCS, the more likely the neuron is to fire. Hence even when not presented with its optimal stimulus, the neuron remains faithful to the orchestra of which it is a member and is most likely to fire when the orchestra plays its favorite tune,

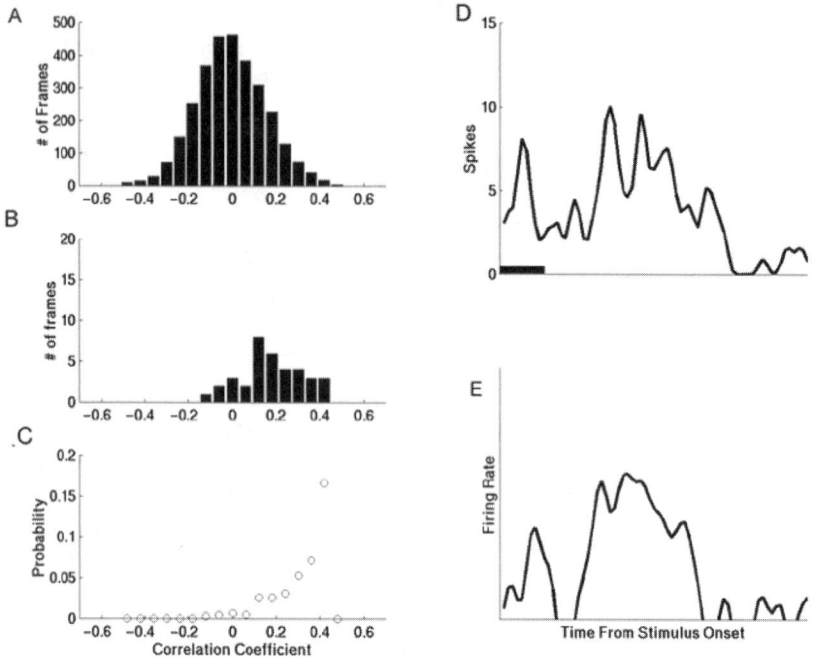

Figure 9-7. Predicting evoked activity of a single neuron when the stimulus is orthogonal to its optimal one. (A) The histogram of the distribution of correlation coefficients between each time frame and the map obtained using a stimulus orthogonal to the one used in the recording session. The histogram this time has no tail in the positive domain, as in figure 9-5A, and in fact, is indeed slightly shifted toward the negative domain, as we would expect. (B) The subset of the histogram in (A), including only times that corresponds to spike times. The shift toward positive values indicates that spikes were still most likely to occur when the correlation coefficients were high, namely, when the pattern of population activity was similar to the neuron's PCS. (C) Predicting the probability of spike occurrence per correlation coefficient by dividing the spikes in histogram (B), by the full histogram (A). (D) The observed PSTH of the neuron. (E) The predicted PSTH of the neuron computed using the probabilities obtained in (C). Kenet et al. unpublished.

namely when the pattern of population activity resembles most its PCS, even when it is the pattern orthogonal to the one evoked by the stimulus.

Single Neurons and Population Behavior during Spontaneous Activities

The last question we asked was, what happens in the spontaneous regime, that is, when the eyes are closed but a given neuron is firing, what do other members of the orchestra do? What would the conductor hear? We followed the same procedure here as we did in the evoked regime. Namely, we again evaluated the similarity between every instantaneous spatial pattern of population activity (obtained in the absence of stimulation) and the neuron's PCS. Note that the PCS in fact was derived from the evoked regime, and hence the procedure we use here directly tests if there is a tight link between spontaneous activity and the functional architecture as measured in the evoked regime. A sample trace of the resulting correlation coefficients time course is shown in figure 9-8A. When we look at the distribution of the correlation coefficients over many spontaneous activity recording sessions, that is, draw the resulting histogram (figure 9-8B), we see that this time there is no bias or tail in any direction. We then repeated what we did before and took only a subset of the frames, namely the frames that coincided with action potentials. The spikes that occurred during the trace shown in figure 9-8A are drawn below it in figure 9-8C. We observed a tendency for these spikes to occur during periods of high correlation coefficients. The histogram of frames that correspond only to spike times is shown in figure 9-8D and indeed quantifies well this observation. We see that in the subset there is a definite bias towards positive correlation coefficients. In fact, 80 percent of spontaneous spikes occurred when the instantaneous state of population activity was positively correlated with the neuron's PCS (a number similar to that obtained in the evoked regime). This means that even in the absence of a stimulus, the neuron still tended to fire when the instantaneous cortical activity pattern was most similar to the neuron's PCS. To determine when does the neuron tend to fire spontaneously, we once again computed the probability for the neuron to fire given a particular correlation coefficient, by dividing the two histograms. We found that also the resulting function (figure 9-8F) is surprisingly similar to the one obtained when an optimal stimulus was used (figure 9-5F), namely a monotonically increasing function past a certain threshold. To try and predict the observed spike train we then used these computed values, as before. The observed (figure 9-8C) and predicted (figure 9-8E) spike trains were similar; the bursts concurred well with the upswings in the value of the correlation between these two patterns, which indicates the similarity between the spontaneous patterns and the PCS. Thus, the spontaneous activity of the neuron could be reconstructed to a large extent form the time course of the similarity between the spontaneous population activity and the neuron's PCS. It is important to emphasize that the PCS of the neuron was computed from an evoked recording session.

These results imply that the spatial pattern revealed by computing the PCS of a neuron and the spontaneous cortical state obtained by averaging all single frames that coincided with spontaneous action potentials should be similar. Figure 9-9A and figure 9-9B illustrate one case where these two spatial patterns are indeed very similar. Since the PCS obtained during activation

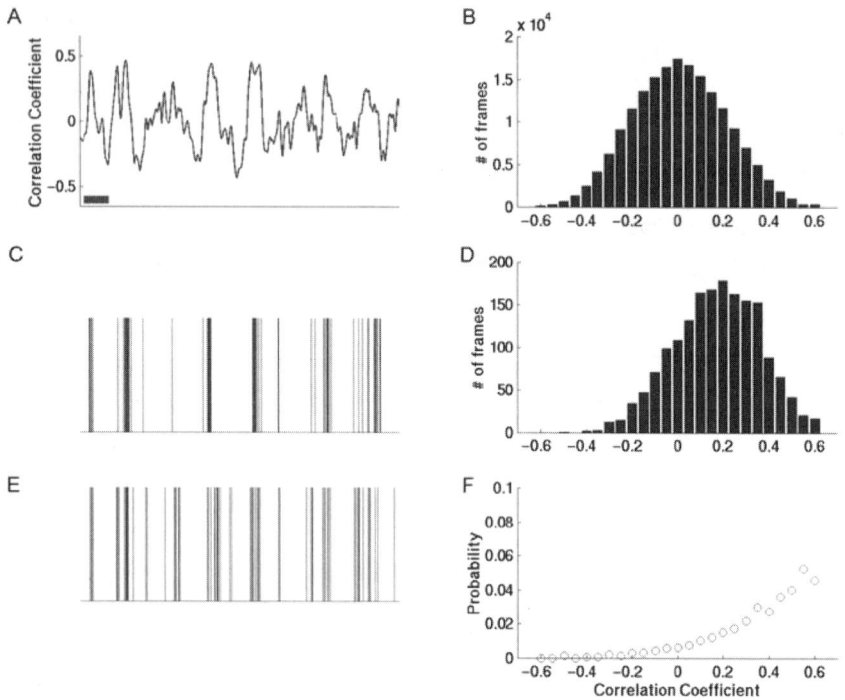

Figure 9-8. Predicting the spontaneous single neuron's spike train from patterns of ongoing population activity and its evoked PCS. (A) The time course of the correlation coefficients in a time window from a spontaneous recording session. (B) The histogram of distribution of correlation coefficients over all spontaneous recording sessions from five hemispheres. This time, unlike the histograms from evoked recording sessions, the histogram has no bias. Indeed, we expect none, since we do not expect any cortical state to be preferred over other cortical states in the absence of stimulation. (C) Spikes that occurred in the same window as plotted in (A). (D) The subset of the histogram plotted in (B), which includes only frames that coincide with spike times. We see that once again spikes occurred mostly when the instantaneous pattern of population activity was similar to the PCS. This time the result is most surprising, because it means this state, which is usually associated with an evoked map, occurred in the absence of stimulation. (E) A prediction of the observed spike train of the neuron, from the similarity of the instantaneous ongoing pattern of population activity to the PCS. (F) We compute the probability for a neuron to fire an action potential at any given value of the correlation coefficient by dividing the bottom histogram (D) by the top one (B). This yields a steadily increasing function past a certain threshold. At low values of correlation coefficients, the firing rate is practically zero, and above a certain value it increases monotonically. This graph implies that a neuron's instantaneous firing rate is higher the more similar the instantaneous population activity pattern is to its functional architecture. Thus, this result is true even in the absence of stimulation. Modified from Tsodyks et al. (1999).

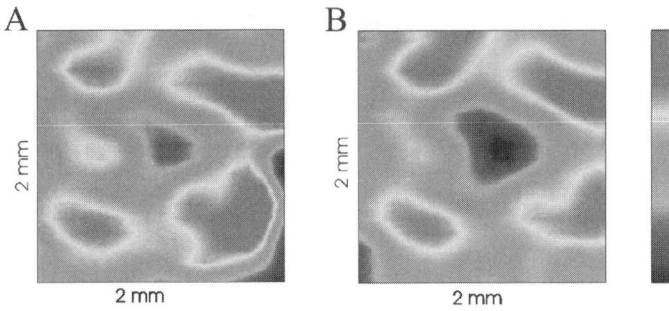

Figure 9-9. Evoked versus spontaneous cortical states. See color insert.

is similar to the functional architecture (figure 9-5), these results indicate that the spontaneous activity of cortical neurons depends on the underlying functional architecture related to their tuning properties (Tsodyks et al. 1999). The fact that the same cortical states appeared during both spontaneous and evoked activity of the neurons where the evoked spiking activity is connected to a specific cortical spatial pattern of activity even when the stimulus is not optimal for the neuron (figure 9-7) also suggests that even in the presence of sensory inputs the response of cortical neurons is strongly affected by the instantaneous cortical states and is not only merely a direct reflection of the input.

Soloists or Obedient Players?

The fact that nearly the same population state occurs during both spontaneous and evoked activity of a cortical neuron suggests that in the absence of stimulation the cortical network wanders through various states represented by coherent firing of different neuronal assemblies (see Kenet et al. 2003). When the network activity happens to be in a particular state, neurons with a preference for this state will have the highest firing rate. When a stimulus is presented, it will quickly push the network from whichever state it was in into the neuron's PCS, which in turn represents the stimulus. The similarity between the underlying functional architecture and the neuron's PCS suggests not only a significant degree of coherent activity in a given cortical orientation column at the recording site but also coherent activation of distant functional domains with similar tuning properties over a large cortical area. This synchronicity during spontaneous activity is presumably mediated by the long-range horizontal cortical connections (Gilbert 1993; Malach et al. 1993; Fitzpatrick 1996; Singer 1994). The results presented here also suggest that orientation domains tuned to different orientations will show a lower level of spike synchronicity during ongoing activity. In the absence of a stimulus, the transitions between cortical states may be affected by feedback inhibition (Sillito et al. 1980; Tsodyks and Sejnowski 1995; Ben-Yishai, Hansel, and Sompolinsky 1997), the dynamic properties of cortical connections (Thomson and Deuchars 1994; Tsodyks and Markram 1997; Abbott et al. 1997), or spike frequency adaptation (Connors and Gutnick 1990; Ahmed et al. 1998; Hansel and Sompolinsky 1998).

It is important to comment on the feasibility of demonstrating that such cortical states occur also in the awake animal and do serve as cortical representation. We know that the technical difficulties involved in such an experiment are immensely greater (Grinvald et al. 1999) and therefore it is likely that in the awake animal these states will not be as readily observable unless very extensive averaging can be done. While in the anesthetized animal there is high cortical synchronicity in the ongoing activity, indicating that large groups of neurons are coherently active, in the awake animal the degree of synchronicity is much lower, indicating that each coherent assembly contains a much smaller number of neurons and/or that many more independent assemblies perform many more computational tasks. Furthermore, because in the awake state the activity in higher cortical area is far larger and hence the feedback far stronger, such feedback from higher brain areas (Felleman and Van Essen 1991; Salin and Bullier 1995; Ullman 1995) should affect the cortical states in lower areas greatly and subdivide the neuronal assemblies even further. Thus in the awake state a neuron's PCS may embody many dynamic representations of higher cortical processing by a smaller groups of coherently active neurons, rather than mostly the representations of orientations which we observed in the anesthetized cat. Therefore, distinguishing between different states is likely to become a formidable task requiring much more averaging. For example, if in the awake cat there are only tenfold more coherent assemblies, then the recording duration should be 100-fold longer to be able to visualize the neurons' PCSs.

One may also wonder what cortical states may look like in other, higher, cortical areas. Areas seventeen and eighteen are the lowest cortical areas in visual processing, and hence they are, luckily for us, relatively easy to map and associate cortical states with. It is likely that as the cortical hierarchy scale is climbed, the degree of complexity of the definition of a state will rise along with the complexity of the parameters represented in that area. States, and hence neurons with individual preferred cortical states, do exist also in higher areas (Tanaka 2000). Yet, as with the awake animal, the degree of their complexity makes their detection a challenging task for future work.

To sum up, our results indicate that the spontaneous firing of single neurons is tightly linked to the cortical networks in which they are embedded. The idea of a network is a central concept in theoretical brain research (e.g., Amit 1989) and it is now finally possible to directly visualize the cortical networks and their states in action, at high spatiotemporal resolution. It appears that in addition to the fruitful studies of single neurons, it is also important to study their behavior in the context of the entire neuronal network. Exploring cortical states is likely to reveal new and fundamental principles about neural strategies for cortical processing and cortical representations of objects, memories, context, or expectations, among other things. Particularly rewarding may be the exploration of the interplay between dynamically switching internal cortical states and the representation of sensory input in primary sensory areas.

References

Abbott L. F., Varela J. A., Sen K., and Nelson S. B. 1997. "Synaptic depression and cortical gain control," *Science* 275:220–224.

Ahmed B., Anderson J. C., Douglas R. J., Martin K. A., and Whitteridge D. 1998. "Estimates of the net excitatory currents evoked by visual stimulation of identified neurons in cat visual cortex," *Cereb. Cortex* 8:462–476.

Amit D. J. and Brunel N. 1997. "Model of global spontaneous activity and local structured activity during delay periods in the cerebral cortex," *Cereb. Cortex* 7:237–252.

Arieli A., Shoham D., Hildesheim R., and Grinvald. A. 1995. "Coherent spatiotemporal pattern of ongoing activity revealed by real time optical imaging coupled with single unit recording in the cat visual cortex," *J. Neurophysiol.* 73: 2072–2093.

Arieli A., Sterkin A., Grinvald A., and Aertsen A. 1996a. "Dynamics of ongoing activity: Explanation of the large in variability in evoked cortical responses," *Science* 273:1868–1871.

Arieli A., Donchin O., Aertsen A., Bergman H., Gribova A., Grinvald A., and Vaadia E. 1996b. "The impact of ongoing cortical activity on evoked potentials and behavioral responses in the awake behaving monkey," *Soc. Neurosci.*, 22:2022.

Ben-Yishai R., Hansel D., and Sompolinsky H. 1997. "Traveling waves and the processing of weakly tuned inputs in a cortical network module," *J Comput Neurosci.* 4:57–77.

Bonhoeffer T. and Grinvald A. 1991. "Iso-orientation domains in cat visual cortex are arranged in pinwheel-like patterns," *Nature* 353:429–431.

Bosking W. H., Zhang Y., Schofield B., and Fitzpatrick D. 1997. "Orientation selectivity and the arrangement of horizontal connections in tree shrew striate cortex," *J. Neurosci.* 17:2112–2127.

Connors B. W. and Gutnick M. J. 1990. "Intrinsic firing patterns of diverse neocortical neurons," *Trends Neurosci.* 13:99–104.

Felleman J. and Van Essen D. C. 1991. "Distributed hierarchical processing in the primate cerebral cortex," *Cereb. Cortex* 1:1–47.

Fitzpatrick D. 1996. "The functional organization of local circuits in visual cortex: insights from the study of tree shrew striate cortex," *Cereb. Cortex* 6:329–341.

Gilbert C. D. 1993. "Circuitry, architecture, and functional dynamics of visual cortex," *Cereb. Cortex* 3:373–386.

Grinvald A. 1984. "Real time optical imaging of neuronal activity: from single growth cones to the intact brain," *TINS* 7:143–150.

Grinvald A., Cohen L. B., Lesher S., and Boyle M. B. 1981. "Simultaneous optical monitoring of activity of many neurons in invertebrate ganglia, using a 124 element 'Photodiode' array," *J. Neurophysiol.* 45:829–840.

Grinvald A., Hildesheim R., Farber I. C., and Anglister L. 1982. "Improved fluorescent probes for the measurement of rapid changes in membrane potential," *Biophys. J.* 39:301–308.

Grinvald A., Manker A., and Segal M. 1982. "Visualization of the spread of electrical activity in rat hippocampal slices by voltage sensitive optical probes," *J. Physiol.* 333:269–291.

Grinvald A., Salzberg B. M., and Cohen L. B. 1977. "Simultaneous recordings from several neurons in an invertebrate central nervous system," *Nature* 268:140–142.

Grinvald A., Shoham D., Bonhoeffer T., Ts'o D., Frostig R. D., Shmuel A., Glaser D., Vanzetta I., Shtoyerman E., Sterkin A., Slovin H., Wijnbergen C., Hildesheim R., and Arieli A. 1999. *Modern Techniques in Neuroscience Research*, chapter 16, W. Windherst and H. Johansson, Eds. (Berlin: Spinger-Verlag).

Hansel D. and Sompolinsky H. 1998. *Methods in Neuronal Modeling*, chapter 13, C. Koch and I. Segev, Eds. (Cambridge, Mass.: MIT Press).

Hebb D. O. 1949. *The Organization of Behaviour* (New York: Wiley).

Hubel D. H., Wiesel T. N., and Stryker M. P. 1977. "Orientation columns in macaque monkey visual cortex demonstrated by the 2-deoxyglucose autoradiographic technique," *Nature* 269:328–330.

Hubel D. H. and Wiesel T. N. 1962. "Receptive fields, binocular interactions and functional architecture in the cat's visual cortex," *J. Physiol.* 160:106–154.

Kenet T., Bibitchkov D., Tsodyks M., Grinvald A., and Arieli A. 2003 "Spontaneously emerging cortical representations of visual attributes," *Nature* 425:954–956.

Konnerth A. and Orkand R. K. 1986. "Voltage sensitive dyes measurpotential changes in axons and glia of frog optic nerve," *Neuroscience Lett.* 66:49–54.

Lev-Ram R. and A. Grinvald. 1986. "K^+ and Ca^{2+} dependent communication between myelinated axons and oligodendrocytes revealed by voltage-sensitive dyes," *Proc. Natl. Acad. Sci. USA* 83:6651–6655.

Malach R., Amir Y., Harel M., and Grinvald A. 1993. "Relationship between intrinsic connections and functional architecture revealed by optical imaging and *in vivo* targeted biocytin injections in primate striate cortex," *Proc. Natl. Acad. Sci. USA* 90:10469–10473.

Malach R., Tootell R. B. H., and Malonek D. 1994. "Relationship between orientation Domains, Cytochrome Oxidase Stipes and intrinsic horizontal connections in Squirrel monkey area V2," *Cereb. Cortex* 4:151–165.

Orbach H. S., Cohen L. B., and Grinvald A. 1985. "Optical mapping of electrical activity in rat somatosensory and visual cortex," *J. Neurosci.* 5:1886–1895.

Salin P. A. and Bullier J. 1995. "Corticocortical connections in the visual system: structure and function," *J. Physiol. Rev.* 75:107–540.

Salzberg B. M., Davila H. V., and Cohen L. B. 1973. "Optical recording of impulses in individual neurons of an invertebrate central nervous system," *Nature* 246:508–509.

Salzberg B. M., Grinvald A., Cohen L. B., Davila, H. V., and Ross W. N. 1977. "Optical recording of neuronal activity in an invertebrate central nervous system; simultaneous recording from several neurons," *J. Neurophys.* 40: 1281–1291.

Shadlen M. N. and Newsome W. T. 1994. "Noise, neural codes and cortical organization," *Curr. Opin. Neurobiol.* 4:569–579.

Shmuel A. and Grinvald A. 1996. "Relationships between functional organization for direction of motion and for orientation selectivity in cat area 18," *J. Neurosci.* 16:6945–6964.

Shoham D., Glaser D. E., Arieli A., Kenet T., Wijnbergen C., Toledo Y., Hildesheim R., and Grinvald A. 1999. "Imaging cortical dynamics at high spatial and temporal resolution with novel blue voltage-sensitive dyes," *Neuron* 24:791–802.

Sillito A. M., Kemp J. A., Milson J. A., and Berardi N. 1980. "A re-evaluation of the mechanisms underlying simple cell orientation selectivity" *Brain Res.* 194:517–520.

Singer W. 1994. "Coherence as an organizing principle of cortical functions," *Int. Rev. Neurobiol.* 37:153–183.

Softky W. R. and Koch C. 1993. "The highly irregular firing of cortical cells is inconsistent with temporal integration of random EPSPs." *J. Neurosci.* 13: 334–350.

Tanaka K. 2000. "Mechanisms of visual object recognition studied in monkeys," *Spat. Vis.* 13:147–163.

Thomson M. and Deuchars J. 1994. "Temporal and spatial properties of local circuits in neocortex," *TINS* 17:119–126.

Tsodyks M. and Markram H. 1997. "The neural code between neocortical pyramidal neurons depends on neurotransmitter release probability," *Proc. Natl. Acad. Sci.* 94:719–723.

Tsodyks M., Kenet T., Grinvald A., and Arieli A. 1999. "Linking spontaneous activity of single cortical neurons and the underlying functional architecture," *Science* 286:1943–1946.

Tsodyks M., Sejnowski T. 1995. "Rapid switching in Balanced Cortical Network Models," *Network* 6:1–141

Ullman S. 1995. "Sequence seeking and counter streams: a computational model for bidirectional information flow in the visual cortex," *Cereb. Cortex* 5:1–11.

Van Vreeswijk C. and Sompolinsky H. 1996. "Chaos in neuronal networks with balanced excitatory and inhibitory activity," *Science* 274:1724–1726.

10

What Is the Other 85 Percent of V1 Doing?

Bruno A. Olshausen and David J. Field

Introduction

The primary visual cortex (area V1) of mammals has been the subject of intense study for at least four decades. Hubel and Wiesel's original studies in the early 1960s created a paradigm shift by demonstrating that the responses of single neurons in the cortex could be tied to distinct image properties such as the local orientation of contrast. Since that time, the study of V1 has become something of a miniature industry, to the point where the annual Society for Neuroscience meeting now routinely devotes multiple sessions entirely to V1 anatomy and physiology. Given the magnitude of these efforts, one might reasonably expect that we would by now have a fairly concrete grasp of how V1 works and its role in visual system function. However, as we shall argue here, there still remains so much unknown that, for all practical purposes, we stand today at the edge of the same dark abyss as did Hubel and Wiesel forty years ago.[1]

It may seem surprising to some that we should take such a stance. V1 does, after all, have a seemingly ordered appearance—a clear topographic map and an orderly arrangement of ocular dominance and orientation columns. Many neurons are demonstrably tuned for stimulus features such as orientation, spatial frequency, color, direction of motion, and disparity. And there has now even emerged a fairly well agreed upon standard model for V1, in which simple cells compute a linearly weighted sum of the input over space and time (usually a Gabor-like function) and the output is passed through a pointwise nonlinearity, in addition to being subject to contrast gain control to avoid response saturation (figure 10-1). Complex cells are similarly explained in terms of summing the outputs of a local pool of simple cells with similar tuning properties but different positions or phases. The net result is to think of V1 roughly as a

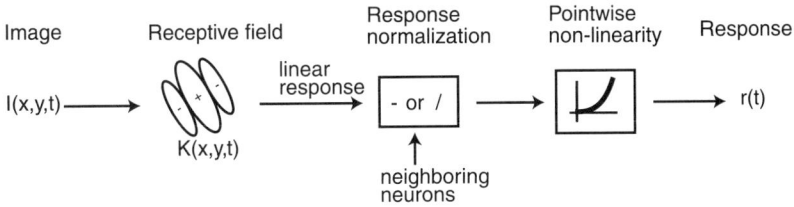

| Image | Receptive field | Response normalization | Pointwise non-linearity | Response |

Figure 10-1.   Standard model of V1 simple cell responses. The neuron computes a weighted sum of the image over space and time, and this result is normalized by the responses of neighboring units and passed through a pointwise nonlinearity (see, e.g., Carandini, Heeger, and Movshon 1997).

"Gabor filter bank." There are now many papers showing that this basic model fits much of the existing data well, and many scientists have come to accept this as a working model of V1 function (see, e.g., Lennie 2003a).

But behind this picture of apparent orderliness, there lies an abundance of unexplained phenomena, a growing list of untidy findings, and an increasingly uncomfortable feeling among many about how the experiments that have led to our current view of V1 were conducted in the first place. The main problem stems from the fact that cortical neurons are highly nonlinear. They emit all-or-nothing action potentials, not analog values. They also adapt, so their response properties depend upon the history of activity. Cortical pyramidal cells have highly elaborate dendritic trees, and realistic biophysical models suggest that each thin branch could act as a nonlinear subunit, so that any one neuron could be computing many different nonlinear combinations of its inputs (Hausser and Mel 2003), in addition to being sensitive to coincidences (Softky and Koch 1993; Azouz and Gray 2000, 2003). Everyone knows that neurons are nonlinear, but few have acknowledged the implications for studying cortical function. Unlike linear systems, where there exist mathematically tractable, textbook methods for system identification, nonlinear systems can not be teased apart using some straightforward, reductionist approach. In other words, there is no general method for characterizing nonlinear systems.[2]

The reductionist approach has formed the bedrock of V1 physiology for the past four decades. Indeed, it would seem necessary, given the stunning complexity of the brain, to tease apart one chunk at a time. But whether or not the reductionist approach tells you anything useful depends entirely on how you reduce. Some modes of interaction may be crucial to the operation of the system, and so cutting them out—either in theories or experiments—may give a misleading picture of how the system actually works. Obviously, if one knew in advance what the important modes of interaction were then one could choose to reduce appropriately. But when it comes to the brain we really haven't a clue. V1 physiologists have for the most part chosen one particular way to reduce complexity by using highly simplified stimuli and recording from only one neuron at a time, and from this body of experiments has emerged the standard model which forms the basis for our conceptual understanding of V1. But whether or not the physiologists chose correctly is anyone's guess. The best-case scenario is that they did, and that the standard model is more or less correct. The worst-case scenario, which we lean toward, is that they chose inappropriately in many cases, that the standard model is but one small part of the full story, and that we still have much to learn about V1.

In this chapter we lay out the reasons for our skepticism by identifying five fundamental problems with the reductionist approach that have led us to the current view of V1 as a Gabor filter bank. In addition, we attempt to quantitate the level of our current understanding by considering two important factors: an estimate of the fraction of V1 neuron types that have actually been characterized and the fraction of variance explained in the responses of these neurons under natural viewing conditions. Together, these lead us to conclude that at present we can rightfully claim to understand only 10 to 20 percent of how V1 actually operates under normal conditions. Our aim in pointing out these things is not to simply tear down the current framework. Indeed, we ourselves have attempted to account for some aspects of the standard model in terms of efficient coding principles (sparse coding), so obviously we buy into at least part of the story. Rather, our goal is to make room for new theories that we believe are essential for understanding V1 and its relation to perception, and we shall present a few candidates in the second part of this chapter. A central conclusion that emerges from this exercise is that we need to begin seriously studying how V1 behaves using natural scenes. Based on these observations, we will then be in a more informed position when it comes to making choices about how to reduce complexity to tease apart the fundamental components of the system.

Five Problems with the Current View

Biased Sampling of Neurons

The vast majority of our knowledge about V1 function has been obtained from single-unit recordings in which a single microelectrode is brought into close proximity with a neuron in cortex. Ideally, when doing this one would like to obtain an unbiased sample from any given layer of cortex. But some biases are unavoidable. For instance, neurons with large cell bodies will give rise to extracellular action potentials that have larger amplitudes and propagate over larger distances than neurons with small cell bodies. Without careful spike sorting, the smaller extracellular action potentials may easily become lost in the background when in the vicinity of neurons with large extracellular action potentials. This creates a bias in sampling that is not easy to dismiss.

Even when a neuron has been successfully isolated, detailed investigation of the neuron may be bypassed if it does not respond "rationally" to the investigator's stimuli or fit the stereotype of what the experimenter believes the neuron should do. This is especially true for higher visual areas like area V4, but it is also true for V1. Such neurons are commonly regarded as "visually unresponsive." It is difficult to know how often such neurons are encountered because oftentimes either they simply go unreported or it is simply stated that only visually responsive units were used for analysis.

While it is certainly difficult to characterize the information-processing capabilities of a neuron that seems unresponsive, it is still important to know in what way these neurons are unresponsive. What are the statistics of activity? Do they tend to appear bursty or tonic? Do they tend to be encountered in particular layers of cortex? And most importantly, are they merely unresponsive to bars and gratings, or are they also equally uninterpretable in their responses to a wider variety of stimuli, such as natural images? A seasoned experimentalist who has recorded from hundreds of neurons would probably have some feel for these things. But for the many readers not directly involved in collecting the data, there is no way of knowing these unreported aspects of V1 physiology. It is possible that someone may eventually come up with a theory that could account for some of these unresponsive neurons, but this can't happen if no one knows they are there.

Another bias that arises in sampling neurons, perhaps unintentionally, is that the process of hunting for neurons with a single microelectrode will almost invariably steer one towards neurons with higher firing rates. This is especially disturbing in light of recent analyses showing that based on estimates of energy consumption, the average firing rates of neurons in cortex must be rather low—that is, less than 1 Hz (Attwell and Laughlin 2001; Lennie 2003b). On the other hand, one finds many neurons in the literature for which even the spontaneous or background rates are well above 1 Hz, suggesting that they are likely to be substantially overrepresented (Lennie 2003b). What makes matters worse is that if we assume an exponential distribution of firing rates, as has been demonstrated for natural scenes (Baddeley et al. 1997), then a mean firing rate of 1 Hz would yield the distribution shown in figure 10-2a. Under natural conditions, then, only a small fraction of neurons would exhibit the sorts of firing rates normally associated with a robust response. For example, the total probability for firing rates of even 5 Hz and above is 0.007, meaning that one could wait one to two minutes before observing a one-second interval containing five or more spikes. It seems possible that such neurons could either be missed altogether or else purposely bypassed because they do not yield enough spikes for data analysis. For example, the overall mean firing rate of V1 neurons in the Baddeley et al. study was 4.0 Hz (standard deviation 3.6 Hz), suggesting that these neurons constitute a subpopulation that were perhaps easier to find but

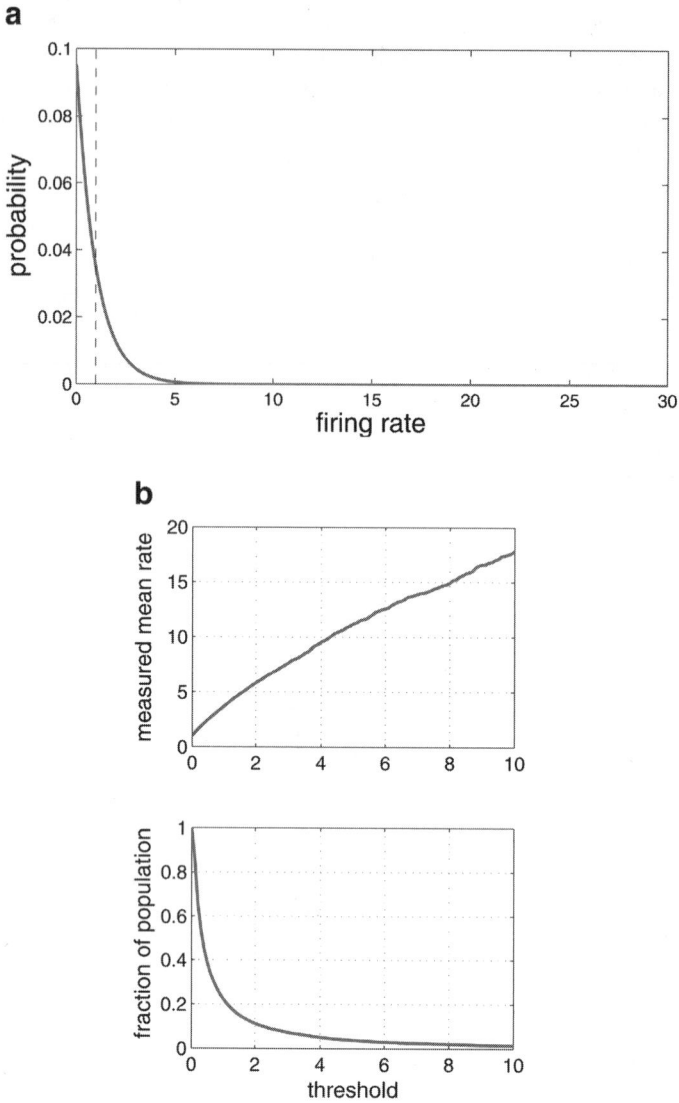

Figure 10-2. (a) Probability distribution of firing rates, assuming an exponential distribution with a mean of 1 Hz (dashed line). (b) Resulting overall mean-rate and fraction of the population captured as a result of recording from neurons only above a given mean firing-rate (threshold).

not necessarily representative of the population as a whole. In fact, the authors point out that even this rate is considered low (which they attribute to anesthesia), as previous studies (Legendy and Salcman 1985) report the mean firing rate to be 8.9 Hz (standard deviation 7.0 Hz).

Presumably V1 contains a heterogeneous population of neurons with different mean firing rates. If we assume some distribution over these rates, then it is possible to obtain an estimate of the fraction of the population characterized and the subsequent observed mean rate, if one were to selectively record from neurons only above a certain mean rate. The result of such an analysis, assuming a log-normal distribution of mean rates so as to yield an overall mean of 1 Hz, is shown in figure 10-2b. As one can see, an overall mean of 4 Hz implies that the selection criterion was somewhere between 1–2 Hz, which would capture less than 20 percent of the population.

Neurophysiological studies of the hippocampus provide an interesting lesson about the sorts of biases introduced by low firing rates. Prior to the use of chronic implants, in which the activity of neurons could be monitored for extended periods while a rat explored its environment, the granule cells of the dentate gyrus were thought to be mostly high rate "theta" cells (e.g., Rose, Diamond, and Lynch 1983). But it eventually became clear that the majority are actually very low-rate cells (Jung and McNaughton 1993), and that for technical reasons only high-rate interneurons were being detected in the earlier studies (Skaggs, personal communication). In fact, Thompson and Best (1989) found that nearly two-thirds of all hippocampal neurons that showed activity under anesthesia became silent in the awake, behaving rat. This overall pattern appears to be upheld in macaque hippocampus, where the use of chronic implants now routinely yields neurons with overall firing rates below 0.1 Hz (Barnes et al. 2003), which is in stark contrast to the "low baseline rates" of 8.1 Hz reported by Wirth et al. (2003) using acutely implanted electrodes.

The dramatic turn of events afforded by the application of chronic implants combined with natural stimuli and behavior in the hippocampus can only make one wonder what mysteries could be unraveled when the same techniques are applied to the visual cortex. What is the natural state of activity during free-viewing of natural scenes? What are the actual average firing rates and other statistics of activity in layers 2 and 3? What are the huge numbers of granule cells in macaque layer 4, which outnumber the geniculate fiber inputs by thirty to one, doing? Do they provide a sparser code than their geniculate counterparts? And what about the distribution of actual receptive field sizes? Current estimates show that most parafoveal neurons in V1 have receptive field sizes on the order of 0.1 degree. But based on retinal anatomy and psychophysical performance one would expect to find a substantial number of neurons with receptive fields an order of magnitude smaller, approximately 0.01 degree (Olshausen and Anderson 1995). Such receptive field sizes are extremely rare, if not nonexistent, in macaque V1 (De Valois, Albrecht, and Thorell 1982; Parker and Hawken 1988).

Overall, then, one can identify at least three different biases in the sampling of neurons: (1) neurons with large cell bodies, (2) "visually responsive" neurons, and (3) neurons with high firing rates. So where does this leave us? If we assume that 5 to 10 percent of neurons are missed because they have weak extracellular action potentials, another 5 to 10 percent are discarded because they are not visually responsive, and 40 to 50 percent are missed because of low firing rates (as demonstrated in figure 10-2), then even allowing for some overlap among these populations would yield the generous estimate that 40 percent of the population has actually been characterized.

Biased Stimuli

Much of our current knowledge of V1 neural response properties is derived from experiments using reduced stimuli. Oftentimes these stimuli are ideal for characterizing linear systems—such as spots, white noise, or sine wave gratings—or else they are designed around preexisting notions of how neurons should respond. The hope is that the insights gained from studying neurons using these reduced stimuli will generalize to more complex situations—for example, natural scenes. But, of course, there is no guarantee that this is the case. And given the nonlinearities inherent in neural responses, we have every reason to be skeptical.

Sine wave gratings are ubiquitous tools in visual system neurophysiology and psychophysics. In fact, the demand for using these stimuli is so high that some companies produce lab equipment with specialized routines designed for this purpose (e.g., Cambridge Research Systems). But sine waves are special only because they are eigenfunctions of linear, time- or space-invariant systems. For nonlinear systems, they bear no particular meaning, nor do they occupy any special status. In the auditory domain, sine waves could be justified from the standpoint that many natural sounds are produced by oscillating membranes. However, in the visual world there are few things that naturally oscillate either spatially or temporally. One is thus led to the unavoidable conclusion that there is no principled reason for using sine waves to study vision.

White noise, M-sequences, and spots suffer from similar problems. They are informative stimuli only to the extent that the system is linear. Otherwise, they are no more valid than any other stimulus. Although it is true that an orthonormal basis (which could comprise any of these stimuli) can fully describe any pattern, characterizing the responses to each basis function in isolation is pointless when the system is nonlinear.

What about bars of light or Gabor patches? The use of these stimuli also makes some assumptions about linearity. However, this approach primarily assumes that neurons are selective or tuned to localized, oriented, bandpass structure and that the appropriate parameters for characterizing them are properties such as position, length and width, orientation, spatial-frequency, and so on. This may seem reasonable given the fact that images also contain localized, oriented structure (i.e., edges), but how do we really know this is the right choice?

The use of reduced stimuli is sometimes justified by the fact that one would actually like to know how a neuron responds to a single point of light in an image, or to a specific spatial-frequency. In this case, one does need to construct a controlled stimulus. But we would argue that such questions are misplaced to begin with. Given the nonlinearities inherent in real neurons, there is every reason to believe that neurons are selective to certain combinations of pixel values or spatial-frequencies, and so probing the system's response to one element at a time will not necessarily tell you how it responds to particular combinations. Of course, we will never know this until it is tested, and that's precisely the problem—the central assumption of the elementwise, reductionist approach has yet to be thoroughly tested.

The brute force solution would be to exhaustively search the stimulus space. However, even an 8×8 patch with 6 bits of gray level requires searching $2^{384} > 10^{100}$ possible combinations. Needless to say, this is far beyond what any experimental method could explore. Therefore, the hope is that the nonlinearities are smooth enough to allow predictions from a smaller set of stimuli.

We believe that the solution to these problems is to turn to natural scenes. Our intuitions for how to reduce stimuli should be guided by the sorts of structure that occur in natural scenes, not arbitrary mathematical functions or stimuli that are simple to think about or happen to be easy to generate on a monitor. If neurons are selective to specific combinations of stimuli, then we will need to explore their responses to the sorts of combinations that occur in natural scenes. And at the same time, we will need to put more effort into mathematically characterizing the structure of natural scenes to better understand the forms of structure contained in them. No matter what stimuli one uses to characterize the response of a neuron, the true test that the characterization is correct is to demonstrate that one can predict the neuron's behavior in ecological conditions.

Biased Theories

Currently in neuroscience there is an emphasis on "telling a story." This often encourages investigators to demonstrate when a theory explains data, not when a theory provides a poor model. We therefore have theories of grating responses and line weighting functions which may predict grating responses and line responses with some success, but which appear to provide poor predictions for natural scene data (as noted in the section "Ecological Deviance" below). In addition, editorial pressures often encourage one to make a tidy picture out of data that may actually be quite messy. This of course runs the risk of forcing a picture that does not actually exist. Theories then emerge that are centered around explaining a particular subset of published data or which can be conveniently proven, rather than being motivated by functional considerations—that is, how does this help the brain to solve difficult problems in vision?

Hubel and Wiesel introduced the classification of neurons into categories of simple, complex, and hypercomplex or end-stopped based on their investigations using stimuli composed largely of bars of light. Simple cells are noted for having oriented receptive fields organized into explicit excitatory and inhibitory subfields, whereas complex cells are tuned for orientation but are more relatively insensitive to position and the sign of contrast (black-white edge vs. white-black edge). Hypercomplex cells display more complex shape selectivity and some appear most responsive to short bars or the terminations of bars of light (so-called end-stopping). Are these categories real or a result of the way neurons were stimulated and the data analyzed?

A widely accepted theory that accounts for the distinction between simple and complex cells is that simple cells compute a (mostly linear) weighted sum of image pixels, whereas complex cells compute a sum of the squared and half-rectified outputs of simple cells of the same orientation—that is, the so-called energy model (Adelson and Bergen 1985). This theory is consistent with measurements of response modulation in response to drifting sine wave gratings, otherwise known as the "F1/F0 ratio" (Skottun et al. 1991). From this measure one finds clear evidence for a bimodal distribution of neurons, with simple cells having ratios greater than one, and complex cells having ratios less than one. Recently, however, it has been shown that this particular nonlinear measure tends to exaggerate or even introduce bimodality rather than revealing an actual, intrinsic property of the data (Mechler and Ringach 2002; Priebe et al. 2004). When receptive fields are characterized instead by the degree of overlap between zones activated by increments or decrements in contrast, one still obtains a bimodal distribution, in addition to a clear separation according to layer, but the F1/F0 ratio does a poor job at predicting which cells are complex according to the overlap criterion (Kagan, Gur, and Snodderly 2002). In addition, the energy model of complex cells does a poor job accounting for complex cells with a partial overlap of activating zones. Thus, the way in which response properties are characterized can have a profound effect on the resulting theoretical framework that is adopted to explain the results.

The notion of end-stopped neurons introduces even more questions when one considers the structure of natural images. Most natural scenes are not littered with line terminations or short bars—see for example figure 10-3 *left*. Indeed, at the scale of a V1 receptive field, the structures in this image are quite complex and they defy the simple, line drawing–like characterization of a "blocks world." Where in such an image would one expect an end-stopped neuron to fire? By asking this question, one could possibly be led to a more ecologically relevant theory of these neurons than suggested by simple laboratory stimuli.

Another theory bias often embedded in investigations of V1 function is the notion that simple cells, complex cells, and hypercomplex cells are actually coding for the presence of edges, corners, or other two-dimensional shape features in images. But years of research in computer vision have now

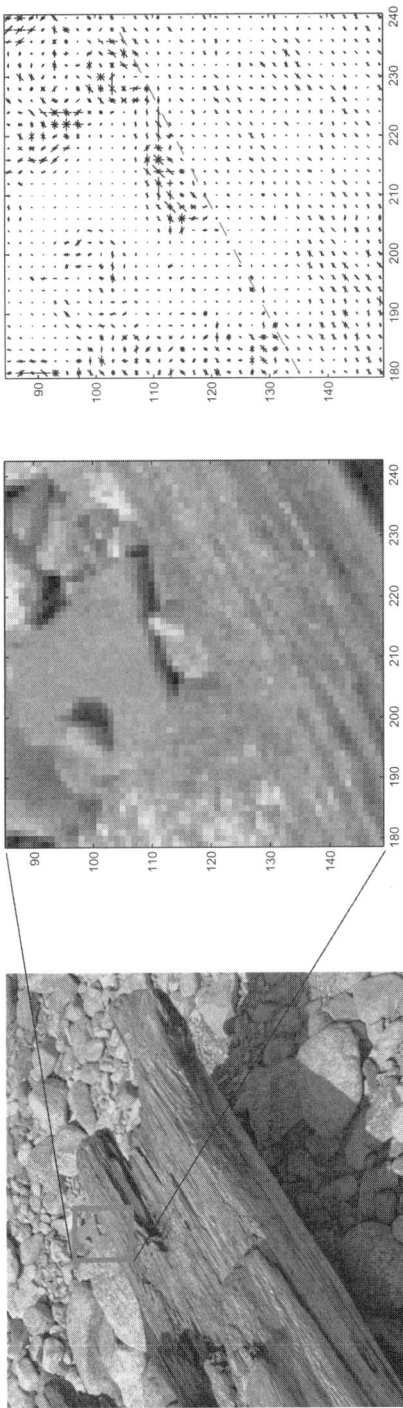

Figure 10-3. A natural scene (*left*), and an expanded section of it (*middle*). Far right shows the information conveyed by an array of complex cells at four different orientations. The length of each line indicates the strength of response of a model complex cell at that location and orientation. The dashed line shows the location of the boundary of the log in the original image.

conclusively demonstrated that it is impossible to compute the presence even of simple edges in an image in a purely bottom-up fashion (i.e., a filter such as a simple or complex cell model). As an example, figure 10-3 demonstrates the result of processing a natural scene with the standard energy-model of a complex cell. Far from making contours explicit, this representation creates a confusing array of activity from which it would be quite difficult to make sense of what is going on. Our perception of crisp contours, corners, and junctions in images is largely a post hoc phenomenon that is the result of massive inferential computations performed by the cortex. In this sense our initial introspections about scene structure may actually be a poor guide as to the actual problems faced by the cortex.

In order to properly understand V1 function, we will need to go beyond bottom-up filtering models and think about the "priors" used by V1 or fed back from higher areas for interpreting images (Olshausen 2003; Lee and Mumford 2003). Our theories need to be guided by functional considerations and an appreciation for the ambiguities contained in natural images rather than appealing to simplistic notions of feature detection that are suggested by a select population of recorded neurons using reduced stimuli.

Interdependence and Contextual Effects

It has been estimated that roughly 5 percent of the excitatory input in layer 4 of V1 arises from the lateral geniculate nucleus (LGN), with the majority resulting from intracortical input (Peters and Payne 1993; Peters, Payne, and Budd 1994). Thalamocortical synapses have been found to be stronger, making them more likely to be effective physiologically (Ahmed et al. 1994). Nevertheless, based on visually invoked membrane potentials, Chung and Ferster (1998) have argued that the geniculate input is responsible for just 35 percent of a layer-4 neuron's response. This leaves 65 percent of the response determined by factors outside of the direct feedforward input. Using optical imaging methods, Arieli et al. (1996) show that the population ongoing activity can account for 80 percent of an individual V1 neuron's response variance. Thus, we are left with the real possibility that somewhere between 60 and 80 percent of the response of a V1 neuron is a function of other V1 neurons or inputs other than those arising from LGN.

It should also be noted that recent evidence from the early blind has demonstrated that primary visual cortex has the potential for a wide range of multimodal input. Sadato et al. (1996) and Amedi et al (2003) demonstrated that both tactile Braille reading and verbal material can activate the visual cortex in those who have been blind from an early age, even though no such activation occurs in those with normal sight. This implies that in the normal visual system, the primary visual cortex has the potential for interactions with quite high-level sources of information.

That V1 neurons are influenced by context—that is, the spatiotemporal structure outside of the classical receptive field (CRF)—is by now well known and has been the subject of many investigations over the past decade. Knierim and Van Essen (1992) showed that many V1 neurons are suppressed by a field of oriented bars outside the classical receptive field of the same orientation, and Sillito et al. (1995) have shown that one can introduce quite dramatic changes in orientation tuning based on the orientation of gratings outside the CRF (see figure 10-4). But these investigations have likely tapped only a portion of the interdependencies and contextual effects that actually exist.

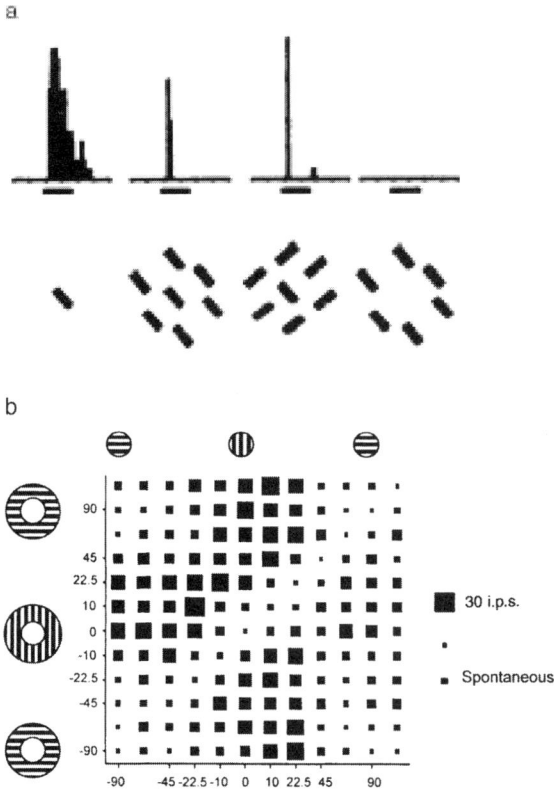

Figure 10-4. (a) Knierim and Van Essen (1992) showed that many neurons in V1 of the alert macaque monkey are inhibited by stimuli outside their classical receptive field, even though these stimuli by themselves elicit little or no response from the neuron. Shown for each configuration is the peristimulus time histogram of the neuron's response (bar denotes stimulus duration, which was 500 milliseconds). (b) Sillito et al. (1995) showed that the presence of a grating stimulus outside the classical receptive field (CRF) could have a profound influence on the orientation tuning to a grating inside the CRF. Shown are the responses—denoted by the size of the square—to different combinations of grating orientation in the center and surround.

The problem in teasing apart contextual effects in a reductionist fashion is that one faces a combinatorial explosion in the possible spatial/featural configurations of surrounding stimuli such as bars or gratings. What we really want to know is how neurons respond within the sorts of context encountered in natural scenes. For example, given the results of Knierim and Van Essen (1992), or Sillito et al. (1995), what should we reasonably expect to result from the sorts of context seen in the natural scene of figure 10-3? Indeed, it is not even clear whether one can answer the question since the contextual structure here is so much richer and more diverse than that which has been explored experimentally. Some of the initial studies exploring the role of context in natural scenes have demonstrated pronounced nonlinear effects that tend to sparsify activity in a way that would have been hard to predict from the existing reductionist studies (Vinje and Gallant 2000). More studies along these lines are needed, and most importantly, we need to understand how context is doing this.

Another striking form of interdependence exhibited by V1 neurons is in the synchrony of activity. Indeed, the fact that one can even measure large-scale signals such as the local field potential or electroencephalogram (EEG) implies that large numbers of neurons must be acting together. Gray et al. (1989) demonstrated gamma-band synchronization between neurons in cat V1 when bars moved through their receptive fields in similar directions, suggesting that synchrony is connected to a binding process. More recently, Worgotter et al. (1998) have shown that receptive field sizes change significantly with the degree of synchrony exhibited in the EEG, and Maldonado et al. (2004) have shown that periods of synchronization preferentially occur during periods of fixation as opposed to during saccades of drifts. However, what role synchrony plays in the normal operation of V1 neurons is entirely unclear, and it is fair to say that this aspect of response variance remains a mystery.

Ecological Deviance

Publishing findings only in conditions when a particular model works would be poor science. It is important to know not only where the current models can successfully predict neural behavior, but also under what conditions they break down and why. And as we have emphasized above, it is most important to know how they fare under ecological conditions. If the current models fail to predict neural responses under such conditions, then the literature should reflect this.

The Gallant lab at UC Berkeley has for the past several years attempted to see how well one can predict the responses of V1 neurons to natural stimuli using a variety of different models. However, assessing how well these models fare and what this implies about our current understanding of V1 is difficult for two reasons. One is that responses to repeated trials vary, and so if one wishes to attribute the variability to noise, then one must measure the intertrial variability and discount this from the unexplained variance. The other problem is that whatever model is chosen, one is always subject to the criticism that it is not sufficiently elaborate, and thus any inability to predict the neuron's response is simply due to some missing element in the model.

For example, David, Vinje, and Gallant (1999) have explored two different types of models—a linear model in which the neurons response is a weighted sum of the image pixels, and a "Fourier power" model in which the neuron's response is a weighted sum of the local power spectrum (which is capable of capturing the position/phase invariance nonlinearity of a complex cell). After correcting for intertrial variability, which is only approximate due to the limited number of trials, these models can explain approximately 20 to 30 percent of the response variance. It is possible that with more trials and with the addition of other nonlinearities such as contrast normalization, adaptation, and response saturation the fraction of variance explained would rise considerably. However, our own view is that these are well-established models that have been given a fair run for their money, and the addition of simple response nonlinearities such as these is unlikely to improve matters much. Moreover, we would contend that one can not easily dismiss inter-trial variability as "noise"—that is, it could well be due to internally generated activity that plays an important role in information processing that we simply do not as yet understand (Arieli et al. 1996; see also the section "Dynamical Systems and the Limits of Prediction"). Given these results with both linear and Fourier power models, we conjecture the best-case scenario is that the percentage of variance explained is likely to asymptote at 30 to 40 percent with the standard model.

One of the reasons for our pessimism is due to the way in which these models fail. For example, figure 10-5 shows data collected from the Gray lab at Montana State University, Bozeman, in which the responses of V1 neurons to repeated presentations of a natural movie are recorded using tetrodes. Shown (figure 10-5a) is the peristimulus time histogram of one neuron in addition to the predicted response generated by convolving the neuron's space-time receptive field (obtained from an M-sequence) with the movie. Needless to say, the fit is not good despite the fact that the receptive field as measured from the M-sequence kernel is a "normal," Gabor-like function that translates over time (i.e., space-time inseparable). Moreover, the responses of neighboring neurons having similar receptive fields are extremely heterogeneous, much more so than expected from a bank of similarly tuned filters (figure 10-5b). It quickly becomes clear from looking at these and many other such neurons that adding further pointwise nonlinearities or simple network nonlinearities such as contrast normalization is going to be of limited use. What seems to be suggested by this behavior is a complex network nonlinearity that involves fairly specific interactions among neurons.

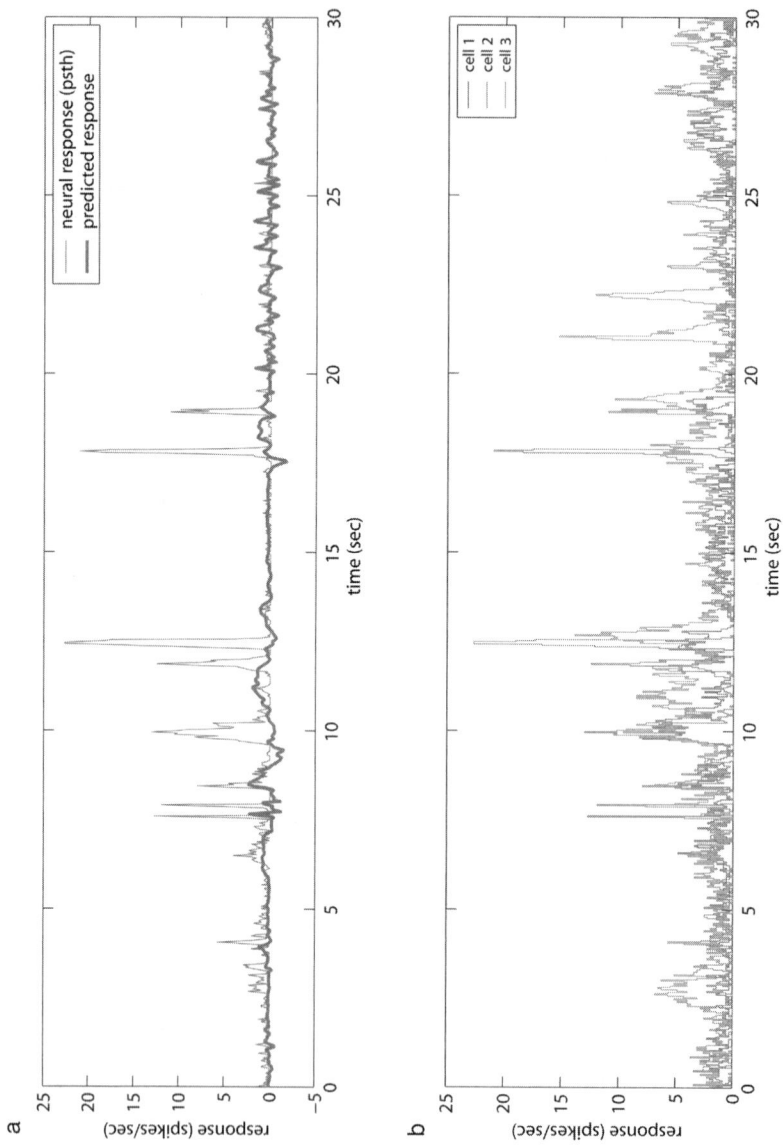

Figure 10-5. Activity of V1 neurons in anaesthetized cat in response to a natural movie. See color insert.

Unfortunately, journals are often unprepared to publish results when a study demonstrates the failure of a model unless the study also presents a competing model that works well. But even without a competing model, these sorts of data are crucial to presenting a complete picture of V1 function. And given the magnitude of the task before us, it could take years before a good model emerges. In the meantime, what would be most helpful is to accumulate a database of single-unit or multiunit data (stimuli and neural responses) that would allow modelers to test their best theory under ecological conditions.

Finally, it should be noted that much better success has been obtained in using receptive field models to predict the responses of neurons to natural scenes in the LGN (Dan et al. 1996). This would seem to suggest that the difficulty in predicting responses in cortex has to do with the effects of the massive, recurrent intracortical circuitry that is engaged during natural vision.

Summary

Table 10-1 presents a summary of the five problems we have identified with the current, established view of V1 as a "Gabor filter bank." Given these factors, is it possible to quantitate how well we currently understand V1 function? We attempt to estimate this as follows:

$$[\text{fraction understood}] = [\text{fraction of variance explained from neurons recorded}] \times [\text{fraction of population recorded}]$$

If we consider that roughly 40 percent of the population of neurons in V1 has actually been recorded from and characterized together with what appears to be 30 to 40 percent of the response variance of these neurons that is explained under natural conditions using the currently established models, then we are left to conclude that we can currently account for approximately 12 to 16 percent of V1 function. Thus, 85 percent of V1 function has yet to be explained.

Table 10-1   Five Problems with the Current View of V1

Biased sampling	Biased stimuli	Biased theories	Interdependence and context	Ecological deviance
large neurons, visually responsive neurons, neurons with high firing rates	use of reduced stimuli such as bars, gratings, and spots	simple/complex cells; data-driven theories verses functional theories	influence of intracortical input, context in natural scenes, synchrony	responses to natural scenes deviate from predictions of standard models

New Theories

Given the above observations, it becomes clear that there is so much unexplored territory that it is very difficult to rule out theories at this point (although there are some obvious bounds dictated by neural architecture—e.g., anatomical convergence, etc.). In the sections below, we discuss some of the theories that are plausible given our current data. However, it must be emphasized that considering that there may exist a large family of neurons with unknown properties, and given the low level of prediction for the neurons studied there is still considerable room for theories dramatically different from those theories presented here.

Dynamical Systems and the Limits of Prediction

Imagine tracking a single molecule within a hot gas as it interacts with the surrounding molecules. The particular trajectory of one molecule will be erratic and fundamentally unpredictable without knowledge of all other molecules with potential influence. Even if we presumed the trajectory of the particular molecule was completely deterministic and following simple laws, in a gas with large numbers of interacting molecules one could never provide a prediction of the path of a single molecule except over very short distances.

In theory, the behavior of single neurons may have similar limitations. To make predictions of what a single neuron will do in the presence of a natural scene may be fundamentally impossible without knowledge of the surrounding neurons. The nonlinear dynamics of interacting neurons may put bounds on how accurately the behavior of any neuron can be predicted. And at this time, we cannot say where that limit may be.

What is fascinating in many ways then is that neurons are as predictable as they are. For example, work from the Gallant lab has shown that under conditions where a particular natural scene sequence is repeated to a fixating macaque monkey, a neuron's response from trial to trial is fairly reliable (e.g., Vinje and Gallant 2000). This clearly suggests that the response of the neuron is dependent in large part on the stimulus, certainly more so than a molecule in the "gas model." So how do we treat the variability that is not explained by the stimulus? We may find that the reliability of a local group of neurons is more predictable than a single neuron, which would then require multielectrode recording to attempt to account for the remaining variance. For example, Arieli et al. (1996) have shown that much of the intertrial variability may be explained in terms of large-scale fluctuations in ongoing activity of the surrounding population of neurons measured using optical recording. However, what role these large-scale fluctuations play in the normal processing of natural scenes has yet to be investigated.

Sparse, Overcomplete Representations

One effort to explain many of the nonlinearities found in V1 is to argue that neurons are attempting to achieve some degree of gain control (e.g., Heeger 1992). Because any single neuron lacks the dynamic range to handle the range of contrasts in natural scenes, it is argued, the contrast response must be normalized. Here we provide a different line of reasoning to explain the observed response nonlinearities of V1 neurons (further details are provided by Olshausen and Field 1997 and Field and Wu 2004). We argue that the spatial nonlinearities primarily serve to reduce the linear dependencies that exist in an overcomplete code, and, as we shall see, this leads to a fundamentally different set of predictions about the population activity.

Consider the number of vectors needed to represent a particular set of data with dimensionality D (e.g., an 8×8 pixel image patch would have $D = 64$). No matter what form the data takes, such data never requires more than D vectors to represent it. A system where data with dimensionality D is spanned by D vectors is described as "critically sampled." Such critically sampled systems (e.g., orthonormal bases) are popular in the image coding community as they allow any input pattern to be represented uniquely, and the transform and its inverse are easily computed. The wavelet code, for example, has seen widespread use, and wavelet-like codes similar to that of the visual system have been shown to provide very high efficiency, in terms of sparsity, when coding natural scenes (e.g., Field 1987). Some basic versions of ICA also attempt to find a critically sampled basis which minimizes the dependencies among the code outputs, and the result is a wavelet-like code with tuning much like the neurons in V1 (Bell and Sejnowski 1997; van Hateren and van der Schaaf 1998).

However, the visual system is not using a critically sampled code. In cat V1, for example, there are twenty-five times as many output fibers as there are input fibers from the LGN, and in macaque V1 the ratio is on the order of 50 to 1. Such overcomplete codes have one potential problem: the vectors are not linearly independent. Thus, if neurons were to compute their output simply from the inner-product between their weight vector and the input, their responses will be correlated.

Figure 10-6a shows an example of a two-dimensional data space represented by three neurons with linearly dependent weight vectors. Even assuming the outputs of these units are half-rectified so they produce only positive values, the data are redundantly represented by such a code. The only way to remove this linear dependence is through a nonlinear transform. One of the nonlinear transforms that will serve this goal is shown in figure 10-6b. Here, we show the isoresponse curves for the same three neurons. This curvature represents an unusual nonlinearity. For example, consider the responses of a unit to two stimuli: the first stimulus aligned with the neuron's weight vector, and a second

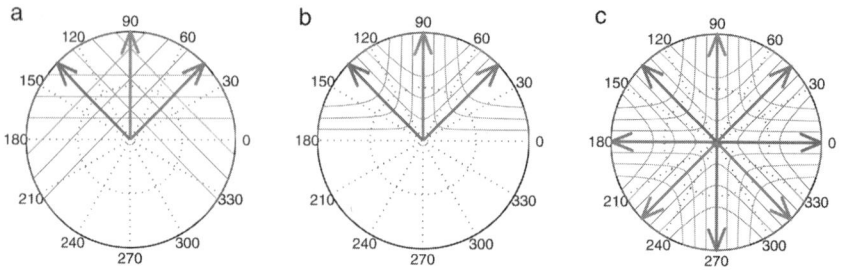

Figure 10-6. Overcomplete representation. See color insert.

stimulus separated by 90 degrees. The second stimulus will have no effect on the neuron on its own since its vector is orthogonal to that of the neuron. However, when added to the first vector, the combined stimulus will be on a lower iso-response curve (i.e., the neuron will have reduced its activity). In other words, the response curvature of the neuron results in a nonlinearity with the characteristic "nonclassical," suppressive behavior: stimuli which on their own have no effect on the neuron (stimuli orthogonal to the principal direction of the neuron) can modulate the behavior of an active neuron. This general nonlinearity comes in several forms and includes end-stopping and cross-orientation inhibition and is what is typically meant by the term *nonclassical surround*. Indeed, as Zetzsche, Krieger, and Wegmann (1999) note, this curvature is simply a geometric interpretation of such behaviors. With the addition of a compressive nonlinearity this curvature results in the behavior described as "contrast normalization."

In contrast to the gain control or divisive normalization theory, we argue that the nonlinearities observed in V1 neurons are present primarily to allow a large (overcomplete) population of neurons to represent data using a small number of active units, a process we refer to as *sparsification*. The goal is not to develop complete independence because the activity of any neuron partially predicts the lack of activity in neighboring neurons. However, the code allows for expanding the dimensionality of the representation without incurring the linear dependencies that would be present in a population with non-orthogonal weight vectors.

Importantly, this model predicts that the nonlinearities are a function of the angle between the neuron's weight vector and those surrounding it. Future multielectrode recordings may provide the possibility to test this theory. From the computational end, we have found that our sparse coding network (Olshausen and Field 1996, 1997) produces nonlinearities much like those proposed. It seems possible, then, that the family of nonlinearities found in V1 can eventually be explained within one general framework of efficient coding.

Contour Integration

If the contrast normalization model were a complete account of V1 neurons, then we might expect the surround suppression to be relatively unspecific. However, the physiological and anatomical evidence implies that V1 neurons have a rather selective connection pattern both within layers and between layers. For example, research investigating the lateral projections of pyramidal neurons in V1 has shown that the long-range lateral connections project primarily to regions of the cortex with similar orientation columns as well as to similar ocular dominance columns and cytochrome oxidase blobs (Malach et al. 1993; Yoshioka et al. 1996). The short-range projections, by contrast, do not show such specificity. Early studies exploring the horizontal connections in V1 discovered that selective long-range connections extend laterally for two to five millimeters parallel to the surface (Gilbert and Wiesel 1979), and studies on the tree shrew (Rockland and Lund 1983; Bosking et al. 1997), primate (e.g., Malach et al. 1993; Sincich and Blasdel 2001), ferret (Ruthazer and Stryker 1996), and cat (e.g., Gilbert and Weisel 1989) have all demonstrated significant specificity in the projection of these lateral connections.

A number of neurophysiological studies also show that colinearly oriented stimuli presented outside of the classical receptive field have a facilitory effect (Kapadia et al. 1995; Kapadia, Westheimer, and Gilbert 2000; Polat et al. 1998). The results demonstrate that when a neuron is presented with an oriented stimulus within its receptive field, a second colinear stimulus will sometimes increase the response rate of the neuron while the same oriented stimulus presented orthogonal to the main axis (displaced laterally) will produce inhibition or at least less facilitation.

These results suggest that V1 neurons have an orientation- and position-specific connectivity structure beyond what is usually included in the standard model. One line of research suggests that this connectivity helps resolve the ambiguity of contours in scenes and is involved in the process of contour integration (e.g., Field, Hayes, and Hess 1993). This follows from work showing that the integration of local oriented elements provides an effective means of identifying contours in natural scenes (Parent and Zucker 1989; Sha'ashua and Ullman 1988). This type of mechanism could work in concert with the sparsification nonlinearities mentioned above since the facilitatory interactions would primarily occur among elements that are nonoverlapping—that is, receptive fields whose weight vectors are orthogonal.

An alternative theoretical perspective is that the effect of the orientation- and position-specific connections should be mainly suppressive, with the goal of removing dependencies among neurons that arise due to the structure in natural images (Schwartz and Simoncelli 2001). In contrast to the contour integration hypothesis, which proposes that the role of horizontal connections is to amplify the structure of contours, this model would attempt to attenuate the presence of any structure in the V1 representation. Although this may be a desirable outcome in terms of redundancy reduction, we would argue that the cortex has objectives other than redundancy reduction per se (Barlow 2001). Chief among these is to provide a meaningful representation of image structure that can be easily read and interpreted by higher-level areas.

Finally, it is important to note, with respect to the discussion in the previous section, that the type of redundancy we are talking about here is due to long-range structure in images beyond the size of a receptive field, not that which is simply due to the overlap among receptive fields. Thus, we propose that the latter should be removed via sparsification, while the former should be amplified by the long-range horizontal connections in V1.

Surface Representation

We live in a three-dimensional world, and the fundamental causes of images that are of behavioral relevance are surfaces, not two-dimensional features such as spots, bars, edges, or gratings. Moreover, we rarely see the surface of an object in its entirety—occlusion is the rule, not the exception, in natural scenes. It thus seems quite reasonable to think that the visual cortex has evolved effective means to parse images in terms of the three-dimensional structure of the environment—that is, surface structure, foreground-background relationships, and so on. Indeed, there is now a strong body of psychophysical evidence showing that three-dimensional surfaces and figure-ground relationships constitute a fundamental aspect of intermediate-level representation in the visual system (Nakayama, He, and Shimojo 1995; see also figure 10-7).

Nevertheless, it is surprising how little V1 physiology has actually been devoted to the subject of three-dimensional surface representation. Recently, a few studies have begun to yield interesting findings about surface representation in the extrastriate cortex (Nguyenkim and DeAngelis 2003; Zhou, Friedman, and von der Heydt 2000; Bakin, Nakayama, and Gilbert 2000). But V1's involvement in surface representation remains a mystery. Although many V1 neurons are disparity selective, this by itself does not tell us how surface structure is represented or how figure-ground relationships of the sort depicted in figure 10-7 are resolved.

At first sight it may seem preposterous to suppose that V1 is involved in computing three-dimensional surface representations. But again, given how little we actually do know about V1, combined with the importance of three-dimensional surface representations for guiding behavior, it is a plausible hypothesis to consider. In addition, problems such as occlusion demand resolving figure-ground relationships in a relatively high-level representation where topography is preserved (Lee and Mumford 2003). There is now beginning to emerge physiological evidence supporting this idea. Neurons in V1 have been shown to produce a differential response to the figure versus

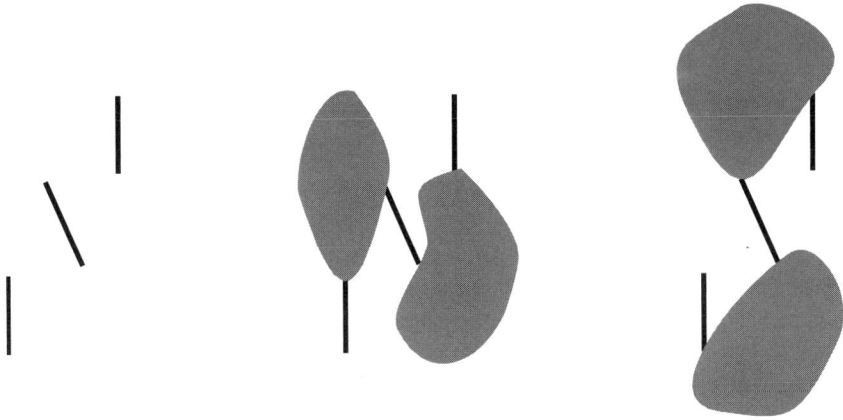

Figure 10-7. The three line-strokes at left are interpreted as vastly different objects depending on the arrangement of occluders. Thus, pattern completion depends on resolving figure-ground relationships. At what level of processing is this form of completion taking place? Since it would seem to demand access to high-resolution detail in the image, it cannot simply be relegated to high-level areas.

background in a scene of texture elements (Lamme 1995; Zipser, Lamme, and Schiller 1996), and a substantial fraction of neurons in V1 are selective to border ownership (Zhou, Friedman, and von der Heydt 2000). In addition, Lee et al. (1998) have demonstrated evidence for a medial axis representation of surfaces in which V1 neurons become most active along the skeletal axis of an object. It seems quite possible such effects are just the tip of the iceberg, and there could be even more effects lurking.

Top-Down Feedback and Disambiguation

Although our perception of the visual world is usually quite clear and unambiguous, the raw image data that we start out with is anything but that. Looking back at figure 10-3, one can see that even the presence of a simple contour can be ambiguous in a natural scene. The problem is that information at the local level is insufficient to determine whether a change in luminance is an object boundary, simply part of a texture, or a change in reflectance. Although boundary junctions are also quite crucial to the interpretation of a scene, a number of studies have now shown that human observers are quite poor judges of what constitutes a boundary or junction when these features are shown in isolation (Elder, Beniaminov, and Pintilie 1999; McDermott 2003). Thus, the calculation of what forms a boundary is dependent on the context, which provides information about the assignment of figure and ground, surface layout, and so forth.

Arriving at the correct interpretation of an image, then, constitutes something of a chicken-egg problem between lower and higher levels of image analysis. The low-level shape features that are useful for identifying an object—edges, contours, surface curvature, and the like—are typically ambiguous in natural scenes, so they cannot be computed directly based on a local analysis of the image. Rather, they must be inferred based on global context and higher-level knowledge. However, the global context itself will not be clear until there is some degree of certainty about the presence of low-level shape features. A number of theorists have argued from this that recognition depends on information circulating through corticocortical feedback loops in order to disambiguate representations at both lower and higher levels in parallel (Mumford 1994; Ullman 1995; Lewicki and Sejnowski 1997; Rao and Ballard 1999; Lee and Mumford 2003).

An example of disambiguation at work in the visual cortex can be seen in the resolution of the aperture problem in computing direction of motion. Because receptive fields limit the field of a view of a neuron to just a portion of an object, it is not possible for any one object to signal, in a purely bottom-up fashion, the true direction of the object. Pack, Berezovskii, and Born (2001) have shown that the initial phase of response of neurons in MT signals the direction of motion directly orthogonal to a contour and that the latter phase of the response reflects the actual direction of the object that the contour is part of, presumably from the interaction with other neurons viewing other parts of the object. It is interesting to note that this effect does not occur under anesthesia. A similar delayed response effect has been demonstrated in end-stopped V1 neurons as well (Pack et al. 2003).

Recent evidence from fMRI (functional magnetic resonance imaging) points to a disambiguation process occurring in V1 during shape perception (Murray et al. 2002). Subjects viewed a translating diamond that was partially occluded so that the vertices are invisible, resulting in a bistable percept in which the line segments forming the diamond are seen moving independently in one case and coherently in the direction of the object motion in the other case. When subjects experience the coherent motion and shape percept, activity in the lateral occipital complex (LOC) increases while activity in V1 decreases. This is consistent with the idea that when neurons in LOC are representing the diamond, they feed this information back to V1 to refine the otherwise ambiguous representations of contour motion. If the refinement of activity attenuates the many incorrect responses while amplifying the few that are consistent with the global percept, the net effect could be a reduction as seen in the BOLD signal measured by fMRI. An alternative interpretation for the reduction in V1 is based on the idea of predictive coding, in which higher areas actually subtract their predictions from lower areas.

There exists a rich set of feedback connections from higher levels into V1, but little is known about the computational role of these connections. Recent experiments in which higher areas are cooled to look at the effect upon activity in lower areas seem to suggest that these connections play a role in enhancing the salience of stimuli (Hupé et al. 1998). But we would argue that feedback has a far more important role to play in disambiguation, and as far as we know, no one has yet investigated the effect of feedback using such cooling techniques under normal conditions that would require disambiguation. (See also Young [2000] for arguments similar to those presented here, in addition to chapter 6, this volume.)

Dynamic Routing

A challenging problem faced by any visual system is that of forming object representations that are invariant to position, scale, rotation, and other common deformations of the image data. The currently accepted, traditional view is that complex cells constitute the first stage of invariant representation by summing over the outputs of simple cells whose outputs are half-rectified and squared—that is, the classical "energy model" (Adelson and Bergen 1985). In this way, the neuron's response changes only gradually as an edge is passed over its receptive field. This idea forms the basis of so-called Pandemonium models, in which a similar feature extraction/pooling process is essentially repeated at each stage of visual cortex (see Tarr [1999], in addition to chapter 16, this volume, for a review).

However, the Pandemonium model leaves much to be desired—namely, there is no provision for how phase or information about relative spatial relationships is preserved. Clearly, though, we have conscious access to this information. In addition, resolving figure-ground relationships and occlusion demands it.

One of us has proposed a model for forming invariant representations that preserves relative spatial relationships by explicitly routing information at each stage of processing (Olshausen et al. 1993). Rather than passively pooling, information is dynamically linked from one stage to the next by a set of control neurons that progressively remaps information into an object-centered reference frame. Thus, it is proposed that there are two distinct classes of neurons—those conveying image and feature information and those controlling the flow of information. The former corresponds to the invariant part, the latter to the variant part. The two are combined multiplicatively, so mathematically this constitutes a bilinear model (e.g., Tenenbaum and Freeman 2000; Grimes and Rao 2003).

Is it possible that dynamic routing occurs in V1 and underlies the observed shift-invariant properties of complex cells? If so, there are at least two things we would expect to see: (1) that complex cells are actually selecting at any given moment which simple cells they are connected to and (2) that there should also exist control neurons to do the selection. It is interesting that the observed invariance properties of complex cells are just as consistent with the idea of routing as they are with pooling. What could possibly distinguish between these models is the population activity: if the complex cell outputs are the result of passive pooling, then one would expect a dense, distributed representation of contours among the population of complex cells, whereas if information is routed then the representation at the complex cell level would remain sparse. The control neurons, on the other hand, would look something like contrast normalized simple cells, which represent phase independent of magnitude (Zetzsche and Rohrbein 2001).

One of the main predictions of the dynamic routing model is that the receptive fields of the invariant neurons would be expected to shift depending on the state of the control neurons. Connor et al. (1997) showed that some neurons in V4 shift their receptive fields depending on where the animal is directing its attention, and Motter and Poggio (1990) showed that some complex cells in V1 appear to shift their receptive fields in order to compensate for the small eye movements that occur during fixation (but see also Gur and Snodderly 1997). However, more experiments are needed to properly characterize the invariance properties of visual neurons under normal viewing conditions.

Conclusions

Our goal in this chapter has been to point out that there are still substantial gaps in our knowledge of V1 function, and more importantly that there is more room for new theories to be considered than the current conventional wisdom might allow. We have identified five specific problems with the current view of V1, emphasizing the need for using natural scenes in experiments in addition to multiunit recording methods, in order to obtain a more representative picture of V1 function. While the single-unit, reductionist approach has been a useful enterprise for getting a handle on basic response properties, we feel that its usefulness as a tool for investigating V1 function has nearly been exhausted. It is now time to dig deeper, using richer experimental paradigms and developing theories that can help to elucidate how the cortex performs the computationally challenging problems of vision.

As we explore the response properties of V1 neurons using natural scenes, we are likely to uncover some interesting new phenomena that defy explanation with current models. It is at this point that we should be prepared to revisit the reductionist approach in order to tease apart what is going on. Reductionism does have its place, but it needs to be motivated by functionally and ecologically relevant questions, similar to the European tradition in ethology.

At what point will we actually "understand" V1? This is obviously a difficult question to answer, but we believe at least three ingredients are required: (1) an unbiased sample of neurons of all types, firing rates, and layers of V1, (2) the ability to observe simultaneously the activities of hundreds of neurons in a local population, and (3) the ability to predict, or at least qualitatively model, the responses of the population under natural viewing conditions. Given the extensive feedback connections into V1 in addition to the projections from pulvinar and other sources, it seems unlikely that we will ever understand V1 in isolation. Thus, our investigations must also be guided by how V1 fits into the bigger picture of thalamocortical function.

Acknowledgments We thank Bill Skaggs for discussions on hippocampal physiology, Charlie Gray and Jonathan Baker for sharing preliminary data, and Jack Gallant for clarifying the issues involved in predicting neural responses.

Notes

1. We do not mean to imply here that nothing has been learned, but rather that what has been learned is but a small fraction of what lies ahead and still needs to be understood.

2. The Volterra series expansion is often touted as a general approach for characterizing nonlinear systems, but it has been of little practical value in analyzing neural systems because it requires estimating many higher-order moments. In addition, it is an overly general, "black box" approach that does not easily allow one to incorporate prior knowledge about the types of nonlinearities known to exist in the nervous system.

References

Adelson EH, Bergen JR (1985) Spatiotemporal energy models for the perception of motion. J. Opt. Soc. Am. A, 2, 284–299.

Ahmed B, Anderson JC, Douglas RJ, Martin KA, Nelson JC (1994) Polyneuronal innervation of spiny stellate neurons in cat visual cortex. J. Comp. Neurol., 341, 39–49.

Amedi A, Raz N, Pianka P, Malach R, Zohary E (2003) Early "visual" cortex activation correlates with superior verbal memory performance in the blind. Nat. Neurosci., 6, 758–766.

Arieli A, Sterkin A, Grinvald A, Aertsen A (1996) Dynamics of ongoing activity: Explanation of the large variability in evoked cortical responses. Science, 273, 1868–1871.

Attwell D, Laughlin SB (2001) An energy budget for signaling in the grey matter of the brain. J. Cereb. Blood Flow Metab., 21, 1133–1145.

Azouz R, Gray CM (2000) Dynamic spike threshold reveals a mechanism for synaptic coincidence detection in cortical neurons in vivo. Proc. Natl. Acad. Sci. USA, 97, 8110–8115.

Azouz R, Gray CM (2003) Adaptive coincidence detection and dynamic gain control in visual cortical neurons in vivo. Neuron, 37, 513–523.

Baddeley R, Abbott LF, Booth MCA, Sengpiel F, Freeman T, Wakeman EA, Rolls ET (1997) Responses of neurons in primary and infererior temporal visual cortices to natural scenes. Proc R. Soc. Lond. B, 264, 1775–1783.

Bakin JS, Nakayama K, Gilbert CD (2000) Visual responses in monkey areas V1 and V2 to three-dimensional surface configurations. J. Neurosci., 20, 8188–8198.

Barlow HB (2001) Redundancy reduction revisited. Network, 12, 241–253.

Barnes CA, Skaggs WE, McNaughton BL, Haworth ML, Permenter M, Archibeque M, Erickson CA (2003) Chronic recording of neuronal populations in the temporal lobe of awake young adult and geriatric primates. Program No. 518.8. Abstract Viewer/Itinerary Planner. Washington, DC: Society for Neuroscience.

Bell AJ, Sejnowski TJ (1997) The independent components of natural images are edge filters. Vision Res., 37, 3327–3338.

Bosking WH, Zhang Y, Schofield B, Fitzpatrick D (1997) Orientation selectivity and the arrangement of horizontal connections in Tree Shrew striate cortex. J. Neurosci., 17, 2112–2127.

Carandini M, Heeger DJ, Movshon JA (1997) Linearity and normalization in simple cells of the macaque primary visual cortex. J. Neurosci., 17, 8621–8644.

Chung S, Ferster D (1998) Strength and orientation tuning of the thalamic input to simple cells revealed by electrically evoked cortical suppression. Neuron, 20, 1177–1189.

Connor CE, Preddie DG, Gallant JL, Van Essen DC (1997) Spatial attention effects in macaque area V4. J. Neurosci., 17, 3201–3214.

Dan Y, Atick JJ, Reid RC (1996) Efficient coding of natural scenes in the lateral geniculate nucleus: experimental test of a computational theory. J Neurosci., 16, 3351–3362.

David SV, Vinje WE, Gallant JL (1999) Natural image reverse correlation in awake behaving primates. Society for Neuroscience Abstracts, 25, 1935.

De Valois RL, Albrecht DG, Thorell LG (1982) Spatial frequency selectivity of cells in macaque visual cortex. Vision Res., 22, 545–559.

Elder JH, Beniaminov D, Pintilie G (1999) Edge classification in natural images. Invest. Ophthalmol. Vis. Sci., 40, S357.

Field DJ (1987) Relations between the statistics of natural images and the response properties of cortical cells. J. Opt. Soc. Am., A, 4: 2379–2394.

Field DJ, Hayes A, Hess RF (1993) Contour integration by the human visual system: evidence for a local "association field." Vision Res., 33, 173–193.

Field DJ, Wu M (2004) An attempt towards a unified account of non-linearities in visual neurons. J Vision, 4, 283a.

Gilbert CD, Wiesel TN (1979) Morphology and intracortical projections of functionally characterised neurones in the cat visual cortex. Nature, 280, 120–125.

Gilbert CD, Wiesel TN (1989). Columnar specificity of intrinsic horizontal and corticocortical connections in cat visual cortex. J. Neurosci., 9, 2432–2442.

Gray CM, Konig P, Engel AK, Singer W (1989) Oscillatory responses in cat visual cortex exhibit inter-columnar synchronization which reflects global stimulus properties. Nature, 338, 334–337.

Grimes DB, Rao RP (2005) Bilinear sparse coding for invariant vision. Neural Computation, 17, 47–73.

Gur M, Snodderly DM (1997) Visual receptive fields of neurons in primary visual cortex (V1) move in space with the eye movements of fixation. Vision Res., 37, 257–265.

van Hateren JH, van der Schaaf A (1998) Independent component filters of natural images compared with simple cells in primary visual cortex. Proc. R. Soc. Lond. B, 265, 359–366.

Hausser M, Mel B (2003) Dendrites: bug or feature? Curr. Opin. Neurobiol., 13, 372–383.

Heeger DJ (1992) Normalization of cell responses in cat striate cortex. Vis. Neurosci., 9, 181–198.

Hupé, JM, James, AC, Payne, BR, Lomber, SG, Girard, P, Bullier, J (1998) Cortical feedback improves discrimination between figure and background by V1, V2 and V3 neurons. Nature, 394, 784–787.

Jung MW, McNaughton BL (1993) Spatial selectivity of unit activity in the hippocampal granular layer. Hippocampus, 3, 165–182.

Kagan I, Gur M, Snodderly DM (2002) Spatial organization of receptive fields of V1 neurons of alert monkeys: Comparison with responses to gratings. J. Neurophys., 88, 2557–2574.

Kapadia MK, Ito M, Gilbert CD, Westheimer G (1995) Improvement in visual sensitivity by changes in local context: parallel studies in human observers and in V1 of alert monkeys. Neuron, 15, 843–856.

Kapadia MK, Westheimer G, Gilbert CD (2000) Spatial distribution of contextual interactions in primary visual cortex and in visual perception. J. Neurophys., 84, 2048–2062.

Knierim JJ, Van Essen DC (1992) Neuronal responses to static texture patterns in area V1 of the alert macaque monkey. J. Neurophys., 67, 961–980.

Lamme VA (1995) The neurophysiology of figure-ground segregation in primary visual cortex. J. Neurosci., 15, 1605–1615.

Lee TS, Mumford D, Romero R, Lamme VA (1998) The role of the primary visual cortex in higher level vision. Vision Res., 38, 2429–54.

Lee TS, Mumford D (2003) Hierarchical Bayesian inference in the visual cortex. J. Opt. Soc. Am. A, 20, 1434–1448.

Legendy CR, Salcman M (1985) Bursts and recurrences of bursts in spike trains of spontaneously active striate cortex neurons. J. Neurophys. 53, 926–939.

Lennie P (2003a) Receptive fields. Curr. Biol., 13, R216–R219.

Lennie P (2003b) The cost of cortical computation. Curr. Biol., 13, 493–497.

Lewicki MS, Sejnowski TJ (1996) Bayesian unsupervised learning of higher order structure. In: Advances in Neural Information Processing Systems, 9, Mozer MC, Jordan MI, Petsche T, eds., Cambridge, MA: MIT Press, pp. 529–535.

Malach R, Amir Y, HarelM, Grinvald A (1993). Relationship between intrinsic connections and functional architecture revealed by optical imaging and in vivo targeted biocyting injections in primate striate cortex. Proc. Natl. Acad. Sci. USA, 90, 10469–10473.

Maldonado P, Babul C, Singer W, Rodriguez E, Grun S (2004). Synchrony and oscillations in primary visual cortex of monkeys viewing natural images. Submitted.

McDermott J (2004) Psychophysics with junctions in real images. Perception, 33, 1101–1127.

Mechler F, Ringach DL (2002) On the classification of simple and complex cells. Vision Res., 42, 1017–1033.

Motter BC, Poggio GF (1990) Dynamic stabilization of receptive fields of cortical neurons (VI) during fixation of gaze in the macaque. Exp. Brain Res., 83, 37–43.

Mumford D (1994) Neuronal architectures for pattern-theoretic problems. In: Large Scale Neuronal Theories of the Brain, Koch C, Davis, JL, eds., Cambridge, MA: MIT Press, pp. 125–152.

Murray SO, Kersten D, Olshausen BA, Schrater P, Woods DL (2002) Shape perception reduces activity in human primary visual cortex. Proceedings of the National Academy of Sciences, USA, 99(23), 15164–15169.

Nakayama K, He ZJ, Shimojo S (1995) Visual surface representation: a critical link between lower-level and higher level vision. In: An Invitation to Cognitive Science, Kosslyn SM, Osherson DN, eds., Cambridge, MA: MIT Press, pp. 1–70.

Nguyenkim JD, DeAngelis GC (2003) Disparity-based coding of three-dimensional surface orientation by macaque middle temporal neurons. J. Neurosci., 23(18), 7117–7128.

Olshausen BA (2003) Principles of image representation in visual cortex. In: The Visual Neurosciences, Chalupa LM, Werner JS, eds. Cambridge, MA: MIT Press, pp. 1603–1615.

Olshausen BA, Anderson CH (1995) A model of the spatial-frequency organization in primate striate cortex. The Neurobiology of Computation: Proceedings of the Third Annual Computation and Neural Systems Conference. Bower JM, Ed., Boston: Kluwer Academic Publishers, pp. 275–280.

Olshausen BA, Anderson CH, Van Essen DC (1993) A neurobiological model of visual attention and invariant pattern recognition based on dynamic routing of information. J Neurosci, 13, 4700–4719.

Olshausen BA, Field DJ (1996) Emergence of simple-cell receptive field properties by learning a sparse code for natural images. Nature, 381, 607–609.

Olshausen BA, Field DJ (1997) Sparse coding with an overcomplete basis set: A strategy employed by V1? Vision Res., 37, 3311–3325.

Pack CC, Berezovskii VK, Born RT (2001) Dynamic properties of neurons in cortical area MT in alert and anaesthetized macaque monkeys. Nature, 414, 905–908.

Pack CC, Livingstone MS, Duffy KR, Born RT (2003) End-stopping and the aperture problem: Two-dimensional motion signals in macaque V1. Neuron, 39, 671–680.

Parent P, Zucker S (1989) Trace inference, curvature consistency and curve detection. IEEE Transactions on Pattern Analysis and Machine Intelligence, 11, 823–839.

Parker AJ, Hawken MJ (1988) Two-dimensional spatial structure of receptive fields in monkey striate cortex. J. Opt. Soc. Am. A, 5, 598–605.

Peters A, Payne BR (1993) Numerical relationships between geniculocortical afferents and pyramidal cell modules in cat primary visual cortex. Cereb. Cortex, 3, 69–78.

Peters A, Payne BR, Budd J (1994) A numerical analysis of the geniculocortical input to striate cortex in the monkey. Cereb. Cortex, 4, 215–29.

Polat U, Mizobe K, Pettet MW, Kasamatsu T, Norcia AM (1998) Collinear stimuli regulate visual responses depending on cell's contrast threshold. Nature, 391, 580–584.

Priebe NJ, Mechler F, Carandini M, Ferster D (2004) The contribution of spike threshold to the dichotomy of cortical simple and complex cells. Nat. Neurosci., 7, 1113–1122.

Rao RP, Ballard DH (1999) Predictive coding in the visual cortex: a functional interpretation of some extra-classical receptive-field effects. Nat. Neurosci., 2, 79–87.

Rockland KS, Lund JS (1983) Intrinsic laminar lattice connections in primate visual cortex. J. Comp. Neurol., 216, 303–18.

Rose G, Diamond D, Lynch GS (1983) Dentate granule cells in the rat hippocampal formation have the behavioral characteristics of theta neurons. Brain Res., 266, 29–37.

Ruthazer ES, Stryker MP (1996). The role of activity in the development of long-range horizontal connections in area 17 of the ferret. J. Neurosci., 16, 7253–7269.

Sadato N, Pascual-Leone A, Grafman J, Ibanez V, Deiber MP, Dold G, Hallett M (1996) Activation of the primary visual cortex by Braille reading in blind subjects. Nature, 380, 526–528.

Schwartz O, Simoncelli EP (2001) Natural signal statistics and sensory gain control. Nat. Neurosci., 4, 819–825.

Sha'ashua A, Ullman S (1988) Structural saliency. Proceedings of the International Conference on Computer Vision, Tampa, Florida, 482–488.

Sillito AM, Grieve KL, Jones HE, Cudeiro J, Davis J (1995) Visual cortical mechanisms detecting focal orientation discontinuities. Nature, 378, 492–496.

Sincich LC, Blasdel GG (2001) Oriented axon projections in primary visual cortex of the monkey. J. Neurosci., 21, 4416–4426.

Skottun BC, De Valois RL, Grosof DH, Movshon JA, Albrecht DG, Bonds AB (1991) Classifying simple and complex cells on the basis of response modulation. Vision Res., 31, 1079–1086.

Softky WR, Koch C (1993) The highly irregular firing of cortical cells is inconsistent with temporal integration of random EPSPs. J. Neurosci., 13(1), 334–50.

Tarr MJ (1999) News on views: pandemonium revisited. Nat. Neurosci., 2, 932–935.

Tenenbaum JB, Freeman WT (2000) Separating style and content with bilinear models. Neur. Comp. 12, 1247–1283.

Thompson LT, Best PJ (1989) Place cells and silent cells in the hippocampus of freelybehaving rats. J. Neurosci., 9, 2382–2390.

Ullman S (1995) Sequence seeking and counter streams: a computational model for bidirectional information flow in the visual cortex. Cereb. Cortex, 5, 1–11.

Vinje WE, Gallant JL (2000) Sparse coding and decorrelation in primary visual cortex during natural vision. Science, 287, 1273–1276.

Wirth S, Yanike M, Frank LM, Smith AC, Brown EN, Suzuki WA (2003) Single neurons in the monkey hippocampus and learning of new associations. Science, 300, 1578–81.

Worgotter F, Suder K, Zhao Y, Kerscher N, Eysel UT, Funke K (1998) State-dependent receptive-field restructuring in the visual cortex. Nature, 396, 165–168.

Yoshioka T, Blasdel GG, Levitt JB, Lund JS (1996). Relation between patterns of intrinsic lateral connectivity, ocular dominance, and cytochrome oxidase-reactive regions in macaque monkey striate cortex. Cereb. Cortex 6, 297–310.

Young MP (2000). The architecture of visual cortex and inferential processes in vision. Spatial Vision, 13, 137–146.

Zetzsche C, Krieger G, Wegmann B (1999) The atoms of vision: Cartesian or polar? J. Opt. Soc. Am. A., 16, 1554–1565.

Zetzsche C, Rohrbein F (2001) Nonlinear and extra-classical receptive field properties and the statistics of natural scenes. Network, 12, 331–350.

Zhou H, Friedman HS, von der Heydt R (2000). Coding of border ownership in monkey visual cortex. J. Neurosci., 20, 6594–6611.

Zipser K, Lamme VA, Schiller PH (1996) Contextual modulation in primary visual cortex. J. Neurosci., 16, 7376–7389.

PART IV

WHAT CAN BRAINS COMPUTE?

11

Which Computation Runs in Visual Cortical Columns?

Steven W. Zucker

Introduction

Behavior results from the solution of problems: we plan activities; we sense our environment; we accomplish motor tasks. Each of these vague statements actually implies a universe of specific problems, such as object identification, trajectory planning, and motor control. But even these problems need further specification before they can be examined in the laboratory: using vision to describe one limit of this specification, data (e.g., spike trains) characterize the response of a particular neuron in V1 to a moving bar stimulus at a particular orientation, contrast, and velocity, to take one (still partially specified) example. To understand this neuron further, synaptic arrangements can be measured, and neurotransmitters identified. How can such data be put together into a theory of visual cortex?

The task normally ascribed to theory is to develop models that predict data such as that above, and a close relationship between modeling and experiment (Jennings and Aamodt 2000) is sought. For example, one might model synaptic facilitation and nonlinearities. This encourages a "devil's in the details" perspective. The hunt is on for "the basic unit of computation" (Zador 2000), which, according to many, may reside in synaptic nonlinearities. Little, if any, analysis is aimed toward models at a global, abstract level, and few connections, if any, are attempted between the capabilities of the model and the problem that originally motivated it.

There is no question that such data and such models are a necessary part of understanding information processing in the brain. The question is whether they are sufficent. Consider what could happen with success. The trend toward specifics would create the danger of burying theory under mountains of detail. An experimental program that provided direct measurements for all the neurons in the visual system, even limiting physiology to receptive-field characteristics, anatomy to a listing of synaptic interactions, and stimuli to small image patches, would result in a suffocating amount of data. In the limit, modeling would become simulation, and the effective complexity of the simulation would be of the same order as the data (or worse). How could it be explained; generalized; applied to other problems? With such high dimensionality, how could it be related to the original problem? Returning to the above example to be concrete: how much of the edge detection problem does the moving bar experiment capture? To answer questions such as these requires abstraction, and our goal in this chapter is to raise questions about the nature of this abstraction. To enable our pursuit to be somewhat concrete and to provide a few exemplars for what we seek, the focus is on early vision. In order of specificity, the questions are as follows.

Perhaps the most natural place to seek abstraction is in a model for neural computation, and the basic question we ask is, what does *computation* mean in the phrase *neural computation*? Digital computers immediately suggest an answer. In brief, since computers consist of logical circuits, it follows that if certain primitive functions (such as AND, OR, NOT, etc.) could be implemented in neural circuitry, then any computation could be implemented by arranging these primitives in an appropriate fashion. This view is attractive because of its universality, and it provides a foundation for building computations out of neural components (Shepard 1990; Koch 1999).

I do not believe that such a strategy is enough because a successful pairing between logical functions and synapses would, in the end, get buried exactly as above. There could be about as many "logical gates" as there are dendritic interactions (maybe more). In effect, we would be no better off than the designers of VLSI circuits, and we know how difficult it is to design them (Lengauer 1991). There are limits to how far a designer can go without tools to help with abstraction, in which complex circuits at the silicon level become "units" at the design level. Nevertheless, the most complex circuits constructed are still many orders of magnitude simpler than the visual cortex. "Silicon neurons" can be designed into circuits providing there is symmetry in the design, either in layout (e.g., spatially regular arrays such as those that arise in retinal models) or in interconnection (e.g., complete recriprical connections (Hahnloser et al. 2000), but it is unclear how these simplifications relate to information processing. Often the temptation is to model a unit in a complex but without the interactions that define the complex (for example, those models for a single cortical column that include no inter-columnar interactions [Hansel and Sompolinsky 1998]) or to design the interactions with no detail on the units (Mumford 1994; Ullman 1994). Surely both are required: understanding the behavior of a single ball is clearly part of understanding the game of billiards, but it is also clear that emergent properties of the game simply do not exist at the single-ball level.

Thus we claim that while local models are important, it is as important to explore the other way around; that is, to find the right types of abstraction for characterizing the cortical machine. This will relate to the classes of problems being solved on it, and it will highlight their emergent properties. To be useful, abstraction must help in framing problems, loosely in the way high-level programming languages provide a framework for conceptualization. (VLSI design environments share this feature with programming languages.) But our task is not to seek a programming language in the standard sense because our abstraction must articulate those aspects of the physiology that are constraining, a point in direct contrast to the way in which programming languages often hide the computer's architecture from the programmer. From this perspective the separation of levels in (Marr 1982) needs to be reconsidered; there are constructive senses in which the "problem," the "computational" level, and the "implementation" level are intimately related. One would not expect visual area V1, with its highly articulated structure (discussed in the section "Functional Architecture of the Visual Cortex" below), to be Turing universal. We ask, instead: what V1 is good for; that is, which tasks are naturally formulated (encoded) and solved (computed) in it?

Underlying this chapter is an assumption that a deep connection exists between the functional architecture of cortical areas and the computational abstractions they support. This may not be the case. It could be argued, for example, that columnar architectures exist because of evolutionary accidents and not because of emergent computations. This is not to say that there is no genetic component to brain evolution, which would be incorrect, but to raise the possibility that, during evolution, certain genes that, say, control body segmentation processes became encapsulated in the brain's developmental sequence, and they simply remained in place since then. To our knowledge no evidence currently exists to support this accidental position (Allman 1999). Or it may be argued that there is no deep connection between global computations and function; rather function emerges as a result of learning (Koch, 1997), and no satisfying global explanation exists. If so, then the comments above about generalization and applicability apply again. Showing that a basic computational abstraction exists would argue against both of these positions, and could, in particular, provide an explicit criterion for the evaluation of learning strategies.

To underline the subtlety of our task, we return to the question of whether the problem level can be separated from the implementation level. Circuits in silicon have been developed to the point that the quantal nature of holes and electrons need not be considered. This does not appear to be the case for neural circuits, however, in which the quantal nature of synapses and the timing of spikes is decidedly probabilistic (Mainen and Sejnowski 1995; Cowan, Sudhoff, and Stevens 2001). How does such stochastic structure inform neural computation? (We shall return to this question at the end of the chapter.)

To introduce the relationship between problems and computations, in the next section we begin with the abstract hierarchy developed in theoretical computer science. Our treatment of the hierarchy is very informal, and it is not altogether perfectly suited for questions in computational neuroscience. Nevertheless, it does suggest two notions that will be fundamental for what follows: (1) how can a problem statement be related to a computation; and (2) given a problem, how can an instance of it be encoded? But this hierarchy is too general, so we then focus on vision, and in particular on early vision, where specific cell properties, functional architecture, and behavioral constraints all exist. This allows us to be more specific, and we eventually arrive at the question, what is the natural abstraction for computation in cortical columns? We illustrate one style of analysis that could answer such questions, and a possible solution opens the door to considering the delicacy of cortical dynamics and spike trains. Thus very abstract analysis can suggest very concrete, testable results, a topic that relates back to experimental formulations (see opening paragraph).

Problem Abstractions and Computational Abstractions

Intuitively, when one thinks about solving problems by neural computation, one thinks about what to represent and how to use it. "How to" implies a procedure, and procedures run on problem instances. These notions are interrelated, and we now deal with both.

The difficulty in solving a problem can be formalized in terms of algorithms for solving it. To illustrate, imagine a traveling salesman who seeks a tour among the cities in his territory. Knowing the distance between each pair of cities, it is natural for the salesman to attempt to minimize his travel. Starting from home, an impetuous salesman might strategize that always driving to the next closest city until all are visited will result in a shortest path. This "greedy" strategy has the advantage that it involves little planning, but it has the disadvantage that it will not necessarily find the minimum (shortest tour). A compulsive salesman might attempt to evaluate the distance for every possible tour and then sort them to choose the shortest. While this is guaranteed to give the right answer, the combinatorics make it totally impractical. For only hundreds of cities, a computer might have to work for centuries to do a total enumeration!

Because enumeration seems so natural for such problems, one might guess that only discrete combinatorial solutions are possible. However, this guess biases one's thinking in too constrained a fashion; "analog" approaches to the traveling salesman problem exist (Durbin and Willshaw 1987). Neural networks provide another approximation (Hopfield and Tank 1985). Neither, however, is guaranteed to find the shortest tour. As with the greedy approach, "how to" questions force us to consider limits: how well can the salesman do on the traveling salesman problem, given the amount and type of his resources?

The second issue has to do with the domain over which the problem is defined. Intercity distances are fundamental to the salesman because transportation abstracts part of his job. Emergent structure at that level allows him to relate intercity distances to minimizing fuel costs, pollution, and guaranteeing that all customers are served. Although this example is artificial, consider how an experimentalist might evaluate the salesman's behavior. The given "experimental data" is an ordered list of cities visited. What is his strategy? Although our context suggests minimizing total distance, the experimentalist might try to establish that the salesman is trying to alternate cities according to size of purchase, so two large orders in consecutive cities do not overload his production staff.

Our problem in building an abstraction for vision is to find, in a manner of speaking, the equivalent of a theory of transportation. This should abstract over images as transportation abstracts over customers, and in effect it is a theory by which we should be able to calculate the (vision analog to) intercity distances. As with inferences about the salesman, different abstractions are possible, and we shall illustrate two of them. Before doing so, however, we continue the general discussion.

Complexity of Computations

How well can the salesman do on the traveling salesman problem? For a small number of cities—say, fewer than five—the difficulty of the problem hardly matters, since the total enumeration can be done easily. Computers could handle maybe twenty or thirty cities by brute force, but after that the situation becomes intractable. Clearly the traveling salesman problem is more difficult than, say, counting all of the cities, but how much more difficult? Should the salesman attempt to find an optimal solution to his problem or should he settle for something less?

When we talk about problems, normally we mean a particular instance of a problem; for example, the traveling salesman problem (TSP) over all McDonald's restaurants in Woods Hole, Massachusetts, or in Canada. Complexity theorists seek solutions by general methods that will work for *any* instance of a problem, assuming computing or other resources are available. These general procedures become well-specified algorithms when a particular machine model is assumed, such as a deterministic Turing machine or a well-defined programming language. With such a machine model, it is possible to specify the *time complexity function* or the number of steps required to solve the problem. We know from the above example that this will be a function of the particular instance of the problem, and the input to the problem specifies the instance. For the TSP this input includes the number of cities as well as the distances between pairs of cities.

With this structure we can define classes of problem complexity in terms of time complexity: those problems whose complexity grows no faster than a linear function of the size of the input (such as counting the number of cities) are said to be easy, while those that grow faster than a polynomial function in the size of the input are hard (such as finding the salesman's tour). An important class of hard problems are NP-complete (Garey and Johnson 1979), meaning that they can be solved in polynomial time on a "nondeterministic Turing machine" (a machine that can attempt an unbounded number of independent computational streams in parallel), and that they can be reduced to one another. Thus a solution to one problem makes others in this class easy, in the sense that it can be used to solve another in polynomial time. The famous open question in computer science, "does $P = NP$?" asks whether there exist polynomial solutions for nondeterministic polynomial problems. The traveling salesman problem is NP-complete, which is why heuristics, such as the greedy approach, are attempted for it. Even more difficult decision problems than TSP exist, as do some totally undecidable problems (e.g., the famous "halting problem").

The TSP involves a mix of discrete notions (the set of cities and the roads connecting them) and continuous notions (the distances between cities are real numbers). Abstractly the cities can be thought of as nodes in a graph, existing roads as edges in this graph, and distances as continuous weights attached to the edges. The traveling salesman problem was to find a tour through this graph such that the sum of the weights attached to the edges along the tour was minimal. (How such graphs and numbers are encoded is a key part of the representation question.) This mixture of continuous and discrete quantities leads to very deep questions mathematically and computationally (Grotschel, Lovasz, and Schrijver 1993); it is important that we keep such distinctions in mind for this chapter, because the neural computational literature separates them with the idea that continuous or analog computations are multiplications and amplifications of signals, while discrete computations involve bits (Koch 1997).

A type of parallelism entered through nondeterminism, and the combinatorics derived from the number of tours through the graph. The solution was still obtained in a very general class of machines. In comparison to this, the problems that we face in abstracting visual function are already highly constrained, and parallelism arises in a somewhat different guise. As we review in the next section, primary visual cortex has a special structure, and we believe this structure casts light on the class of problems for which it is naturally suited. As with the TSP, it involves a mix of discrete and continuous notions. Understanding it will lead us to suggest a "cortical columnar machine" and a characterization of the computations it can support. One of the more curious aspects of this machine is that both continuous and discrete views of it are possible.

Functional Architecture of the Visual Cortex

The visual cortex is organized largely around orientation, that is, around selective responses to local oriented bars. In a classical observation (Hubel and Wiesel 1977), recordings along a tangential penetration encounter a sequence of cells with regular shifts in orientation preference, while normal penetrations reveal cells with similar orientation and position preferences but different receptive field sizes. Together they define an array of orientation columns, and combined with eye of origin, these columns provide a representation for visual information processing. In effect these columns represent an instance of the cortical columnar machine specialized for problems in vision. Our goal is to untangle the machine from the specific problem encoding.

The widespread appearance of columns is normally explained as a packing problem: since an array of different features is calculated for each point in the retinotopic array, including orientation, scale, eye-of-origin, and so on, more than a three-dimensional arrangement of information is required. If there were only one feature, say, orientation, then a columnar architecture would not be required, since this feature could be arranged along the third dimension of cortex, with retinotopic position the other two. (This would simply be a scalar map.) A minimal wiring length constraint then predicts that different features will be clustered, and arranged as closely as possible to their retinotopic coordinates (Mitchison 1991; Koulakov and Chklovskii 2001) provided uniform coverage of the retinotopic array is maintained (Swindale et al. 2000).

It is essential to note that the above discussion—multiple superimposed feature maps and minimal wiring length—does not directly address questions of problem encoding, of computation, or of information processing: what are the relevant features and their interactions, and how should they be computed? (Machine models that include communication costs have been developed; Pippenger, 1993.) A functional abstraction suitable for the domain is required, and we know at least since Helmholtz that, for vision, inferencing will be required. We need, then, to specify visual inferences as well-defined computations.

If the domain is taken to be images, and statistical regularities over images are sought, then a coding view in which receptive fields model filters arises (Field, Hayes, and Hess 1993; Barlow and Blakemore 1989). Information is processed by composing these filters (e.g., Volterra series), and the layer-to-layer variation with scale amounts to scale-spaces of filters (Simoncelli et al. 1992). The computation here is convolution, and linear operators, wavelets, and related abstractions are applied. Channel models (DeValois and DeValois 1988) are simplified versions of this.

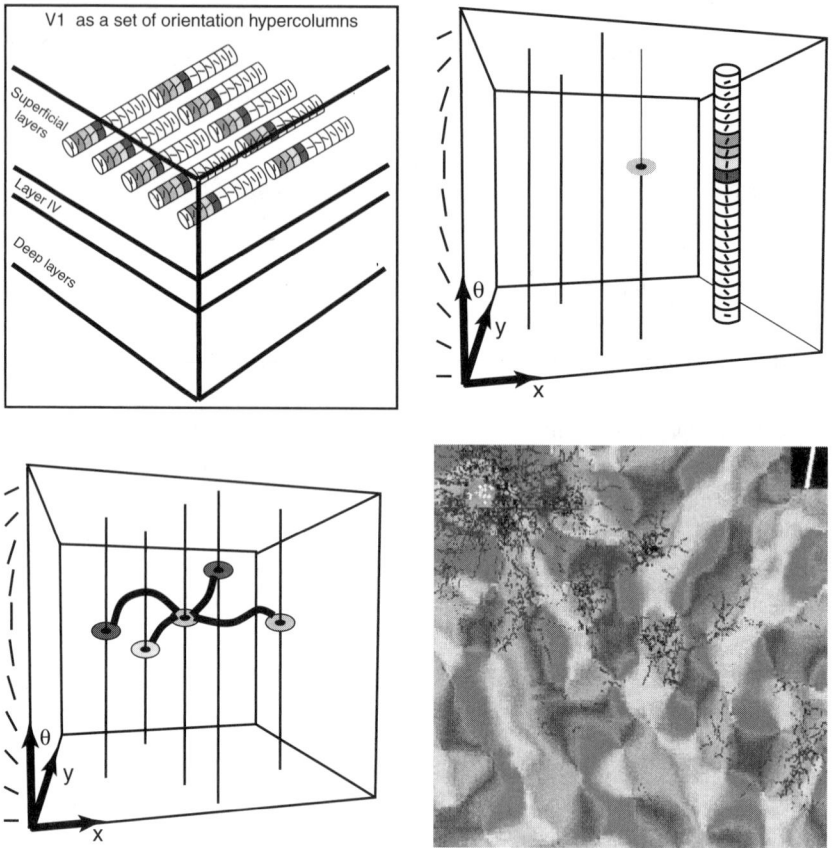

Figure 11-1. Abstracting the functional architecture of visual cortex. See color insert.

But note that as we are reminded by Helmholtz, there is still no inference. Filters provide an enhancement; they map images into images. For example, lateral inhibitory networks can implement filters for contrast enhancement, and Mach first noticed the importance of this for emphasizing edges. Maass, Natschlger, and Markram (2004) have studied which maps are possible from spike trains to spike trains. But if such filters are playing a role in edge detection, as is normally supposed, then an additional computational requirement arises: *detection* requires a nonlinear decision, a selection process that maps filter outputs into discrete classes, say 1 (to signal that an edge is present at a particular location and with a particular orientation) and 0 (for edge absent). Filtering followed by selection is now a composite computation, a local form of inference, and different ways to implement it have been studied (Nowlan and Sejnowski 1995). This generalizes, in the spirit of artificial neural networks, to independent "winner-take-all" operations among filter outputs at the same position.

The filtering abstraction leads to emergent notions of "energy" in the response, and the composition of filters followed by selection of, for example, the maximal energy response can clearly be done in columns. But does the filtering abstraction capture the limits of what can be done in a columnar architecture? Is this edge detection? These are the instantiated versions of our earlier questions about computation and representation.

That this is not yet a complete model for columnar computation follows from the interdependence between columns. The anatomical and physiological evidence for intercolumnar interactions in the superficial layers is classical (Szentagothai) and continues to accumulate (Bosking, Zhang, Schofield, and Fitzpatrick 1997; Kapadia et al. 1995; Nelson and Frost 1985; Malach et al. 1993; Douglas and Martin 2004). Decisions at each location are not independent: they are coupled. Decisions are a function of neighbors, and neighbors' decisions are functions of their neighbors. (This coupling will be fundamental to answering the title question of this chapter.) Returning to edge detection, such neighbor interaction is commonly explained in terms of perceptual organization: co-aligned facilitation could enhance long, straight edges. Theorists suggested this nearly twenty years ago (Mitchison and Crick 1982), and it serves as a prime instance of interactions beyond the classical receptive field (Allman, Miezin, and McGuinness, 1985). Thus the physiology appears to be consistent with function, which provides a hypothesis worth testing.

To implement and test this model for edge detection—filtering followed by selection—requires specifying the filters, their interactions, and the detection mechanism. Research in computer vision has considered these issues and has settled on a model precisely in this class (Canny 1986). While performance seems reasonable (figure 11-2), detailed examination reveals fundamental problems. Without implementation, however, it would have been difficult (if not impossible) to foresee them.

The problem of edge detection is not philosophical; it is constructive. Given the component operations (filtering, detection), how should they be put together? Cortical columns arrange the filters for efficent evaluation, but what interactions across position and scale? Can the results in figure 11-2 be evaluated in nonsubjective terms? The more such questions are asked, the more arise: In which step of the above model—the filtering, the decision, or both—do the problems arise? Canny implemented a type of hysteresis in the decision process, a variable-strength form of co-aligned facilitation. Does the problem lie in the hysteresis, or in the co-aligned facilitation? Many examples in the literature (e.g., Kapadia et al. 1995, figure 10), show facilitation between cells with up to 50 degrees orientation differences. Are such exceptions just a physiological "smear" in the connections, or is there more going on? How can these ideas be extended to deal with stereo (recall the ocular dominance bands) or shading? In short, how can visual information processing be structured on orientation hypercolumns?

Figure 11-2. An illustration of the geometric problems in interpreting standard approaches to edge detection. *Left*, the original image. The edge map (*center*) is obtained from the Canny operator. Matlab implementation, scale = 3. This operator is designed with a hysteresis stage in effect to enforce co-aligned facilitation and normalization so that the "best" edge orientation can be selected for each position. Thus, it effectively realizes the two basic postulates for filters. While the result seems solid at first glance, a detail from the shoulder/arm region (*right*) shows the incorrect topology; neither the detail of the shoulder musculature nor the separation from the arm are localized correctly. A more careful examination of the full edge map (*center*) now confirms the incorrect topology in many locations. Any attempt to follow along the right shoulder would get lost in the arm. We contend that a sufficient approach to edge detection should be topologically correct, which casts doubt on the claim that the Canny operator is doing edge detection.

Defining the problem of edge detection is more involved than defining the traveling salesman problem. The natural "distance" measure introduced by co-aligned facilitation is deviation from straightness. How might we discover better "distance" measures?

Abstracting the Curve Inference Problem

We are now at the point where we need to build an abstraction for vision, that is, in the senses discussed earlier, an analog to transportation. To illustrate a possible approach, we review a computational model that is under development.

We take orientation to be fundamental. Let visual orientation selectivity be a substrate for representing those tangents that approximate the curves that bound objects, that define highlights and other surface markings, and that group into sets of curves to provide texture flows (e.g., hair, fur), and other visual patterns. As suggested in the section "Problem Abstractions and Computational Abstractions" above, horizontal interactions between orientations are thought to reduce the errors inherent in locally estimating orientation so that orientation change can be used to localize corners and discontinuities (as occur at the point where one object occludes another in depth, for example), and for grouping and completing contour fragments obscured by highlights and specularities (Zucker, Dobbins, and Iverson 1989). But how can we derive the connections between tangents: by what rules do they constrain one another? We suggest that differential geometry is the natural abstraction to adopt because it lies intermediate between images and objects and because it provides a theoretical level whose constructs emerge as the natural constraints on interaction.

Curve Inference and Tangent Maps

Local measurements of orientation are inherently ambiguous; while often correct (but coarse), sometimes local intensity structure, or noise, can conspire to affect the maximal local response. To introduce global constraints, we adopt the mathematics of differential geometry, which dictates that any such interactions must involve curvature (Parent and Zucker 1989). The formal question is how to transport a tangent at one location to a nearby location, and the analysis is not unlike driving a car, in the following sense. At each instant of time the axis of the car defines its (tangent) orientation, and the relationship between the orientation of the car at one instant with that at the next depends on how much the road curves; in operational terms, it depends on how much the steering wheel has to be turned during transport. This requires that curvature must be represented systematically with respect to orientation in the cortex, and we (and others) have established that another property of cortical neurons—*endstopping*—is sufficent for achieving this (Dobbins, Zucker, and Cynader 1987). The majority of superficial, interblob orientation selective cells in V1 are also endstopped to some extent, and these biselective dimensions of orientation and endstopping are precisely what is required to represent tangent and curvature. We have used such notions of transport to derive the strength of horizontal interactions (Ben-Shahar and Zucker 2003), which agree with available data for straight situations (curvature $= 0$) but generalize as well to explain the non–co-linear data; see figure 11-3.

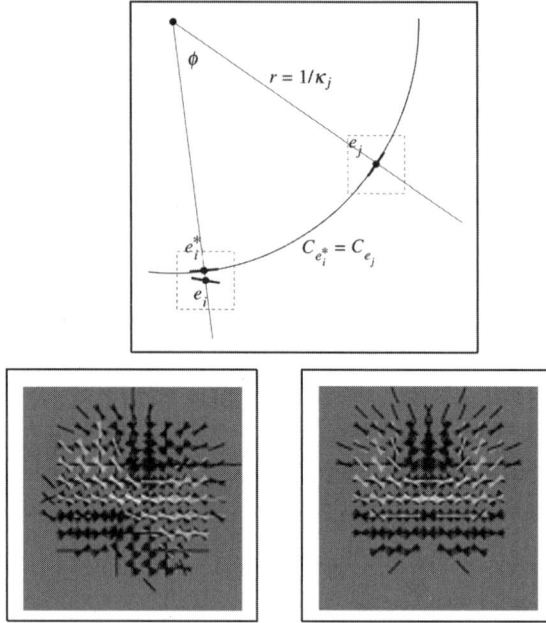

Figure 11-3. The geometry of intercolumnar interactions. *Top*, cocircularity indicates how consistent a neighboring tangent e_j is with a given tangent e_i. Since the positions i and j are close, the actual edge curve can be approximated by its osculating circle, and curvature is approximately constant. The neighboring tangent can be transported along the osculating circle, and the mismatch in position and orientation (between e_i^* and e_i) provides a (distance) measure between them. The larger the mismatch, the bigger the distance. Networks can be designed to select those tangents that minimize such distances (Hummel and Zucker 1983; Miller and Zucker 1999). *Bottom*, in network implementations, the transport results are precomputed and embedded in the connections. A very small mismatch results in an excitatory connection, and a larger mismatch in an inhibitory one. Two examples of the compatibilities derived from cocircularity are shown for curve inference. Think of a sampling of layer 2–3 pyramidal cells. The bars indicate orientation preference, and all compatibility fields are with respect to the central neuron (shown at maximal brightness). The brightness for each bar is the strength of the synapse with the central tangent; positive contrast is excitatory and negative contrast is inhibitory. Multiple bars at the same position indicate several cells in the same orientation hypercolumn. The connections are intended to model long-range horizontal interactions. Analogous connections can be defined for hue, texture, and shading interactions (Zucker, Dobbins, and Iverson 1998).

We emphasize that transport and curvature are emergent concepts at the geometric level. They predict a system of interconnections that appears roughly co-linear, with some smearing; this smearing is not noise, but is the result of curvature. The curvature-based models predict both the shape of this spread and its second-order (variance) statistics.

An analysis of the results in figure 11-4 is also possible at the abstract level. Rather than make comparisons with "intuitive" notions of edge, we appeal again to the basic mathematics of the situation. Whitney has classified maps from smooth surfaces into smooth surfaces (Golubitsky and Guillemin 1973) and has shown that only two situations can occur generically (i.e., without changing under small changes in viewpoint): the fold and the cusp (the position where the fold disappears into the surface); see figure 11-5.

Folds clearly indicate boundaries when viewed from a given position; in fact, the word implies that the tangent plane to the surface "folds" away from the viewer's line of sight. They thus become singular, the two-dimensional tangent space collapses to a one-dimensional tangent, and this is the object we seek.

The Position-Orientation Representation

The geometry just developed begins to articulate the relationship between problems and representations developed in the introduction. The requirement is a space of tangents at each position, which a redrawing of the classical "ice

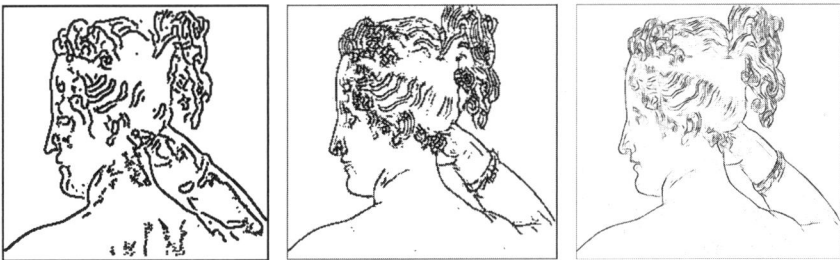

Figure 11-4. Performance of our model for boundary and edge detection. *Left*, the Canny output at a scale larger than shown in figure 11-2 (*center*). Note how the topological problems remain despite the scale variation. This places doubt on standard scale-space models for edge detection. *Center*, the tangent map obtained from our logical/linear operators (Iverson and Zucker 1995). Note differences in the edge topology, with the shoulder musculature clearly indicated and proper T-junctions around the neck and chin. Such T-junctions signal orientation discontinuities. *Right*, the result of our relaxation process using the cocircularity compatibilities (five iterations of Hummel and Zucker 1983). Note how the isolated responses through the hair have been removed and how the details in high curvature regions (such as the ear) have been improved.

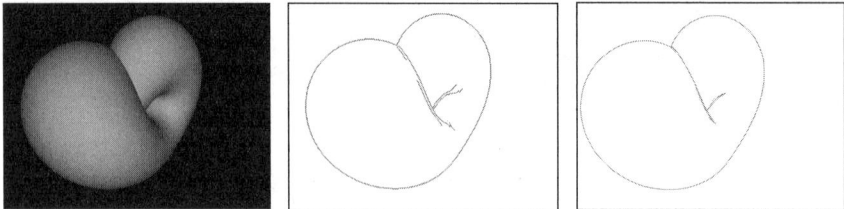

Figure 11-5. The motivation from differential topology for early vision. The image of the Klein bottle (*left*) shows how T-junctions can arise from occlusion relationships (e.g., at the top of the figure), and how certain interior edges can end (e.g., where the fold smoothly joins the body). The Canny edge structure shown (*center*) is inconsistent with both of these topological observations. Notice how the boundary T-junction is not connected, how it smoothes the outline, and how the interior folds blur into the shading. *Right*, the output of the logical/ linear operator. Notice how the T-junctions are maintained and how the contours end at cusps. Such configurations resemble the shoulder musculature in the Paolina image. Dark tangents signify lines; gray ones, edges; see text for a description of how this structure is loaded into the columnar machine.

cube" model elaborates (figure 11-1). This can be viewed as a structure "on top of" the image, with retinotopic (x,y) coordinates extended into a third dimension ("height") corresponding to orientation. But the orientation axis is somewhat different than the length axis because orientation wraps around 2π, or the circle S^1. Thus we speak of this space not as (x,y,z), but as (x,y,θ), where θ is the tangent angle. A point in this space lives in $R^2 \times S^1$. In differential geometry this space is related to the unit tangent bundle.

It is instructive to consider a few examples of how curves in the plane lift into $R^2 \times S^1$, to underline the uses of this encoding within a cortical columnar machine. First, a straight line in the plane lifts to a "horizontal" straight line in $R^2 \times S^1$, whose "height" depends only on the angle θ. A smooth, closed curve in the plane, say an ellipse, lifts into a smooth, closed curve in $R^2 \times S^1$. Discontinuities in orientation lift into broken curves (see figure 11-6).

Encoding a Problem Instance

We have been considering the superficial layers as a kind of geometric machine, and we will be formalizing this shortly. Before doing that, however, we demonstrate another use for the emergent notion of tangent: we can examine the intensity profiles in the tangent and the normal directions separately. Note that linear receptive fields would average these together. We observe immediately that

- *Normal direction:* The fold condition can take on a different intensity profile for a bounding edge (which involves a dark-to-light transition) from an interior fold (which often involves a light-to-dark-to-light transition, or two edges very close together) or vice versa. This latter profile is often

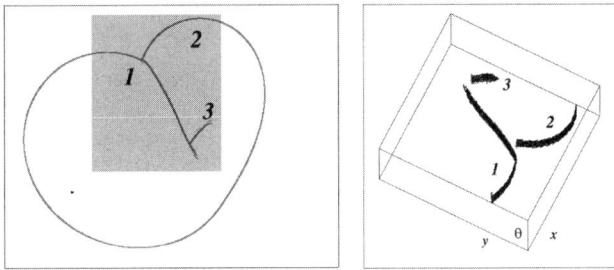

Figure 11-6. An illustration of the lift into position (x,y), orientation (θ) space. Three contour fragments from the tangent map in figure 11-5 (*right*) are highlighted. Notice how the discontinuity in orientation at the 1-2 T-junction is separated, highlighting multiple orientations at the same position, the natural columnar representation for orientation discontinuities.

called a line. Standard linear operators, such as those underlying the Canny, can confuse these conditions.

• *Tangential direction:* The definition of a tangent demands that continuity conditions exist (that is, that the limit of one point approaching another must exist). This corresponds to continuity constraints on the intensity pattern.

A necessary condition for a tangent to exist is that the continuity condition be satisfied for both lines and edges, and we have developed a class of non-linear local operators, called logical/linear operators (Iverson and Zucker 1995) that use Boolean conditions to test whether the above structural criteria are met; if so, they return the average; if not, they veto to zero. "Edge operators" are separated from "line operators," and lines can arise either in light-dark-light conditions (typical of a crack or a crease) or dark-light-dark conditions (typical of a highlight). These nonlinearities are different from the compatibility fields above, because they are taking place at much smaller spatial scales. In effect this processing can be used to "load" the orientation and curvature information into the columnar machine. For this we recall the differences in receptive field size from layer to layer in V1, which suggests the following sketch:

• the initial development of orientationally selective responses at a small scale (layer 4C);
• the development of orientationally selective responses over a larger scale (layers 5 and 6); shunting or other inhibition can implement the logical/ linear nonlinearities (Borg-Graham, Monier, and Fregnac 1998);
• the combination of the above into orientation/endstopped (or orientation/ curvature) covarying responses; this builds up an orientation column (projection from layer 6 up to superficial layers via inhibitory interneuron);
• the refinement of responses in the orientation/curvature column (horizontal interactions in layers 2–3) by computing with cliques (discussed next).

Thus our suggestion, in crude terms, is that intracolumnar processing is used to extract initial data for the full columnar machine to process. While it is impossible to know at this time whether the above sketch is correct, notice how the nonlinearities, which derive from emergent requirements such as tangent continuity, correspond with layer segregation.

Texture, Color, and Surface Properties

The emergence of differential geometry in curve inference implicates V1 in computing surface descriptions. We already saw an example in figure 11-6 of this, in which boundary discontinuities typically provide a monocular occlusion clue; see also Lehky and Sejnowski (1988).

Texture and color are two different aspects of surface coverings. Ben-Shahar and Zucker (2003) show that oriented textures and the hue component of color (2004) are both formal possibilities. Since certain aspects of shading fall into this same formal class, it should be included as well (Huggins and Zucker 2001).

More important, from the surface perspective, are monocular clues about qualitative surface structure that are available from combinations of the above cues. These include information about whether the surface is smoothly folding away from the viewer or is abruptly cut off (Huggins and Zucker 2001); whether abrupt intensity changes correspond to object or shading boundaries (Ben-Shahar and Zucker 2003), whether hue changes correspond to cast shadows or mutual illumination (Ben-Shahar and Zucker 2004), and so on. We shall have more to say about surfaces shortly.

The Columnar Machine

The (position, orientation) representation for early vision suggests an abstraction, in loose analogy with the traveling salesman problem. Discrete objects, such as orientations (analogous to cities), are connected via neighbor relations between orientations, loosely analogous to the roads connecting cities in the traveling salesman problem. The continuous distances associated with roads now take two forms: a measure of activity associated with each discrete object (which has no analog in the TSP) and a measure of consistency between objects (lengths of roads, which might be both positive and negative in this generalization). But a difference arises because the discrete objects are further organized into columns, and conditions can be enforced differently along and between the columns. For example, while two orientations along a column can signal an orientation discontinuity, all orientations at every position should not be allowed. Thus one would want constraints on activity over neighborhoods. The notation to capture these abstractions is shown in figure 11-7.

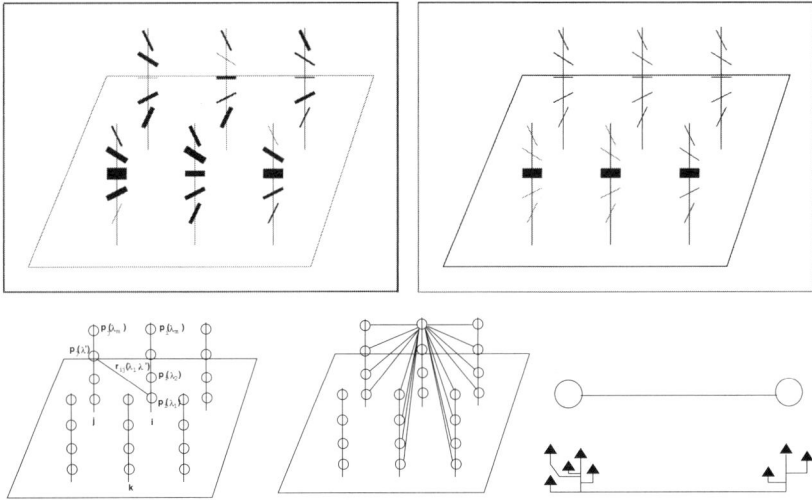

Figure 11-7. Deriving an abstract model for columnar computation based on the redrawn columns in figure 11-1. The thickness of the oriented bars denotes activity of cells; initial values in time (*top left*) and then later (*top right*) denote processing in time, which may be viewed as the result of a selection process along each fiber. Thus there are temporal as well as spatial and orientation dimensions to visual information processing; dynamics are not shown. The abstraction into mathematical symbols (*bottom left*) is then reduced to densely interconnected networks of pyramidal cells (*bottom right*). These are now described in detail. In the abstract orientation columns in figure 11-1, the individual tokens representing a "cell" whose receptive field exhibits a particular orientation preference are represented by a label (λ); a set of such labels is defined for every (columnar) location $i, j, \ldots, n$. Attached to each label at each position is a number, the probability of label λ at node i. Interactions between labels at neighboring positions are weighted by synaptic coefficients, or compatibilities $r_{i,j}(\lambda,\lambda')$, that capture the influence that λ' at j has on λ at i. For the curve detection example the geometrically derived $r_{i,j}(\lambda,\lambda')$ were shown in figure 11-3. *Bottom center*, each label at each node has a network of connections to labels at neighboring nodes; these model, for example the horizontal interactions in layers 2–3 of V1. *Bottom right*, the abstract model is reduced to a network of pyramidal cells by realizing each O–O abstract complex as a group of cells with rich excitatory interactions. Although only a handful of cells are shown in this drawing, calculations in Miller and Zucker (1999) suggest cliques of thirty to forty cells are sufficient for orientation resolution at the level of hyperacuity. The polymatrix game abstraction further suggests dynamics in which about five spikes in twenty-five milliseconds signal the clique that corresponds to the game equilibrium.

We now develop a computational model consistent with this architecture. The computations are suffcient to use the geometric constraints derived in the previous section, and were in fact used (formulated as a relaxation) in obtaining the results shown. Although it embodies aspects of constraint satisfaction and energy models, there are substantial differences. The computation takes form as a game, in a technical sense, which can then be specialized to networks of neurons in various ways. (We describe one.) To illustrate the formal style, this section is more mathematical than previous ones.

One might have expected Bayesian decision theory (Knill and Richards 1996) to arise here, given that decisions are being made (e.g., is there an edge at location [x,y]?) with probabilistic information. But the intercolumnar connections suggest that decisions are coupled across positions, with the question of whether there is an edge at (x,y) depending on whether there are edges near (x,y), and vice versa. Such coupled systems of decisions are games in the formal sense.

There are three steps to the development in this section. First, we introduce the concept of a polymatrix game and show how this leads to a natural dynamical system. (Analog systems are related to dynamical systems.) We then rewrite the game as a linear complementarity problem and note discrete (vertex pivoting) algorithms for solving it. Thus it can be related to complexity classes, such as NP-complete. But most importantly, it is this discrete view that suggests a very different way of obtaining a solution—computation via cliques of neurons—that provides a very rapid answer. Such models should be compared with other attempts to formulate spike-based processing, where the pattern recognition problem has been simplified significantly (Hopfield 1995).

Polymatrix Games

An *n-person game* (Nash 1951) is a set of *n players*, each with a set of *m pure strategies*. Player *i* has a real-valued payoff function $s_i(\lambda_1, \ldots, \lambda_n)$ of the pure strategies $\lambda_1, \ldots, \lambda_n$ chosen by the *n* players. Players can adopt *mixed* strategies, or probability distributions on each player's *m* pure strategies. A player *i*'s payoff for a mixed strategy is the expected value of *i*'s pure strategy payoff given all players choose according to their mixed strategies. Notice how mixed strategy payoffs are only meaningful in terms of all players' simultaneous actions, which is an important generalization from the filtering/local detection model developed earlier. It captures the interactive component across position implied by the overlaping neuronal connections. In terms of the cortical machine (figure 11-7), you might think of each columnar position as a player and each label as a pure strategy. $p_i(\lambda)$ is the probability distribution that defines player *i*'s mixed strategy. (We shall consider other games shortly.)

A *competitive* or *Nash equilibrium* is a collection of mixed strategies for each player such that no player can receive a larger expected payoff by changing his/her mixed strategy given the other players stick to their mixed strategies. Nash showed that such equilibria always exist.

A *polymatrix game* is an *n*-person game in which each payoff to each player *i* in pure strategy is of the form

$$s_i(\lambda_1, \ldots, \lambda_n) = \sum_j r_{ij}(\lambda_i, \lambda_j)$$

where for all i, $r_{ii}(\lambda_i, \lambda_i') = 0$. We may interpret r_{ij} as i's payoff from j given their respective pure strategies λ_i and λ_j. This implies a payoff to i in mixed strategies of the form

$$\sum_{\lambda_i, j, \lambda_j} p_i(\lambda_i) r_{ij}(\lambda_i, \lambda_j) p_j(\lambda_j), \tag{11.1}$$

where $p_i(\lambda_i)$ is the probability player i chooses strategy λ_j.

The quadratic form of equation 11.1 can be generalized to include a penalty function in i's payoff (equation 11.1) that is a convex quadratic function of i's mixed strategy. It is also possibile to include a constant payoff term $c_i(\lambda_i)$ for each strategy λ_i of each player i, regardless of the other players' choices. We make the $c_i(\lambda_i)$ explicit, since these will correspond to bias terms that are important below. Therefore our payoff (11.1) to i is of the more general form:

$$\frac{1}{2} \sum_{\lambda_i, \hat{\lambda}_i} p_i(\lambda_i) r_{ii}(\lambda_i, \hat{\lambda}_i) p_i(\hat{\lambda}_i) \;+\; \sum_{\lambda_i, j \neq i, \lambda_j} p_i(\lambda_i) r_{ij}(\lambda_i, \lambda_j) p_j(\lambda_j) \;+\; \sum_{\lambda_i} c_i(\lambda_i) p_i(\lambda_i).$$
$$\tag{11.2}$$

The elements of a polymatrix game can be specified with an $mn \times mn$ consistency matrix R and an mn bias vector c given by

$$R = \begin{bmatrix} [r_{11}(\lambda_1, \hat{\lambda}_1)] & \cdots & [r_{1n}(\lambda_1, \lambda_n)] \\ \vdots & \ddots & \vdots \\ [r_{n1}(\lambda_n, \lambda_1)] & \cdots & [r_{nn}(\lambda_n, \hat{\lambda}_n)] \end{bmatrix} \quad c = \begin{bmatrix} [c_1(\lambda_1)] \\ \vdots \\ [c_n(\lambda_n)] \end{bmatrix}. \tag{11.3}$$

For each player i it will also be convenient to refer to $[c_i(\lambda_i)]$ as c_i, to i's vector of mixed strategies $[p_i(\lambda_i)]$ as p_i, and to the vector of p_i's as p. If we let A be the $n \times mn$ matrix

$$\begin{bmatrix} -1 \ldots -1 \cdots 0 \ldots 0 \\ \vdots & \ddots & \vdots \\ 0 \ldots \cdots -1 \ldots -1 \end{bmatrix}$$

and let $q^\top$ be the n-vector $(-1, \ldots, -1)$, then p is a vector of all players' mixed strategies if and only if

$$Ap = q, \quad p \geq 0. \tag{11.4}$$

We express the gradient of equation 11.2 as

$$\sum_{j=1}^{n} \left[r_{ij}(\lambda_i, \hat{\lambda}_j) \right] p_j + c_i. \tag{11.5}$$

Assume p satisfies equation 11.4 and is fixed except for player i, whose payoff is given by equation 11.2. Since this function is concave, and the constraint set (a simplex) is convex, a given mixed strategy for i will have a maximum payoff if and only if i's gradient (equation 11.5) has a vanishing projection onto the constraint set (equation 11.4). Miller and Zucker show that an alternative characterization of the competitive equilibria of the poly-matrix game (equation 11.3) in terms of the equilibria of the dynamical system

$$p' = Rp + c,$$
$$Ap = q, \quad p \geq 0. \tag{11.6}$$

In other words, these equilibria are precisely the points at which the vector field of (6) vanishes. If R is symmetric then p' is the gradient of

$$1/2 p^\top R p + c^\top p. \tag{11.7}$$

The first term in equation 11.7 corresponds to the *average local potential* in relaxation labeling (Hummel and Zucker 1983), which is the form extremized in the example computations in this chapter.

Although the above system seems extremely abstract, methods for solving it via gradient descent (when the r_{ij} are symmetric) are classical (Hopfield 1984; Hummel and Zucker 1983) Statistical methods, such as simulated annealing (Kirkpatrick, Gelatt, and Vecchi 1983), Gibbs sampling, and Markov Chain Monte Carlo, can also be applied. However, such methods can be extremely delicate numerically, and it is difficult to imagine how they can be applied to neurons over short time periods (Nowak and Bullier 1997). The obvious idea of relating iterations of the gradient descent process to neuronal firing is questionable, because what does "firing rate" mean over such short time periods (Rieke et al., 1996)?

We thus seek different ways of writing this system that permits a proper reduction to biophysical models of neurons. Using the Kuhn-Tucker theorem (Kuhn and Tucker 1951) it can be shown that that p is an equilibrium for equation 11.6 if and only if there also exist vectors y, u, v such that p, y, u, v satisfy the system

$$\begin{bmatrix} R & -I_n \\ I_n & 0 \end{bmatrix} \begin{bmatrix} I_n & 0 \\ 0 & I_n \end{bmatrix} \begin{bmatrix} p \\ y \\ u \\ v \end{bmatrix} = \begin{bmatrix} -c \\ e \end{bmatrix} + \delta \begin{bmatrix} -\tilde{c} \\ 0 \end{bmatrix} \tag{11.8}$$

$$p, y, u, v \geq 0$$
$$p^\top u + y^\top v = 0$$

Here I_n is the $n \times n$ identity matrix.

The above system of equations is an example of a *linear complementarity problem*, which in general is NP-complete. Identifying this abstract structure immediately opens connections to several important special cases, including linear and convex quadratic programming (Garey and Johnson 1979), in which it is polynomial. Two-person zero sum games are equivalent to linear programs, and the selection problems discussed earlier are in this special class; some decision problems can be defined as "games against nature," or games against a completely random opponent.

Computing with Cliques of Neurons

Connections between different problems are not the only advantage to emerge at the abstract level. Insight into possible cortical dynamics which differ fundamentally from gradient descent also emerge by capitalizing on the different perspectives provided by continuous (analog) and discrete (vertex pivoting) computations.

An important technique for linear complementarity problems is called *Lemke's algorithm* (Cottle, Pang, and Stone 1992), a vertex pivoting algorithm that uses bias to find a solution. While we shall not discuss it in detail, we do return to the bias terms c in equation 11.8, and observe that these terms appear with a kind of switch δ indicating whether the bias is on or not.

This bias interpretation can be applied to neurons as follows. Think of neurons as players and pure strategies as whether a neuron should depolarize (spike) or hyperpolarize (Miller and Zucker 1992, 1999). Then, if neurons are modeled as piecewise-linear amplifiers, synapses (compatibilities) as conductances, and so on, they can be placed in the above form. The key idea behind our model is to consider groups of tightly interconnected excitatory neurons capable of bringing themselves to saturation feedback response following a modest initial afferent bias current, like a match igniting a conflagration (in the phrase of Douglas and Martin, 1992).

The basic computation is in two phases and builds upon the observed regular spiking behavior of pyramidal cells. Prior to phase I the cells in a patch of cortex are initially quiescent. Among these hundreds of thousands of cells are several times that number of cortical *cliques*. (Miller and Zucker calculate each clique contains about thirty-three highly interconnected cells.) A "computation" amounts to activating the cells in a single clique, but no others, to saturation feedback response levels. In phase I ($\delta = 1$ above): afferent stimulation produces a single spike in a majority of the cells in one clique (whose cells can be distributed among many different iso-orientation areas), as well as a certain number of other cells outside the clique (noise). Because the clique has a sufficient level of excitatory interconnections, all its cells drive themselves to saturation response levels of about five spikes in twenty-five milliseconds (end of phase I), whereas the initially activated cells outside the clique return to their resting membrane potentials and do not spike further (end of phase II; $\delta = 0$). Thus the clique has been "retrieved" through a parallel analog computation, and it is this clique that defines the equilibrium point of the system (equation 11.8).

The realization of a clique of orientationally selective cells would be a family of receptive fields; one can predict that such a family provides a stable, high-resolution (hyperacuity) representation of short curve segments (figure 11-8).

Learning, Codes, and the Games Neurons Play

Game theory is an abstract model for a class of computations natural for and applicable to problems in early vision. While the previous discussions were rather formal, they did suggest a strikingly different setting for neural computation than the neuron versus synapse framework discussed in the introduction to this chapter. This at least raises questions about the current dogma in several different areas. We now simply list several of them.

First: mixed strategies are probabilistic and easily related to the probability of neurotransmitter release. Thus the stochastic property of neurons (Mainen and

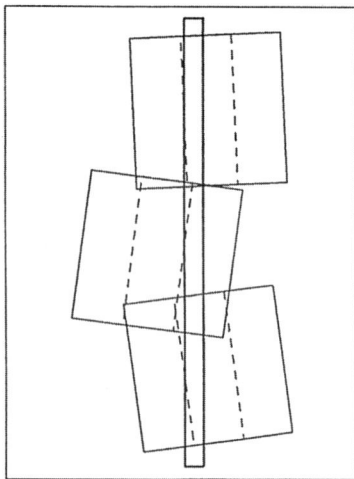

Figure 11-8. Distributed representation for a thin line contour derives from a family of receptive fields covering it. Each of these receptive fields comes from a single cortical neuron, and a clique consists of about thirty-three neurons. In this example receptive fields are represented by rectangles, and a white slit contour stimulus (heavy black outline) excites a highly interconnected clique of simple cortical (S) cells to maximal saturated feedback response by crossing, in the appropriate direction and within a narrow time interval, the edge response region of a sufficiently large proportion of the clique's cells. Biophysical simulations suggest the clique is signaled by about five spikes in about twenty-five milliseconds. Three such receptive fields, out of the approximately thirty-three required, are illustrated here.

Sejnowski 1995; Cowan, Sudhoff, and Stevens 2001) that seemed so curious and different from silicon circuits in the introduction now takes on a hypothetical role that could be central to understanding neural computation.

Second: sparse coding is now widely accepted as a model for V1 neuronal receptive fields (Olshausen and Field 1997). However, each clique of neurons in the computation above can be viewed as an atom in a code, and the output of each computation could be viewed as selecting a sparse representation inferred from the image. Thus it is not at all clear which is the correct level for specifying the neural code.

Third: there is a classical identification between reliability and redundancy. Neural computations are notoriously reliable, but simply building them in a redundant fashion would make little use of cortical architecture and would be energy inefficent (compare Attwell and Laughlin 2001). The clique computation suggests a new direction to explore for the redundant portion of the computation: it is aimed at increased accuracy. (This was demonstrated for hyperacuity.) Is achieving accurate/reliable configurations a general design principle for neural computation?

Fourth: perceptual learning occurs over several different time scales. At the short scale it is curious to observe that spike timing dependent plasticity (Dan and Poo 2004) occurs at precisely the same time scale as the clique computation. How universal is the twenty millisecond epoch?

Fifth: for longer time scales learning from the game theoretic standpoint involves changing (or selecting from among different) Nash equilibria. Evolutionary game theory (Weibull 1995) provides one model for this; many others arise in behavioral game theory (Camerer, 2003). Are these models relevant for perceptual learning?

Columnar Computations Beyond V1

Although the discussion of columnar computations has been concentrated in the superficial layers of the first visual area, many other structures exhibit columnar organization with long-range patchy connections between them. Perhaps the most natural to consider next is V2, because the stereo computation is richly elaborated into it (Hubel and Livingstone 1987).

The Geometry of Stereo Correspondence

So far we have considered the input from only one eye; we now consider both. Abstractly this implies an important construction for the columnar machine: the "product" of two machines, one for the left eye and the other for the right eye. Mathematically this suggests working in $(R^2 \times S^1) \times (R^2 \times S^1)$ and compatability fields that also take the product form. We have developed this

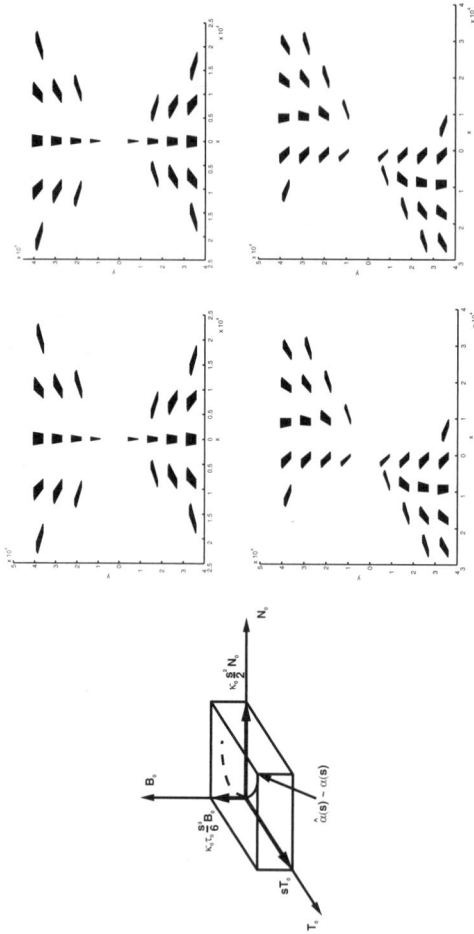

Figure 11-9. The local approximation of a curve in three space involves its projection in the left and right images. *Far left*, a curve in three space can be described by the relationships between its tangent, normal and binormal. As the curve moves across depth planes, there exists a positional disparity between the projection of the curve in the left image and the projection in the right image, as well as higher order disparities, for example disparities in orientation. *Top left and top right*, for the stereo correspondence problem, we are given two edge maps (one for the left eye and one for the right); each of these will be consistent (in the sense that they satisfy the transport constraint); our goal now is to make them consistent with a local approximation to the space curve from which they project. The osculating object for space curves is a helix. Two examples of positive discrete compatibility fields are shown. Note how they incorporate both position and orientation disparity; this is especially evident in the pair of compatibility fields shown in the bottom left and bottom right panels.

238

Figure 11-10. The depth map associated with a stereo lily pair with only the flower shown in detail. Each point in the lily image is a tangent in three-space that is geometrically consistent with its neighbors. See color insert.

product structure into an algorithm for computing stereo correspondences (Alibhai and Zucker 2000; Li and Zucker 2005) that generalizes the tangent fields for plane curves to those for general space curves. A curve in space can be described by the relationships between its tangent, normal and binormal at each point. As the curve moves across depth planes, there exists a positional disparity between the projection of the curve in the left image and the projection in the right image. But most importantly, there also exist higher-order disparities, for example disparities in orientation, that occur. It is these types of relationships that can be capitalized upon when solving the correspondence problem. Rather than correlating left/right image pairs, these algorithms require the existence of a curve in three-space whose projection into the left and right image planes is commensurate with the locus of tangent pairs in a neighborhood of the proposed match.

Since stereo is elaborated into V2, a full series of questions arise immediately about which computations the long-range horizontal connections within V2 implement? How do the problems represented in V2 differ from those in V1? Why, in general, are computations elaborated from one area to the next? The geometry of stereo just sketched suggests one possibility—that the computation of orientation disparity requires a finer resolution for orientation than those computations implemented in V1—but this possibility is at best a vague hypothesis today.

Columnar Computations in Other Areas

In a recent review, Douglas and Martin (2004) list superficial lateral connections in cat, tree shrew, and monkey somatosensory, motor, and visual areas, as well as in the prefrontal cortex and inferotemporal cortex. Whether the game-theoretic model in its abstract form can be extended to problems in all of these areas remains an open question. To our knowledge, however, no other formal model has been proposed that could cover all of them.

The existence of feedforward and feedback projections raises all of the above questions about intercolumnar interaction to questions about interareal interaction. In particular, which of the feedforward and feedback connections serve analagous computational roles to the long-range horizontal connections within an area, and which are implementing a qualitatively different computation? While the feedforward connections are thought to be patchy, a necessary question is whether the feedback connections to layer 1 are diffuse (as limited data currently suggest) or are also patchy remains central.

Summary and Conclusions

There is a substantial divide in computational neuroscience, separating the modeling of neurons at a biophysical level from the modeling of function at an information processing level. We concentrated in this chapter on the second, information processing level, and raised a series of questions about it.

Although our current understanding of this level is quite primitive, theoretical computer science provides some guidance. To be concrete we concentrated on early vision, and representational structure emerged regarding tangents, curvatures, and continuity. That is, problem instances were formulated using the structure of differential geometry. This is not unexpected, given that differential geometry is the natural mathematics of surfaces and objects.

For computation, linear complementarity emerged as a generalization of columnar operations. Algorithms for solving linear complementarity problems provided new insight into finding fast solutions by neural mechanisms. In all of the above cases it was clear that the resulting networks are quite plausible; but starting only with network components, it seems implausible that all of the above abstract functions would have been inferred.

Much remains to be done to obtain a deep understanding of what *computation* means in the phrase neural computation. We focused on a particularly elegant neural architecture—the long-range lateral connections in the superficial layers of the visual cortex—and found a model in game theory. Whether these techniques can be extended beyond this architecture remains unclear. The situation is, in a sense, complementary to the one that Hilbert faced a century ago in mathematics. For him the language and context were clear, and the questions could be posed crisply. It was the answers that were difficult. In computational neuroscience, we understand how to formulate neither the questions nor the answers in general. Perhaps progress will accompany more precise formulations.

Acknowledgments Research supported by AFOSR, DARPA, NIH, and ONR. S. Alibhai and G. Li collaborated on the stereo computations, L. Iverson computed the cocircular compatibilities, O. ben-Shahar the texture and color fields, and P. Huggins the Klein bottle examples. I thank J. Allman, A. Hasenstaub, and P. Huggins for comments, D. Fitzpatrick and M. Stryker for figures.

References

S. Alibhai and S. W. Zucker. 2000. Contour-based correspondence for stereo. In *Computer Vision—ECCV 2000, Lecture Notes in Computer Science 1842*, June.

J. Allman. 1999. *Evolving Brains*. New York: W.H. Freeman.

J. Allman, F. Miezin, and E. McGuinness. 1985 Stimulus specific responses from beyond the classical receptive field: Neurophysiological mechanisms for local-global comparisons in visual neurons. *Ann. Rev. Neurosci.*, 8:407–430.

D. Attwell and S. B. Laughlin. 2001. Art energy budget for signalling the grey matter of the brain. *J. Cereb. Blood Flow Metab.*, 21:1133–1145.

H. Barlow and C. Blakemore, editors. 1989. *Visual Coding and Efficiency*. Cambridge: Cambridge University Press.

O. Ben-Shahar and S. Zucker. 2003. Geometrical computations explain projection patterns of long-range horizontal connections in visual cortex. *Neural Comput.*, 16:445–476.

O. Ben-Shahar and S. Zucker. 2004. Hue geometry and horizontal connections. *Neural Netw.*, 17:753–771.

L. Borg-Graham, C. Monier, and Y. Fregnac. 1998. Visual input evokes transient and strong shunting inhibition in visual cortical neurons. *Nature*, 393: 369–373.

W. H. Bosking, Y. Zhang, B. Schofield, and D. Fitzpatrick. 1997. Orientation selectivity and the arrangement of horizontal connections in the tree shrew striate cortex. *J. Neurosci.*, 17(6):2112–2127.

C. F. Camerer. 2003. *Behavioral Game Theory*. Princeton, NJ: Princeton University Press.

J. Canny. 1986. A computational approach to edge detection. *IEEE Trans. Pattern Analysis and Machine Intelligence*, 8:679–698.

R. W. Cottle, J.-S. Pang, and R. Stone. 1992. *The Linear Complementary Problem*. San Diego, CA: Academic Press.

W. M. Cowan, T. C. Sudhoff, and C. F. Stevens, editors. 2001. *Synapses*. Baltimore, MD: Johns Hopkins University Press.

Y. Dan and M.-M. Poo. 2004. Spike timing-dependent plasticity of neural circuits. *Neuron*, 44:23–30.

K. DeValois and R. L. DeValois. 1998. *Spatial Vision*. New York: Oxford University Press.

A. Dobbins, S. W. Zucker, and M. S. Cynader. 1987. Endstopped neurons in the visual cortex as a substrate for calculating curvature. *Nature*, 329:438–441.

R. J. Douglas and K. A. C. Martin. 2004. Neuronal circuits of the neocortex. *Annu. Rev. Neurosci.*, 27:419–451.

R. J. Douglas and K. A. C. Martin. 1992. Exploring cortical microcircuits: A combined physiological and computational approach. In T. McKenna, J. Davis, and S. Zornetzer, editors, *Single Neuron Computation*, 381–412, New York: Academic Press.

R. Durbin and D. Willshaw. 1987. Art analogue approach to the traveling salesman problem using an elastic net method. *Nature*, 343:698–671.

D. Field, A. Hayes, and R. Hess. 1993. Contour integration by the human visual system: evidence for a local association field. *Vision Res.*, 33:173–193.

M. Garey and D. Johnson. 1979. *Computers and Intractability*. New York: W.H. Freeman.

M. Golubitsky and V. Guillemin. 1973. *Stable Mappings and their Singularities*. New York: Springer-Verlag.

M. Grotschel, L. Lovasz, and A. Schrijver. 1993. *Geometric Algorithms and Combinatorial Optimization*. New York: Springer-Verlag.

R. H. Hahnloser, R. Sapreshkar, M. Mahowald, R. Douglas, and H. Seung. 2000. Digital selection and analogue amplification coexist in a cortex-inspired silicon circuit. *Nature*, 405:947–951.

D. Hansel and H. Sompolinsky. 1998. Modeling feature selectivity in local cortical circuits. In C. Koch and I. Segev, editors, *Methods in Neuronal Modeling* 467–499, Cambridge, MA: MIT Press.

J. J. Hopfield. 1984. Neurons with graded response have collective computational properties like those of two-state neurons. *Proc. Natl. Acad. Sci. USA*, 81:3088–3092.

J. J. Hopfield. 1995. Pattern recognition computation using action potential timing for stimulus representation. *Nature*, 376:33–36.

J. J. Hopfield and D. W. Tank. 1985. "Neural" computation of decisions in optimatization problems. *Biol. Cybern.*, 52:141–152.

D. H. Hubel and M. S. Livingstone. 1987. Segregation of form, color, and stereopsis in primate area 18. *J. Neurosci.*, 7(11):3378–3415.

D. H. Hubel and T. N. Wiesel. 1977. Functional architecture of macaque monkey visual cortex. *Prod. R. Soc. Lond. B*, 198:1–59.

P. Huggins and S. W. Zucker. 2001. Folds and cuts: How shading flows into edges. In *Eighth International Conference on Computer Vision*, 153–158, Washington, DC: Computer Society Press.

R. Hummel and S. W. Zucker. 1983. On the foundations of relaxation labeling processes. *IEEE Trans. Pattern Analysis and Machine Intelligence*, 6:267–287.

L. A. Iverson and S. W. Zucker. 1995. Logical/linear operators for image curves. *IEEE Trans. Pattern Analysis and Machine Intelligence*.

C. Jennings and S. Aamodt. 2000. Computational approaches to brain function. *Nat. Neurosci.*, 3(Suppl.):1160.

M. Kapadia, M. Ito, C. Gilbert, and G. Westheimer. 1995. Improvement in visual sensitivity by changes in local context: Parallel studies in human observers and in v1 of alert monkeys. *Neuron*, 15:843–856.

S. Kirkpatrick, J. C. Gellatt, and M. Vecchi. 1983. Optimization by simulated annealing. *Science*, 220:671–680.

D. C. Knill and W. Richards. 1996. *Perception as Bayesian Inference*. Cambridge: Cambridge University Press.

C. Koch. 1997. Computation and the single neuron. *Nature*, 385:207–210.

C. Koch. 1999. *Biophysics of Computation*. Oxford: Oxford University Press.

A. A. Koulakov and D. B. Chklovskii. 2001. Orientation preference patterns in mammalian visual cortex: a wire length minimization approach. *Neuron*, 29:519–527.

H. W. Kuhn and A. W. Tucker. 1951. Nonlinear programming. In J. Neyman, editor, *Second Berkeley symposium on mathematical statistics and probability*, 481–492, Berkeley: University of California Press.

S. R. Lehky and T. J. Sejnowski. 1988. Network model of shape-from-shading: neural function arises from both receptive and projective fields. *Nature*, 333(2):452–454.

T. Lengauer. 1991. Vlsi theory. In J. van Leeuwen, editor, *Handbook of Theoretical Computer Science, vol A.*, 835–868. Cambridge, MA: MIT Press.

G. Li and S. Zucker. 2005. Contour-based binocular stereo: Inferencing coherence in stereo tangent space. *Int. J. Computer Vision*, page to appear.

W. Maass, T. Natschlger, and H. Markram. 2004. On the computational power of circuits of spiking neurons. *J. of Physiology (Paris)*, in press.

Z. Mainen and T. Sejnowski. 1995. Reliability of spike timing in neocortical neurons. *Science*, 268:1502–1506.

R. Malach, Y. Amir, M. Harel, and A. Grinvald. 1993. Relationship between intrinsic connections and functional architecture revealed by optical imging and in vivo targeted biocytin injections in primate striate cortex. *Proc. Natl. Acad. Sci. USA*, 90:10469–10473.

D. Marr. 1982. *Vision*. San Fransisco, CA: W.H. Freeman.

D. A. Miller and S. W. Zucker, 1992. Efficent simplex-like methods for equilibria of nonsymmetric analog networks. *Neural Comput.*, 4:167–190.

D. A. Miller and S. W. Zucker. 1999. Computing with self-excitatory cliques: A model and an application to hyperacuity-scale computation in visual cortex. *Neural Comput.*, 11:21–66.

G. Mitchison. 1991. Neuronal branching patterns and the economy of cortical wiring. *Proc. Royal Soc. (Lond.)*, 245:151–158.

G. Mitchison and F. Crick. 1982. Long axons within the striate cortex: Their distribution, orientation, and patterns of connection. *Proc. Natl. Acad. Sci. (USA)*, 79:3661–3665.

D. Mumford. 1994. Neuronal architectures for pattern-theoretic problems. In C. Koch and J. Davis, editors, *Large-Scale Neuronal Theories of the Brain*, 125–152. Cambridge, MA: MIT Press.

J. F. Nash. 1951. Noncooperative games. *Ann. Math.*, 54:286–295.

J. Nelson and B. Frost. 1985. Intracortical facilitation among co-oriented, co-axially aligned simple cells in cat striate cortex. *Exp. Brain Res.*, 61:54–61.

L. Nowak and J. Bullier. 1997. The timing of information transfer in the visual system. In K. Rockland, J. Kaas, and A. Peters, editors, *Cerebral Cortex*, vol. 12, 125–152. New York: Plenum Press.

S. J. Nowlan and T. J. Sejnowski. 1995. A selection model for motion processing in area mt of primages. *J. Neuro.*, 15:1195–1214.

B. A. Olshausen and F. D. Field. 1997. Sparse coding with an overcomplete basis set: A strategy employed by v1? *Vision Research*, 37:3311–3325.

P. Parent and S. W. Zucker. 1989. Trace inference, curvature consistency and curve detection. *IEEE Trans. Pattern Analysis and Machine Intelligence*, 11(8):823–839.

N. Pippenger. 1993. Communication networks. In J. van Leeuwen, editor, *Handbook of Theoretical Computer Science*, vol. A., 805–834. Boston: Kluwer.

F. Rieke, D. Warland, R. van Steveninck, and W. Bialek. 1996. *Spikes: Exploring the neural code*. Cambridge, MA: MIT Press.

G. Shepard. 1990. *The Synaptic Organization of the Brain*. New York: Oxford University Press.

W. T. Simoncelli, E. P. Freeman, E. H. Adelson, and D. J. Heeger. 1992. Shiftable multiscale transforms. *IEEE Trans. on Infor. Theory*, 38:587–607.

N. V. Swindale, D. Shoham, A. Grinvald, T. Bonhoeffer, and M. Hubener. 2000. Visual cortex maps are optimized for uniform coverage. *Nat. Neuro.*, 3:822–826.

J. Szentagothai. 1975. The "module" concept in cerebral cortex architecture. *Brain Research*, 95:475–496.

S. Ullman. 1994. Sequence seeking and counter streams: a model for bi-directional information flow in the cortex. In C. Koch and J. Davis, editors, *Large-Scale Neuronal Theories of the Brain*, 257–270. Cambridge, MA: MIT Press.

J. W. Weibull. 1995. *Evolutionary Game Theory*. Cambridge, MA: MIT Press.

A. Zador. 2000. The basic unit of computation. *Nat. Neurosci.*, 3(Suppl.):1167.

S. W. Zucker, A. Dobbins, and L. Iverson. 1989. Two stages of curve detection suggest two styles of visual computation. *Neural Comput.*, 1:68–81.

12

Are Neurons Adapted for Specific Computations? Examples from Temporal Coding in the Auditory System

C. E. Carr, S. Iyer, D. Soares, S. Kalluri, and J. Z. Simon

Introduction

Are neurons adapted for specific computations? Evolution has led to the appearance of specialized neurons, such as the neurons in the auditory system that encode temporal information with great precision (Trussell 1997; Oertel 1999). Nevertheless, it is not clear whether all neurons are adapted for particular computations or even whether specialized computational units are desirable under all circumstances. Some neurons may have more general responses. Other neuron types change their responses under the action of some modulator, but these might be regarded as being adapted for several computations, rather than for some general input-output function (see Golowasch et al. 1999; Stemmler and Koch 1999; Turrigiano, Abbot, and Marder 1994).

It is important to understand the functions of single neurons. Johnston et al (1996) wrote, "Before one can hope to understand systems of neurons fully, one must be able to describe the function of the basic unit of the nervous system, that is, the single neuron and its associated dendritic tree." To make the case that neurons may be adapted for particular tasks, we will use the example of temporal coding cells in the vertebrate auditory system because their function is well known. This allows us to tie physiological and morphological observations to function.

Encoding Temporal Information

In the auditory system, precise encoding of temporal information has direct behavioral relevance. The timing of firing of auditory neurons carries information used for both localization and interpretation of sound. Psychophysical studies support a role of a timing code for localization and pitch detection, and there is good evidence that localization of interaural phase differences falls off with frequency in the same way that temporal encoding falls off (see Hafter and Trahiotis 1997 for review). Therefore, those features of auditory neurons that lead to improved temporal processing should experience positive selection.

Cellular Specializations for Encoding Time: Quality of Input

Sound coming from one side of the body reaches one ear before the other, and the auditory system uses these time differences to localize the sound source. The auditory system encodes the phase of the auditory signal and then uses interaural phase differences to compute sound location (Heffner and Heffner 1992). Nocturnal predators such as the barn owl and mammals that use auditory information to direct their visual foveas towards a sound source all have well-developed abilities to localize sound (Heffner and Heffner 1992). The barn owl's ability to detect small phase or time differences is acute, and the owl is able to catch mice on the basis of auditory cues alone (Konishi 1973). Accurate and precise processing of the auditory stimulus is required for this detection. Auditory nerve fibers phase lock to the waveform of the acoustic stimulus, and this information is preserved and improved in the brain. Two lines of evidence support the idea that accurate temporal coding is important. First, measurements of the vector strength of the auditory nerve signal (calculated from the variability in the timing of action potentials with respect to the phase of the acoustic stimulus) show an improvement in high-frequency phase locking in the owl as compared to other animals by an octave or more (Koppl 1997). Second, models of coincidence detection perform better when the vector strength of the inputs improves (Simon, Carr, and Shamma 1999; Colburn, Han, and Culotta 1990; Grau-Serrat, Carr, and Simon 2003).

Presynaptic Specializations for Encoding Temporal Information

In the bird, auditory nerve afferents divide into two with one branch to the cochlear nucleus angularis (NA), a structure that codes for changes in sound level, and the other branch to the cochlear nucleus magnocellularis (NM) that codes for phase (Sullivan and Konishi 1984; Takahashi, Moiseff, and Konishi 1984). In mammals, similar cell types receiving auditory nerve input are contained in a single nucleus, the ventral cochlear nucleus (see Ryugo 1991). The termination of the auditory nerve onto the somas of avian NM neurons and mammalian bushy cells take the form of a specialized calyceal or endbulb terminal while avian NA neurons and mammalian stellate cells are contacted through bouton-like synapses (figure 12-1A; Jhaveri and Morest 1982; Brawer and Morest 1974; Ryugo and Fekete 1982). The endbulb terminals envelop the postsynaptic cell body and are characterized by numerous release sites. They therefore form a secure and effective connection for the precise relay of the phase-locked discharges of the auditory nerve fibers to their postsynaptic targets. Physiological measures show that phase-locking abilities are correlated with the morphology of the nerve terminals so that phase locking is preserved in the neurons of the NM, and lost at higher frequencies in the noncalyceal projection to the NA (Koppl 1997). In the cat, there is a slight improvement in phase locking between the nerve and the bushy cells of the cochlear nucleus presumably due to monaural coincidence of auditory nerve fibers (Joris, Smith, and Yin 1994; Rothman, Young, and Manis 1993), while in the barn owl, there is a slight decrease (Koppl 1997).

Endbulb terminals are not essential for transmission of phase-locked spikes at low frequencies. The very low best frequency cells of the NM receive large bouton terminals from the auditory nerve and can also phase lock to frequencies below ~1 kHz (Koppl 1997). The task of encoding temporal information precisely becomes more difficult with increasing frequency. The

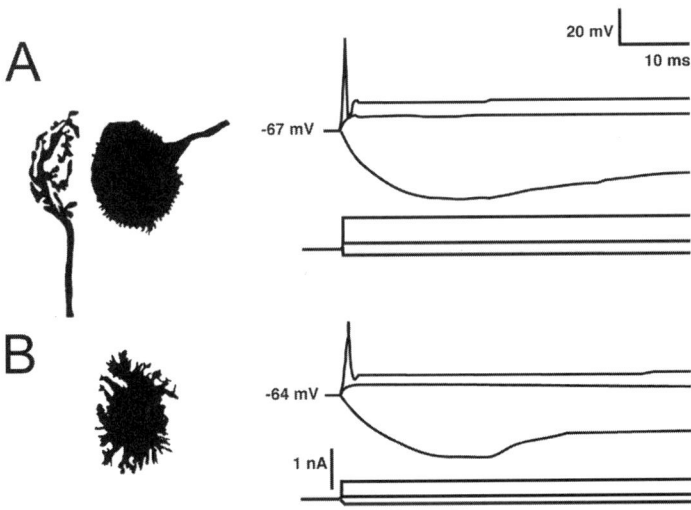

Figure 12-1.   Time coding neurons in the bird brain exhibit a suite of physiological and morphological features suited to their function. (A) Auditory nerve endbulb terminals and magnocellular neurons in barn owl (*left*) and current clamp recordings from chicken NM neuron (*right*; from Reyes, Rubel, and Spain 1994). (B) Laminaris neuron in barn owl (*left*) and current clamp recordings from chicken NL (*right*; from Reyes, Rubel, and Spain 1996).

reason for this is clear when one considers that the absolute temporal precision required for phase locking to high frequencies is greater than that needed for low frequencies, that is, the same variation in temporal jitter of spikes translates to greater variation in terms of degrees of phase for high frequencies. Hill, Stange, and Mo (1989) estimated phase locking in the auditory fibers of the pigeon in terms of the commonly used synchronicity index (vector strength) as well as by measuring temporal dispersion. Vector strength of phase locking decreased for frequencies above 1 Khz. Temporal dispersion, however, also decreased with frequency, indicating enhanced temporal synchrony as frequency increased. The upper frequency limit of phase locking, therefore, appears to depend on irreducible jitter in the timing of spikes (see Carr and Friedman 1999). Thus, endbulb terminals may have emerged as an adaptation for transmission of phase information for frequencies above 1kHz, perhaps associated with the development of hearing in land vertebrates (Rubel and Fritzsch 2002).

The invasion of the presynaptic action potential into the calyx leads to the synchronous release of quanta at many endbulb release sites giving this synapse a high safety factor of transmission (Isaacson and Walmsley 1996; Taschenberger et al. 2002). The invading presynaptic action potential is extremely narrow, being about 250 µsec at 35°C in postnatal day 8–10 animals (Borst, Egelhaff, and Haag 1995; Taschenberger and von Gersdorff 2000) probably due to rapid repolarization mediated by specific potassium conductances. Calcium influx into the presynaptic terminal is also brief and occurs only during the falling phase of the presynaptic action potential (Borst and Sakmann 1996). Because the action potential is narrow, its downstroke occurs quickly, as does calcium influx, reducing the synaptic delay. In addition, the brief period of calcium influx produces a confined and phasic period of neurotransmitter release, which also increases the temporal precision of transmission across the synapse (Sabatini and Regehr 1999).

Transmitter release becomes more precise during development, which leads to less desensitization of the postsynaptic alpha amino-3-hydroxy-5-methylisooxazole-4-propionic acid (AMPA)-type glutamate receptors (Brenowitz and Trussell 2001, Taschenberger et al. 2002). In the medial nucleus of the trapezoid body (MNTB), vesicle pool size, exocytotic efficiency, and the number of active zones increase with age. These changes lead to active zones that are less prone to multivesicular release, reducing AMPA receptor saturation and desensitization (Taschenberger et al. 2002). Similarly, endbulb synapses on chicken NM showed synaptic maturation around the time of hatching with an increased pool of synaptic vesicles, lower release probability, larger transmitter quanta, and reduced AMPA receptor desensitization (Brenowitz and Trussell 2001). These factors improve the ability of avian endbulb synapses and mammalian MNTB calyces to provide an accurate representation of high-frequency firing in mature synapses.

Postsynaptic Specializations for Encoding Temporal Information

Both avian and mammalian time-coding cells possess a number of morphological and physiological specializations that make them well suited to preserve the temporal firing pattern of auditory nerve inputs. In addition to the specialized synaptic arrangement, large cell bodies and reduced dendritic arbors serve to keep the cells electrically compact. Time-coding neurons possess a particular combination of synaptic and intrinsic membrane properties, including fast AMPA receptors and specific K^+ conductances. These features lead to a single or a few well-timed spikes in response to a depolarizing stimulus (figure 12-1A; for reviews see Oertel 1999 and Trussell 1997, 1999). A similar suite of physiological and morphological features also characterizes the neurons of the medial nucleus of the trapezoid body and the type-II neurons of the ventral nucleus of the lateral lemniscus, both of which receive endbulb synapses (Brew and Forsythe 1995; Wu 1999).

Activation of AMPA receptors at endbulb synapses generates extremely brief but large synaptic currents (Raman and Trussell 1992; Zhang and Trussell 1994; Isaacson and Walmsley 1996). The brevity of EPSCs in these neurons depends not only on the time course of release but also on the specific properties of the postsynaptic AMPA receptors. AMPA receptors in time coding auditory neurons have fast kinetics and very rapid desensitization rates such that the duration of miniature EPSCs in auditory neurons are among the shortest recorded for any neuron (Raman and Trussell 1992; Geiger et al 1995; Gardner, Trussell, and Oertel 1999). These receptors are also characterized by high Ca^{2+} permeability (Otis, Raman, and Trussell 1995). AMPA receptors in auditory neurons have low levels of GluR1 and perhaps GluR2 subunits and high levels of GluR3 and GluR4 subunits with the majority being of the flop isoform (reviewed by Trussell 1999; Ravindranathan, Parks, and Rao 1996; Parks 2000). These results are consistent with expression studies showing that AMPA receptors containing GluR4 subunits gate rapidly and that flop variants desensitize most quickly (Mosbacher et al. 1994; Geiger et al. 1995).

Although brief EPSCs underlie the rapid synaptic potential changes seen in time coding neurons, the intrinsic electrical properties of these neurons also shape the synaptic response as well as the temporal firing pattern. Of particular interest are the voltage sensitive K^+ conductances. The importance of these conductances in sculpting the response properties of auditory neurons was first demonstrated by Manis and Marx (1991) who showed that differences in the electrical responses of bushy cells and stellate cells in the mammalian cochlear nucleus can be attributed to a distinct complement of outward K^+ currents in each cell type. At least two K^+ conductances underlie phase locked responses in auditory neurons: a low threshold conductance

(LTC) and a high threshold conductance (HTC) (Manis and Marx 1991; Brew and Forsythe 1995; Reyes, Rubel, and Spain 1994; Rathouz and Trussel 1998; Wang et al. 1998).

The LTC activates at potentials near rest and is largely responsible for the outward rectification and nonlinear current voltage relationship around the resting potential seen in a number of auditory neurons (figure 12-1A; see Oertel 1999 for review). Activation of the LTC leads to a short active time constant so that the effects of excitation are brief and do not summate in time (Oertel 1999). Only large EPSPs reaching threshold before significant activation of the LTC would produce spikes with short latencies, whereas small EPSPs which depolarize the membrane more slowly would allow time for LTC activation to shunt the synaptic current and prevent action potential generation and thus long latency action potentials. Blocking the LTC elicits multiple spiking in response to depolarizing current injection (Manis and Marx 1991; Rathouz and Trussel 1998) or synaptic activation (Brew and Forsythe 1995). K^+ channels underlying the LTC appear to be composed of Kv1.1 and Kv1.2 subunits. Both subunits are expressed in auditory neurons although the subcellular distribution is unknown (Grigg, Brew, and Tempel 2000). Consistent with a role for Kv1.1 subunits in the LTC, synaptic activation of MNTB neurons in Kv1.1 null mice produce action potentials with more jitter compared to wild type (Brew, Hallows, and Tempel 2003; Kopp-Scheinpflug et al. 2003).

The HTC is characterized by fast kinetics and an activation threshold around −20 mV (Brew and Forsythe 1995; Rathouz and Trussell 1998; Wang et al. 1998). These features of the HTC result in fast spike repolarization and a large but brief afterhyperpolarization without influencing input resistance, threshold, or action potential rise time. Thus, the HTC can keep action potentials brief without effecting action potential generation. In addition, the HTC minimizes Na^+ channel inactivation allowing cells to reach firing threshold sooner, facilitating high frequency firing. Relatively specific pharmacological blockade of the HTC broadens action potentials and reduces the fast afterhyperpolarization (Brew and Forsythe 1995). Furthermore, blockade of the HTC diminishes the ability of MNTB neurons to follow high-frequency stimuli in the range of 300–400 Hz but had little effect on responses to low frequency stimulation (<200 Hz; Wang et al. 1998).

Elimination of the Kv3.1 gene in mice results in the loss of the HTC and failure of MNTB neurons to follow high-frequency stimulation (Macica et al. 2003). Neurons that fire fast, including many auditory neurons, express high levels of Kv3 mRNA and protein, although it should be noted that not all neurons that express Kv3 subunits have fast firing abilities (Perney and Kaczmarek 1997; Parameshwaran, Carr, and Perney 2001; Li, Kaczmarek, and Perney 2001). Interestingly, in several auditory nuclei including avian NM and NL (Parameshwaran, Carr, and Perney 2001), rat MNTB (Li, Kaczmarek, and Perney 2001), Kv3.1 protein expression varied along the tonotopic map such that mid to high best frequency neurons are most strongly immunopositive, while neurons with very low best frequencies are only weakly immunopositive. A high to low frequency gradient of Kv3.3 expression has also been observed in electrosensory lateral line lobe of a weakly electric fish (Rashid et al. 2001). These results suggest that the electrical properties of higher-order auditory neurons may vary with frequency tuning. Since no differences in either spontaneous or driven rates have been observed across the tonotopic axis, however, Kv3 channels may be functioning as more than just a facilitator of high frequency firing and may also enhance the temporal precision of spike discharges.

Distribution of Kv3.1 protein in auditory neurons is largely somatic and/or axonal, consistent with its role in spike repolarization (Perney and Kaczmarek 1997; Li, Kaczmarek, and Perney 2001; Parameshwaran, Carr, and Perney 2001). EM studies have shown that Kv3.1 is present in the membranes of endbulb terminals onto MNTB neurons suggesting that Kv3.1 channels may be at least partially responsible for the extremely brief action potential seen at this terminal. Kv3.1 protein is also present in the NM axons innervating the NL in owl but not chicken (Parameshwaran, Carr, and Perney 2001). The increased levels of HTC associated with Kv3.1 expression in owl NM axons would reduce the width of the action potential invading the NM terminals and thus the amount of neurotransmitter released. Modeling of coincidence detector neurons suggest that an increase in the width of the input EPSC could impair ITD coding (Simon, Carr, and Shamma 1999; Grau-Serrat, Carr, and Simon 2003). Thus, the selective increase of Kv3.1-like currents in the NM delay line axons in owl may contribute to the temporal synchrony necessary for accurate phase locking.

Coincidence Detection

In birds and mammals, precisely timed spikes encode the timing of acoustic stimuli, and interaural acoustic disparities propagate to binaural processing centers such as the avian nucleus laminaris (NL) and the mammalian medial superior olive (MSO; Young and Rubel 1983; Carr and Konishi 1990; Joris, Smith, and Yin 1998). The projections from the NM to NL and from mammalian spherical bushy cells to MSO resemble the Jeffress model for encoding interaural time differences (Jeffress 1948). The Jeffress model has two elements: delay lines and coincidence detectors. A Jeffress circuit is an array of coincidence detectors, every element of which has a different relative delay between its ipsilateral and contralateral excitatory inputs. Thus, ITD is encoded into the position (a *place code*) of the coincidence detector whose delay lines best cancels out the acoustic ITD (for reviews, see Joris, Smith, and Yin 1998 and Konishi

1991). Neurons of NL and MSO phase lock to both monaural and binaural stimuli but respond maximally when phase-locked spikes from each side arrive simultaneously, that is, when the difference in the conduction delays compensates for the ITD (Goldberg and Brown 1969; Yin and Chan 1990; Carr and Konishi 1990; Overholt, Rubel, and Hyson 1992; Pena et al. 1996).

Delay Line-Coincidence Detection Circuits

The barn owl is capable of great accuracy in detecting time differences, and its auditory system is hypertrophied in comparison to birds like the chicken whose auditory systems are less specialized. The details of delay line circuit organization vary between species (figure 12-2). In the chicken, NL is composed of a monolayer of bipolar neurons that receive input from ipsi- and contralateral cochlear nucleus onto their dorsal and ventral dendrites, respectively (Rubel and Parks 1975). These dendrites increase in length with decreasing best frequency. Only the projection from the contralateral cochlear nucleus acts as a delay line, while inputs from the ipsilateral cochlear nucleus arrive simultaneously at all neurons (Overholt, Rubel, and Hyson 1992). This pattern of inputs creates a single map of interaural time difference (ITD) in any tonotopic band in the mediolateral dimension of NL (Overholt, Rubel, and Hyson 1992). In the barn owl, magnocellular axons from both cochlear nuclei act as delay lines (Carr and Konishi 1988; Carr and Konishi 1990). They convey the phase of the auditory stimulus to NL such that axons from the ipsilateral NM enter NL from the dorsal side, while axons from the contralateral NM enter from the ventral side. Recordings from these interdigitating ipsilateral and contralateral axons show regular changes in delay with depth in NL (Carr and Konishi 1990). Thus these afferents interdigitate to innervate dorsoventral arrays of neurons in NL in a sequential fashion and

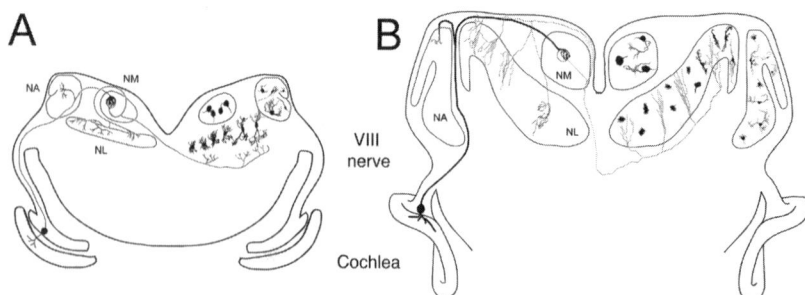

Figure 12-2. Schematic of a coronal section through the brainstem of (A) chicken and (B) owl. The medial branch of the auditory nerve innervates NM. NL receives bilateral projections from NM. The cells are not drawn to scale. (A) modified from Rubel and Parks (1988).

produce multiple representations of ITD within the nucleus. Despite the differences in organization of NL in owls and chickens, interaural time differences are detected by neurons that act as coincidence detectors in both species (Sullivan and Konishi 1984; Joseph and Hyson 1993; Pena et al 1996; Kubke, Massoglia, and Carr 2002). Very similar principles apply to the mammalian superior olive (Goldberg and Brown 1969; Yin and Chan 1990).

An important feature of both avian and mammalian coincidence detectors is that they share physiological features with NM neurons and mammalian bushy cells. Coincidence detectors exhibit specific K^+ conductances that lead to a single or a few well-timed spikes in response to a depolarizing stimulus in vitro (figure 12-1B; Reyes, Rubel, and Spain 1996; Smith 1995; Kuba, Koyano, and Ohmori 2002). The LTC channels should decrease the effective membrane time constant, that is, the average membrane time constant for a cell receiving and processing in vivo rates of EPSPs, which will be much shorter than the passive membrane time constant (Softky 1994; Mainen and Sejnowski 1995; Gerstner et al. 1996; Grau-Serrat, Carr, and Simon 2003). These fast conductances may be critical to coincidence detection—the models described in the next section of this chapter suggest that they are instrumental in keeping the firing rate near zero when the inputs are completely out of phase but allowing nonzero firing rate when the inputs are monaural.

Coincidence detector neurons in birds and mammals may display similar conductances and bipolar morphologies, but they are not identical. In mammals, MSO neurons do not express either Kv3.1 mRNA or protein (Grigg, Brew, and Tempel 2000; Li, Kaczmarek, and Perney 2001). They do, however, express high levels of Kv3.3 message (Grigg, Brew, and Tempel 2000; Li, Kaczmarek, and Perney 2001). Thus, differences in Kv3.1 expression between NL and MSO structures may reflect species differences in the expression of Kv3 subfamily members. We do not know whether this variation in expression also represents a significant physiological difference. A second substantial difference is in inhibitory inputs. In mammals the MSO receives well-timed inhibitory input from the medial and lateral nucleus of the trapezoid body (Cant and Hyson 1992; Kuwabara and Zook 1992; Grothe and Sanes 1994; Grothe 2003). These inhibitory inputs may enhance coincidence detection in several ways. First, by producing a somatic shunt during coincidence detection to decrease the membrane time constant (Brughera, Stutman, and Carney 1996; Thompson, Rowland, and Spirou 2004). Second, in the Mongolian gerbil, a small mammal with low frequency hearing, precisely timed glycine-controlled inhibition in the MSO appears to shift the ITD curve so that the peak change in firing rate falls within the physiologically relevant range of ITDs (Brand et al. 2002). In birds, inhibitory inputs in NL are more diffuse and appear to decrease excitability through a gain control mechanism (Monsivais, Yang, and Rubel 1999; Funabiki, Koyano, and Ohmori 1998; Yang, Monsivais, and Rubel 1999; Pena et al. 1996).

Models of Coincidence Detection Relate Dendritic Structure to Detection of Interaural Time Differences

A singular feature of the coincidence detectors in mammals and of low best frequency NL cells in birds is their common morphological organization. Both are bitufted neurons with inputs from each ear segregated on the dendrites (figure 12-3). Modeling studies have shown that this dendritic organization improves coincidence detection (Agmon-Snir, Carr, and Rinzel 1998; Grau-Serrat, Carr, and Simon 2003). Thus the cell morphology and the spatial distribution of the inputs enriches the computational power of these neurons beyond that expected from point neurons. How does the dendritic structure of the coincidence detectors enhance their computational ability? An ITD discriminator neuron should fire when inputs from two independent neural sources coincide (or almost coincide) but not when two inputs from the same neural source (almost) coincide. A neuron that sums its inputs linearly would not be able to distinguish between these two scenarios. To understand this mechanism, we constructed a biophysically detailed model of coincidence detector neurons using NEURON (Simon et al 1999; Grau-Serrat, Carr, and Simon 2003).

Two dendritic nonlinearities aid coincidence detection. First, synaptic inputs arriving at the same dendritic compartment sum non-linearly because the driving force decreases with depolarization (Agmon-Snir, Carr, and Rinzel 1998). Hence, the net synaptic current from several inputs arriving simultaneously at nearby sites on the same dendrite is smaller than the net current generated if these inputs are distributed on different dendrites. As a result, the conductance threshold, or minimum synaptic conductance needed to trigger a somatic action potential, is higher when the synaptic events are on the same dendrite compared to when they are split between the bipolar dendrites. Second, each dendrite acts as a current sink for inputs on the other dendrite, consequently increasing the voltage change needed to trigger a spike at the soma when inputs arrive only on one side. This effect is boosted by the presence of a low threshold K^+ conductance similar to that found in NM and bushy neurons so that out of phase inputs are subtractively inhibited (Grau-Serrat, Carr, and Simon 2003). With only monaural input, the LTC in the opposite dendrite is somewhat activated, producing a mild current sink. When, however, there are recent EPSPs in the opposite dendrite due to out-of-phase inputs, the LTC is strongly activated and acts as a large current sink suppressing spike initiation. Thus, the model predicts the experimental finding (Goldberg and Brown 1969; Yin and Chan 1990; Carr and Konishi 1990) that the monaural firing rate while lower that the binaural in-phase rate, is higher than the binaural out-of-phase rate.

One dendritic effect diminishes with increasing stimulus frequency. When typical chick-like parameters are used, sublinear summation in the dendrites only improves coincidence detection below 2kHz, after which discrimination

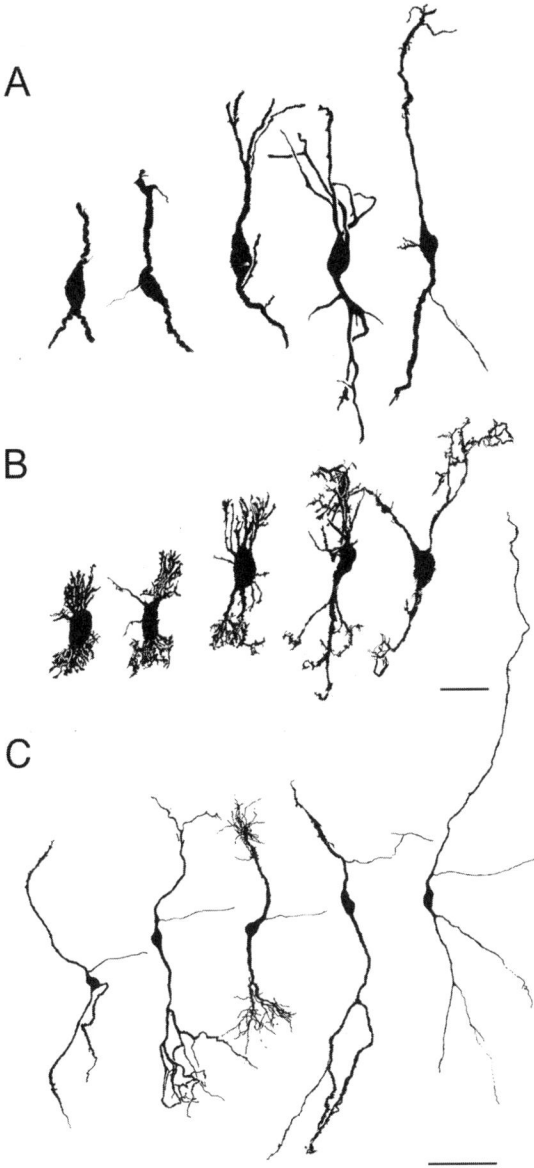

Figure 12-3. Coincidence detectors share bitufted morphology. Alligator (A),
avian (B), and mammalian (C) low-frequency coincidence detector neurons.
(A, B) The stimulus frequency of alligator and chicken nucleus laminaris (NL) cells
increases from left to right (chicken data adapted from Jhaveri and Morest 1982,
alligator NL neurons from Carr and Soares unpublished results). (C) The dendritic
morphology of the principal cells of the medial superior olive from the guinea pig
(adapted from Smith 1995) differs somewhat from the chicken and alligator, and a
frequency gradient is not apparent. Nevertheless, the bipolar architecture and the
segregation of the inputs arriving from both ears is common to both mammalian
and archosaur coincidence detectors with low best frequencies.

between in-phase and out-of-phase inputs is poor (Agmon-Snir, Carr, and Rinzel 1998). This is consistent with observation from rabbit MSO neurons, where ITD sensitivity has only been observed for sounds at or below 2kHz (Batra, Kuwada, and Fitzpatrick 1997). The second dendritic nonlinearity, subtractive inhibition of out-of-phase inputs, improves coincidence detection all frequencies (Grau-Serrat, Carr, and Simon 2003) and may therefore be most significant in avian coincidence detectors between 2 and 8kHz. It is also clear that the quality of phase-locked inputs has some bearing on coincidence detection: typical chick-like parameters but with barn owl–like phase locking allow ITD discrimination up to 4–6 kHz (Grau-Serrat, Carr, and Simon 2003). The benefits conveyed by the neuronal structure of the coincidence detectors allows us to argue that selective forces have directed the evolution of coincidence detectors in the bird NL and mammalian MSO, perhaps in parallel (Carr and Soares 2002).

Encoding Onsets

Both birds and mammals have neurons that respond preferentially to onsets, or transients in sound. Onsets play an important role in theories of speech perception (Stevens 1995), music perception, sound localization (Zurek 1987), and segregation and grouping of sound sources (Bregman 1990). The computational importance of encoding onsets may be also inferred by the parallel evolution of onset coding in bird cochlear nuclei (figure 12-4; Warchol and Dallos 1990; Sullivan and Konishi 1984; Koppl and Carr 2003). Mammals differ from birds, however, in that they have a specialized *octopus cell* pathway for onset coding in addition to onset responses in other cell types (see Oertel et al. 2000 for review). Octopus cells integrate auditory nerve inputs across a range of frequencies and encode the time structure of stimuli with great precision (Kim, Rhode, and Greenberg 1986; Golding, Robertson, and Oertel 1995; Ferragamo and Oertel 2002; Oertel et al. 2000).

Octopus Cells Transform Auditory Nerve Inputs to Produce Onset Responses

How does the transformation of the auditory nerve response to onset code occur? Octopus cells have a few thick dendrites emanating from one end of the cell body (see Oertel et al. 2000). These dendrites are perpendicular to entering auditory nerve fibers, enabling them to sample nerve inputs spanning a broad range of frequencies (Kane 1973; Golding et al. 1995). The relatively broad tuning of octopus cells may be a reflection of this anatomy. Many of the electrical and anatomical properties of octopus cells resemble those that help bushy and magnocellular cells encode the time structure of stimuli. Like bushy cells, octopus cells appear to exhibit little dendritic filtering. They have

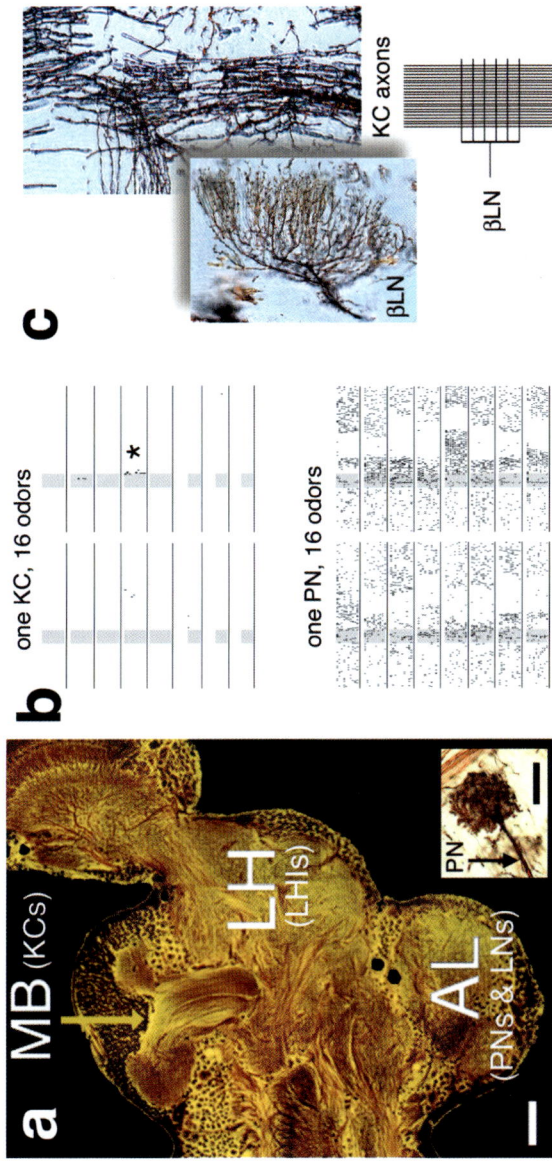

Figure 1-1. The locust olfactory circuits and the transformation of response properties between first and second relay. (a) Bodian stain of a locust brain (transverse section) showing the antennal lobe (AL), mushroom body (MB), and lateral horn (LH). Inset: terminal dendrite of one projection neuron (PN) in one glomerulus; in locusts, each PN sends dendrites in ~15 of ~1,000 glomeruli. Scale bar: 80μm (20μm, inset). (b) *Bottom*, tetrode recording from one PN in the AL and its responses to sixteen different odors (ten trials each, 1-second odor pulses indicated by shaded areas). Note the high baseline rates, high probability of response, and odor-specific temporal patterning. *Top*, tetrode recording from one Kenyon cell (KC) in the MB and its responses to the same sixteen odors as the PN. Note very low baseline rate, high specificity, and brevity of response. Details in Perez-Orive et al. 2002. (c) Golgi stains of KC axon tracks in the MB beta lobe and of β lobe neuron (βLN) dendrites, sampling across those axons (see diagram in inset). Golgi stains kindly provided by Sarah Farivar.

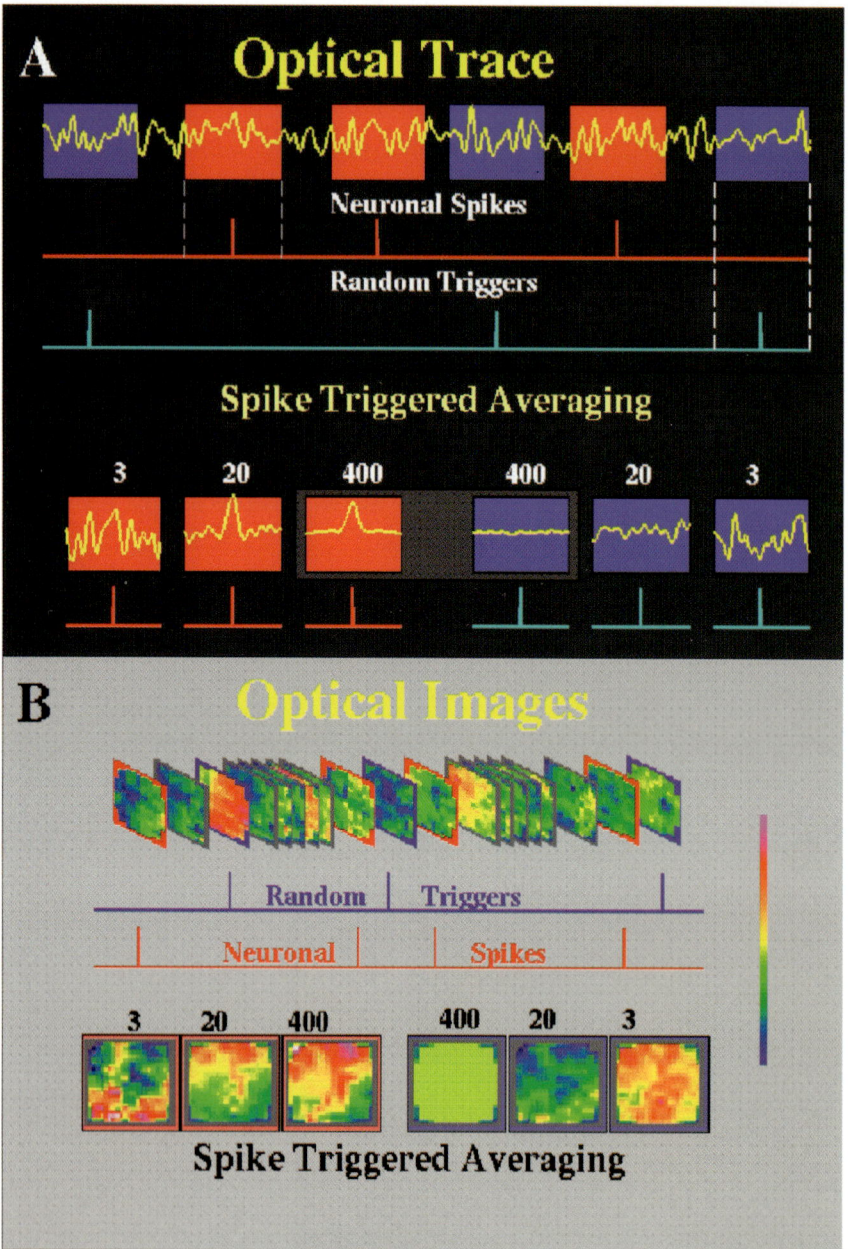

Figure 9-2. The procedure for selective visualization of the dynamics of coherent neuronal assemblies. (A) Time course of the optical signal obtained by spike trigger averaging. The top yellow trace shows the amplitude of the optical signal reflecting compound electrical activity from a given cortical site, measured for eight seconds. The red trace below shows the simultaneously recorded action potentials from the reference neuron. The long recording session was subdivided into one-second time segments (red windows on the top trace), each centered on the timing of the action potential. The blue trace below shows random virtual spikes that are used as a control for the procedure (blue windows). The bottom

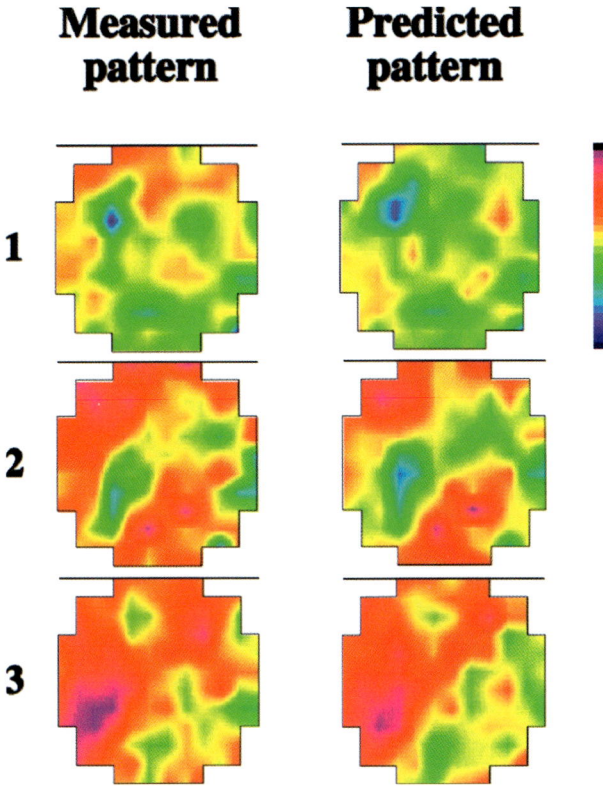

Measured pattern　　**Predicted pattern**

1

2

3

Figure 9-4.　Predicting cortical evoked responses in spite of their large variability. Three examples of comparing predicted and measured responses are shown. A single-trial response to a stimulus was predicted by summing the reproducible response component (estimated by averaging many trials of evoked responses) and the ongoing network dynamics (approximated by the initial state, i.e., the ongoing pattern just prior to the onset of the evoked response). Left columns, measured responses. Right columns, predicted responses, obtained by adding the initial state in each case. Modified from Arieli et al. (1996).

traces in the red windows show the time course of the spike-triggered averaged signal after averaging three, twenty, and four hundred time segments during which action potential occurred. The traces in blue windows shows the results obtained from averaging the control virtual spikes. A clear coherent activity is detected already after averaging twenty events. (B) Spatial patterns of movies obtained by spike triggered averaging. The top shows a series of images in the form of a movie instead of shown the activity in a single cortical site depicted in (A). The two traces below show the timing of simultaneously recorded action potential and virtual action potential that served as a control. The bottom frames show the spatial pattern observed at a given time resulting from spike-triggered averaging. Note that the control patterns are rather flat already after averaging twenty random events without real action potentials (three blue windows at the bottom right). Modified from Sterkin et al. (1999) and Grinvald et al. (1999).

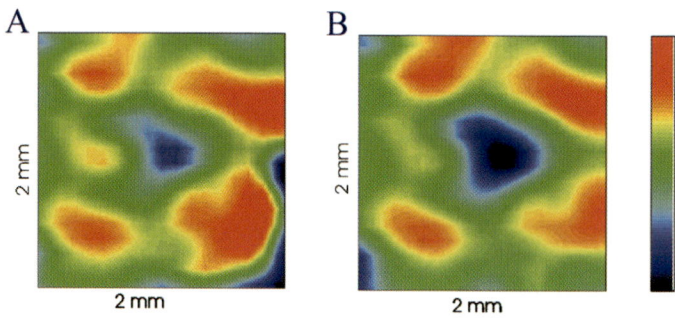

Figure 9-9. Evoked versus spontaneous cortical states. (A) The pattern obtained by averaging all frames that correspond to spike times during an evoked recording session. (B) The pattern obtained by averaging all frames that correspond to spike times during a spontaneous recording session using the same neuron. Surprisingly, we see that the two patterns are extremely similar. Modified from Tsodyks et al. (1999).

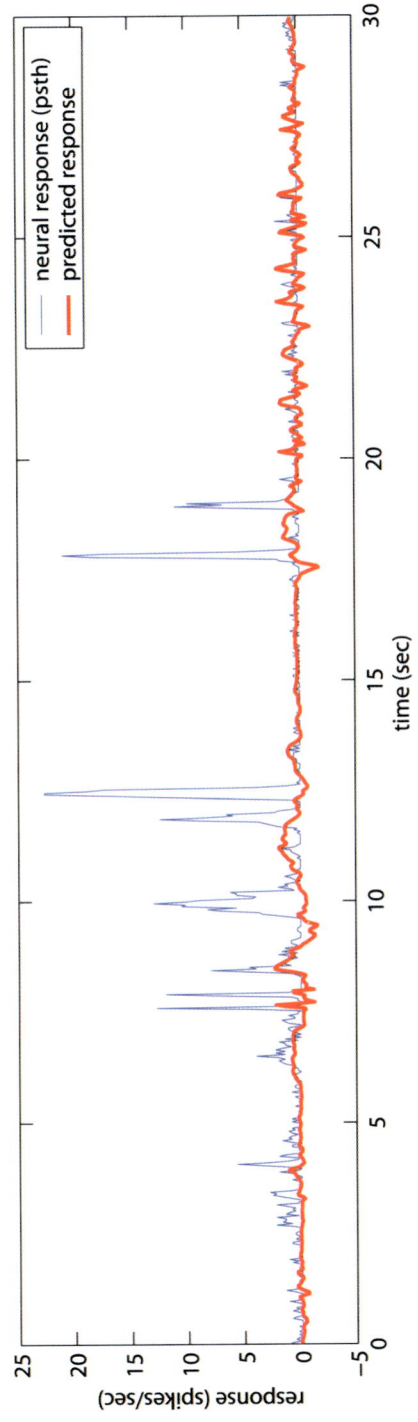

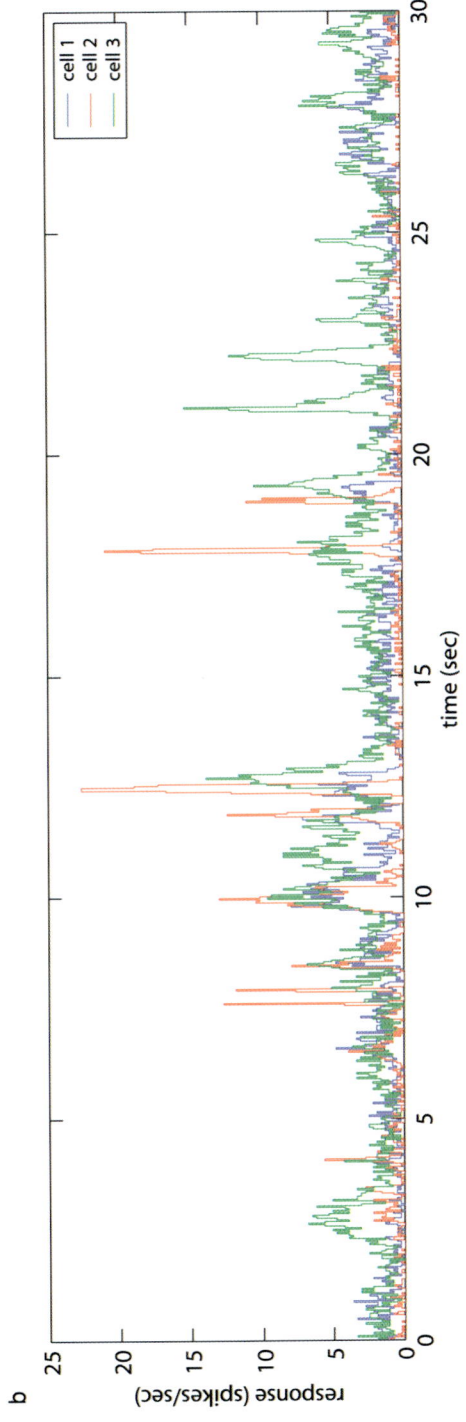

Figure 10-5. Activity of V1 neurons in anaesthetized cat in response to a natural movie. (a) The peristimulus time histogram of one neuron's response (blue), together with the predicted response (red) generated from convolving the space-time receptive field with the movie. (b) Simultaneous responses of three different neurons, with similar receptive fields, to the same movie.

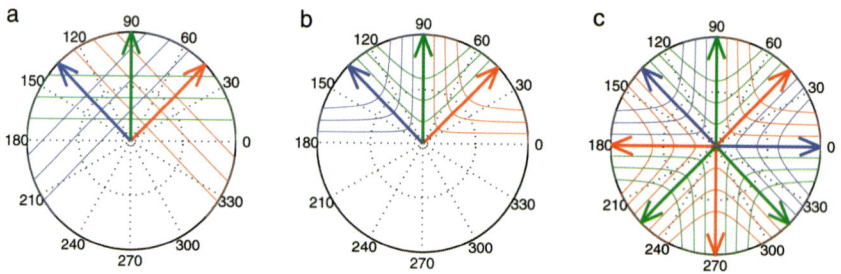

Figure 10-6. Overcomplete representation. (a) Shown are the isoresponse contours of linear neurons with linearly dependent weight vectors. A stimulus falling anywhere along a given contour will result in the same response from the neuron. (b) Curving the response contours removes redundancy among these neurons. (c) Response contours of an entire overcomplete tiling of the stimulus space.

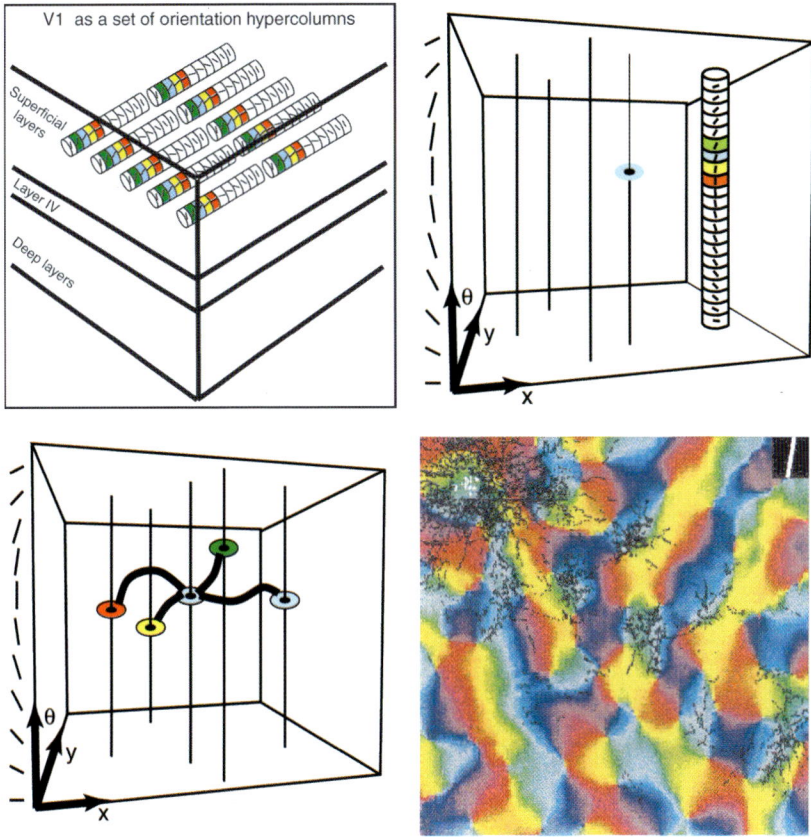

Figure 11-1. Abstracting the functional architecture of visual cortex. *Top left,* the standard Hubel-Wiesel "ice cube" model, which, although it is a cartoon, expresses the fundamental observation that each local retinotopic area is covered by receptive fields that span a range of orientations (several tangential penetrations are shown, each of which corresponds to cells with spatially overlapping receptive fields), sizes (normal penetration), and eye-of-origin. *Top right,* a redrawing of the ice cube model, emphasizing a retinotopic array (the tilted plane in [x,y] coordinates) and columns of superficial-layer cells (one drawn as in the penetration) to show their orientation preference in the vertical direction. *Bottom left,* the axon from one cell is drawn to show how it forms long-range horizontal connections with cells selective for some orientation at nearby positions. When this cartoon is "flattened" onto the retinotopic array, the groups of cells form orientation domains. *Bottom right,* data (courtesy of David Fitzpatrick) from which long-range connection statistics can be obtained. Different colors represent orientation preference estimated from optical imaging; and black dots are axonal tracers. (The projection is from one eye only; receptive field scale is not shown.) When organized in this fashion, a geometric view of processing emerges in which the fiber of orientations at each position in the retinotopic array abstracts the orientation column, and the arrangement of neighboring fibers suggests an architecture that would support interaction between orientations.

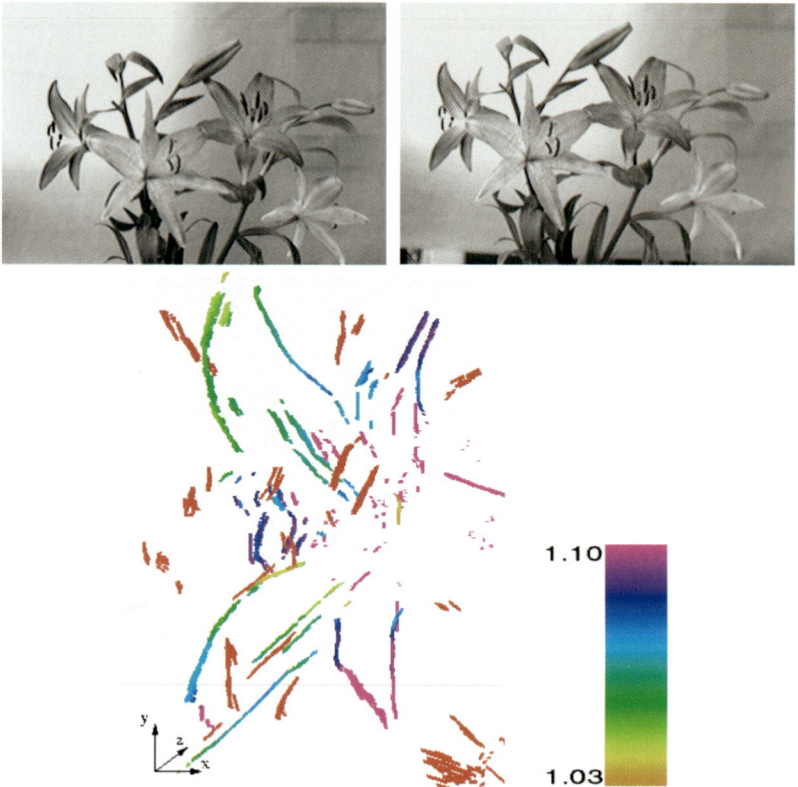

Figure 11-10. The depth map associated with a stereo lily pair with only the flower shown in detail. Each point in the lily image is a tangent in three space that is geometrically consistent with its neighbors.

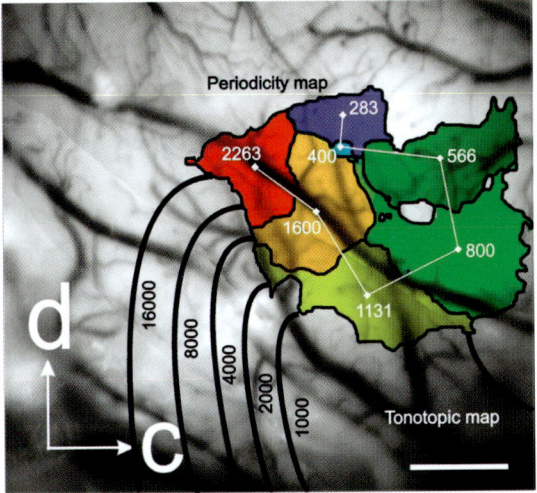

Figure 17-1. Topography of periodicity map in gerbil AI. Cyclic periodicity map of best-stimulus representations from one animal obtained using optical recording of intrinsic signals with amplitude modulated stimuli with different periodicities (f_m; blue to red codes for increasing periodicity, white numbers). The map is superimposed on an image of the cortical surface, including a sketch of the tonotopic map (representation of tone frequency, black numbers). The white line connects the centers of gravity of best-stimulus representations to reveal the cyclic, horseshoe-like gradient of the periodicity map. d, dorsal; c, caudal. Scale bar, 0.5 mm.

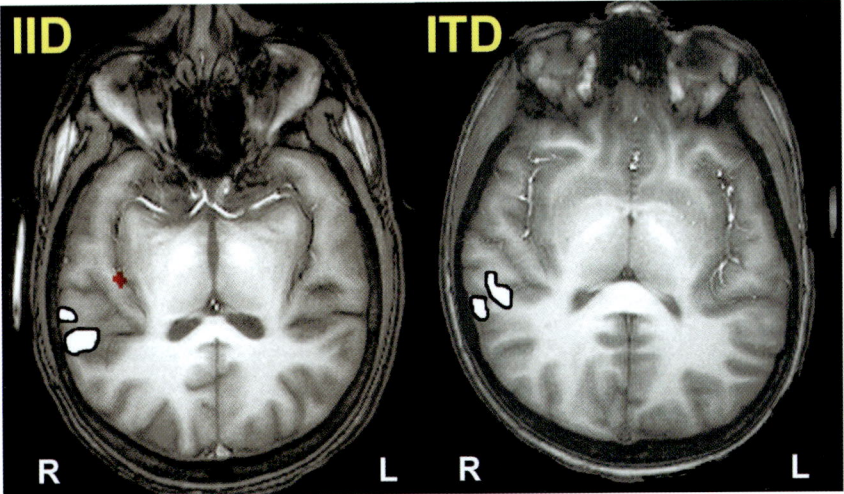

Figure 17-3. Parameter-independence of movement-selective fMRI activation in human auditory cortex. The activations in the yellow areas in the right planum temporale in two subjects were obtained during perceived motion of an auditory object generated by interaural phase shifts of envelops of an amplitude-modulated sound (interaural intensity difference, *left panel*) or by changing interaural onset delays of a pure tone (interaural time difference, *right panel*), respectively.

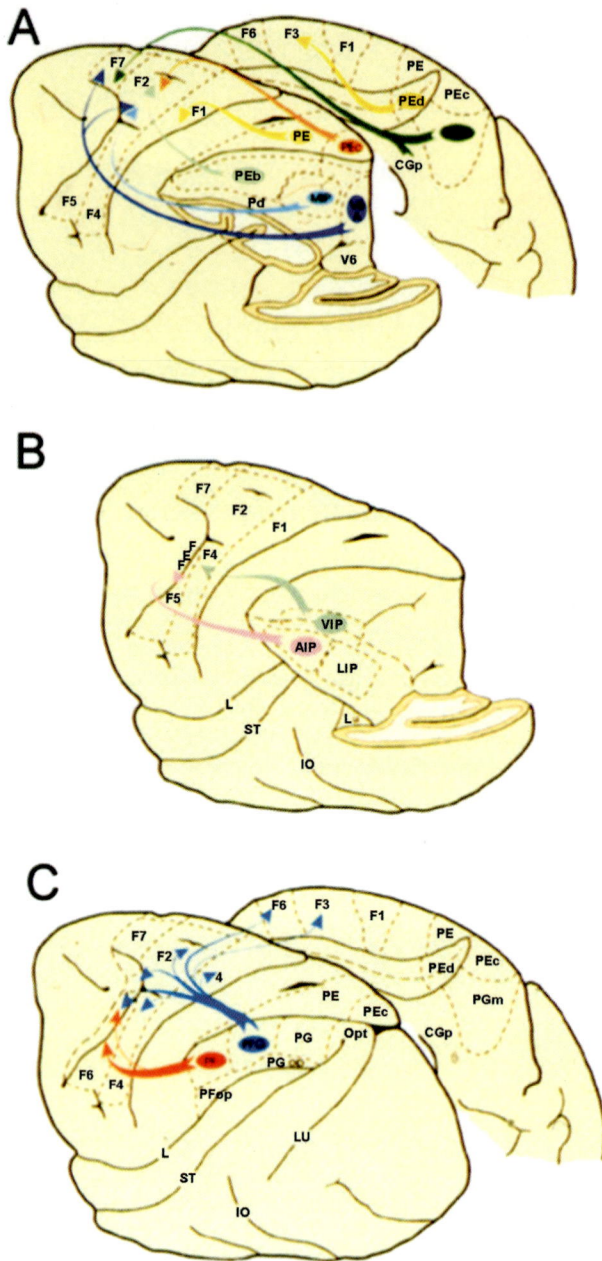

Figure 18-2. Schematic diagrams showing the extrinsic afferent connections of the parietodependent motor areas. (A) Parietal projections from areas located in the superior parietal lobule. In this view of the brain the inferior parietal lobule and the occipital lobe have been removed in order to show the areas located in the medial bank of the intraparietal sulcus and in the anterior bank of the parieto-occipital sulcus, respectively. (B) Parietal projections from areas located in the lateral bank and in the fundus of the intraparietal sulcus. In order to show these areas, the intraparietal sulcus has been opened and the occipital lobe removed. (C) Parietal projections from areas located on the convexity of the inferior parietal lobule. Modified from Rizzolatti et al. (1998).

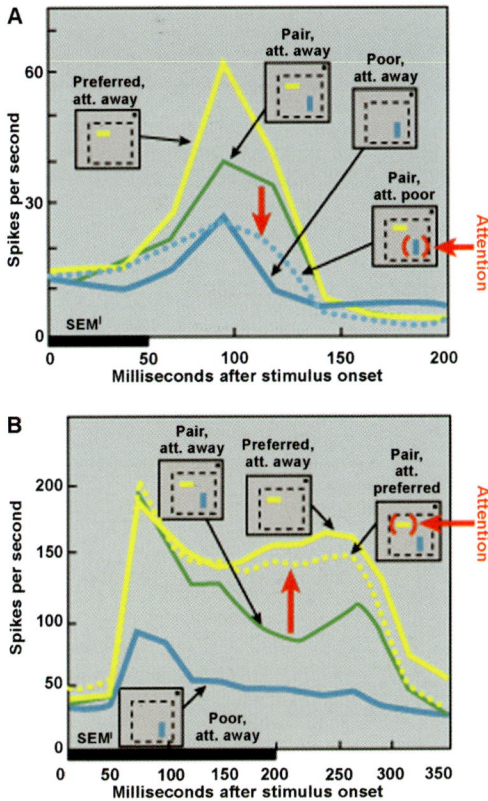

Figure 20-2. Attention to one stimulus of a pair filters out the effect of the ignored stimulus. (A) The x axis shows time (in milliseconds) from stimulus onset, and the thick horizontal bar indicates stimulus duration. Small iconic figures illustrate sensory conditions. Within each icon, the dotted line indicates the RF, and the small dot represents the fixation point. The location of attention inside the RF is indicated in red. Attention was directed away from the RF in all but one condition. The preferred stimulus is indicated by a horizontal yellow bar and the poor stimulus by a vertical blue bar. In fact, the identity of both stimuli varied from cell to cell. The yellow line shows the response to the preferred stimulus. The solid blue line shows the response to the poor stimulus. The green line shows the response to the pair with attention away from the RF. The dotted blue line shows the response to the pair when attention was directed to the poor stimulus. The addition of the poor stimulus suppressed the response to the preferred stimulus. Attention to the poor stimulus (red arrow) magnified its suppressive effect and drove the response down to a level that is similar to the response elicited by the poor stimulus alone. (B) Response of a second V2 neuron. The format is the same as in (A). As in the neuron in (A), the response to the preferred stimulus was suppressed by the addition of the poor stimulus. Attention directed to the preferred stimulus filtered out this suppression, returning the neuron to a response similar to the response that was elicited when the preferred stimulus appeared alone inside the RF. Adapted from Reynolds, Chalazzi, and Desimone (1999), copyright 1999 the Society for Neuroscience.

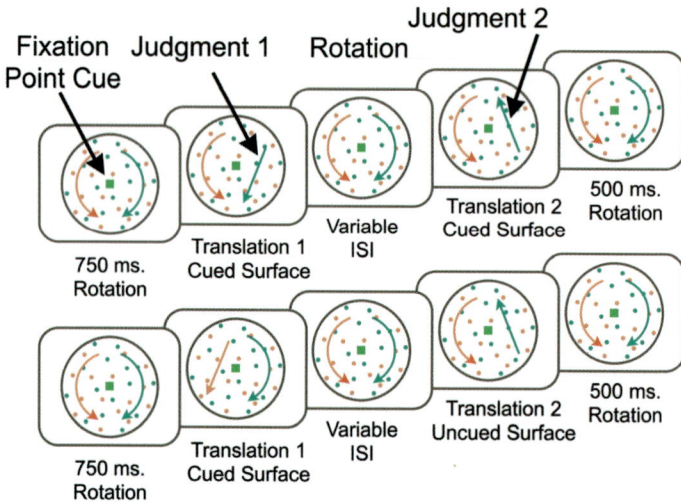

Figure 20-4. Object-based attention task introduced by Valdes-Sosa and colleagues. Panels are arranged from left to right according to the sequence of events in each trial. On half of all cued trials, the fixation point was green (upper panels), indicating that the green surface would be the first to translate. On the remaining cued trials, the fixation point was red (lower panels), indicating that the red surface would translate first. The observer began each trial with a key press, resulting in a period of 750 milliseconds in which the two surfaces rotated around the fixation point in opposite directions. The cued surface then translated for 150 milliseconds in one of the eight cardinal directions, while the other surface continued to rotate. Following this first translation, the two surfaces continued to rotate for a variable delay of 150–1050 milliseconds, at which point one of the two surfaces, with equal probability, shifted for 150 milliseconds. After this second shift, both surfaces continued to rotate for an additional 500 milliseconds. Observers had to maintain fixation throughout the trial and report the direction of each shift. Adapted from Mitchell et al. (2003) copyright 2003, with permission from Elsevier.

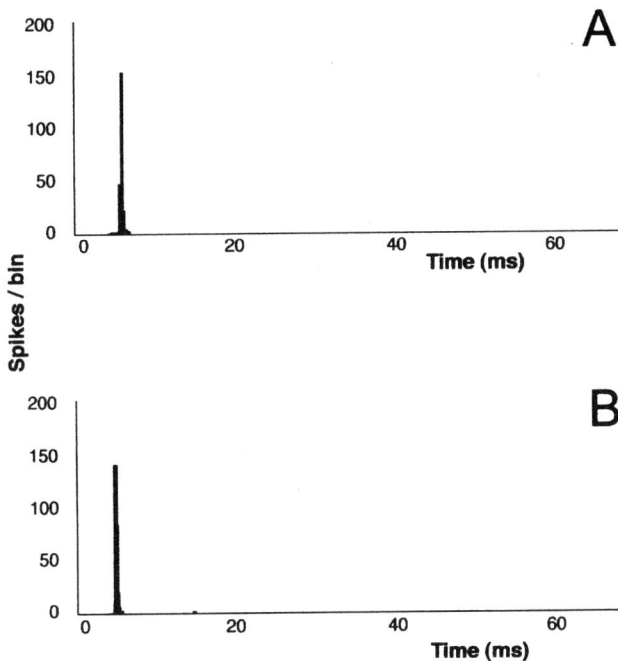

Figure 12-4.   Onset responses in birds and mammals can show a prominent response at the onset of a tonal stimulus, followed by little or no sustained activity. (A) Onset response from an octopus cell (from Winter and Palmer 1995). (B) Onset responses from the nucleus angularis in the barn owl (from Koppl and Carr 2003).

large spherical cell bodies and even though their dendrites are fairly long (120 to 180 μm), they are also thick (Kane 1973; Brawer and Morest 1974; Golding, Robertson, and Oertel 1995; Golding, Ferragamo, and Oertel 1999). The spherical cell body and thick dendrites make the cell electrically compact (Kane 1973; Cai, Walsh, and McGee 1997). Moreover, the many weak auditory nerve inputs are on the soma and proximal dendritic surfaces (Kane 1973), and miniature synaptic currents measured in octopus cells are brief (Gardner, Trussell, and Oertel 1999). Such brief synaptic responses should preserve the time pattern of discharges in the corresponding presynaptic auditory nerve fibers (Kalluri and Delgutte 2003a).

The brevity of the small synaptic responses in octopus cells requires the coincident activation within one millisecond of enough auditory nerve inputs to produce sufficient depolarization to bring the cell to threshold (Golding, Robertson, and Oertel 1995; Oertel et al. 2000). The intrinsic biophysical properties of octopus cells support this coincidence detection role. Like other neurons in the auditory brainstem that preserve the time patterns of discharge in their inputs, octopus cells have a very low input resistance (2–7 MΩ just below resting voltage, and the membrane time constant is near 200 μs, the smallest of any cochlear nucleus neuron (Golding, Ferragamo, and Oertel 1999). The low input resistance of octopus cells is determined in part by two voltage-dependent conductances that are active at rest: a hyperpolarization-activated, mixed-cation conductance, g_h, and a depolarization-activated, low-threshold potassium (K^+) conductance (see Oertel et al. 2000). Like the coincidence detectors in NL and MSO, the low-threshold K^+ conductance in octopus cells allows them to be sensitive to coincident activation of their inputs by making the membrane sensitive to fast transients in the synaptic input (Golding, Ferragamo, and Oertel 1999). A rapidly rising input, such as that arising from the synchronous activation of synapses, can depolarize the membrane to threshold before the relatively slow low-threshold K^+ conductance is activated (2–3 ms time constant). In contrast, a slower input would fail to drive the membrane voltage to threshold because it could not outpace this conductance (Cai, Walsh, and McGee 1997; Kalluri and Delgutte 2003a,b).

Onset Responses Have Evolved in Parallel in Birds

The computational importance of encoding onsets, or rapid fluctuations, may be inferred by the parallel evolution of onset coding in the bird cochlear nucleus angularis (NA). In the barn owl, the chicken, and the blackbird, some NA neurons exhibit onset responses, while others have primary-like, chopper, and Type IV responses (figure 12-4B; Warchol and Dallos 1990; Sullivan and Konishi 1984; Koppl and Carr 2003). Nevertheless, there does not seem to be an avian counterpart of the octopus cell. NA onset cells have relatively narrow frequency tuning curves, unlike the octopus cell (Rhode, Oertel, and Smith 1983; Koppl and Carr 2003). Furthermore, Golgi analyses of barn owl NA and intracellular labeling of chicken NA neurons in brain slices have not revealed cells with thick dendrites that extend across the incoming auditory nerve inputs (Soares and Carr 2001, Soares et al. 2002).

The presence of onset units argues for NA's involvement in temporal processing. In mammals, onset responses are found in several cell types in the cochlear nucleus and may encode temporal features such as broadband transients (Oertel et al. 2000; Kalluri and Delgutte 2003b). NA onset neurons may serve a similar function. Thus NM may mediate coding of the temporal information used for the computation of ITDs, while one of the cell types in NA may encode other temporal features of the stimulus (Koppl and Carr 2003).

Neuronal Structure and Function

When compared with a simple integrate-and-fire unit, the auditory neurons that phase lock, detect coincidences, and encode temporal patterns all exhibit a suite of physiological and morphological adaptations that suit them for their task. Other neuronal systems exhibit similarly well-equipped neural circuits. The blowfly has an array of direction-selective, motion-sensitive cells that conform to the Reichardt model of motion detection (Borst, Helmchen, and Sakmann 1995). An array of Reichardt motion detectors projects onto the lobular plate tangential cell to create a response to both the direction and velocity of pattern motion. The geometry of the tangential cell dendrites supports this computational task in visual motion control because they are aligned with the direction of motion.

The question remains whether all neurons are adapted for specific computations. Neurons in the auditory brainstem and fly motion detectors appear to be, and a similar case may be made for phase coding neurons in weakly electric fish (Matsushita and Kawasaki 2004; for reviews see Carr and Friedman 1999; Kawasaki 2000). In other body tissues, cells appear to have a precise function, and it could be argued that the same should be true for brain, once its functions are understood. Nevertheless, the brain must be able to respond to changing and disparate stimuli, so it would be not be advantageous to have all cells and neural circuits restricted in their responses. Turrigiano, Abbott, and Marder (1994) have shown that there are activity-dependent changes in the intrinsic properties of cultured neurons, so neurons could be equipped with a suite of features suited for particular computations, but also retain the ability to modify these over time (Desai, Rutherford, and Turrigiano 1999).

Acknowledgments   This work was supported by NIH DC00436 to C. E. C., by T32 DC00046 to S. K., by NIH R03 DC04382 and NSF 972033 to J. Z. S., and by NIH P30 DC0466 to the University of Maryland Center for the Evolutionary Biology of Hearing.

References

H. Agmon-Snir, C.E. Carr, J. Rinzel. (1998) The role of dendrites in auditory coincidence detection *Nature* 393: 268–272.

R. Batra, S. Kuwada, D.C. Fitzpatrick. (1997) Sensitivity to interaural temporal disparities of low- and high-frequency neurons in the superior olivary complex. I. Heterogeneity of responses. *J Neurophysiol* 78: 1222–1236.

A. Borst, M. Egelhaaf, J. Haag. (1995) Mechanisms of dendritic integration underlying gain control in fly motion-sensitive interneurons. *J Comput Neurosci* 2: 5–18.

J.G. Borst, F. Helmchen, B. Sakmann. (1995) Pre- and postsynaptic whole-cell recordings in the medial nucleus of the trapezoid body of the rat. *J Physiol* 489: 825–840.

J.G. Borst, B. Sakmann. (1996) Calcium influx and transmitter release in a fast CNS synapse. *Nature* 383: 431–434.

A. Brand, O. Behrend, T. Marquardt, D. McAlpine, B. Grothe. (2002) Precise inhibition is essential for microsecond interaural time difference coding. *Nature* 417: 543–547.

J.R. Brawer, D.K. Morest. (1974) Relations between auditory nerve endings and cell types in the cat's anteroventral cochlear nucleus seen with Golgi method and Nomarski optics. *J Comp Neurol* 160: 491–506.

A.S. Bregman. (1990) Auditory Scene Analysis. Cambridge, Mass.: MIT Press.

S. Brenowitz, L.O. Trussell. (2001) Maturation of synaptic transmission at endbulb synapses of the cochlear nucleus. *J Neurosci* 21: 9487–9498.

H.M. Brew, J.L. Hallows, B.L. Tempel. (2003) Hyperexcitability and reduced low threshold potassium currents in auditory neurons of mice lacking the channel subunit Kv1.1. *J Physiol* 548: 1–20.

H.M. Brew, I.D. Forsythe. (1995) Two voltage-dependent K+ conductances with complementary functions in postsynaptic integration at a central auditory synapse. *J Neurosci* 15: 8011–8022.

A.R. Brughera, E.R. Stutman, L.H. Carney. (1996) A model with excitation and inhibition for cells in the medial superior olive. *Auditory Neurosci* 2: 219–233.

Y. Cai, E. J. Walsh, J. McGee. (1997) Mechanisms of onset responses in octopus cells of the cochlear nucleus: implications of a model. *J Neurophys* 78: 872–883.

N.B. Cant, R.L. Hyson. (1992) Projections from the lateral nucleus of the trapezoid body to the medial superior olivary nucleus in the gerbil. *Hear Res.* 58: 26–34.

C.E. Carr, M.A. Friedman. (1999) Evolution of time coding systems. *Neural Comput* 11: 1–20.

C.E. Carr, M. Konishi. (1988) Axonal delay lines for time measurement in the owl's brainstem. *Proc Natl Acad Sci* 85: 8311–8315.

C.E. Carr, M. Konishi. (1990) A circuit for detection of interaural time differences in the brainstem of the barn owl. *J Neurosci* 10: 3227–3246.

C.E. Carr, D. Soares. (2002) Evolutionary convergence and shared computational principles in the auditory system. *Brain Behav Evol* 59: 294–311.

H.S. Colburn, Y. Han, C.P. Culotta. (1990) Coincidence model of MSO responses. *Hear Res.* 49: 335–355.

N.S. Desai, L.C. Rutherford, G.G. Turrigiano. (1999) Plasticity in the intrinsic excitability of cortical pyramidal neurons. *Nat Neurosci* 2: 515–520.

M.J. Ferragamo, D. Oertel. (2002) Octopus cells of the mammalian ventral cochlear nucleus sense the rate of depolarization. *J Neurophysiol* 87: 2262–2270.

K. Funabiki, K. Koyano, H. Ohmori. (1998) The role of GABAergic inputs for coincidence detection in the neurones of nucleus laminaris of the chick. *J Physiol* 508: 851–869.

S. Gardner, L. Trussell, D. Oertel. (1999) Time course and permeation of synaptic AMPA receptors in cochlear nucleus neurons correlate with input. *J Neurosci* 19: 8721–8729.

J.R.P. Geiger, T. Melcher, D.-S. Koh, B. Sakmann, P.H. Seeburg, P. Jonas, H. Monyer. (1995) Relative abundance of subunit mRNAs determines Gating and Ca^{2+} permeability of AMPA receptors in principal neurons and interneurons of rat CNS. *Neuron* 15: 193–204.

W. Gerstner, R. Kempter, J.L. van Hemmen, H. Wagner. (1996) A neuronal learning rule for sub-millisecond temporal coding. *Nature* 383: 76–81.

J.M. Goldberg, P.B. Brown. (1969) Response of binaural neurons of dog superior olivary complex to dichotic tonal stimuli: Some physiological mechanisms of sound localization. *J Neurophysiol* 32: 613–636.

N.L. Golding, M.J. Ferragamo, D. Oertel. (1999) Role of intrinsic conductances underlying responses to transients in octopus cells of the cochlear nucleus. *J Neurosci* 19: 2897–2905.

N.L. Golding, D. Robertson, D. Oertel. (1995) Recordings from slices indicate that octopus cells of the cochlear nucleus detect coincident firing of auditory nerve fibers with temporal precision. *J Neurosci* 15: 3138–3153.

J. Golowasch, M. Casey, L.F. Abbott, E. Marder. (1999) Network stability from activity-dependent regulation of neuronal conductances. *Neural Comput* 11: 1079–1096.

V. Grau-Serrat, C.E. Carr, J.Z. Simon. (2003) Modeling coincidence detection in nucleus laminaris. *Biol Cybern* 89: 388–396.

J.J. Grigg, H.M. Brew, B.L. Tempel. (2000) Differential expression of voltage-gated potassium channel genes in auditory nuclei of themouse brainstem. *Hear Res* 140: 77–90.

B. Grothe. (2003) New roles for synaptic inhibition in sound localization. *Nat Rev Neurosci* 4: 540–550.

B. Grothe, D.H. Sanes. (1994) Synaptic inhibition influences the temporal coding properties of medial superior olivary neurons: an *in vitro* study. *J Neurosci* 14: 1701–1709.

E.R. Hafter, C. Trahiotis. (1997) Functions of the binaural system. In: Handbook of Acoustics, edited by M. Crocker, 1461–1480. New York: Wiley.

R. S. Heffner, H.E. Heffner. (1992) Evolution of sound localization in mammals. In: The Evolutionary Biology of Hearing, edited by DB Webster, RR Fay, AN Popper, 691–716. New York: Springer-Verlag.

K.G. Hill, G. Stange, J. Mo. (1989) Temporal synchronization in the primary auditory response in the pigeon. *Hear Res* 39: 63–74.

J.S. Isaacson, B. Walmsley. (1996) Amplitude and time course of spontaneous and evoked excitatory postsynaptic currents in bushy cells of the anteroventral cochlear nucleus. *J Neurophysiol* 76: 1566–1571.

L.A. Jeffress. (1948) A place theory of sound localization. *J Comp Physiol Psych* 41: 35–39.

S. Jhaveri, K. Morest. (1982) Sequential alterations of neuronal architecture in nucleus magnocellularis of the developing chicken: A Golgi study. *Neurosci* 7: 837–853.

D. Johnston, J.C. Magee, C.M. Colbert, B.R. Christie. (1996) Active properties of neuronal dendrites. *Ann Rev Neurosci.* 19: 165–186.

P.X. Joris, P.H. Smith, T.C.T. Yin. (1998) Coincidence detection in the auditory system: 50 years after Jeffress. *Neuron* 21: 1235–1238.

A.W. Joseph, R.L. Hyson. (1993) Coincidence detection by binaural neurons in the chick brain stem. *J Neurophysiol* 69: 1197–1211.

S. Kalluri, B. Delgutte. (2003a) Mathematical models of cochlear nucleus onset neurons: I. Point neuron with many weak synaptic inputs. *J Comput Neurosci* 14: 711–790.

S. Kalluri, B. Delgutte. (2003b) Mathematical models of cochlear nucleus onset neurons: II. Model with dynamic spike-blocking state. *J Comput Neurosci* 14: 91–110.

E.C. Kane. (1973) Octopus cells in the cochlear nucleus of the cat: Heterotypic synapses upon homeotypic neurons. *Intern J Neuroscience* 5: 251–279.

M. Kawasaki. (2000) Phylogenetic evolution of computational algorithms. *Nonparametric Approach to Knowledge Discovery* 8: 77–80.

D. Kim, W. Rhode, S. Greenberg. (1986) Responses of cochlear nucleus neurons to speech signals: neural encoding of pitch, intensity, and other parameters. In: Auditory Frequency Selectivity, edited by B. Moore and R. Patterson, 281–288. New York: Plenum Press.

M. Konishi. (1973) How the owl tracks its prey. *Am Sci* 61: 414–424.

M. Konishi. (1991) Deciphering the brain's codes. *Neural Computation* 3: 1–18.

C. Koppl. (1997) Phase locking to high frequencies in the auditory nerve and cochlear nucleus magnocellularis of the barn owl, *Tyto alba. J Neurosci* 17: 3312–3321.

C. Koppl, C.E. Carr. (2003) Computational diversity in the cochlear nucleus angularis of the barn owl. *J Neurophysiol* 89: 2313–2329.

C. Kopp-Scheinpflug, K. Fuchs, W.R. Lippe, B.L. Tempel, R. Rübsamen. (2003) Decreased temporal precision of auditory signaling in KCNA1-null mice: An electrophysiological study in vivo. *J Neurosci* 23: 9199–9207.

H. Kuba, K. Koyano, H. Ohmori. (2002) Development of membrane conductance improves coincidence detection in the nucleus laminaris of the chicken. *J Physiol* 540: 529–542.

M.F. Kubke, D.P. Massoglia, C.E. Carr. (2002) Developmental changes underlying the formation of the specialized time coding circuits in barn owls (*tyto alba*). *J Neurosci* 22: 7671–7679.

N. Kuwabara, J.M. Zook. (1992) Projections to the medial superior olive from the medial and lateral nuclei of the trapezoid body in rodents and bats. *J Comp Neurol* 324: 522–538.

W. Li, L.K. Kaczmarek, T.M. Perney. (2001) Localization of two high-threshold potassium channel subunits in the rat central auditory system. *J Comp Neurol* 437: 196–218.

C.M. Macica, C.A. von Hehn, L.Y. Wang, C.S. Ho, S. Yokoyama, R.H. Joho, L.K. Kaczmarek. (2003) Modulation of the kv3.1b potassium channel isoform adjusts the fidelity of the firing pattern of auditory neurons. *J Neurosci* 23: 1133–1141.

Z.F. Mainen, T.J. Sejnowski. (1995) Reliability of spike timing in neocortical neurons. *Science* 268:1503–6

P.B. Manis, S.O. Marx. (1991) Outward currents in isolated ventral cochlear nucleus neurons. *J. Neurosci* 11: 2865–2880.

A. Matsushita, M. Kawasaki. (2004) Unitary giant synapses embracing a single neuron at the convergent site of time-coding pathways of an electric fish, *Gymnarchus niloticus. J Comp Neurol* 472: 140–155.

P. Monsivais, L. Yang, E.W. Rubel. (2000) GABAergic inhibition in nucleus magnocellularis: implications for phase locking in the avian auditory brainstem. *J Neurosci* 20: 2954–2963.

J. Mosbacher, R. Schoeper, H. Monyer, N. Burnashev, P.H. Seeburg, J.P. Ruppersberg. (1994) A molecular determinant for submillisecond desensitization in glutamate receptors. *Science* 266: 1059–1062.

D. Oertel. (1999) The role of timing in the brainstem auditory nuclei. *Ann Rev Physiol* 61: 497–519.

D. Oertel, R. Bal, S. Gardner, P. Smith, P. Joris. (2000) Detection of synchrony in the activity of auditory nerve fibers by octopus cells of the mammalian cochlear nucleus. *Proc Nat Acad Sci* 97: 11773–11779.

T.S. Otis, I.M. Raman, L.O. Trussell. (1995) AMPA receptors with high Ca2+ permeability mediate synaptic transmission in the avian auditory pathway. *J Physiol (Lond)* 482: 309–315.

E.M. Overholt, E.W. Rubel, R.L. Hyson. (1992) A circuit for coding interaural time differences in the chick brain stem. *J Neurosci* 12: 1698–1708.

S. Parameshwaran, C.E. Carr, T.M. Perney. (2001) Expression of the Kv3.1 potassium channel in the avian auditory brainstem. *J Neurosci* 21: 485–494.

T.N. Parks. (2000) The AMPA receptors of auditory neurons. *Hear Res* 147: 77–91.

J.L. Pena, S. Viete, Y. Albeck, M. Konishi. (1996) Tolerance to sound intensity of binaural coincidence detection in the nucleus laminaris of the owl. *J Neurosci* 16: 7046–7054.

T.M. Perney, L.K. Kaczmarek. (1997) Localization of a high threshold potassium channel in the rat cochlear nucleus. *J Comp Neurol* 386: 178–202.

I.M. Raman, L.O. Trussell. (1992) The kinetics of the response to glutamate and kainate in neurons of the avian cochlear nucleus. *Neuron* 9: 173–186.

A.J. Rashid, E. Morales, R.W. Turner, R. J. Dunn. (2001) The contribution of dendritic Kv3 K+ channels to burst threshold in a sensory neuron. *J Neurosci* 21: 125–35.

M. Rathouz, L.O. Trussell. (1998) A characterization of outward currents in neurons of the nucleus magnocellularis. *J Neurophysiol* 80: 2824–2835.

A. Ravindranathan, T.N. Parks, M.S. Rao. (1996) Flip and flop isoforms of chick brain AMPA receptor subunits: cloning and analysis of expression patterns. *Neuroreport* 7: 2707–2711.

A.D. Reyes, E.W. Rubel, W.J. Spain. (1994) Membrane properties underlying the firing of neurons in the avian cochlear nucleus. *J Neurosci* 14: 5352–5364.

A.D. Reyes, E.W. Rubel, W.J. Spain. (1996) *In vitro* analysis of optimal stimuli for phase-locking and time-delayed modulation of firing in avian nucleus laminaris neurons. *J Neurosci* 16: 993–1007.

W.S. Rhode, D. Oertel, P.H. Smith (1983) Physiological response properties of cells labeled intracellularly with horseradish peroxidase in cat ventral cochlear nucleus. *J Comp Neurol* 213: 448–463.

J.S. Rothman, E.D. Young, P.B. Manis. (1993) Convergence of auditory nerve fibers onto bushy cells in the ventral cochlear nucleus: implications of a computational model. *J Neurophysiol* 70: 2562–83.

E.W. Rubel, T.N. Parks. (1975) Organization and development of brainstem auditory nuclei of the chicken: Tonotopic organization of *N. magnocellularis* and *N. laminaris. J Comp Neurol* 164: 411–434.

E.W. Rubel, T.N. Parks. (1988) Organization and development of the avian brainstem auditory system. In: Brain Function, edited by G.M. Edelman, W.E. Gall, M.W. Cowan, 3–92. New York: Wiley.

E.W. Rubel, B. Fritzsch. (2002) Auditory system development: Primary auditory neurons and their targets. *Ann Rev Neurosci* 25: 51–101.

D.K. Ryugo, D.M. Fekete. (1982) Morphology of primary axosomatic endings in the anteroventral cochlear nucleus of the cat: A study of the endbulbs of Held. *J Comp Neurol* 210: 239–257.

D.K. Ryugo. (1991) The auditory nerve: peripheral innervation, cell body morphology, and central projections. In: The Mammalian Auditory Pathway: Neuroanatomy, edited by D.B. Webster, A.N. Popper, R.R. Fay, 23–65. New York: Springer-Verlag.

B.L. Sabatini, W.G. Regehr. (1999) Timing of synaptic transmission. *Annu Rev Physiol* 61: 521–542.

J.Z. Simon, C.E. Carr, S.A. Shamma. (1999) A dendritic model of coincidence detection in the avian brainstem. *Neurocomputing* 26–27: 263–269.

P.H. Smith. (1995) Structural and functional differences distinguish principal from nonprincipal cells in the guinea pig MSO slice. *J Neurophysiol* 73: 1653–1667.

D. Soares, C.E. Carr. (2001) The cytoarchitecture of the nucleus angularis of the barn owl (*Tyto alba*). *J Comp Neurol* 429: 192–205.

D. Soares, R.A. Chitwood, R.L. Hyson, C.E. Carr. (2002) Intrinsic neuronal properties of the chick nucleus angularis. *J Neurophysiol* 88: 152–162.

W. Softky. (1994) Sub-millisecond coincidence detection in active dendritic trees. *Neuroscience* 58: 13–41.

M. Stemmler, C.K. Koch. (1999) How voltage-dependent conductances can adapt to maximize the information encoded by neuronal firing rate. *Nat Neurosci* 2: 521–527.

K.N. Stevens. (1995) Applying phonetic knowledge to lexical access. In: 4th European Conference on Speech Communication and Technology, vol. 1. Madrid, Spain, pp. 3–11.

W.E. Sullivan, M. Konishi. (1984) Segregation of stimulus phase and intensity coding in the cochlear nucleus of the barn owl. *J Neurosci* 4: 1787–1799.

T. Takahashi, A. Moiseff, M. Konishi. (1984) Time and intensity cues are processed independently in the auditory system of the owl. *J Neurosci* 4: 1781–1786.

H. Taschenberger, R.M. Leao, K.C. Rowland, G.A. Spirou, H. von Gersdorff. (2002) Optimizing synaptic architecture and efficiency for high-frequency transmission. *Neuron* 36: 1127–1143.

H. Taschenberger, H. von Gersdorff. (2000) Fine-tuning an auditory synapse for speed and fidelity: developmental changes in presynaptic waveform, EPSC kinetics, and synaptic plasticity. *J Neurosci* 20: 9162–9173.

J.M. Thompson, K.C. Rowland, G.A. Spirou. (2004) Cellular basis for ITD sensitivity in the MSO. *ARO Abstract* 1169.

L.O. Trussell. (1997) Cellular mechanisms for preservation of timing in central auditory pathways. *Curr Opin Neurobiol* 7: 487–492.

L.O. Trussell. (1999) Synaptic mechanisms for coding timing in auditory neurons. *Annu Rev Physiol* 61: 477–496.

G. Turrigiano, L.F. Abbott, E. Marder. (1994) Activity-dependent changes in the intrinsic properties of cultured neurons. *Science* 264: 974–977.

L.Y. Wang, L. Gan, I.D. Forsythe, L.K. Kaczmarek. (1998) Contribution of the Kv3.1 potassium channel to high-frequency firing in mouse auditory neurones. *J Physiol (Lond)* 509: 183–194.

M.E. Warchol, P. Dallos. (1990) Neural coding in the chick cochlear nucleus. *J Comp Physiol* 166: 721–734.

I. Winter, A. Palmer. (1995) Level dependence of cochlear nucleus onset unit responses and facilitation by second tones or broadband noise. *J. Neurophysiol* 73: 141–159.

S.H. Wu. (1999) Physiological properties of neurons in the ventral nucleus of the lateral lemniscus of the rat: Intrinsic membrane properties and synaptic responses. *J Neurophysiol* 2872–2874.

L. Yang, P. Monsivais, E.W. Rubel. (1999) The superior olivary nucleus and its influence on nucleus laminaris: a source of inhibitory feedback for coincidence detection in the avian auditory brainstem. *J Neurosci* 19: 2313–2325.

T.C.T. Yin, J.C.K. Chan. (1990) Interaural time sensitivity in medial superior olive of cat. *J Neurophysiol* 64: 465–488.

S.R. Young, E.W. Rubel. (1983) Frequency-specific projections of individual neurons in chick brainstem auditory nuclei. *J Neuroscience* 7: 1373–1378.

S. Zhang, L.O. Trussell. (1994) A characterization of excitatory postsynaptic potentials in the avian nucleus magnocellularis. *J Neurophysiol.* 72:705–718.

P.M. Zurek. (1987) The precedence effect. In: Directional Hearing, edited by W.A. Yost and G. Gourevitch, 85–105. New York: Springer-Verlag.

13

How Is Time Represented in the Brain?

Andreas V. M. Herz

Introduction

Behaviorally relevant sensory signals are not constant in time but vary on many scales. A sudden or gradual change in the patterns of a visual scene or the precise time course of an acoustic communication signal contains information that is of greatest importance for an organism, not the mean illumination level or mean sound intensity. To process and interpret time-varying signals in a meaningful way, environmental signals have to be integrated over time. As a consequence, temporal relations between and within external stimuli cannot just be encoded in a one-to-one fashion by the nervous system.

These observations trigger a general question: How is time represented in the brain? Given the large variety and intriguing temporal complexity of many natural pattern sequences, sophisticated neural representations are likely to have been invented during the course of evolution. No simple, universal answer is therefore to be expected to the question. Progress in the understanding of neural coding in the time domain may nevertheless be achieved if one concentrates on a more specific problem: Which types of representations best support flexible and robust computations of temporal relations?

Of particular importance are algorithms that are compatible with naturally occurring signal variations such as a change of the stimulus intensity or a change of the duration of all signal components, also known as *time warp*. In the following, a collection of basic computing principles is presented. The focus is on algorithms that deal with sensory pattern sequences that vary over time scales from a few to a few hundred milliseconds.

Spatial Representation of Temporal Sequences

Axonal, synaptic, and dendritic delays provide an ideal neuronal substrate to integrate temporal information over short time scales (Caianiello 1961), as is illustrated in figure 13-1. Well-known examples are time-comparison circuits (Jeffress 1948) that allow computation of the location of objects from binaural acoustic signals. This is of particular importance for animals hunting at night, such as barn owls (Carr and Konishi 1988; Konishi 1992). Similar circuits are also used by electric fish (Carr, Heiligenberg, et al. 1986; Carr, Maler, et al., 1986; Heiligenberg 1991; Konishi 1992) and constitute in their most simple form the central comparison unit of Reichardt-type velocity detectors (Reichardt 1965).

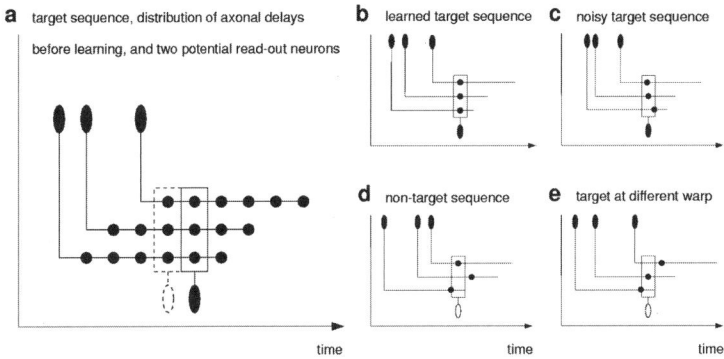

Figure 13-1. Schematic illustration of a neural network with a broad delay distribution. (a) Input neurons (three are shown as large ovals) are activated at different times by the target sequence. Action potentials travel along the axons (black lines) and reach synapses (small circles) at successive delay times τ. Hebbian plasticity as described by equation 13.2 is based on the correlation of delayed presynaptic activity and present activity of postsynaptic neurons (two are shown) within a narrow time window (rectangle). Note that in this drawing, the horizontal axis represents time, not real space. Furthermore, the vertical direction has no physical meaning and only a small group of neurons is depicted. Finally, within a feedback network, every neuron would function as both pre- and postsynaptic units. (b) A subset of synapses has been strengthened during the learning process. When the system is presented with the target sequence during a recall phase, the very same delay lines are activated and trigger dendritic currents in postsynaptic neurons (only one is shown for simplicity). (c) If the input sequence is corrupted by small amounts of noise, the postsynaptic neuron still receives sufficient activation and correctly recognizes the sequence. (d) At large noise levels or for non-target sequences, however, the postsynaptic neuron does no longer respond (open ellipse). (e) The same is true if the target sequence is stretched or compressed in time—systems with fixed transmission delays do not generalize with respect to time warp.

Time lags may also play an important role in other neural systems. The striking geometric layout of the cerebellum—long parallel fibers intersecting the dendritic trees of Purkinje cells at a right angle—has been hypothesized to be a "clocking device in the millisecond range" (Braitenberg 1967) or to support resonant "tidal waves" (Braitenberg 1997) that are generated by external sensory inputs whose individual arrival times match the internal signal propagation along the parallel fibers.

In essence, fixed delay lines provide a means to map the temporal domain into a spatial dimension. Delay lines thus facilitate a broad variety of computations that involve comparisons between signals received at different times spread out as much as the longest available transmission delay. One application is the associative recall of temporal sequences in neural networks with delayed feedback (Kleinfeld 1986; Sompolinsky and Kanter 1986; Amit 1988, Riedel, Kühn, and van Hemmen 1988). In such networks, synaptic plasticity may be used to *learn* the time structure of a target sequence in a Hebbian manner (Herz et al. 1989).

As an example, consider a model network with N graded-response neurons,

$$\tau_{RC}\frac{d}{dt}u_i(t) = -u_i(t) + I_i(t) + \sum_{j=1}^{N}\int d\tau T_{ij}(\tau)g[u_j(t-\tau)]. \qquad (13.1)$$

In this type of model, the membrane potential $u_i(t)$ of each neuron i follows a leaky-integrator dynamics with time constant τ_{RC}. The neuron receives both a time-dependent external input signal $I_i(t)$ and delayed feedback from the other neurons, whose output activity is described by a short-time averaged firing rate that depends through the sigmoid nonlinearity g on the membrane potential. To include discrete axonal time lags as well as continuous delay distributions due to synaptic transmission and dendritic integration processes, the coupling strength of a synapse from neuron j to neuron i is not simply a scalar variable T_{ij} as in traditional autoassociative neural networks but is described by a function $T_{ij}(\tau)$ whose argument denotes the time required by a signal to travel from neuron j to neuron i along the specific connection path. Within this class of model, the contribution from a single axonal delay τ_a corresponds to a delta function $T_{ij}(\tau) \sim \delta(\tau-\tau_a)$; additional synaptic delays lead to a functional dependence such as $T_{ij}(\tau) \sim (\tau-\tau_a) \cdot \exp[(\tau_a-\tau)/\tau_{syn}] \cdot \theta(\tau-\tau_a)$ or similarly shaped delay distributions.

Nonzero synaptic connections $T_{ij}(\tau)$ represent the network topology and overall delay structure. The particular synaptic strengths determine the network dynamics and thus the network's computational capabilities. Hebb's postulate for synaptic plasticity (1949) provides an algorithm to store both static objects and temporal associations such as tunes and rhythms: "When an axon of cell A is near enough to excite cell B and *repeatedly* or *persistently* takes part in firing it, some growth process or metabolic change takes place in one or both cells such that A's efficiency, as one of the cells firing B, is increased." Within the present framework, this postulate can be implemented by synaptic modifications of the type

$$\Delta T_{ij}(\tau) = F[h_i(t), u_j(t - \tau)]. \qquad (13.2)$$

Here $h_i(t)$ denotes the total current driving neuron i, that is, the last two terms in equation 13.1. The learning rule implies that synaptic coupling strengths change according to the correlations of the pre- and postsynaptic activity as measured at the synaptic site (Herz et al. 1989; Herz 1995). For example, a synapse located at the end of a long axon (large transmission delay) encodes time-lagged correlations between pre- and postsynaptic activity, while a synapse located near the presynaptic soma encodes correlations at approximately equal times. The first type of synapse therefore represents specific temporal features of the target sequence whereas the second type of synapse represents individual "snapshots" of the same sequence. In either case, synaptic strengths facilitate the storage and associative replay of temporal sequences by properly concentrating information in time, as shown in figure 13-1. Recent electrophysiological results by Markram et al. (1997) and Bi and Poo (1998) support the view that synaptic plasticity requires precisely timed pre- and postsynaptic activity.

As demonstrated by computer simulations (Herz et al. 1989) and analytical studies (Herz, Li, and van Hemmen 1991), the learning rule (equation 13.2) allows model networks to function as content-addressable memories for spatiotemporal patterns. Perhaps surprisingly, broad unspecific delay distributions *improve* the associative capabilities as long as a sufficient number of connections are provided whose time lags exceed the typical transition time between successive patterns within one target sequence. Using the example of a cyclic sequence containing three patterns, this robustness with respect to details of the delay distribution is shown in figure 13-2.

Learning is successful if the structure of the learning task matches both the network architecture and the learning algorithm. In the present context, the task is to store spatiotemporal target objects such as stationary patterns and temporal sequences. A successful internal representation of these objects is guaranteed by a broad distribution of time lags τ in conjunction with a high connectivity. The representation itself is accomplished by a Hebbian rule so that correlations of the target objects in both space (ij) and time (τ) are measured and stored. The dynamics of the neural network, operating with the very same delays, are able to extract the spatiotemporal information encoded in the $T_{ij}(\tau)$. As a consequence of this "happy triadic relation" (Minsky 1986) between learning task, network architecture and learning rule, retrieval is extremely robust.

This robustness should be compared with the chaotic behavior typically exhibited by systems of nonlinear delay-differential equations (Mackey and Glass 1977, Glass and Mackey 1988, Riedel, Kühn, and van Hemmen 1988).

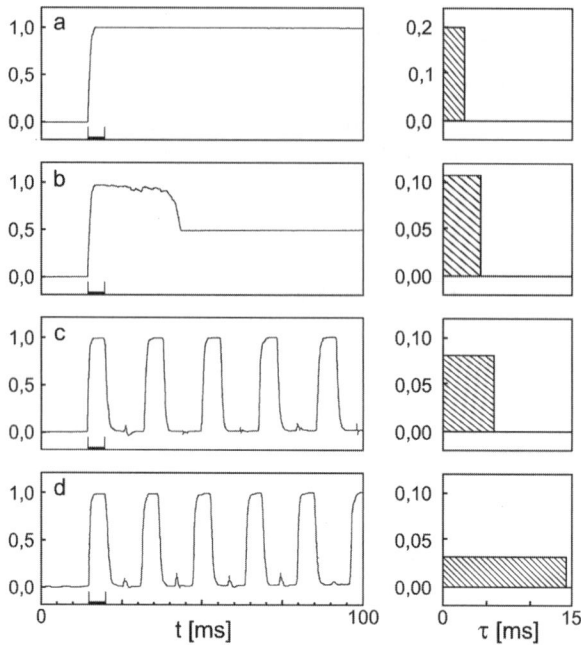

Figure 13-2. Performance of the model network as a function of the delay distribution. Each track represents the time evolution of the "overlap" (*left panels*) with the first pattern of a cyclic target sequence consisting of three static patterns, for a given distribution of axonal delays (*right panels*). The overlap measures the similarity between the current network state and a stored pattern and takes values between one (network state is identical with the target pattern) and minus one (network state is identical with the inverted target pattern). Values around zero imply that the current network state is uncorrelated with the target pattern. All delay distributions are discrete, with a spacing of one-half millisecond. To trigger the retrieval of the target sequence, the first pattern of the sequence is presented as external input $I(t)$ to the system, between $t = 15$ ms and $t = 20$ ms, as is illustrated by the black horizontal bars. In (a), all delays are shorter than the duration (5 ms) of a static pattern within the target sequence. The first pattern is retrieved as shown by the transition of the network, but the desired target sequence is not triggered due to the lack of appropriately long delays. In (b), additional delays destabilize the first pattern and a static mixture of the three patterns is reached, where the overlap with each pattern is approximately half. For larger maximal time lags (c and d), the entire target sequence is replayed and its period depends only marginally on the specific shape of the delay distribution. Note that a stable cycle can be produced even in the absence of synapses with short delays that would stabilize a single pattern (data not shown, but see Herz et al. 1989) if enough delays are provided that are longer than the duration of an elementary static pattern. Redrawn from Herz (1990).

270

The present networks are, however, endowed with a broad delay distribution where Hebbian learning automatically *selects* the connections most suitable to stabilize the time course of a target sequence. In other words, Hebbian learning "tames" otherwise chaotic systems so that they can operate as useful computational devices.

Firing Neurons, Derailed Spike Trains

To investigate the development of interaural-time-difference maps (Jeffress 1948) within a framework that is more realistic from a biophysical point of view, the approach presented in the last section has been extended with great success to model networks with integrate-and-fire neurons (Gerstner, Kempter, et al. 1996). Integrate-and-fire neurons capture the essential dynamics of most biological neurons—integration of synaptic inputs and subsequent generation of action potentials. In particular, the time evolution of a (leaky) integrate-and-fire model neuron is described by equation 13.1 as long as the membrane potential u remains below a certain firing threshold θ. If, however, u reaches θ the neuron instantaneously generates an action potential or "spike" modeled as a δ-function and u is reset to some value $u_{reset} < \theta$.

Increasing the coupling strength of a synapse causes a shift of the spike times of the postsynaptic neuron. Within feedback networks, synaptic plasticity may therefore lead to large rearrangements of the temporal sequence of action potentials generated by a neuron; such a sequence is often also called a *spike train*. If information is represented on the level of time-averaged firing rates, these rearrangements will in general not change the information conveyed by the spike train in a significant manner. If, on the other hand, information is encoded with high precision on the level of interspike intervals or individual spike times, synaptic plasticity may completely alter the information contained in a spike train. Unless carefully controlled, synaptic plasticity may therefore jeopardize the computational capabilities of a neural network by "derailing" spike trains.

Sequences of precisely timed spike patterns across many neurons, also known as "synfire chains" (Abeles 1982, 1991), have been suggested to play an important role for cortical information processing. Numerical simulations and theoretical investigations support the principal feasibility of this concept (see, for example, Diesmann, Gewaltig, and Aertsen 1999). There have also been numerous attempts to store target synfire chains as dynamical attractors of model networks with spiking neurons (see, for example, Hertz and Prugel-Bennett 1996). The limited success of these attempts is directly related to the difficulty to stabilize the desired "synfire chains" by a learning rule. Keeping the *total* synaptic input to a given neuron constant when changing individual synaptic strengths might be a key for solving this problem.

Time Warp and Analog Match

Combined with Hebbian learning schemes, a broad distribution of transmission delays can be used to "concentrate information in time" in that stimuli that were originally spread out over a large time interval are grouped together. This mechanism has also been implemented in several technical applications such as artificial speech recognition (Unnikrishnan, Hopfield, and Tank 1992). For acoustic communication systems, one natural limitation of these algorithms arises from the trial-to-trial variability of communication signals which is of particular importance for poikilothermic animals whose body temperature strongly fluctuates. For example, the length of single "syllables" of the mating songs of grasshoppers varies up to 300 percent as a function of the ambient temperature (von Helversen and von Helversen 1994). Large time warps are, however, also common in human speech, where the duration of a word spoken by a given speaker may change by up to 100 percent depending on the specific circumstance (see, e.g., Unnikrishnan, Hopfield, and Tank 1992). Psychophysical data show that in both systems, even such large variations are tolerated with ease by the receiver.

Time-warp-invariant sequence recognition is a major challenge for any recognition system based on fixed transmission delays: how should a neural system such as the feedback network described by equations 13.1 and 13.2 classify the two temporal sequences *AAABBBCCCCCCDDD* and *ABCCD* as "equal" (up to an overall threefold time warp) and yet discriminate the two sequences from a third sequence, for example *ABBBBCDDDD*? This task is a special case of the so-called analog-match problem—the task to recognize a multidimensional input vector $(x_1, \ldots, x_N)$ in a scale-invariant manner, or to put it differently, the task to classify all vectors $(\lambda x_1, \ldots, \lambda x_N)$ with $\lambda \in R$ as equal. In our special case the analog vector to be recognized is the vector $(T - t_1, \ldots, T - t_N)$ of onset *times* of the elementary components of a target sequence, measured relative to the end point T of this sequence.

If the input vector $(x_1, \ldots, x_N)$ is constant or varies only slowly in time, the analog-match problem can be solved by using encoding neurons that receive an additional periodic subthreshold input whose strength and time course is tuned such that the nth encoding cell generates its action potentials at a time $\log(x_n)$ before the next maximum of the underlying rhythm (Hopfield 1995). In this scheme, the firing pattern of the N encoding neurons represents the analog input as a periodic activity pattern. Increasing the overall stimulus intensity by a factor λ results in a global time advance of the firing pattern by $\log(\lambda)$ without any changes of the relative firing times.

To *decode* this distributed representation, the encoding neurons are connected with one or several read-out neurons. Each of the read-out neurons is programmed to function as a "grandmother neuron" and recognizes only one specific target pattern, for example the pattern $(y_1, \ldots, y_N)$. To do so, the connection between the nth encoding neuron and the read-out neuron responsible for pattern $(y_1, \ldots, y_N)$ contains a time lag of length $\log(y_n)$ and the read-out neuron itself is a coincidence detector. If the sensory input vector is $(y_1, \ldots, y_N)$, spikes generated by the encoding neurons will converge at the *same* time at the read-out neuron and trigger a spike, signaling the successful recognition of the input pattern. Apart from changing the overall response time, multiplicative scale transformations $x_i \rightarrow \lambda x_i$ of the input pattern have no effect on the recognition event. By using a high firing threshold for the read-out neuron, only input patterns very similar to the target pattern are recognized, choosing a somewhat lower threshold allows the recognition of noisy and incomplete patterns albeit at the price of an increased probability of false alarms.

Based on this framework, the original time-warp problem may be solved if it can be rephrased as an analog-match problem. Neural responses that decay in time after a transient stimulus offer an ideal mechanism to achieve this goal. For example, the firing rate of an adapting sensory neuron, say neuron i, that is triggered at time t_i by the onset of a step input is a decaying analog variable $r_i(t)$ with stereotypical time course, $r_i(t) = f_i(t - t_i)$ for $t > t_i$. A comparison of the present firing rate with the initial firing rate—at stimulus onset for on-cells and stimulus offset for off-cells—therefore provides a direct measure for the time elapsed since the triggering event. In general, the function f_i will depend on stimulus intensity. For simplicity of the argument, we will mostly consider target sequences with fixed intensity, that is, we concentrate on symbolic sequences.

Readout I: Subthreshold Oscillations and Coincidence Detection

To map the time-warp problem onto the analog-match problem, the decaying output activity $r(t)$ of each sensory neuron is applied as an input for one encoding neuron of the type used for the analog-match problem. If the time course of the decaying firing rate of the sensory neurons and the shape of the subthreshold oscillation of the encoding neurons are properly matched, a situation can be achieved the ith encoding neuron generates its action potentials at times $\log(t - t_i)$ before the next maximum of the subthreshold oscillation (for technical details, see Hopfield, Brody, and Roweis 1998).

Following a presentation of the target sequence, the relations between the components of the analog pattern $(r_1(t), \ldots, r_N(t))$ constantly change as time evolves—and cause subsequent changes of the relative firing times of the encoding neurons. Around $t = T$, however, the relative firing times of each of the encoding neurons approach the logarithm of the corresponding components of the target pattern, $(\log(T - t_1), \ldots, \log(T - t_N))$. Provided that signals between the encoding neurons and the read-out neuron are delayed by $\tau_I = \log(T - t_i)$, the spikes of all encoding neurons reach the read-out neuron at the same time and trigger a spike, as in the original analog-match problem. It follows that the recognition process is largely insensitive to global time warps $(T - t_i) \rightarrow \lambda (T - t_i)$ of the external stimulus (Hopfield 1996).

In its basic form, this approach has two intrinsic constraints—(1) it does not tolerate intensity changes of the external stimulus and (2) a target sequence already stored cannot be extended by additional elements because all time intervals are measured and represented relative to the sequence's end. The first limitation can be overcome by using feature detectors for signal preprocessing (Hopfield, Brody, and Roweis 1998), the second problem can be solved by more elaborate hierarchical encoding schemes (Hopfield 1996).

Readout II: Transient Synchronization

Coupled integrate-and-fire neurons tend to synchronize if their firing rates are approximately equal (Gerstner, Ritz, and van Hemmen 1993, Tsodyks and Sompolinsky 1993). In fact, synchronization is established very rapidly if the summed synaptic strengths of each neuron (as measured at its input side) are equal and if all neurons receive the same constant external input (Herz and Hopfield 1995, Hopfield and Herz 1995). Time-varying inputs will therefore cause transient synchronization if over some time interval, most inputs have about the same strength.

This input-dependent transient synchronization can be exploited to recognize temporal sequences independent of global time-warps (Hopfield and Brody 2000, 2001). To do so, receptor neurons with a broad distribution of firing-rate decay times are connected one-to-one to neurons in a second network layer that is endowed with lateral feedback connections. As shown in figure 13-3 for a target sequence with three patterns, the activities of the receptor neurons converge at some time T if one selects appropriately long decay times for cells encoding early components of the sequence and increasingly shorter decay times for cells encoding later components. The two circles in the figure illustrate that the choice of T is arbitrary as long as decay times can be found such that most or all input currents converge at one point in time.

If the decay time of the receptor neurons exceeds the time needed for synchronization by the recurrently connected layer, presenting the target sequence to the network will lead to a transient synchronization at around $t = T$. This event can easily be read out by a down-stream grandmother neuron. Noisy input sequences cause somewhat weaker synchronization; nontarget sequences do not result in any synchronized activity. Although based on an entirely different mechanism, this network thus concentrates information in time in a way that qualitatively resembles the delay-line network discussed

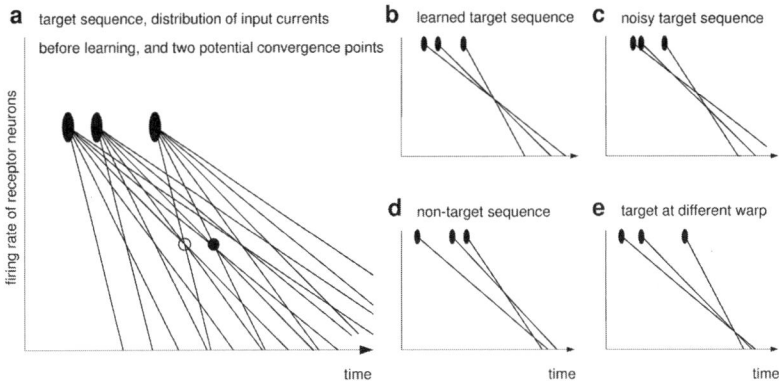

Figure 13-3. Schematic illustration of an associative neural network with a broad distribution of input decay times. (a) Input currents from a total of 3*6 receptor neurons that are activated at three different times (shown as large ovals) by the target sequence. The two circles denote points in time where three currents converge, one for each feature of the target sequence. The three currents of one such set are selected as inputs for three neurons in the second network layer that is endowed with feedback connections. Note that, unlike figure 13-1, the vertical direction has now a physical meaning. (b) When the system is presented with the target sequence during a recall phase, the very same decaying currents are activated and trigger synchronous spikes in the feedback layer. Only one set of three selected currents is shown. (c) If the input sequence is corrupted by small amounts of noise, the three currents do not converge exactly, but the network still receives sufficient activation and correctly recognizes the sequence. (d) At large noise levels or for nontarget sequences, the selected currents do no longer converge. (e) If the target sequence is stretched or compressed in time, the three currents still converge: in contrast to fixed delays, firing-rate adaptation can be used to recognize temporal sequences independent of global time warp (after Hopfield and Brody 2001).

earlier. This similarity is also evident from a comparison of figure 13-3 with figure 13-1. Unlike networks with fixed time lags, however, the present system tolerates a time warp because the decaying input currents still converge at one point in that case. Note also that information about the time warp's magnitude is not lost but encoded in the frequency of the triggered oscillation.

Multiple sequences can be embedded in the network structure if the cross-talk between different sequences is not too large. This can be achieved by balancing excitatory and inhibitory connections within the set of neurons representing one sequence in the second layer. In this scenario, the firing *rates* of all neurons are primarily determined by the input currents from the receptor neurons, both before and after learning a new sequence, and thus cannot trigger oscillations that represent other pattern sequences. In essence, this recipe is reminiscent of learning rules in traditional firing-rate based attractor neural networks that aim to minimize interference effects due to nonorthogonal target patterns (see, for example, Hertz, Krogh, and Palmer 1991).

The basic scheme sketched above can be extended in several directions. Hopfield and Brody (2001) discuss two enhancements that allow (1) covering sequences where the same symbol occurs multiple times within a target sequence and (2) including inputs that provide evidence *against* a particular target sequence. A further extension of the algorithm is suggested by hardware considerations: In networks with fixed delay times, a broad delay distribution can be achieved by using the multiple synapses of each neuron (figure 13-1). A broad distribution of decay times, however, requires one neuron for each current (figure 13-3) and is much more expensive. This problem can be overcome by utilizing synaptic depression and facilitation (Markram and Tsodyks 1996, Abbott et al. 1997) and sensory neurons with or without decaying firing rates. Within this alternative scheme different decay times can be realized at different synapses of one neuron, in direct analogy to the different time lags of different synapses along one axon. Supported by suitable learning algorithms, each synapse might then operate as one elementary timing device.

Readout III: Shunting Inhibition

Time-warp-invariant sequence recognition is a computational operation of utmost importance. More complex signal processing tasks may require that temporal aspects of a stimulus such as its duration are held transiently in short-term memory. This raises the question how temporal aspects can be mapped into a nontemporal response dimension independent of stimulus intensity. Neural circuits with fixed delay lines and coincidence detectors (Reichardt 1965, Braitenberg 1967, Konishi 1992) are one possibility, decaying neural activity together with shunting inhibition offers a simple alternative on the single-cell level, as discovered recently by the author.

In what follows, we briefly sketch this mechanism and discuss its scope and limitations. Consider, for simplicity, an isolated stimulus A that increases from zero to a constant intensity at time $t_{on} = 0$ before it returns to zero at time $t_{off} = D$. The stimulus is encoded by onset and offset cells whose firing rates decay exponentially with the same time constant τ. If the maximal firing rates of the onset and offset cells are a and b, respectively, their time evolutions are

$$f_{on}(t) = 0 \quad \text{for } t < 0, \qquad f_{on}(t) = ae^{-t/\tau} \quad \text{for } t \geq 0 \qquad (13.3)$$

and

$$f_{off}(t) = 0 \quad \text{for } t < D, \qquad f_{off}(t) = be^{-(t-D)/\tau} \quad \text{for } t \geq D. \qquad (13.4)$$

If the offset cell excites a read-out neuron which is also receiving shunting inhibition from the onset cell, the total input of the read-out neuron is essentially given by

$$x(t) = f_{off}(t)/[f_{on}(t) + c] \qquad (13.5)$$

where the parameter c measures the input resistance of the read-out neuron. For times $t < D$, the input vanishes, for $t > D$, it is given by

$$x(t) = (b/a)e^{D/\tau}[1 + (c/a)e^{t/\tau}]^{-1}. \tag{13.6}$$

For strong shunting inhibition, that is for $a \gg c$, the maximal activation of the read-out neuron becomes

$$x_{max} \cong (b/a)e^{D/\tau}. \tag{13.7}$$

The maximal activation thus encodes the stimulus duration D and does *not* depend on stimulus intensity since it only involves the *ratio* of a and b. Furthermore, the activation stays approximately constant at its maximal value for times $t \ll \tau \log (a/c)$.

Equation (6) can be rewritten as

$$x(t) = (b/a)e^{D/\tau}[1 + ce^{(t-\tau \log a)/\tau}]^{-1}. \tag{13.8}$$

This form illustrates that varying the stimulus intensity (scaling both a and b) has only a single consequence for their combined effect $x(t)$, namely a time shift of the response curve in proportion to the logarithm of the stimulus intensity. Shunting inhibition may thus be used to encode stimulus duration as response intensity and, vice versa, stimulus intensity as response duration. The latter property offers the additional feature that more salient, high-intensity signals remain available for longer periods of (processing) time.

Based on this elementary computational building block, pattern sequences can be processed in several ways. If the beginning and end of each feature of a sequence is encoded by onset and offset detectors, computations are insensitive to variations in the intensity of these individual components. The two sequences *AABBBCDD* and *AAbbbCDD* are thus classified as equal, where for graphical illustration, patterns with low intensity have been denoted by lower-case symbols. Similarly, recognition is not affected by pattern exchange, $ABC = BAC = \dots$. If only on-detectors are used so that the end of one pattern is actually encoded as the beginning of the next pattern, the system does not tolerate local intensity changes but still recognizes a sequence independent of global intensity shifts, $AABBBCDD = aabbbcdd$. By the same token, pattern exchanges are no longer possible.

As shown by these examples, combining exponential decay processes through a divisive operation naturally generates a neural representation where the duration of external stimuli is mapped on neural activity, and stimulus intensity is encoded by response duration. Firing-rate adaptation and shunting inhibition are two particular biophysical realizations of this principle. Alternative mechanisms to implement slowly decaying input currents include synaptic depression, bursting and (post-)synaptic processes with long time constants. Stimulus durations could also be memorized for longer times if the output of the comparison units was used as an input to a feedback network. The internal representation of temporal relations could then also be

Table 13-1 Overview of Several Schemes to Represent Time in Neural Systems

Neural Representations of Time	Mechanisms and Goals	Typical Time Scales	Time-warp Sensitivity	Biological Learning	Fine-tuning Required for
Pairwise delay lines within a feedforward system (Jeffress 1948; Gerstner et al. 1996)	Coincidence detection for time comparison; auditory localization	Micro- to millisecond	Very high	Hebbian	—
Broad distribution of delay lines in a feedforward system (Braitenberg 1997)	Generation of resonant "tidal waves"; detection of movement patterns	Few milliseconds	Very high	Hebbian	—
Broad distribution of delay lines within a feedback system (Herz et al. 1989)	Hebbian learning and associative retrieval of pattern sequences	Few milliseconds	High	Hebbian	—
Activity decay, subthreshold oscillations, delay lines, coincidence detection (Hopfield 1996)	Map: elapsed time on relative spike times; warp-insensitive recognition	Millisecond to second	Low	Hebbian	Shape of the subthreshold oscillations
Activity decay and feedback networks with spiking neurons (Hopfield and Brody 2000)	Transient collective synchronization; warp-insensitive recognition	Millisecond to second	low	—	—
Activity decay and shunting inhibition (Herz 2001)	Division by single cells; map: stimulus intensity on time, duration on activity	Millisecond to second	High	—	Decay times of excitation and inhibition

modulated in a controlled manner by systematic variations of the overall firing rates through higher-level command signals.

Conclusions

At first sight, transmission delays, input currents that decay in time, and synaptic short-time dynamics such as depression or facilitation might be regarded as a nuisance to associative neural computation. As the various examples summarized in table 13-1 demonstrate, these biophysical processes do, however, support interesting calculations in the time domain that would otherwise require much more elaborate architectures and algorithms. In addition, the broad variety of dynamical processes sketched here also shows that there are many more possibilities to represent time in a neural system than just by itself.

This chapter focused on the learning and associative recognition of target sequences that represent sensory stimuli. Acoustic communication served as an example to illustrate that neural representations that are invariant with respect to naturally occurring signal variations are of particular biological significance. At higher levels of information processing, other aspects will also come into play. It might, for example, be advantageous to compress temporal sequences in time for further processing or permanent storage. Indeed, compressed neural activity patterns have been observed during sleep phases in the hippocampus (Skaggs et al. 1996). In addition, spatial and movement-related information could be represented in the temporal domain, as by phase encoding (O'Keefe and Recce 1993). Time scales that are much longer than those considered in the present article could also involve various types of more traditional clock-counter and interval-timing representations (Gibbon 1977, Meck 1983, Killeen and Fetterman 1988, Miall 1989, Matell and Meck 2000).

It is thus most likely that we have only begun to grasp the complexity of time representations in the brain. Further computing principles will be found that, like the ones already known, enable an organism to concentrate information over multiple time scales and for a whole range of different purposes. One general goal is probably common to all of these algorithms, independently of the specific modality or species considered: to predict the future from past experience and to facilitate appropriate behavioral actions—right on time.

Acknowledgments I would like to thank Martin Stemmler for stimulating discussions, Hartmut Schütze for technical support with the figures, and Laurenz Wiskott and Tim Gollisch for critical comments on the manuscript. This work has been supported by the DFG through the Innovationskolleg Theoretische Biologie.

References

M. Abeles (1982): Local cortical circuits. Springer Verlag, Berlin.

M. Abeles (1991): Corticonics: Neural circuits of the cerebral cortex. Cambridge University Press, Cambridge.

L.F. Abbott, J.A. Varela, K. Sen, S.B. Nelson (1997): Synaptic depression and cortical gain control. *Science* **275**: 220–224.

D.J. Amit (1988): Neural networks counting chimes. *Proc. Natl. Acad. Sci. USA* **86**, 7871–7875.

G.Q. Bi and M.M. Poo (1998) *J. Neurosci.* **18**, 10464–10472.

V. Braitenberg (1967): Is the cerebellar cortex a biological clock in the millisecond range? *Prog. Brain Res.* **25**, 334–346.

V. Braitenberg (1997): The detection and generation of sequences as a key to cerebellar function: experiments and theory. *Behav. Brain Sci.* **20**, 229–245.

E. Caianiello (1961): Outline of a theory of thought processes and thinking machines. *J. Theoretical Biology* **1**, 204–235.

C.E. Carr, W. Heiligenberg, and G.J. Rose (1986): A time-comparison circuit in the electric fish midbrain. I. Behavior and physiology. *J. Neurosci.* **6**, 107–119.

C.E. Carr and M. Konishi (1988): Axonal delay lines for time measurement in the owl's brainstem. *Proc. Natl. Acad. Sci. USA* **85**, 8311–8315.

M. Diesmann, M.O. Gewaltig, and A. Aertsen (1999): Stable propagation of synchronous spiking in cortical neural networks. *Nature* **402**, 529–533.

C.R. Gallistel and J. Gibbon (2000): Time, rate and conditioning. *Psychological Review* **107**, 289–344.

W. Gerstner, R. Kempter, J.L. van Hemmen, and H. Wagner (1996): A neuronal learning rule for sub-millisecond temporal coding. *Nature* **383**: 76–81.

W. Gerstner, R. Ritz, and J.L. van Hemmen (1993): A biologically motivated and analytically soluble model of collective oscillations in the cortex. I Theory of weak locking. *Biol. Cybern.* **68**: 363–374.

J. Gibbon (1977): Scalar expectancy theory and Weber's law in animal timing. *Psychological Review* **84**, 279–325.

L. Glass and M. Mackey (1988): From clocks to chaos. Princeton University Press, Princeton, N.J.

D.O. Hebb (1949): The organization of behavior. Wiley, New York.

W.F. Heiligenberg (1991): Neural nets in electric fish. MIT Press, Cambridge, Mass.

O. von Helversen and D. von Helversen (1994): Forces driving coevolution of song and song recognition in grasshoppers. In: Neural Basis of Behavioral Adapations, K. Schildberger and N. Elsner (Eds), *Fortschritte der Zoologie* **39**, 253–284.

J. Hertz, A. Krogh, and R. Palmer (1991): Introduction to the theory of neural computation. Addison-Wesley, Redwood City, Calif.

J. Hertz and A. Prugel-Bennett (1996): Learning synfire chains: turning noise into signal. *Int. J. Neural Syst.* **7**, 445–450.

A.V.M. Herz (1990): Untersuchungen zum Hebbschen Postulat: Dynamik und statistische Physik raum-zeitlicher Assoziation. Ph.D. thesis, Universität Heidelberg.

A.V.M. Herz (1995): Global analysis of recurrent neural networks. In: E. Domany, J.L. van Hemmen and K. Schulten (eds):Model of neural networks III: Association, generalization and representation. Springer, New York, 1–53.

A.V.M. Herz and J.J. Hopfield (1995): Earthquake cycles and neural reverberations: colletive oscillations in systems with pulse-coupled threshold elements. *Phys. Rev. Lett.* **75**, 1222–1225.

A.V.M. Herz, Z. Li, and J.L. van Hemmen (1991): Statistical mechanics of temporal association in neural networks with transmission delays. *Phys. Rev. Lett.* **66**, 1370–1373.

A.V.M. Herz, B. Sulzer, R. Kühn, and J.L. van Hemmen (1989): Hebbian learning reconsidered: Representation of static and dynamic objects in associative neural networks. *Biol. Cybern.* **60**, 457–467.

J.J. Hopfield (1995): Pattern recognition computation using action potential timing for stimulus representation. *Nature* **376**, 33–36.

J.J. Hopfield (1996): Transforming neural computations and representing time. *Proc. Natl. Acad. Sci. USA* **93**, 15440–15444.

J.J. Hopfield, C.D. Brody, and S. Roweis (1998): Computing with action potentials. *Adv. Neural Inf. Processing* **10**, 166–172.

J.J. Hopfield and C.D. Brody (2000): What is a moment? "Cortical" sensory integration over a brief interval. *Proc. Natl. Acad. Sci. USA* **97**, 13919–13924.

J.J. Hopfield and C.D. Brody (2001): What is a moment? Transient synchrony as a collective mechanism for spatiotemporal integration. *Proc. Natl. Acad. Sci. USA.* **98**, 1282–1287.

J.J. Hopfield and A.V.M. Herz (1995): Rapid local synchronization of action potentials: toward computation with coupled integrate-and-fire neurons. *Proc. Natl. Acad. Sci. USA* **92**, 6655–6662.

L.A. Jeffress (1948): A place theory of sound localization. *J. Comp. Physiol. Psychol.* **41**, 35–39.

P.R. Killeen and J.G. Fetterman (1988): A behavioral theory of timing. *Psychological Review* **95**, 274–295.

D. Kleinfeld (1986): Sequential state generation by model neural networks. *Proc. Natl. Acad. Sci. USA* **83**, 9469–9473.

M. Konishi (1992) Similar algorithms in different sensory systems and animals. *Cold Spring Harbor Symposia on Quantitative Biology* **55**, 575–584.

M. Mackey and L. Glass (1977): Oscillations and Cha in phiological control systems. *Science* **197**, 287–289.

H. Markram, J. Lübke, M. Frotscher, and B. Sakmann (1997): Regulation of synaptic efficacy by coincidence of postsynaptic APs and EPSPs. *Science* **275**, 213–215.

H. Markram and M. Tsodyks (1996): Redistribution of synaptic efficacy between neocortical pyrmidal neurons. *Nature* **382**, 807–810.

M.S. Matell and W.H. Meck (2000): Neuropsychological mechanisms of interval timing behavior. *BioEssays* **22**, 94–103.

W.H. Meck (1983): Selective adjustmnt of the speed of internal clock and memory processes. *J. Exp. Pschology* **9**, 171–201.

C. Miall (1989): The storage of time intervals using oscillating neurons. *Neural Computation* **4**, 108–119.

M. Minsky (1986): The society of mind. Simon and Schuster, New York.

J. O'Keefe and M.L. Recce (1993): Phase relationship between hippocampal place units and the EEG theta rhythm. *Hippocampus* **3**, 317–330.

W. Reichardt (1965): On the theory of lateral inhibition in the complex eye of Limulus. *Prog. Brain Res.* **17**, 64–73.

U. Riedel, R. Kühn, and J.L. van Hemmen (1988): Temporal sequences and chaos in neural nets. *Phys. Rev. A* **15**, 1105–1108.

W.E. Skaggs, B.L. McNaughton, M.A. Wilson, and C.A. Barnes (1996): Theta phase prescession in hippocampal neuronal populations and the compression of temporal sequences. Hippocampus **6**, 149–172.

H. Sompolinsky and I. Kanter (1986): Temporal association in asymmetric neural networks. *Phys. Rev. Lett.* **55**, 304–307.

M. Tsodyks and H. Sompolinsky (1993): Patterns of synchrony in inhomogeneous networks of oscillators with pulse interactions. *Phys. Rev. Lett.* **71**, 1280–1283.

K. Unnikrishnan, J.J. Hopfield, and D. Tank (1992): Speaker-independent digit recognition using a neural network with time-delayed connections. *Neural Computation* **4**, 108–119.

14

How General Are Neural Codes in Sensory Systems?

David McAlpine and Alan R. Palmer

Is there anything of which it may be said "See, this is new"?
—Ecclesiastes 1:10

Within the central nervous system, sensory information and other brain states are represented in the form of codes contained in the neural discharge patterns. Fundamental questions in neuroscience concern the nature of these codes. How are they established? Why is one coding strategy implemented in the nervous system in preference to other, apparently equally plausible, alternatives? Such questions are addressed in diverse manners by experimental neuroscientists, for example, by comparing coding strategies across species or by comparing coding strategies across stimulus modalities. Equally, however, fundamental questions remain concerning the *acceptance* of any particular neural representation or coding strategy as being appropriate for general consideration. In particular, why does a given explanation concerning some aspect of sensory processing become established in the literature as a general principle, applicable to sensory processing in other species or stimulus modalities? Although it is beyond the scope of this chapter to address all of these questions in the detail they undoubtedly warrant, we will explore some of the issues they raise by comparing the representation of auditory-spatial cues in the central auditory nervous systems of barn owls and mammals. The barn owl is usually considered the *de facto* model to explain localization of sound-sources in space, often to the extent that explicit reference is made to the similarity between barn owl and *human* sound localization abilities and mechanisms (e.g., Saberi et al. 1998, 1999). Certain lines of evidence from the nonspecialized mammalian auditory system suggest that caution should be exercised in drawing too close an analogy with auditory processing carried out by auditory "specialists" such as barn owls and echo-locating bats. This is perhaps understandable in the case of bats, since they generally operate in the ultrasound range, actively interrogating their environment with sound and processing auditory signals in a manner unlike that envisaged, or observed, in terrestrial mammals. However, unlike bats, the audible range of the barn owl approximates that of many mammals, including humans, extending from a few hundred hertz up to, at most, a few tens of kilohertz. Furthermore, compelling evidence exists that barn owls and mammals utilize the same acoustic cues in order to localize sound sources in space. For example, barn owls are sensitive to *interaural time differences* (ITDs) (e.g., Moiseff and Konishi 1981), one of the two *binaural* cues utilized in sound-source localization in humans and other mammals. The question remains, however, is this sufficient reason to consider that the sensory representation of these cues and the means by which they are used within the central nervous system to create the percept of acoustic space, are similar or even identical? One area in which barn owls and mammals differ is in the *range* of acoustic frequencies over which ITD cues are utilized. Barn owls use ITDs for localizing sounds in the frequency range 3–9 kHz, whereas mammals use ITDs for localizing sounds exclusively below 2 kHz. How important is this difference? Is it simply a nuance, of interest only to the *cognoscenti*, or does it have wider implications for both the sensory representation and the potential coding strategy adopted by barn owls compared with mammals? In this chapter we show that careful attention to the frequency range over which ITD-processing is carried out has important consequences both for the form of the sensory representation of ITDs and the means by which this sensory representation is translated in the central nervous system. Cross-species comparisons have proved to be a powerful tool in furthering our understanding of the brain. However, in some instances, even when basic neural mechanisms appear completely analogous, care needs to be exercised before accepting that this implies a complete commonality.

Analogies, Models, and Generalities in the Nervous System

The 1913 edition of Webster's Dictionary defines *analogy* as "a resemblance of relations; an agreement or likeness between things in some circumstances or effects, when the things are otherwise entirely different." Many successful explanations of complex systems depend to some extent on analogy, whether explicit or implied, with other systems. Often this is because certain relationships they possess truly do resemble, or appear to resemble in some aspect, relationships in other systems, or because the analogy proves particularly useful in understanding and explaining the complex system. In particular, systems that have been investigated in great detail necessarily are invested with general explanatory power. An obvious example of this is the dominance of the visual system when seeking to understand the functioning brain in general and the auditory system specifically. Mechanisms of visual processing have been appropriated to provide an account of the functional anatomy of auditory cortical circuitry (Kaas and Hackett 1999), a distinction between "magnocellular" and "parvocellular" forms of auditory processing (Stein and Talcott 1999), and a means by which auditory "binding" might be realized (Brosch and Schreiner 1999).

One possible reason for the dominance of the visual system in our understanding of general sensory coding likely relates to the fact that many fundamental issues in visual processing at the neural level have been resolved to a much greater degree than for any other sensory system. In particular, it is undisputed that "We know more about the primary visual cortex…than about any other part of the neocortex, particularly with regard to its functional layout and underlying anatomy" (Blasdel 1989). This sentiment could safely be extended to include many *nonprimary* visual areas also, which are understood better even than *primary* auditory cortex, as well as to many physiological and psychological aspects of visual processing. However, although analogies with and generalizations from visual processing can provide insight into how coding problems might be solved in other sensory modalities, danger resides in overextending these without due concern for the unique problems faced by different sensory systems. In particular, some coding strategies that are considered to be general may be so only at a very superficial level. A specific example of this is the hypothesis that the output of the primary auditory cortex includes a ventral "what" stream and a dorsal "where" stream, in a similar manner to that suggested for the visual system (Romanski et al. 1999). Anatomical evidence suggests that the output of extra-primary auditory cortical areas might well divide ventrally and dorsally (Kaas and Hackett 1999). However, the *functional* analogy is weakened by the consideration of the auditory feature analogous to "where" (position on the primary sensory epithelium) in vision. In the auditory system, position on the primary sensory epithelium indicates sound frequency rather than spatial position (e.g., see critique by Belin and Zatorre 2000). Auditory feature space is most likely encoded in the form of spectrotemporal patterns—as the analysis of the excitation along the length of the basilar membrane, and how it changes over time—rather than the spatial position of a sound source. Spatial position requires computation, which occurs at subcortical levels. In such a case, too close an analogy to visual processing may have been drawn, either because of the dominance of the paradigm or because of the lack of an alternate and cohesive view of auditory cortical processing.

Models of Binaural Hearing

The ability to localize sounds in space is a fundamental attribute of the way that humans and animals perceive their environment. It has obvious survival value in enabling them to determine where their prey or a potential mate is or from where a predator is approaching. Since the late 1940s, after Jeffress's (1948) seminal paper, the dominant model of localization has consisted of a series of delay lines with each frequency component represented in a different delay line and all positions in azimuth represented by different delay taps (see figure 14-1). Crucial aspects of this model are that it is the peak of activation that indicates spatial position and that the majority of delay taps are near zero delay, and certainly within the ecological range of delays which would be experienced by the animal (a few hundreds of microseconds). This theory has been instantiated in several influential computer models (e.g., Stern and Trahiotis 1996) and is sufficiently dominant to be classed as the paradigm. While studies in mammals are generally consistent with this model of binaural hearing (Yin and Kuwada 1983), one reason for its dominance, apart from its inherent elegance, has been the seminal investigations into sound-localization abilities of barn owls by Konishi and his colleagues over the last several decades (Konishi 1973; Carr and Konishi 1990; Saberi et al. 1999). Initiated from an ethological perspective, the study of barn-owl localization behavior and its neurophysiological basis has become the *de facto* model for understanding sound localization. An integral aspect of this model lies in its synthesis of a neural representation of auditory space in the form of a

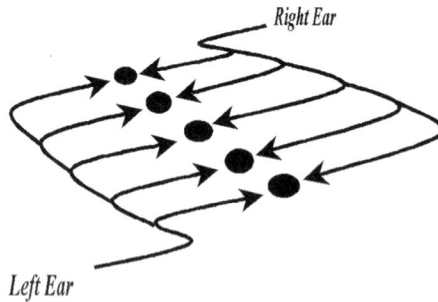

Figure 14-1. The traditional form of the Jeffress model, in which coincidence detectors are innervated by delay lines, axons of varying path-length, from the two ears. The role of this arrangement is to offset external delays that arise as a result of the position of a sound source with equal and opposite internal delays. Each coincidence detector responds maximally at the appropriate ITD (the characteristic delay, or CD) and at ITDs equivalent to the CD ± multiple periods of the stimulating wave form. Evidence from the barn owl brainstem and midbrain suggest that a "Jeffress-like" arrangement exists within each isofrequency laminae extending over the entire frequency range to which the barn owl is sensitive to ITDs.

topological map (Carr and Konishi 1990; Takahashi and Keller 1994). Probably because of this, it is often cited as evidence for the generality of sensory map production as a means of translating inherently nonspatial sensory representations into spatial coordinates (e.g., Lund 1989). However, the extent to which this model of binaural hearing extends to species other than the barn owl and in particular its validity for mammalian hearing is, perhaps surprisingly to the nonspecialist reader, a matter of conjecture. This is despite the seeming comparability of neural mechanisms responsible for generating sensitivity to interaural timing cues in barn owls and mammals.

Comparison of Barn-Owl and Mammalian Binaural Hearing

Barn owls have evolved unique auditory capabilities to equip them for survival in their environment. Their ability to hunt and capture prey using only acoustic cues, in total darkness if necessary, has provided for a powerful narrative account not only of sound localization in this species, but also how such exquisite sensitivity is subject to plastic change during development (e.g., Knudsen 1983). Nonetheless, exquisite auditory-spatial acuity is not restricted to barn owls. Many other avian species (for review, see Klump 2000) and many terrestrial mammals, including humans (e.g., Hafter and De Maio 1975), exhibit spatial acuity comparable to that observed in the barn owl. Despite this, major differences do exist between barn owl and mammalian binaural hearing capabilities, and these differences are integral to understanding where differences in sensory representations and coding strategies might lie. Foremost amongst these differences is the frequency range over which barn owls utilize ITDs for sound localization compared with mammals.

Traditionally, binaural sound localization cues have been divided into low- and high-frequency processes. In their classic description of human pure-tone localization in the free field, Stevens and Newman (1936) observed that human subjects performed best (i.e., showed fewest localization errors for sound sources in azimuth) for frequencies below 1.5 kHz and above 5 kHz. Localization errors were most common for mid-frequency sounds, peaking around 3 kHz. This indicates that humans, and likely other species with acoustic sensitivity spanning a broad frequency range extending down into the subkilohertz range, tap into two different mechanisms when localizing sound sources in the azimuthal plane, one operating at low frequencies and one operating at high frequencies. This dichotomy, now referred to as the duplex theory of binaural hearing, and first proposed by Lord Rayleigh (1907), arises due to the physical dimensions of the acoustic stimulus. The head can create an *acoustic shadow* at the ear furthest from the sound source, with the effect that the sound is less intense at that ear. These *interaural level differences* (ILDs) were, even in the late nineteenth century, widely recognized as a cue for localizing a sound source in space. However, equally, it was recognized that the magnitude of ILDs depends on the spectral content of the stimulus. In particular, the low-pass filter characteristics of the head, which permits the passage of low-frequency sounds around the head by virtue of their longer wavelength, result in ILDs becoming negligible with decreasing sound frequency; the smaller the head size, the higher the cut-off frequency at which this occurs. At that time it was considered that differences in the time of arrival of the sound at the two ears, brought about by the incidence angle of the sound source relative to the head, were too small to be detectable, being, at most, a few hundreds of microseconds. Rayleigh's confirmation of Thompson's (1877, 1878) early work demonstrating that small temporal differences between the ears were indeed detectable and could provide for the ability to localize low-frequency sound sources marked a critical juncture in the development of a theory of spatial hearing, and the duplex theory is now an established tenet of binaural processing. It may come as some surprise, therefore, that the barn owl, the species perhaps most readily associated with binaural sound localization, does not conform to this dichotomy.

Unlike mammals that localize high and low frequencies well but are poor at mid-range frequencies, barn owls localize only very poorly for frequencies below about 3 kHz and above 10 kHz (Knudsen and Konishi 1979; Coles and Guppy 1988), but their localization abilities are exceptionally well developed between these limits. Consistent with this, physiological recordings in a variety of barn-owl auditory nuclei routinely report ITD-sensitive neurons tuned, in terms of their sensitivity to sound frequency, in the ecologically relevant (to the barn owl at least) range 5–9 kHz (e.g., Cohen and Knudsen 1994). The ability in barn owls to utilize temporal information in the stimulus fine-structure at such high frequencies is underpinned by a unique specialization in the processing within their cochleae. Auditory nerve fibers in the barn owl are able to signal the fine-structure of the sound waveform in the timing of their discharges (phase locking) up to 8 or 9 kHz (Koppl 1997). Typically, in mammals, phase locking in the auditory nerve declines above 3 kHz or so and is completely absent by 4 or 5 kHz. Moreover, the upper-frequency limit declines at successive stations in the ascending auditory pathway. Consistent with this, ITD sensitivity in mammals is low-pass and, in humans, is completely absent for frequencies above 1.5 kHz. Other mammals show a similar upper-frequency cut-off, and physiological recordings in a range of mammalian species demonstrate few instances of neural sensitivity to ITDs for frequencies above 2 kHz. The low-pass characteristic of phase locking is thought to arise due to shunting of the capacitance of the hair cell receptor potential with increasing stimulus frequency (Palmer and Russell 1986). The consequence of this is that the AC component of the potential is systematically reduced relative to the DC component, rendering hair cell output incapable of transducing information about stimulus temporal fine-structure into phase-locked activity. The higher-frequency phase locking in barn owl hair cells is presumed to be a result of different hair cell biophysics (Koppl 1997).

Furthermore, unlike mammals, which utilize ILDs for localizing high-frequency sounds in the *azimuthal* plane, barn owls utilize ILDs to localize sounds in the *elevational* plane. This is accomplished by the directionality of the barn owl's ears, the openings of which lie on an interaural plane inclined about twelve degrees to the horizontal, with the left ear opening higher than the right (Coles and Guppy 1988). Amplification of pressure by the external ears is maximal between 3 and 9 kHz, declining sharply above 10 kHz, contributing to the bandpass nature of barn-owl localization abilities.

Anatomical Basis of the Duplex Theory

The dichotomy of binaural hearing into low- and high-frequency processes is strengthened by the existence of apparently dedicated lower-brainstem pathways subserving it. The medial superior olive (MSO) shows a relative overrepresentation of low best frequency (BF) neurons and, in many mammals, is specialized for processing ITD information, whilst the lateral superior olive (LSO), with a relative overrepresentation of high (>2 kHz) BF neurons, processes ILD information. The tonotopic organization at subsequent auditory centers in the auditory pathway maintains this separation. An anatomical dichotomy also exists in the barn owl, but here reflecting a division into pathways encoding interaural time and level differences spanning the same frequency range. The separation of these pathways is maintained until the level of the external nucleus of the inferior colliculus. Here, the integration of the ITD azimuth code and the ILD elevation code provides for a map of auditory space, a topologically organized map, consistent with the concept of a receptive field in the visual system. Despite these differences, however, it is clear that in both barn owls and mammals a dichotomy between ITD and ILD processing exists, both anatomically and physiologically. The major difference appears to lie in the ability of the barn owl to combine the azimuthal ITD cue and the elevational ILD cue to form a topologically ordered space map.

The Neural Basis of Sensitivity to Interaural Time Differences

Similar neural mechanisms have been demonstrated to underlie binaural processing in barn owls and mammals. In both cases, strong evidence exists for a process of binaural coincidence detection underlying sensitivity to ITDs.

Physiological observations indicate that temporal information in the form of action potentials phase-locked to the carrier stimulus waveform converges from each ear onto single neurons in the brainstem to generate interaural-delay sensitivity. Responses of mammalian ITD-sensitive neurons are qualitatively similar to those recorded throughout the barn-owl auditory system. Probing the response of such neurons to pure-tones over a range of ITDs reveals cyclic input/output functions, with response maxima at multiple periods of the stimulus waveform (figure 14-2a). The cyclic nature of the functions has been taken to indicate that binaurally responsive neurons are sensitive to interaural phase disparities rather than to interaural time disparities per se. It is also well established that such neurons often show a characteristic delay (CD), an interaural time difference at which the relative magnitude of the response is independent of stimulus frequency, presumably reflecting the difference in axonal conduction time from the two ears to the binaural coincidence detector. A quantitative measure of the CD is obtained from the relationship between the interaural phase at the peak response and stimulus frequency (figure 14-2b). Many neurons showing CDs have been recorded from auditory nuclei in both barn owls (Takahashi and Konishi 1986) and mammals (e.g., Rose et al. 1966; Yin and Chan 1990). The neuron's response to interaurally delayed noise (figure 14-2c) appears to be a linear summation of the responses to its component frequencies (Yin, Chan, and Irvine 1986) and, as such, confirms the ITD tuning.

Unique Localization Abilities Present Unique Problems

Thus far, it appears that the major difference between mammalian and barn-owl binaural hearing probably lies in the frequency range over which ITD information is utilized. Barn owls appear able to capitalize on their ability to detect ITDs at high frequency by combining ITD information in azimuth and ILD information in elevation over the same frequency range to produce a topologically ordered map of acoustic space. Mammals, conversely, must utilize these binaural cues largely in isolation from each other. This fact aside, are there other consequences of sensitivity to high frequency ITDs that are of interest? In fact, it transpires that this single factor has a profound effect on both the form of the sensory representation of ITD and the likely coding strategy adopted for translating this representation into the localization of a sound source.

Barn owls, utilizing ITDs at high frequencies, are faced with the challenge that although a sound source might hold a single position in azimuth, its location may not be inferred from the output of a single ITD channel in the barn-owl brainstem (figure 14-3). At some frequencies, multiple cycles of the stimulus waveform arrive at the nearer ear before the first arrives at the

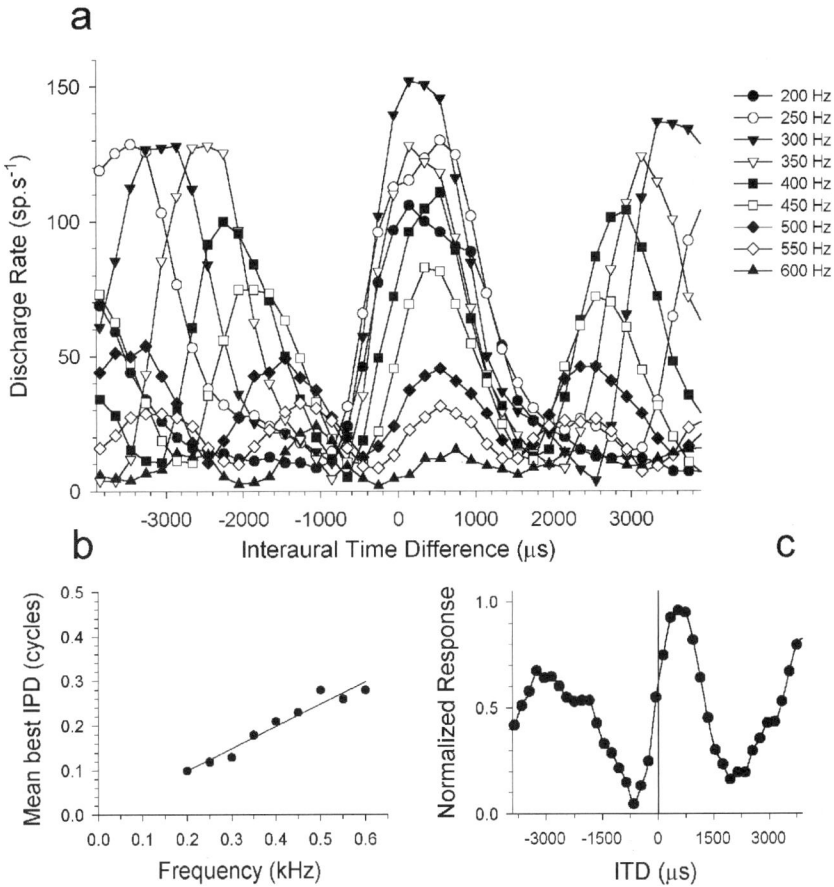

Figure 14-2. (a) Response of an ITD-sensitive neuron, with a BF of 300 Hz, in the IC of the guinea pig. Each pure-tone frequency elicited a periodic ITD function, cycling on the period of the stimulus frequency. (b) The slope of the function plotting best interaural phase difference against stimulus frequency provides an easily quantifiable measure of the characteristic delay. In this case, the characteristic delay occurs at +497 μs, corresponding to a sound leading at the ear contralateral to the inferior colliculus being recorded. Note that this is outside the range of physically-plausible ITDs of the guinea pig (±150 μs). (c) Response of the same neuron to interaurally delayed noise. The solid vertical line indicates an ITD of 0 μs. All of these responses are consistent with the general form of the binaural coincidence envisaged by Jeffress.

farther ear. The action potentials phase locked to each cycle provide binaural coincidence detection across a range of ITDs encompassing the barn owl's head width. This proves problematic for barn owls when they attempt to localize narrow-band sound sources and, predictably, they are susceptible to the perception of phantom auditory images and consequent errors in localization

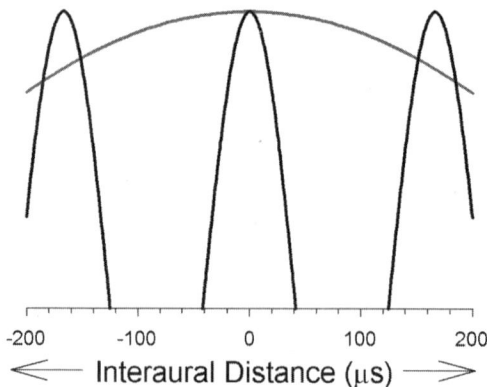

Figure 14-3. The unique problems facing barn owls using high-frequency ITDs. The high-frequency (3–9 kHz) ITD-sensitivity in barn owls provides binaural coincidence detection across a range of ITDs encompassing the barn owl's head width (black curve). This makes barn owls susceptible to the perception of phantom auditory images and consequent errors in localization. Mammals are faced with the opposite problem (gray curve). By virtue of their wavelength, low-frequency sounds can provide for only a very coarse representation of ITD, based on the peak responses of neurons.

(Saberi et al. 1999). Even within individual neurons of the barn owl's brainstem or midbrain auditory nuclei, widening the bandwidth of the stimulus does not disambiguate real from phantom source locations. Filter widths of ITD-sensitive neurons are apparently too narrow to provide for adequate damping of ITD side peaks within individual auditory channels. This problem is overcome in the barn owl by means of a systematic map of ITDs running orthogonal to the primary tonotopic map. Vertical electrode penetrations through the frequency laminae in the central nucleus of the IC of barn owls reveal neurons within those penetrations to be similarly tuned in terms of their characteristic delay (Wagner, Takahashi, and Konishi 1987). Thus, each isofrequency lamina in the barn-owl auditory pathway contains a full representation of the Jeffress model, and broadband signals activate multiple frequency channels across a single "ITD column." Neurons in the external nucleus of the IC integrate the output of this ITD column, such that the frequency-dependent side peaks cancel to leave a single peak at the "true" ITD (Knudsen and Konishi 1978).

In contrast to the barn owl, frequency tuning of ITD-sensitive neurons throughout the mammalian binaural auditory pathway appears sufficiently broad to provide for damping of the side peaks of ITD functions (e.g., Yin, Chan, and Irvine 1986). However, this issue is largely irrelevant for mammals, since they are faced with essentially the opposite problem to that of barn owls. For low-frequency sounds the long wave length, relative to the width of the head, means that it is highly unlikely that more than one cycle of the stimulus waveform will arrive at the nearer ear before the first cycle reaches the farther ear. Phase ambiguities are therefore less likely to occur and unlikely ever to be critical for animals with interaural distances less than about 12 cm—approximately the wavelength of a 1500 Hz sine wave—and that utilize ITDs for exclusively low-frequency localization. The problem facing any species utilizing ITDs to localize low-frequency sound sources is that, simply by virtue of their wavelength, low-frequency sounds can provide for only a very coarse representation of ITD based on the peak responses of neurons. For a hypothetical low-frequency neuron tuned to zero ITD, neural output is largely unmodulated, being essentially maximal, over a wide range of ITDs. Depending on the frequency and exact interaural distance, a maximal response may be elicited by ITDs spanning a range encompassing the entire head width (figure 14-3). Binaural neurons at many levels of the auditory system show essentially half-rectified ITD functions (e.g., Yin and Chan 1990), consistent with a process of half-wave rectification of the output of the cochlea, enhancement of monaural phase locking at the level of the cochlear nucleus (Joris et al. 1994), and cross-correlation of monaural inputs. The important consequence of this is that the stimulus (in the case of narrowband signals) and the frequency-filter properties of the auditory neuron (in the case of broadband signals) defines the precision, in ITD terms, with which binaural neurons encode interaural delays. This means that mammals by virtue of their use of ITDs for low-frequency localization necessarily respond with broad ITD functions, whilst high-frequency neurons, such as those found in barn owls, necessarily respond with sharp ITD functions. So, both broad and narrow ITD functions present processing problems, which have, as we argue below, been solved in different ways.

Distinct Sensory Representations for ITD in Barn Owls and Mammals

The vast majority of ITD functions reported in the barn-owl literature are tuned to ITDs within the physiologically plausible range determined by the width of the head (approx. ±200 μs). They are often close to zero ITD, indicating a relatively high density of spatially tuned neurons encoding spatial positions around midline. This accords with anatomical evidence from the barn-owl *nucleus laminaris* (NL—the avian homologue of the mammalian MSO), suggesting a full compliment of Jeffress-like delay lines within each frequency channel (Carr and Konishi 1990; see figure 14-1]. In contrast, physiological evidence throughout the mammalian auditory pathways reveals that the distribution of ITDs to which mammalian delay-sensitive neurons are tuned lies approximately in the range +200 to +300 μs (see figure 14-4b), outside the physically plausible range of many small mammals and apparently independent of head width (Palmer, Rees, and Caird 1992). Furthermore, convincing anatomical evidence for a systematic arrangement of delay lines in the mammalian brainstem is lacking, despite the attempts of several studies to demonstrate the existence of such an arrangement (Smith, Joris, and Yin 1992; Beckius, Batra, and Oliver 2000). These empirical observations notwithstanding, a tacit assumption in many models of low-frequency binaural hearing in mammals is that an identical (and relatively broad) range of

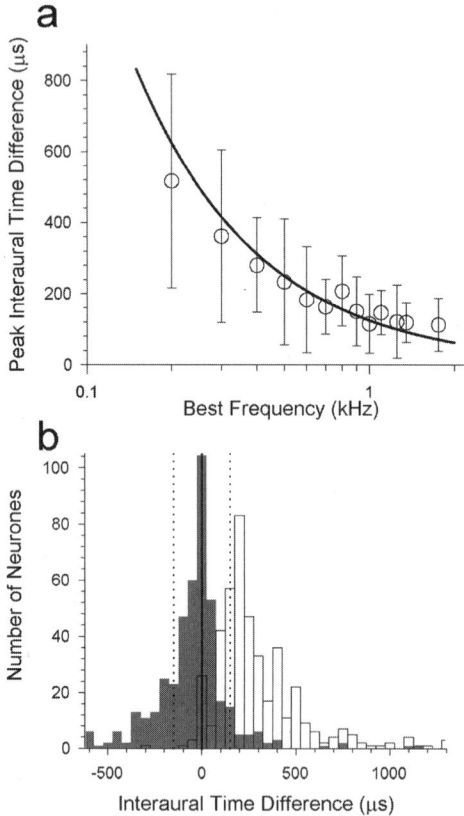

Figure 14-4. (a) Distribution of peak ITDs (± 1 s.d.) as a function of neuronal BF, for 100-Hz bands covering the full frequency range over which mammals are sensitive to ITDs. The solid curve denotes the ITDs corresponding to 45-degree interaural phase difference, indicating that the ITD-sensitive neurons are, on average, distributed around ITDs corresponding to 45-degree IPD. Note that peaks occur at positive ITDs, indicating that neurons are tuned to sounds leading at the ear contralateral to the recording site (the right inferior colliculus, or IC). The representation in the left IC is a mirror-image representation of the representation in the right IC. (b) ITD values at the peak (open bars) and maximum slopes of ITD functions (filled bars) for neurons in the IC of the guinea pig.

peak ITDs is fully represented within each low-frequency channel of the tonotopic gradient for frequencies up to approximately 2 kHz. By analogy with the barn owl, it is has been suggested that there might exist a topological map of peak ITDs running *orthogonal to* the tonotopic axis. It comes as some surprise, therefore, that when the representation of ITD is examined as a function of the tonotopic gradient, a systematic relationship is observed, but one in which the representation of ITD runs *parallel to* the tonotopic axis. Neurons with the lowest BFs tend to have the highest values of peak ITD, whilst neurons with the highest BFs tend to have the lowest values of peak ITD (figure 14-4a). Furthermore, the extent of the representation within each frequency band is not equivalent across BF, with a much wider range of ITDs producing maximum firing in lower-BF than in higher-BF neurons. The consequence of the BF dependence of the peak ITD is that the steepest region of the function relating discharge rate to ITD falls close to midline for all neurons *irrespective of BF* (figure 14-4b). A corollary of this relationship is that both the value and the range of the *interaural phase difference* at the peak within each frequency band are essentially invariant as a function of BF. Peak responses are distributed around 45° IPD (solid curve in figure 14-4a), regardless of neural frequency tuning. These observations do not accord with current models of low-frequency binaural processing and are incompatible with the accepted physiology of the barn-owl auditory system. Given the relationship between BF and peak ITD an obvious question is how such neural activity contributes to low-frequency sound localization?

A Proposed Mammalian Code for Localization

As an alternative model to a labeled line code, where a particular neuron firing maximally indicates the azimuthal position of the source, data from the guinea pig suggest that azimuth could be encoded in the form of a rate code mediated by broadly tuned spatial channels (see figure 14-5). The specific form that such a population code might take, however, is open to question. First, and simplest, is that azimuthal position of a sound source is read from the discharge rate of one of the broadly tuned ITD channels on one side of the brain. Thus, for a sound moving away from the midline, activity will increase in the contralateral hemisphere toward the peak of ITD functions, indicating that the sound source has shifted to a more lateral position. An inherent ambiguity in this model arises, however, from the possibility of confounding changes in stimulus level with changes in spatial position of a sound source. Adjusting either parameter will necessarily influence activity levels within each hemisphere alone. One way of overcoming this potential ambiguity is by reading azimuthal position as a comparison code. This could be achieved either as a comparison of activity between hemispheric channels or as a

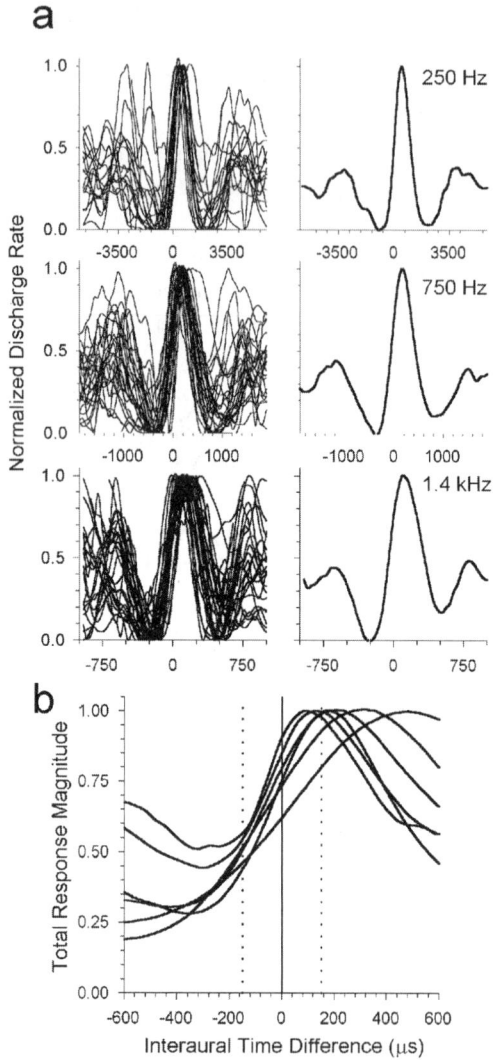

Figure 14-5. (a) Responses to interaurally delayed broadband noise (50 Hz to 5 kHz bandwidth) recorded from IC neurons with BFs around 250 Hz, 750 Hz, and 1.4 kHz, plotted for ITDs over approximately three cycles of neuronal BF and covering the frequency range over which ITD sensitivity is found in mammals. Note the different ITD ranges over which the functions are plotted. Left panels show individual functions, and right panels show averaged functions. (b) Average noise ITD functions plotted over the same ITD range, for neurons with BFs in bands around 250 Hz, 335 Hz, and 425 Hz, 500 Hz, 700 Hz, 1.0 kHz, and 1.4 kHz. The widest NDF with the longest peak ITD is that for the 250-Hz band, and peak ITD decreases systematically as BF increases. The dotted vertical lines indicate the physiological range of ITDs for the guinea pig (±150 μs).

comparison between binaural and monaural activity within the same hemispheric channel. In the two hemispheric channels model, a change in activity brought about by a change in azimuthal position in one channel is accompanied by a change in activity, of opposite sign, in the other hemispheric channel. Thus, as activity in one channel increases, activity in the other decreases, and vice versa. In the single-hemispheric model, changes in binaurally evoked activity are compared with a fixed, azimuth-independent level of monaural activity in the same side of the brain. Both of these comparator models can disambiguate changes in stimulus level from changes in azimuthal position of the sound source. The low-pass filter properties of the head provide no appreciable interaural level difference for low-frequency sounds, irrespective of their position in space. Therefore, a change in stimulus level under free-field listening conditions will produce equal changes in activity within both monaural pathways, and within both binaural pathways. Since both monaural and both binaural channels involved are subject to the same level-dependent change in activity, a *relative* difference between the two binaural channels, or between low-frequency binaural and monaural channels in the same hemisphere will, correctly, be interpreted as a change in spatial position of a sound source.

A final, important point is that none of these models compromises the hypothesis that mammalian ITD sensitivity is realized by a process of binaural cross-correlation. All evidence from the responses of single ITD-sensitive neurons suggests that it is. Nevertheless, it is important to realize that the existence of the binaural cross-correlation process does *not* necessitate the existence of an array of delay lines representing all azimuths at each frequency. Binaural cross-correlation exists within the broader confines of models of coincidence detection, and can be considered without reference to the more specific model in which the neural representation of auditory spatial coordinates is systematically arranged within a topological map.

Developmental Considerations

The representation of ITDs that we describe here has consequences for how such a representation might develop. The barn owl's auditory system has proved to be a powerful tool for studying neural plasticity and development, notably by Knudsen and his colleagues (e.g., Knudsen 1983), but also in the context of neural network models employing Hebbian mechanisms to instruct the range of ITDs encoded (Gerstner et al. 1996). Necessarily this means that the mature representation of ITDs should eventually be constrained to within the physically realizable range, that is, within the head width of the barn owl. However, in the mammal it would appear that whatever mechanism is responsible for establishing the representation of ITDs in the central auditory nervous system operates so as to position the *slopes* of ITD functions, rather than their peaks, through the physiological range. Although it is plausible that the configuration of such a representation during development could arise under the influence of modified Hebbian mechanisms, it is equally plausible that the representation is functionally hardwired. Fixing the peaks of ITD functions to delays that approximate 45° IPD maximizes information capacity within the physiological range and for frontal ITDs *regardless of head size*. Head growth will extend the range of realizable ITDs but is unlikely to influence the positions of the peak ITDs, since these appear unimportant in relation to where the slopes of ITD functions are maximal.

Different Sensory Representations Provide for Different Coding Strategies

The different forms of the sensory representations for ITD in barn owls and mammals raise more general coding issues than simply the frequency range over which barn owls and mammals utilize ITDs for sound localization. Coding sound source position in the barn owl resembles most closely a form of local coding; different spatial positions of a sound source are represented by peak activity of individual neurons. This does not, of course, preclude the notion that many neurons, even within a single isofrequency band, are tuned to the same ITD. However, in mammals, it follows that the relationship between spatial position and peak ITD tuning that we have observed cannot be a form of local coding or, indeed, any form of direct representation, since the majority of neurons are "tuned" to ITD values that can never be experienced. They do not represent a unique spatial position by virtue of their ITD tuning. This potentially liberates us from having to accept a neural code for auditory space that conforms to the concept of a spatial receptive field. It could be argued that constraining auditory spatial representation to topological maps across some physical dimension within the brain is merely to confuse the feature requiring representation with the means by which it is represented. In other words, although it is obvious that within the mammalian brain a *representation* of auditory space is present, there is no reason a priori, why this necessitates the form of a topological map arranged systematically across dimensions of neural tissue. As Middlebrooks et al. (1998) state, "such a model, reduced to a single neuron ... would return us conceptually to the spatially selective neuron that has proven to be so elusive in auditory physiology." This latter sentiment is perhaps surprising to those for whom the barn owl remains their closest encounter with the study of auditory localization. Nevertheless, having failed to find such a representation in the mammalian auditory cortex, Middlebrooks and his colleagues have investigated other codes for spatial position. They have suggested at least one alternative, in the form of a variation in the temporal pattern of discharges with sound-source position of neurons in non-tonotopically organized auditory cortex, rather than in their discharge rate per se (Middlebrooks et al. 1994, 1998). It accords with the view that the brain constructs the auditory scene in the form of a spectrotemporal code. Of course, many plausible potential codes might be posited, each with its own network of supporters. Of ultimate interest to the physiologist, however, is the question, which of these codes is the one that the brain uses to guide behavior? Perhaps the appropriate answer might currently be best summed up in the answer, "the jury is still out."

What Constitutes Generality?

What can the comparison of mechanisms of sound localization in barn owls and mammals teach us concerning the generality of neural codes in sensory systems? We suggest that there are two lessons to learn. First, it is necessary to consider all of the factors that influence information processing in any sensory system since this has implications for how the information may be utilized in constructing a neural code and, therefore, how similar the neural code might be to those in other sensory systems. This lesson may seem so obvious as to warrant instant dismissal. However, it is clearly the case that the stimulus subrange poses barn owls and mammals different problems, both behaviorally and neurally, when encoding ITD information, and that this issue has been ignored in most popular accounts of spatial hearing. Second, conformity of a particular aspect of sensory processing to an established model does not imply general conformity to that model. Again this may seem trite. However, the example of sound localization is a salutary reminder that even *within* seemingly identical stimulus submodalities—and many models of neural coding seek to address issues *across* stimulus modalities—the means by which sensory information is either represented or is translated into a neural code may be different. The Jeffress model, which was originally developed to account for human localization abilities, includes the means by which sensory information from the two ears is brought together *and* the means by which this information is translated into a neural code for auditory space. Although evidence from mammals and barn owls suggests that identical neural mechanisms might underlie the former of these tasks, this cannot be taken to imply that the latter task is also accomplished in an identical manner. Indeed, as we argue above, there is good reason to believe that the representation of acoustic space must, of necessity, differ in barn owls and mammals. To paraphrase Webster's definition, there may an agreement or likeness between the sensory representations and coding strategies in some circumstances, but they are, in fact, otherwise different. In answer to the question "How general are neural codes in sensory systems?" we might therefore appropriate the answer given concerning the nature of the curate's egg and reply "general in parts."

References

Beckius, G. E., Batra, R. and Oliver, D. L. (2000) Axon from anteroventral cochlear nucleus that terminates in medial superior olive of the cat: Observations related to delay lines. *J. Neurosci.* 19: 3146–3161.

Belin, P. and Zatorre, R. J. (2000) 'What,' 'where' and 'how' in auditory cortex. *Nat. Neurosci.* 3: 965–966.

Blasdel, G. G. (1989) Topography of visual function as shown with voltage-sensitive dyes. In J.S. Lund, *Sensory processing in the mammalian brain: Neural substrates and experimental strategies*, Oxford University Press, New York.

Brosch, M. and Schreiner, C. E. (1999) Correlations between neural discharges are related to receptive field properties in cat primary auditory cortex. *Eur. J. Neurosci.* 11: 1–14

Carr, C. E. and Konishi, M. (1990) A circuit for detection of interaural time differences in the brainstem of the barn owl. *J. Neurosci.* 10: 3227–3246.

Cohen, Y.E. and Knudsen, E.I. (1994) Auditory tuning for spatial cues in the barn owl basal ganglia. *J. Neurophysiol.* 72: 285–298.

Coles, R.B. and Guppy, A. (1988) Directional hearing in the barn owl (*Tyto alba*) *J. Comp. Physiol. A*. 163: 117–133.

Gerstner, W., Kempter, R., van Hemmen, J.L., and Wagner H. (1996) A neuronal learning rule for sub-millisecond temporal coding. *Nature* 383: 76–81.

Hafter, E. R. and De Maio, J. (1975) Difference thresholds for interaural delay. *J. Acoust. Soc. Am.* 57: 181–187.

Jeffress, L. A. (1948) A place theory of sound localization. *J. Comp. Physiol. Psychol.* 41: 35–39.

Joris, P. X., Carney, L. H., Smith, P. H., and Yin, T. C. T. (1994) Enhancement of neural synchronization in the anteroventral cochlear nucleus I. Responses to tones at the characteristic frequency. *J. Neurophysiol.* 71: 1022–1036.

Lund J. S. (1989) Mapping strategies of monkey primary visual cortex. In J. S. Lund, *Sensory processing in the mammalian brain: Neural substrates and experimental strategies*, Oxford University Press, New York.

Kaas, J. H. and Hackett, T. A. (1999) What and where processing in auditory cortex. *Nat. Neurosci.* 2: 1045–1047.

Klump, G. M. (2000) Sound localization in birds. In R. J. Dooling, A. N. Popper, and R. R. Fay (eds.), *Comparative hearing: Birds and reptiles*. Springer, New York, Berlin.

Knudsen, E. I. (1983) Early auditory experience aligns the auditory map of space in the optic tectum of the barn owl. *Science* 222: 939–942.

Knudsen, E. I. and Konishi, M. (1978) Space and frequency are represented separately in the auditory midbrain of the owl. *J. Neurophysiol.* 41: 870–884.

Knudsen, E. I. and Konishi, M. (1979) Mechanisms of sound localisation in the barn owl (*Tyto alba*). *J. Comp. Physiol.* 133: 13–21.

Konishi, M. (1973) How the owl tracks its prey. *Am. Sci.* 61: 414–424.

Koppl, C. (1997) Phase locking to high frequencies in the auditory nerve and cochlear nucleus magnocellularis of the barn owl, *Tyto alba*. *J. Neurosci.* 17: 3312–3321.

Middlebrooks, J. C., Clock Eddins, A., Xu, L., and Green, D. M. (1994) A panoramic code for sound location by cortical neurons. *Science* 264: 842–844.

Middlebrooks, J. C., Xu, L., Clock Eddins, A., and Green, D. M. (1998) Codes for sound-source location in nontonotopic auditory cortex. *J. Neurophysiol.* 80: 863–881.

Moiseff, A. and Konishi, M. (1981) Neuronal and behavioural sensitivity to binaural time differences in the owl. *J. Neurosci.* 1: 40–48.

Palmer, A. R. and Russell, I. J. (1986) Phase-locking in the cochlear nerve of the guinea-pig and its relation to the receptor potential of inner hair cells. *Hear. Res.* 24: 1–15.

Palmer, A. R., Rees, A., and Caird, D. (1992) Binaural masking and sensitivity to interaural delays in the inferior colliculus. In R. P. Carlyon, C. J. Darwin and I. J. Russell (eds.), *Processing of complex sounds by the auditory system*, Clarendon Press, Oxford.

Rayleigh, Lord. (1907) On our perception of sound direction. *Philos. Mag.* 13: 214–232.

Romanski, L. M., Tian, B., Fritz, J., Mishkin, M., Goldman-Rakic, P. S., and Rauschecker, J. P. (1999) Dual streams of auditory afferents target multiple domains in the primate prefrontal cortex. *Nat. Neurosci.* 2: 1131–1136.

Rose, J. E., Gross, N. B., Geisler, C. D., and Hind, J. E. (1966) Some neural mechanisms in the inferior colliculus of the cat which may be relevant to localization of a sound source. *J. Neurophysiol.* 2: 288–314.

Saberi, K., Takahashi, Y., Konishi, M., Albeck, Y., Arthur, B. J., and Farahbod, H. (1998) Effects of interaural decorrelation on neural and behavioural detection of spatial cues. *Neuron* 21: 789–798.

Saberi, K., Takahashi, Y., Farahbod, H., and Konishi, M. (1999) Neural basis of an auditory illusion and its elimination in owls. *Nat. Neurosci.* 2: 656–659.

Smith, P. H., Joris, P. X., and Yin, T. C. T. (1992) Projections of physiologically characterized spherical bushy cell axons from the cochlear nucleus of the cat: Evidence for delay lines in the medial superior olive. *J. Comp. Neurol.* 331: 245–260.

Stein, J. F. and Talcott, J. (1999) Impaired neuronal timing in developmental dyslexia—the magnocellular hypothesis. *Dyslexia* 5: 59–78.

Stern, R. M. and Trahiotis, C. (1996) In R. H. Gilkey and T. R. Anderson (eds.), *Binaural and spatial hearing in real and virtual environments*, 499–531, Erlbaum, Mahwah, N.J.

Stevens, S. S. and Newman, E. B. (1936) The localization of actual sources of sound. *Am. J. Psych.* 48: 297–306.

Takahashi, T. and Keller, C. H. (1994) Representation of multiple sound sources in the owl's auditory space map. *J. Neurosci.* 14: 4780–4793.

Takahashi, T. and Konishi, M. (1986) Selectivity for interaural time difference in the owl's midbrain. *J. Neurosci.* 6: 3413–3422.

Thompson, S. (1877) On binaural audition. *Philos. Mag.* 5: 274–276.

Thompson, S. (1878) Phenomena of binaural audition. *Philos. Mag.* 5: 383–391.

Wagner, H., Takahashi, T. T., and Konishi, M. (1987) Representation of interaural time difference in the central nucleus of the barn owl's inferior colliculus. *J. Neurosci.* 7: 3105–3116.

Yin, T. C. T. and Chan, J. C. K. (1990) Interaural time sensitivity in medial superior olive of cat. *J. Neurophysiol.* 64: 465–488.

Yin, T. C. T. and Kuwada, S. (1983) Binaural interaction in low-frequency neurons in inferior colliculus of the cat. III. Effects of changing frequency. *J. Neurophysiol.* 50: 1020–1042.

Yin, T. C. T., Chan, J. C. K., and Irvine, D. R. F. (1986) Effects of interaural time delays of noise stimuli on low-frequency cells in the cat's inferior colliculus. I. Responses to wideband noise. *J. Neurophysiol.* 55: 280–300.

15

How Does the Hearing System Perform Auditory Scene Analysis?

Georg M. Klump

Introduction

In order to identify sources of sounds and recognize composite multipart signals that are common in the natural acoustic environment, the auditory system must be able to group components of sounds originating from one source together (i.e., form auditory objects) and separate them from sounds that are being emitted by other sources. This perceptual process has been termed *auditory scene analysis*, since in many aspects it resembles the analysis of visual scenes or images (Bregman 1990). Bregman suggested that, on the one hand, learning and memory allow the formation of composite-feature detectors in the sensory system that could be involved in selecting objects from an auditory scene, and in humans (and probably other animal species), may thus aid top-down cognitive processing of the incoming sensory information from the acoustic environment. On the other hand, many processes have been identified that do not require learning and memory and can enable the auditory system to single out objects in a bottom-up process of auditory scene analysis. The latter have been termed "primitive processes" by Bregman (1990), and they relate to basic mechanisms of analysis in the auditory system (see also Yost 1991; Griffiths and Warren 2004).

Mechanisms of auditory scene analysis have been studied predominantly in psychoacoustic experiments with human subjects (e.g., see reviews by Bregman 1990; Yost 1991; Griffiths and Warren 2004). However, there is accumulating evidence from behavioral studies of animal subjects that the mechanisms of auditory scene analysis may operate on similar principles in a wide range of animals (see review by Hulse 2002). For example, auditory stream segregation has been demonstrated in different bird species using various stimulus paradigms. Wisniewski and Hulse (1997) and Benney and

Braaten (2000) have shown that European starlings (*Sturnus vulgaris*), zebra finches (*Taeniopygia guttata*), and striated finches (*Lonchura striata domestica*) are capable of isolating a stream of conspecific song from a background of simultaneously presented songs of other species. Stream segregation also extends to more "unnatural" stimuli. Van Noorden (1975, 1977) demonstrated the segregation of tone sequences in human subjects into different auditory streams depending on proximity in time and frequency. MacDougall-Shackleton et al. (1998) have shown that European starlings exhibit the same perceptual segregation of tone sequences into auditory streams. Izumi (2002) observed auditory stream segregation in Japanese monkeys using a paradigm with interleaved tone sequences. The mechanisms of auditory stream segregation also appear to operate in lower vertebrates. Fay (1998) has demonstrated that the goldfish (*Carassius auratus*) is capable of learning physical characteristics of individual components in composite stimuli that bear resemblance to processes of auditory stream segregation (see figure 15-1). He found that the animals would reveal the association of a spectral profile of sound pulses with a specific pulse repetition rate even if the pulses were presented simultaneously. In a generalization test, after conditioning with two simultaneously presented pulse trains, the fish responded differentially to isolated pulse trains with the different spectra that had been present in the mix. Spectral profiles (i.e., the timbre of a sound) are also readily utilized by European starlings in identifying classes of sounds (Braaten and Hulse 1991). Such spectral profiles have been shown to be an effective cue in auditory grouping in psychoacoustic studies in humans (e.g., Handel 1995) and in auditory stream segregation (e.g., Singh and Bregman 1997).

The parallels between human and animal perception of auditory scenes also extend to the observation of the phenomenon of induction, demonstrated by Warren (1970) and Warren et al. (1972). Humans will "fill in" a missing part in an interrupted sound if the auditory system is provided with another stimulus, such as a noise pulse, that bears no resemblance to the signal but provides excitation to the same frequency channels of the auditory system as the interrupted sound. Sugita (1997) first demonstrated that this phenomenon also appears to occur in a behavioral test in the cat. If cats are trained to choose one response upon hearing an interrupted tone and another response upon hearing a continuous tone, they will choose the latter feeder if the temporal gaps between the interrupted tones are filled with noise. Observations in the European starling (Klump et al. 1998) and macaque monkey (Petkov, O'Connor, and Sutter 2003) when they are tested with tone signals suggest that these species experience a similar effect. When European starlings were trained to report exemplars of conspecific (starling) or heterospecific (budgerigar) song elements presented in the middle of starling or budgerigar song, they were more likely to indicate having heard a conspecific song element rather than having heard a heterospecific song element when presented with a pulse

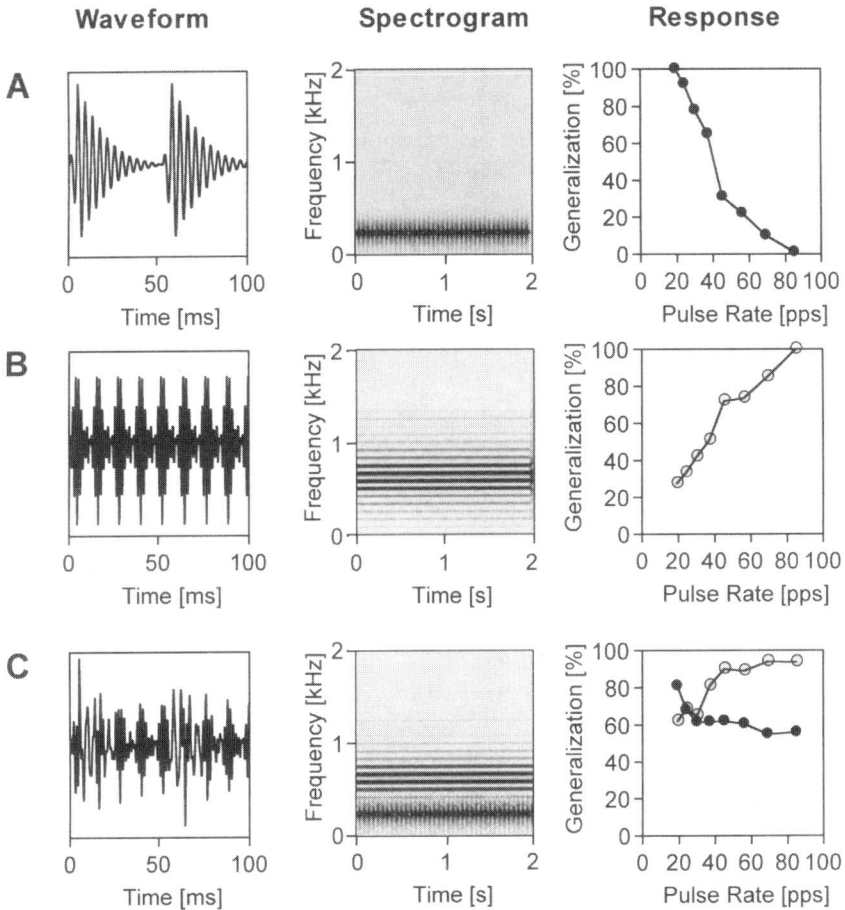

Figure 15-1. Auditory scene analysis in the goldfish (after Fay 1998). The first two columns show a sketch of a part of the waveform and the spectrogram of the training stimuli. The animals were classically conditioned to trains of a 238-Hz pulse presented at a rate of 19 s^{-1} (row A), to trains of a 625-Hz pulse presented at a rate of 85 s^{-1} (row B), or to a signal composed of the two pulse trains presented simultaneously (row C). In the generalization test, the response to a 238-Hz pulse train (filled circles) and a 625-Hz pulse train (open circles) with pulse repetition rates varying between 19 s^{-1} and 85 s^{-1} was determined. If trained with single pulse trains (rows A and B), the animals show the strongest response to the training stimulus, and the response strength decreases with increasing difference between the pulse rates of the training and the test stimuli. If conditioned to a mixture of the 238-Hz and 625-Hz pulse trains, the animals nevertheless exhibit a differential response to the pulse types differing in the spectrum. This differential learning indicates that the animals segregated the two pulse trains in the conditioning stimulus.

of noise replacing the song element (Braaten and Leary 1999). This is interpreted as evidence for induction of the perception of a conspecific song, similar to the effect observed when humans restore a speech sound in a sequence that was replaced by a noise signal (Warren 1970). Restoration of parts of vocalizations also appears to occur in cotton-top tamarin monkeys when a gap introduced in the vocalization is filled with noise (Miller, Dibble, and Hauser 2001). While the induction experiments using tones may be the result of bottom-up processing, it remains unclear whether the experiments using natural vocalizations reflect the influence of top-down processing or can be explained by bottom-up processes (for a review of the analogies and speech and song learning and processing see Doupe and Kuhl 1999).

The animal experiments referred to here indicate that the mechanisms of auditory scene analysis must be common to a wide range of vertebrate species, and they might even be found in invertebrate species that are required to solve the same kind of perceptual problems, for example, when analyzing the mixture of conspecific and heterospecific sound signals in the natural environment or extracting signals from abiotic (e.g., wind-generated) background noise. As I will show in this chapter, many of the behavioral observations of perceptual patterns reflecting mechanisms of auditory scene analysis can be explained by bottom-up processes operating on the sequentially or simultaneously presented sounds, although some of the examples may also reflect top-down processing (e.g., Braaten and Leary 1999). A first review of the basic physiological processes involved in the separation of multiple sound sources in the vertebrate auditory system has been provided by Feng and Ratnam (2000). Extending that review, I will discuss in detail some recent examples of perceptual effects in which both the physiology and the psychophysics of the processes involved in auditory scene analysis have been studied.

Auditory Stream Segregation of Sequential Tone Pulses

Physiological studies can shed light on the processes that may be involved in the segregation of tone sequences into different auditory streams depending on the proximity of the tones in time and frequency. Using multiunit recordings in awake animals, Fishman and colleagues (2001) have shown responses in the primary auditory cortex of the macaque monkey (*Macaca fasciculata*) that reproduce effects of auditory stream segregation. They stimulated the neurons with alternating 25-millisecond tones of two different frequencies, one being the units' best frequency (BF, i.e., the frequency resulting in the strongest excitatory response), and one having a frequency between 10 percent and 50 percent above or below the best frequency (off-BF tone). If trains of alternating tones of the two frequencies were presented to the cortical neurons, then the tones at the BF were likely to suppress the response to the off-BF tones. The amount of suppression depended on the frequency difference between the tones and on the repetition rate, which varied between about 5 s^{-1} and 40 s^{-1}. At slow repetition rates, the units responded well to both the BF tone and the off-BF tone, suggesting a representation of the individual tones or of the train of both tones by the excitatory discharge elicited within the limits of the units' tuning curve. This implies that the tones of both frequencies are encoded in the responses of the same units, which may represent one auditory stream. At fast repetition rates, the units' response depended on the frequency difference between the BF tone and the off-BF tone. When the BF and off-BF tones were presented with a small frequency difference, the response to both was similarly suppressed. When the BF and off-BF tones were presented with a large frequency difference, the response to the off-BF tones was suppressed more than the response to the BF tones. This resulted in a predominant excitatory discharge to BF tones and a reduced response to off-BF tones, that is, the tones of both frequencies were now encoded in the excitation of differing population of units suggesting a separation into different perceptual streams. In a follow-up study Fishman et al. (2004) extended their first experiment using long series of alternating BF and off-BF tones and also varying tone duration between 25 milliseconds and 100 milliseconds. Their results confirmed the findings of the previous study. Kanwal et al. (2003) reported comparable results to those observed by Fishman et al. (2001) when applying a similar stimulus paradigm using 25-millisecond tone signals in the ultrasonic frequency range in a study on bats that are also likely to use auditory streaming when analyzing sequences of echolocation signals (e.g., Moss and Surlykke 2001).

The study of the physiology of auditory stream segregation has been extended to a stimulus paradigm using . . . –ABA–ABA– . . . sequences (– indicates a silent interval that has the same duration as the time interval between the onset of two sequential A and B tones differing in frequency). As has been demonstrated in humans (e.g., van Noorden 1975, 1977) and in the European starling (MacDougall-Shackleton et al. 1998), such a sequence is perceived as one tone series having a galloping rhythm if the A and B tones are analyzed in one stream, or it is perceived as a combination of two separate tone series of A and B tones with an isochronous rhythm each (A tones being presented at half the rate of B tones). When presented with tones of a duration of 100 milliseconds and a tone repetition time that was 100 percent of the tone duration, the starlings reported hearing a galloping rhythm if the tones were separated by 0.9 semitones, and they reported an isochronous rhythm if the tones were separated by 9.3 semitones (MacDougall-Shackleton et al. 1998). The response of neurons in the European starling's auditory forebrain was tested with an ABA– sequence using a range of frequency differences between A and B tones spanning 12 semitones, a range of tone durations from 25 to 100 milliseconds and a range of tone repetition times between 100 percent and 800 percent of the tone duration (Bee and Klump 2004). A change in the relative response magnitude of the neural discharge to A and B tones was observed that resembled the change in the probability of reporting the perception of one stream (figure 15-2). The lines of a constant difference in the normalized response to A and B tones paralleled the perceptual fission boundary in humans (i.e., the combination of frequencies and tone repetition times at which A and B tones cannot be segregated into two streams; see van Noorden 1975), suggesting that the limit for not being able to split the tone series into two streams may be represented by a constant difference in the normalized response magnitude in the auditory system.

Fishman and colleagues (Fishman, Arezzo, and Steinscheider 2001; Fishman et al. 2004) proposed that the time course of suppression being evident in physiological forward masking experiments may provide a basis for explaining auditory stream segregation observed in the physiological experiment in the form of a divergent population response of the units. Similar conclusions have been drawn by Kanwal et al. (2003) and Bee and Klump (2004). Fishman and colleagues (Fishman, Arezzo, and Steinschneider 2001; Fishman et al. 2004)

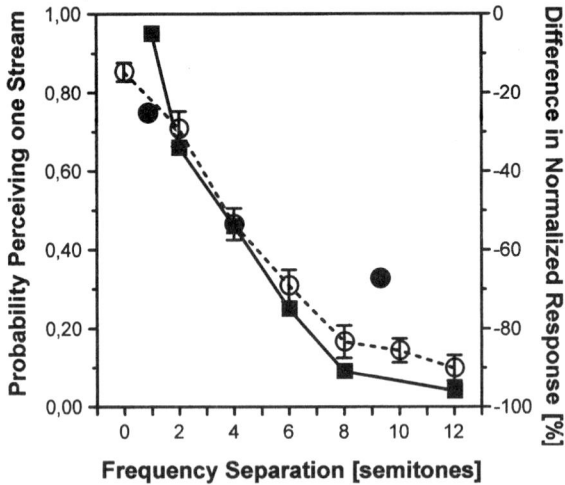

Figure 15-2.   Auditory stream segregation observed in humans and European starlings when stimulated with A and B tones differing in frequency that are presented in an ABA-series (after Bee and Klump 2004, tone durations were 100 or 125 ms and A and B tones were played with a tone repetition time that was identical to the duration). Open circles show human perceptual data reflecting the probability of the subjects to indicate perceiving A and B tones as one stream (data from Anstis and Saida 1985). Filled circles show starling perceptual data obtained by training the birds to distinguish series of tones with a galloping rhythm from series of tones with an isochronous rhythm (data from MacDougall-Shackleton et al. 1998). If the birds indicate a galloping rhythm in an ABA– tone series this is taken as evidence for perceiving the tones as one auditory stream. Filled squares show physiological data obtained from multiunit recordings in field L2 in the starling forebrain. The difference in the neurons' response is an indication of the A and B tones being represented in the excitation of separate populations of neurons.

discuss the role of local inhibitory circuits in the primary auditory cortex in creating the suppression. Inhibition across different frequencies, however, is found on various levels of the auditory pathway starting in the cochlear nucleus (e.g., Rhode and Greenberg 1994). Thus, the suppressive effects observed in the cortex could be the result of across-frequency processing at lower levels of the auditory system. In summary, it appears that some effects observed in auditory stream segregation of two or more interleaved series of sequentially presented tone signals may be the results of very basic processing mechanisms in the auditory system as has been suggested in the computer models developed for explaining this effect (Beauvois and Meddis 1996).

Our knowledge of the physiology of other aspects of auditory stream segregation, however, is rather sparse. Preliminary reports by Micheyl et al. (2003) indicate that in the macaque primary auditory cortex there is also evidence for a neural correlate of the "buildup effect," that is, an increased propensity of subjects perceiving two streams if the alternating tone series lasts longer (e.g., van Noorden 1975). Carlyon et al. (2001) have suggested that attention affects the buildup of streaming in human subjects. So far, no investigations have been conducted to study the underlying physiological mechanism. The recording of the neural response while monitoring the behavioral response will be needed to find evidence for the top-down mechanisms underlying the change of auditory streaming with attention observed in the psychophysical experiments in humans. Further investigations are also needed to elucidate the basis of the temporal coherence boundary (i.e., the combination of frequencies and tone repetition times at which A and B tones cannot be combined into one stream; see van Noorden 1975). It remains unclear what limits the ability of the auditory system to fuse two tone series into one auditory stream. Finally, in reviewing the psychoacoustic literature on auditory stream segregation in humans, Moore and Gockel (2002) pointed out that in addition to the tone frequency a large variety of perceptual differences between signals can result in auditory stream segregation (e.g., differences in the envelope of signals or differences in lateralization resulting from interaural time and intensity differences) if they are sufficiently salient. The study of the neurophysiological correlates of these additional streaming phenomena may offer a deeper insight into the general principles underlying the mechanisms of auditory scene analysis.

Object Formation by Common Modulation: CMR and MDI

Temporally coherent changes in the components of a complex acoustic stimulus are potent cues leading to grouping of the components into an auditory object (Bregman 1990; Hall and Grose 1990). Such changes can be the simultaneous onset and offset of the stimulus components, or more generally speaking, a temporally correlated modulation of the envelope of the stimulus components, or a temporally correlated variation in the frequency of individual components (for a summary of different studies see Bregman 1990). Here I will concentrate on the role of the correlated variation of the envelope of components of a sound. Two effects have been recognized in this context (e.g., see the review by Moore 1992) as possibly reflecting the action of auditory grouping phenomena. The first is the improved detection of a tonal signal in a masker with positively correlated amplitude fluctuations in different frequency bands that can be analyzed separately by the auditory system (termed comodulation masking release, CMR; for a recent review see Verhey, Pressnitzer, and Winter 2003). The second is the impaired analysis of periodic amplitude modulations of one frequency component of a complex signal in the presence of modulations with a similar time course in other stimulus components (termed modulation detection interference or modulation discrimination interference, MDI). It has been suggested that the two effects are related and may reflect the same underlying mechanism (Moore 1992).

There is evidence from psychophysical studies in bird and mammalian species that CMR is a general phenomenon found in the vertebrate auditory system, although some species differences exist. CMR has been measured using two different experimental paradigms. In the first paradigm, one frequency band of masking noise is centered on the frequency of the tone signal that is to be detected, which can either be amplitude modulated or not (in addition to the inherent modulation of the noise envelope). A release from masking has been observed for signals presented in the amplitude-modulated masker, and the size of the release depended on the bandwidth of the masker with maskers of a greater bandwidth resulting in a larger masking release (Hall, Haggard, and Fernandes 1984). In humans, the maximum release from masking ranges from 10 to 15 dB, and it is larger for maskers with slow envelope fluctuations than for maskers with fast envelope fluctuations (e.g., Schooneveldt and Moore 1989). Besides humans, Mongolian gerbils (*Meriones unguiculatus*) are the other mammalian species that exhibits considerable CMR. Given that the gerbil evolved an auditory system that has frequency analysis channels of about the same width as humans (Kittel, Wagner, and Klump 2002) and a temporal resolution that is slightly better than that of humans (Wagner, Klump, and Hamann 2003), it is not surprising to find that the amount of CMR (i.e., the advantage in signal detection gained by coherent modulation of a wide-band acoustic background signal) is relatively similar in the two species (Klump, Kittel, and Wagner 2001). However, in the chinchilla (*Chinchilla laniger*), a closely related rodent species, some types of coherently modulated maskers that result in considerable CMR in the perception of humans, did not elicit notable CMR in a behavioral study (Niemiec, Winter, and Florin 1999). The reason for this discrepancy in the performance of the various mammalian species is not clear. The only bird species studied so far, the European starling, exhibits an even stronger CMR effect than humans (Klump and Langemann 1995). A maximum masking release of between 10 and 20 dB was found in a stimulus paradigm being similar to that applied by Schooneveldt and Moore (1989).

The second stimulus paradigm applied in the study of CMR uses two narrow-band noise maskers with envelopes fluctuating at a slow rate (McFadden 1986). One band of masking noise (the on-frequency band) is centered on the frequency of the signal that needs to be detected. The second band of noise (flanking band) has a mean frequency above or below that of the on-frequency band. If both bands of noise have the same envelope (i.e., the envelopes are perfectly correlated), a considerable release of masking is found compared to cases in which both bands of masking noise have uncorrelated (or weakly correlated) envelopes (e.g., Schooneveldt and Moore 1987; Moore and Schooneveldt 1990). Klump, Kittel, and Wagner (2001), Klump et al. (2001), and Hamann et al. (1999), using similar stimulus conditions to Moore and Schooneveldt (1990), demonstrated that CMR can also be elicited by correlated narrow-band maskers in the Mongolian gerbil and in the starling, respectively. In contrast to the results in humans, however, in these animals a large CMR could be found for on-frequency and flanking bands differing considerably in frequency, whereas in humans the CMR effect was largest if on-frequency and flanking bands were positioned within one auditory filter. In humans it was reduced considerably if on-frequency and flanking bands differed more in frequency.

CMR has been attributed to the operation of two types of processes: (1) mechanisms relying on the analysis of the excitation within a single frequency channel centered on the signal and (2) mechanisms relying on a comparison of excitation across different frequency channels (e.g., Buus 1985; Moore 1992). As indicated by the comparative data reported above, the contribution of within- and across-channel mechanisms may vary between species. Some of the human psychophysical data suggest that in this species within-channel mechanisms may usually dominate the CMR effect (e.g., Schooneveldt and Moore 1987; Verhey, Dau, and Kollmeier 1999; Verhey, Pressnitzer, and Winter 2003), whereas animal studies in the European starling and the Mongolian gerbil suggest a larger contribution of across-channel mechanisms (e.g., Klump et al. 2001; Wagner and Klump 2001).

A few physiological studies have set out to explain the improved detection of signals in amplitude-modulated maskers that exhibit temporally correlated fluctuations in envelope over a wide frequency range and elicit CMR. Mott, McDonald, and Sinex (1990) reported a release from masking by amplitude modulation in chinchilla auditory-nerve fibers responding to the masker in an excitatory fashion. Pressnitzer et al. (2001) and Neuert, Verhey, and Winter (2004) demonstrated CMR-like response patterns in neurons of the ventral and dorsal cochlear nucleus of the guinea pig, respectively. They suggested that an interaction between narrow-band excitation and wide-band inhibition can explain the CMR effect in cochlear nucleus neurons. Henderson, Hongzhe, and Sinex (1999) studied masking release resulting from modulation in the mammalian inferior colliculus. Also in the primary auditory cortex a variation of the neuronal response in relation to the phase of amplitude modulations was observed that could contribute to CMR-like effects (Barbour and Wang 2002). Nelken and colleagues (Nelken, Rotman, and Yosef 1999; Nelken et al. 2001) and Las, Stern, and Nelken (2005) have shown responses in neurons of the cat inferior colliculus, medial geniculate body and the primary auditory cortex that resemble CMR. While in the cat inferior colliculus suppressive effects were observed that were similar to those in the auditory periphery, in the cat medial geniculate body and the primary auditory cortex the mechanism invoked to explain the CMR effect was a suppression of the response to the envelope fluctuations of the masker that was due to the addition of the tone with a steady envelope. In all these studies, the results obtained in the animal experiments were qualitatively similar to results obtained in psychophysical studies in humans applying comparable stimulus paradigms (e.g., Hall, Haggard, and Fernandes 1984; Mott and Feth 1986; Grose and Hall 1989; Schooneveldt and Moore 1989; Delahaye 1999).

On a quantitative level, average neuronal data obtained in the animal experiments demonstrated much less masking release than the psychophysical results obtained in humans (Pressnitzer et al. 2001). In the European starling and the Mongolian gerbil, the only species in which both the physiology and psychoacoustics of CMR have been studied, a similar discrepancy is found (see Foeller, Klump, and Koessel 2001; Klump and Langemann 1995; Klump and Nieder 2001; Kittel, Wagner, and Klump 2000; Langemann and Klump 2001; Nieder and Klump 2001) that cannot be attributed to the fact that physiology and behavioral performance are evaluated in different species. However, in all these studies comparing behavior and physiology in the same species, a considerable fraction of the cells can be found in primary auditory cortical areas exhibiting a degree of masking release in the rate response that is similar or even larger than the masking release observed in psychophysical experiments. It may be possible that the neural system selectively exploits the responses of these most sensitive cells in the behavioral detection performance (see Parker and Newsome 1998 for an in-depth discussion of the issue).

Although CMR effects can be found in studies of humans as well as of other mammals and a bird species, MDI has so far been demonstrated only in humans using psychoacoustic methods (e.g., Yost, Sheft, and Opie 1989). Attempts to show MDI in the European starling and the Mongolian gerbil using different psychoacoustic paradigms have failed so far (Klump et al. 2001; A. Santoso and G. M. Klump, unpublished data). This may indicate that

MDI and CMR do not necessarily reflect common underlying mechanisms as has been suggested (e.g., Moore 1992).

Object Formation by Spatial Processing

Location in space is one cue that characterizes a specific sound source, that is, an auditory object (or a series of objects) comprised of sounds emitted by the same sender (e.g. see Bregman 1990). Auditory image analysis requires the segregation of objects that are presented simultaneously from different positions in space and at the same time suppress the response to echoes. Furthermore, sounds emitted by senders that change their location over time dynamically should be attributed to the same source, that is, be perceived as one auditory stream of objects. Some of the basic processes involved in auditory object formation due to sound source location have been summarized by Feng and Ratnam (2000). They reviewed investigations by Takahashi and Keller (1994) on the representation of signals from simultaneously broadcasting sound sources by cells in the external nucleus of the inferior colliculus (ICx) of the barn owl (*Tyto alba*) that exhibit space-specific responses. Takahashi and Keller (1994) found that sounds (multitone complexes or noise bursts) emitted simultaneously from two speakers at different locations in the horizontal plane could be segregated, if they were sufficiently unique in their frequency contents at least over short time periods (e.g., as is typical for short-time spectra of uncorrelated noise). According to their model, the uniqueness in the short-time spectra of stimuli originating from one location (and thus being characterized by a specific interaural time difference) will lead to a brief column-specific activation of neurons in the ICx. Sounds from different locations (i.e., with differing interaural time delays) will be represented in different columns. Thus, by integrating columnar activity over a longer time (e.g., 100 milliseconds as in the study by Takahashi and Keller 1994), the owl's auditory system could identify simultaneously active but distinct columns that encode spatially separated auditory objects. However, a study by Wagner (1992) indicated that an analysis time window of 100 milliseconds may not be appropriate. Presenting neurons in the inferior colliculus with short periods of binaurally correlated noise (representing a specific interaural time difference to which the neurons were tuned) in long pulses of uncorrelated noise, Wagner found many cells that exhibited a tuned response for time periods of correlation as short as 1 millisecond. The neurons' response could be explained by a leaky-integration process with a time constant of two milliseconds. Such neurons could act more like coincidence detectors than as integrators. Wagner pointed out that a high temporal synchrony of the inputs contributing to this response pattern of IC neurons must be provided to achieve such fast-developing tuned responses. The observation by Takahashi and Keller (1992) that tuning of neurons in the central nucleus of the barn owl IC was sharpened if dynamic binaural cues were presented in a simulated-motion paradigm also suggests that processes with a short integration time provide the basis for spatial auditory object formation. Short integration times may also be a prerequisite for mechanisms of binding of components from spatially separated sound sources and the segregation of simultaneous sounds from different spatial locations that is improved by modulations of the signals (Keller and Takahashi 2001).

The Principle of Good Continuation: Object Formation by Top-Down Processing?

So far, this review has concentrated on bottom-up processes of auditory object formation, which may operate on the basis of very simple mechanisms. The effect of apparent continuity may be, at least in part, an example the of operation of top-down processing. In the experiments by Warren (1970) studying the perceptual restoration of speech interrupted by a noise in humans, nearly all the subjects reported that all speech sounds were present and that the speech sound replaced by the noise was not missing. Studies using auditory evoked-potential measurements and a mismatch-negativity paradigm indicate that the processing of speech sounds depends on sensory memory and learning (see review by Näätänen et al. 2001). For example, using this method in a study of vowel perception, Cheour et al. (1998) found that the neurophysiological response in the cortex to a deviating speech sound presented in a series of reference signals was smaller in subjects speaking a language that does not employ this speech sound than in those speaking a language that does so. Studying the development of this difference in children of 6 and 12 months of age, the authors concluded that it was related to perceptual learning because they found a larger difference between children exposed to different languages at the age of 12 months than at the age of 6 months. This suggests that in the study by Warren (1970), the perceptual restoration of speech may have been supported by memory processes. Experiments in birds investigating song signal perception in the context of auditory scene analysis, similarly indicate a special processing of the species' own song compared to heterospecific songs (Benney and Braaten 2000). Such parallels can be expected given the parallels in the acquisition of birdsong and human speech (Doupe and Kuhl 1999). Studying the physiology of the effect of apparent continuity in the auditory cortex of the cat, Sugita (1997) found neuronal responses to a tone sweep that resembled the perceptual effect. In neurons that responded with a discharge to a frequency-modulated tone but not to the same tone interrupted by a short temporal gap, the response could be restored by presenting a noise in the temporal gap. It appears that feature detectors in the auditory system will respond and elicit the perception of the continuous sound if the appropriate excitation (although being elicited by another stimulus) is present (Warren, Obusek, and Ackroff 1972; Sugita 1997).

Concluding Remarks: What Is the Code for Auditory Objects?

To understand auditory scene analysis by the brain we have to understand how sensory inputs are grouped to form a representation of an auditory object that is independent of that of other auditory objects being formed at the same time (Bregman 1990). The grouping involves linking the neural activity, not only from cells representing one dimension in the parameter space defining the stimulus (such as the frequency spectrum), but also from cells in separate pathways in which the sensory input of other dimensions of the stimulus space is encoded. For example, the auditory system has been shown to process cues characterizing the spatial position of a sound source separate from those de-fining the signal type (i.e., the "where stream" is processed separately from the "what stream"—see for example Rauschecker and Tian 2000). The infor-mation encoded in the spike trains of separate processing units must be combined, otherwise auditory objects with multiple stimulus dimensions could not be depicted. The key question is how this combination is achieved. Most of the physiological studies of mechanisms related to auditory scene analysis that are reported above have taken the mean rate of action potentials integrated over a fixed time window of some 10 to 100 milliseconds as the response measure. On the other hand, physiological studies in the binaural processing of the auditory system have demonstrated that a 1-millisecond time period of correlated binaural stimulation is sufficient to elicit a spatially tuned neural response (Wagner 1992). This suggests that the neural excitation during very short time periods encodes relevant stimulus characteristics.

Studies on the mechanisms of visual scene analysis have proposed that the correlation of activity in distributed populations of cortical neurons, showing precision of the temporal relations in the range of milliseconds, may possibly encode the information of the sensory input (e.g., Gray et al. 1989; Singer and Gray 1995; Singer 1999). While in the visual cortex these responses resulting from internal interactions resulting in oscillations that are not time-locked to the temporal structure of the stimulus (for a report of oscillatory activity in the primate auditory cortex with similar characteristics see Brosch, Budinger, and Scheich 2002), neural discharges that are closely related to the temporal structure of the signal can be found on all levels of the auditory pathway up to the cortex (e.g., Langner 1992; Eggermont 1994; DeCharms and Merzenich 1996; Trussel 1997; Grothe and Klump 2000). Thus, in the auditory pathway already stimulus-based temporal correlations of neural activity across different neurons may have the potential to form the bases for auditory object formation and auditory scene analysis (for an example regarding azimuth sound localization—see Eggermont and Mossop 1998). Furthermore, the discovery that stereotyped temporal patterns of action potentials in the auditory cortex also relate to stimulus characteristics (e.g., in the context of the encoding of the location of a sound source; see Middlebrooks et al. 1994, 1998) provides evidence that the precise time of occurrence of action potentials in distributed neural assemblies may be utilized by the auditory system to encode auditory objects and analyze auditory scenes. Studies exploring temporal correlations of distributed neural activity or temporal patterns of the neural response using stimulus paradigms that have been applied in psychophysical studies of auditory scene analysis are required, before conclusions can be made about the most effective mechanisms for auditory scene analysis by the vertebrate brain.

Acknowledgments   The author's research on primitive principles of mechanisms of sensory processing related to auditory scene analysis is funded by a grant from the DFG (FG 306 "Auditory Objects"). The support of the MPI for the Physics of Complex Systems at Dresden is gratefully acknowledged. The comments of Michael Cherry helped to improve a previous version of the manuscript.

References

S. Anstis and S. Saida, (1985) "Adaptation to auditory streaming of frequency modulated tones," *Journal of Experimental Psychology: Human Perceptual Performance* 11: 257–271.

D.L. Barbour and X. Wang, (2002) "Temporal coherence sensitivity in auditory cortex," *Journal of Neurophysiology* 88: 2684–2699.

M.-C. Beauvois and R. Meddis, (1996) "Computer simulation of auditory stream segregation in alternating-tone sequences," *Journal of the Acoustical Society of America* 99: 2270–2280.

M.A. Bee and G.M. Klump, (2004) "Primitive auditory stream segregation: A neurophysiological study in the songbird forebrain," *Journal of Neurophysiology* 92: 1088–1104.

K.S. Benney and R.F. Braaten, (2000) "Auditory scene analysis in estrildid finches (*Taeniopygia guttata* and *Lonchura striata domestica*): a species advantage for detection of conspecific song," *Journal of Comparative Psychology* 114: 174–182.

R.F. Braaten and S.H. Hulse, (1991) "A songbird, the European starling (*Sturnus vulgaris*), shows perceptual constancy for acoustic spectral structure," *Journal of Comparative Psychology* 105: 222–231.

R.F. Braaten and J.C. Leary, (1999) "Temporal induction of missing birdsong segments in European starlings," *Psychological Science* 10: 162–166.

A.S. Bregman, (1990) *Auditory scene analysis: The perceptual organization of sound* (Cambridge, Mass.: MIT Press).

M. Brosch, E. Budinger, and H. Scheich, (2002) "Stimulus-related gamma oscillations in primate auditory cortex," *Journal of Neurophysiology* 87: 2715–2725.

S. Buus, (1985) "Release from masking caused by envelope fluctuations," *Journal of the Acoustical Society of America* 78: 1958–1965.

R.P. Carlyon, R. Cusack, J.M. Foxton, and I.H. Robertson, (2001) "Effects of attention and unilateral neglect on auditory stream segregation," *Journal of Experimental Psychology: Human Perception and Performance* 27: 115–127.

M. Cheour, R. Ceponiene, A. Lehtokoski, A. Luuk, J. Allik, K. Alho, and R. Näätänen, (1998) "Development of language-specific phoneme representations in the infant brain," *Nature Neuroscience* 1: 351–353.

R.C. DeCharms and M.M. Merzenich, (1996) "Primary cortical representation of sounds by the coordination of action-potential timing," *Nature* 381: 610–612.

R. Delahaye, (1999) "Across-channel effects on masked signal threshold in hearing" (University of Essex, PhD thesis).

A.J. Doupe and P.K. Kuhl, (1999) "Birdsong and human speech: common themes and mechanisms," *Annual Review of Neuroscience* 22: 567–631.

J.J. Eggermont, (1994) "Temporal modulation transfer functions for AM and FM stimuli in cat auditory cortex. Effects of carrier type, modulating waveform and intensity," *Hearing Research* 74: 51–66.

J.J. Eggermont and J.E. Mossop, (1998) "Azimuth coding in primary auditory cortex of the cat. I. Spike synchrony versus spike count representations," *Journal of Neurophysiology* 80: 2133–2150.

R.R. Fay, (1998) "Auditory stream segregation in goldfish (*Carassius auratus*)," *Hearing Research* 120: 69–76.

A.S. Feng and R. Ratnam, (2000) "Neuronal basis of hearing in real world situations," *Annual Review of Psychology* 51: 699–725.

Y.I. Fishman, J.C. Arezzo, and M. Steinschneider, (2001) "Neural correlates of auditory stream segregation in primary auditory cortex of the awake monkey," *Hearing Research* 151: 167–187.

Y.I. Fishman, D.H. Reser, J.C. Arezzo, and M. Steinschneider, (2004) "Auditory stream segregation in monkey auditory cortex: effects of frequency separation, presentation rate, and tone duration," *Journal of the Acoustical Society of America* 116: 1656–1670.

E.R. Foeller, G.M. Klump, and M. Koessl, (2001) "Neural correlates of comodulation masking release in the auditory cortex of the gerbil," *Abstracts Association for Research in Otolaryngology* 24: 177.

C.M. Gray, P. Konig, A.K. Engel, and W. Singer, (1989) "Oscillatory responses in cat visual cortex exhibit inter-columnar synchronization which reflects global stimulus properties," *Nature* 338: 334–337.

T.D. Griffiths and J.D. Warren, (2004) "What is an auditory object?" *Nature Review Neuroscience* 5: 887–892.

J.H. Grose and J.W. Hall, (1989) "Comodulation masking release using SAM tonal complex maskers: Effects of modulation depth and signal position," *Journal of the Acoustical Society of America* 85: 1276–1284.

B. Grothe and G.M. Klump, (2000) "Temporal processing in sensory systems," *Current Opinion in Neurobiology* 10: 467–473.

J.W. Hall and J.H. Grose, (1990) "Comodulation masking release and auditory grouping," *Journal of the Acoustical Society of America* 88: 119–125.

J.W. Hall, M.P. Haggard, and M.A. Fernandes, (1984) "Detection in noise by spectro-temporal pattern analysis," *Journal of the Acoustical Society of America* 76: 50–56.

I. Hamann, G.M. Klump, C. Fichtel, and U. Langemann, (1999) "CMR in a songbird studied with narrow-band maskers," *Abstracts Association for Research in Otolaryngology* 22: 22.

S. Handel, (1995) "Timbre perception and auditory object identification," In: B.J.C. Moore (ed.), *Hearing* (San Diego, Calif., Academic Press), 425–461.

J.A. Henderson, L. Hongzhe, and D.G. Sinex, (1999) "Responses of inferior colliculus neurons to tones in comodulated and uncomodulated noise," *Society of Neuroscience Abstract*: 157.13.

S.H. Hulse, (2002) "Auditory scene analysis in animal communication," *Advances in the Study of Behavior* 31: 163–200.

A. Izumi, (2002) " Auditory stream segregation in Japanese monkeys," *Cognition* 82: 113–122.

J.S. Kanwal, A.V. Medvedev, and C. Micheyl, (2003) "Neurodynamics for auditory stream segregation: tracking sounds in the mustached bat's natural environment. *Network Computation in Neural Systems* 14: 413–435.

C.H. Keller and T.T. Takahashi, (2001) "The neural image of an auditory scene in the owl's midbrain space map," *Abstracts 6th International Congress of Neuroethology*: 134.

M. Kittel, E. Wagner, and G.M. Klump, (2000) "Hearing in the gerbil (*Meriones unguiculatus*): Comodulation masking release," *Zoology* 103, Suppl. III: 68.

M. Kittel, E. Wagner, and G.M. Klump, (2002) "An estimate of the auditory-filter bandwidth in the Mongolian gerbil," *Hearing Research* 164: 69–76.

G.M. Klump, C. Fichtel, I. Hamann, and U. Langemann, (1998) "Mechanisms of object formation in the songbird auditory system," *Verhandlungen Deutsche Zoologische Gesellschaft* 91: 6.

G.M. Klump, M. Kittel, and E. Wagner, (2001) "Comodulation masking release in the Mongolian gerbil," *Abstracts Association for Research in Otolaryngology* 24: 299.

G.M. Klump and U. Langemann, (1995) "Comodulation masking release in a songbird," *Hearing Research* 87: 157–164.

G.M. Klump, U. Langemann, A. Friebe, and I. Hamann, (2001) "An animal model for studying across-channel processes: CMR and MDI in the European starling," In: A.J.M. Houtsma, A. Kohlrausch, V.F. Prijs, and R. Schoonhoven (eds.), *Physiological and psychophysical bases of auditory function* (Maastricht, Shaker Publishing BV), pp. 266–272.

G.M. Klump and A. Nieder, (2001) "Release from Masking in fluctuating background noise is represented in a songbird's auditory forebrain," *NeuroReport* 12: 1825–1829.

U. Langemann and G.M. Klump, (2001) "Signal detection in amplitude-modulated maskers: I. Behavioral auditory thresholds in a songbird," *European Journal of Neuroscience* 13: 1025–1032.

G. Langner, (1992) "Periodicity coding in the auditory system," *Hearing Research* 60: 115–142.

L. Las, E.A. Stern, and I. Nelken, (2005) "Representation of tones in fluctuating maskers in the ascending auditory system," *Journal of Neuroscience* 25: 1503–1513.

S.A. MacDougall-Shackleton, S.H. Hulse, T.Q. Gentner, and W. White, (1998) "Auditory scene analysis by European starlings (*Sturnus vulgaris*): perceptual segregation of tone sequences," *Journal of the Acoustical Society of America* 103: 3581–3587.

D. McFadden, (1986) "Comodulation masking release: Effects of varying the level, duration and time delay of the cue band," *Journal of the Acoustical Society of America* 80: 1658–1667.

C. Micheyl, B. Tian, R.P. Carlyon, and J.P. Rauschecker, (2003) "The neural basis of stream segregation in the auditory cortex," *Abstracts Association for Research in Otolaryngology* 26: 232.

J.C. Middlebrooks, A.E. Clock, L. Xu, and D.M. Green, (1994) "A panoramic code for sound location by cortical neurons," *Science* 264: 842–844.

J.C. Middlebrooks, L. Xu, E. Clock, and D.M. Green, (1998) "Codes for sound location in nontonotopic auditory cortex," *Journal of Neurophysiology* 80: 863–881.

C.T. Miller, E. Dibble, and M.D. Hauser, (2001) "modal completion of acoustic signals by a nonhuman primate," *Nature Neuroscience* 4: 783–784.

B.C.J. Moore, (1992) "Across-channel processes in auditory masking," *Journal of the Acoustical Society of Japan* 13: 25–37.

B.C.J. Moore and H. Gockel, (2002) "Factors influencing sequential stream segregation," *Acta Acustica united with Acustica* 88: 320–332.

B.C.J. Moore and G.P. Schooneveldt, (1990) "Comodulation masking release as a function of bandwidth and time delay between on-frequency and flanking band maskers," *Journal of the Acoustical Society of America* 88: 725–732.

C.F. Moss and A. Surlykke, (2001) "Auditory scene analysis by echolocation in bats," *Journal of the Acoustical Society of America* 110: 2207–2226.

J.B. Mott and L.L. Feth, (1986) "Effects of the temporal properties of a masker upon simultaneous-masking patterns." In: B.C.J. Moore and R.D. Patterson (eds.), *Auditory frequency selectivity: Proceedings of the 7th international symposium on hearing* (London, Plenum Press), 381–386.

J.B. Mott, L.P. McDonald, and D.G. Sinex, (1990) "Neural correlates of psychophysical release from masking," *Journal of the Acoustical Society of America* 88: 2682–2691.

R. Näätänen, M. Tervaniemi, E. Sussman, P. Paavilainen, and I.Winkler, (2001) "'Primitive intelligence' in the auditory cortex," *Trends in Neurosciences* 24: 283–288.

I. Nelken, Y. Rotman, and O.B. Yosef, (1999) "Responses of auditory-cortex neurons to structural features of natural sound," *Nature* 397: 154–157.

I. Nelken, G. Jacobson, L. Ahdut, and N. Ulanovsky, (2001) "Neural correlates of comodulation masking release in auditory cortex of cats." In: A.J.M. Houtsma, A. Kohlrausch, V.F. Prijs, and R. Schoonhoven (eds.), *Physiological and psychophysical bases of auditory function* (Maastricht, Shaker Publishing), 282–289.

V. Neuert, J.L. Verhey, and I.M. Winter, (2004) "Responses of dorsal cochlear nucleus neurons to signals in the presence of modulated maskers," *Journal of Neuroscience* 24: 5789–5797.

A. Nieder and G.M. Klump, (2001) "Signal detection in amplitude-modulated maskers: II. Processing in the songbird's auditory forebrain," *European Journal of Neuroscience* 13: 1033–1044.

A. Niemiec, A. Winter, and Z. Florin, (1999) "Chinchillas do not show masking release in comodulated noise," *Abstracts Association for Research in Otolaryngology* 22: 22.

L.P.A.S. van Noorden, (1975) "Temporal coherence in the perception of tone sequences," Doctoral dissertation, Eindhoven University of Technology.

L.P.A.S. van Noorden, (1977) "Minimum differences of level and frequency for perceptual fission of tone sequences ABAB," *Journal of the Acoustical Society of America* 61: 1041–1045.

A.J. Parker and W.T. Newsome, (1998) "Sense and the single neuron: probing the physiology of perception," *Annual Review of Neuroscience* 21: 227–277.

C.I. Petkov, K.N. O'Connor, and M.L. Sutter, (2003) "Illusory sound perception in macaque monkeys," *Journal of Neuroscience* 23: 9155–9161.

D. Pressnitzer, R Meddis, R. Delahaye, and I.M. Winter, (2001) "Physiological correlates of comodulation masking release in the mammalian ventral cochlear nucleus," *Journal of Neuroscience* 21: 6377–6386.

J.P. Rauschecker and B. Tian, (2000) "Mechanisms and streams for processing of 'what' and 'where' in the auditory cortex," *Proceedings of the National Academy of Science* 97: 11800–11806.

W.S. Rhode and S. Greenberg, (1994) "Lateral suppression and inhibition in the cochlear nucleus of the cat," *Journal of Neurophysiology* 71: 493–514.

G.P. Schooneveldt and B.C.J. Moore, (1987) "Comodulation masking release (CMR): Effects of signal frequency, flanking-band frequency, masker band-width, flanking-band level, and monotic versus dichotic presentation of the flanking band," *Journal of the Acoustical Society of America* 82: 1944–1956.

G.P. Schooneveldt and B.C.J. Moore, (1989) "Comodulation masking release (CMR) as a function of masker bandwidth, modulator bandwidth, and signal duration," *Journal of the Acoustical Society of America* 85: 273–281.

W. Singer, (1999) "Time as coding space?" *Current Opinion in Neurobiology* 9: 189–194.

W. Singer and C.M. Gray, (1995) "Visual feature integration and the temporal correlation hypothesis," *Annual Review of Neuroscience* 18: 555–586.

P.G. Singh and A.S. Bregman, (1997) "The influence of different timbre attributes on the perceptual segregation of complex-tone sequences," *Journal of the Acoustical Society of America* 102: 1943–1952.

Y. Sugita, (1997) "Neuronal correlates of auditory induction in the cat cortex," *NeuroReport* 8: 1155–1159.

T.T. Takahashi and C.H. Keller, (1992) "Simulated motion enhances neuronal selectivity for a sound localization cue in background noise," *Journal of Neuroscience* 12: 4381–4390.

T.T. Takahashi and C.H. Keller, (1994) "Representation of multiple sound sources in the owl's auditory space map," *Journal of Neuroscience* 14: 4780–4793.

L. Trussell, (1997) "Cellular mechanisms for preservation of timing in central auditory pathways," *Current Opinion Neurobiology* 7: 487–492.

J.L. Verhey, T. Dau, and B. Kollmeier, (1999) "Within-channel cues in comodula-tion masking release (CMR): experiments and model predictions using a modulation-filterbank model," *Journal of the Acoustical Society of America* 106: 2733–2745.

J.L. Verhey, D. Pressnitzer, and I.M. Winter, (2003) "The psychophysics and physiology of comodulation masking release," *Experimental Brain Research* 153: 405–417.

E. Wagner and G.M. Klump, (2001) "Comodulation masking release in Mongolian gerbils (*Meriones unguiculatus*) studied with narrow-band maskers." In: N. Elsner and G.W. Kreutzberg, (eds), *Göttingen Neurobiology Report* (Stuttgart and New York: Thieme-Verlag), 415.

E. Wagner, G.M. Klump, and I. Hamann, (2003) "Gap-detection thresholds in the Mongolian gerbil (*Meriones unguiculatus*)," *Hearing Research* 176: 11–16.

H. Wagner, (1992) "On the ability of neurons in the barn owl's inferior colliculus to sense brief appearances of interaural time difference," *Journal of Comparative Physiology A* 170: 3–11.

R.M. Warren, (1970) "Perceptual restoration of missing speech sounds," *Science* 167: 392–393.

R.M. Warren, C.J. Obusek, and J.M. Ackroff, (1972) "Auditory induction: perceptual synthesis of absent sounds," *Science* 176: 1149–1151.

A.B. Wisniewski and S.H. Hulse, (1997) "Auditory scene analysis in the European starling (*Sturnus vulgaris*): discrimination of starling song segments, their segregation from conspecific songs, and evidence for conspecific song categorization," *Journal of Comparative Psychology* 111: 337–350.

W.A. Yost, (1991) "Auditory image perception and analysis: the basis for hearing," *Hearing Research* 56: 8–18.

W.A. Yost, S. Sheft, and J. Opie, (1989) "Modulation interference in detection and discrimination of amplitude modulation," *Journal of the Acoustical Society of America* 86:2138–2147.

16

How Does Our Visual System Achieve Shift and Size Invariance?

Laurenz Wiskott

Introduction

The ease with which we recognize common objects from different distances and perspectives and under different illuminations gives the impression that invariant object recognition is a trivial task. However, an apparently small change in the stimulus can cause a dramatic change in the retinal activity pattern. Assume, for example, you are looking at a zebra and change your gaze by just one width of the zebra stripes. Many responses of retinal sensors will be inverted (change from low to high or vice versa) causing a dramatic change of the neural activity pattern, but you still perceive the same zebra. Thus, in order to achieve invariant recognition, our visual system has to be insensitive to these kinds of dramatic changes in the visual input while being still sensitive to more subtle changes that are perceptually relevant, such as when the zebra turns its head.

The question presented and discussed here is confined to the two apparently simplest geometrical invariances, namely shift and size invariance. The discussion is mainly based on computational models of the visual system (not including the large body of literature on application-oriented systems). This simplifies the discussion in some respects, but it also means that many of the experimental details that have not been modeled yet are left aside. One should also keep in mind that even if a computational model is consistent with all known experimental data, it does not mean that it reveals the actual mechanisms used by the biological system. Despite these and other limitations of the discussion I hope it may nevertheless be useful in addressing the question, how does our visual system achieve shift and size invariance?

The text is structured as follows. "Evidence for Shift and Size Invariance in the Visual System" briefly summarizes psychophysical and neurophysiological evidence for shift and size invariance and its limitations. "Neurobiological Constraints" provides a list of constraints from neuroanatomy and -physiology that need to be taken into account when developing biologically plausible models. From a computational point of view there are two basic approaches by which shift and size invariance can be achieved: normalizing the image or extracting invariant features. These two approaches are briefly discussed in "Two Computational Approaches." "Dynamic Routing Circuit Model" and "Invariant Feature Networks" present models following these two computational approaches and discuss to which extent they can account for different aspects of the visual system. Open questions in modeling shift and size invariance in the visual system are then discussed in "Open Questions." "Other Models" gives a short account of some alternative models not discussed here. Finally, conclusion is given.

Evidence for Shift and Size Invariance in the Visual System

Evidence for shift and size invariance comes from two sides: psychophysics and neurophysiology. Psychophysical experiments with human subjects indicate that recognition of common visual objects is, within the range tested, independent of location and size. For example, in priming experiments Biederman and Cooper (1991, 1992) tested subjects on line drawings of ordinary objects. They found that naming reaction times did not change significantly if the stimuli (4° in diameter) were shifted by 4.8° visual angle. They also found no significant change if stimuli were changed in size from 3.5° to 6.2° in diameter or vice versa. In both experiments stimuli were presented in or near the center of the visual field. In a perceptual learning task with grayscale images of common objects Furmanski and Engel (2000) found no significant performance difference if size changed unexpectedly from 16.4° × 12.7° to 8.2° × 6.4° or vice versa.

Supporting evidence for shift and size invariance also comes from neurophysiological experiments. Tovée, Rolls, and Azzopardi (1994), for instance, recorded from face sensitive neurons in the inferotemporal cortex and the cortex in the banks of the anterior part of the superior temporal sulcus of macaque monkeys. They found the firing rates of the neurons to be largely invariant to a change of fixation point up to the edge of the presented faces and only a small reduction up to 4° beyond the edge. Ito et al. (1995) measured the receptive field properties of neurons in the anterior part of the inferotemporal cortex. They found large receptive fields up to about 50° in diameter for critical features of much smaller size, which indicates large shift invariance; 57 percent of the neurons tested had a stable response within stimulus size ranges larger than two octaves. See Oram and Perrett (1994) for a more detailed overview.

However, psychophysical results also point to some limitations of shift invariance. Nazir and O'Regan (1990), for instance, showed that recognition of random dot patterns of size $0.97° \times 0.86°$ degraded significantly if patterns were trained $2.4°$ to one side of the visual field and then tested in the center or $2.4°$ to the opposite side of the visual field. Dill and Fahle (1998) found similar results in a same-different task for random dot patterns of size $0.5° \times 0.5°$ shifted up to $2°$. The degradation with shift even held when subjects knew the new location beforehand. One possible explanation for these apparently contradicting results might be that our visual system is more invariant for familiar and meaningful objects than for unfamiliar and abstract patterns, like the dot patterns. However, there are also a number of other differences between the experiments pro and contra shift invariance reported here, such as the size of the patterns or the type of task.

Neurobiological Constraints

When developing computational models for shift and size invariant recognition in the visual system, one has to account not only for the invariance properties but also for a number of other aspects of the visual system; see figure 16-1 and Oram and Perrett (1994).

Two pathways: It is generally accepted that the visual system can be divided into two pathways dedicated to different types of processing (Ungerleider and Mishkin 1982), at least from MT and V4 upward (see Merigan and Maunsell 1993 for a review). The ventral pathway, also called form- or *what*-pathway, is dedicated to form perception and object recognition. The dorsal pathway, also called motion- or *where*-pathway, processes motion and other spatial information. Even if the dissociation of *what*- and *where*-processing in the two pathways may not be as strict as generally thought (Merigan and Maunsell 1993), it seems to be clear that computational models have to provide mechanisms for processing these two aspects of visual information in an explicitly accessible fashion.

Layered structure: The visual system has a rich hierarchy of different areas. The ventral pathway, for instance, processes visual information in the following sequence of areas: Retina → lateral geniculate nucleus (LGN) → visual area 1 (V1) → visual area 2 (V2) → visual area 4 (V4) → posterior inferotemporal area (PIT) → central inferotemporal area (CIT) → anterior inferotemporal area (AIT) (→ anterior superior temporal polysensory area [STPa]), with a clear hierarchy given by the connectivity (for reviews see Felleman and Van Essen 1991; Merigan and Mausell 1993; Oram and Perrett 1994). Computational models should reflect this layered structure.

Feedback connections: The majority of connections in the visual system are reciprocal, that is, feedforward connections are mirrored by corresponding feedback connections (Felleman and Van Essen 1991). The role of these feedback connections is fairly unclear. However, a model of the visual system

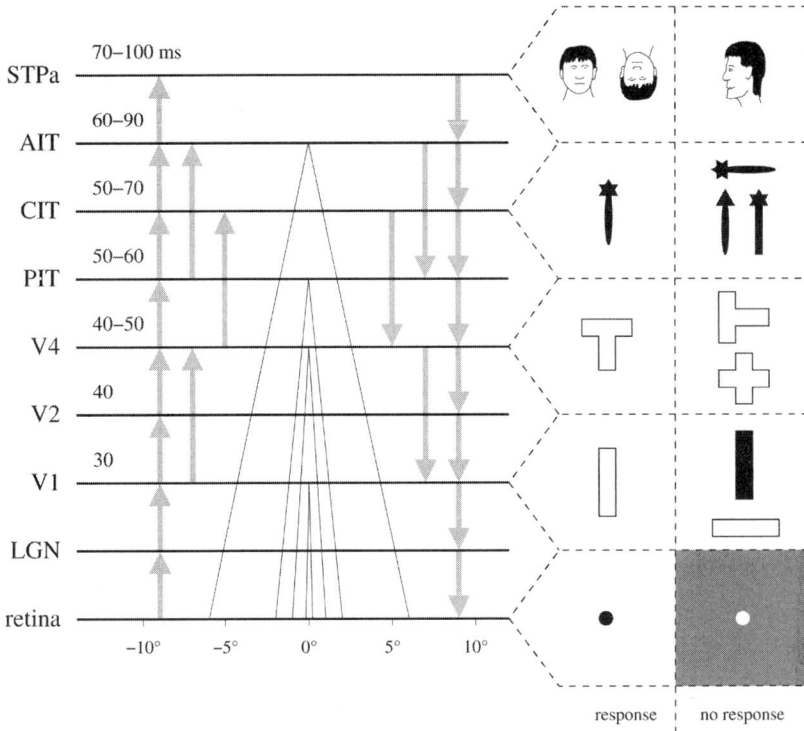

Figure 16-1. Basic properties of the ventral pathway of the visual system (adapted from Oram and Perrett 1994). The pathway has a layered structure (from retina to STPa) with layer-to-layer connections (gray upward arrows). Forward connections are usually mirrored by feedback connections (gray downward arrows). Neurons in different layers respond to features of increasing complexity from bottom to top (sample stimuli to the right). Receptive field size and with it the shift invariance also increase from bottom to top (triangles in the center, tip indicating a neuron and base indicating its receptive field size). Response latencies are very short (numbers on the left).

should at least offer a hypothesis as to what the feedback connections are good for.

Feature hierarchy: Single unit recordings from different areas of the ventral stream indicate that the complexity of the critical features causing a neuron to fire gradually increases from bottom to top (Kobatake and Tanaka 1994; Oram and Perrett 1994).

Invariance hierarchy: Single unit recordings also show that the receptive field sizes gradually increase from bottom to top and with it the amount of shift invariance (Kobatake and Tanaka 1994; Oram and Perrett 1994).

Fast recognition: Measurements of response latencies (see Nowak and Bullier 1997 for a review) as well as electroencephalograph recordings in psychophysical experiments (Thorpe, Fize, and Marlot 1996) show that the visual system performs object recognition very rapidly within about 150 milliseconds.

Attention: It is known from psychophysical as well as neurophysiological studies that visual processing can be modified by attention in various ways (see Desimone and Duncan 1995 for a review). A computational model should offer mechanisms by which attentional selection or biases can be imposed on the processing based on cues such as location, features, or novelty.

Learning: It is infeasible to assume that a complex hierarchical network for invariant object recognition could be genetically predetermined in detail. It is much more likely that the visual system develops through self-organization and unsupervised learning mechanisms from a relatively simple basic structure. A computational model of the visual system thus has to offer ideas about these mechanisms.

Two Computational Approaches

From a computational point of view there are two basic approaches known by which invariances can be achieved, first, by normalization and second, by extracting invariant features. Some principal pros and cons of these approaches are summarized in this section. A discussion as to how consistent these approaches are with what we know about the visual system is given in the succeeding sections in the context of existing neural models.

Normalization

In this approach the image of an object in the visual field is normalized to a standard position and size by an internal transformation. Invariant recognition can then be based on this normalized view. This approach has the following principal advantages ($+$) and disadvantages ($-$).

+ **_Where_-information is made explicit.** Since an explicit normalization is applied, the information about size and location of the object under consideration is available to the system at any time.

− **Recognition requires normalization.** A normalization requires a rough segmentation to determine which part of the visual field contains the object of interest. This is typically a difficult task in natural environments. Thus a sophisticated mechanism of iterating crude recognition, segmentation, normalization, and verification is required to find the correct normalization and recognize the object with certainty. This costs valuable time.

+ **Minimal information loss.** Since shifting and rescaling of a portion of an image is a simple operation, the normalization can be achieved with minimal information loss and great generality. Thus also unfamiliar and unnatural stimuli can be easily represented in an invariant way.

— **No processing toward recognition.** The simplicity of the normalization transformation also implies that no processing toward recognition is achieved. This may be a disadvantage because the time used for the normalization cannot be used for recognition.

Invariant Features

In this alternative approach some features are extracted from the image that are invariant to the location and size of an object in the visual field. Invariant recognition can then be based on these invariant features. This approach has the following principal advantages (+) and disadvantages (−).

— ***Where*-information may be difficult to extract.** Since the point of extracting invariant features is to ignore any positional and size information, this *where*-information may actually be difficult to extract if needed. In any case, it will require additional machinery and probably additional time to do so, with the possible exception of some simple cases (Wiskott 1999).

+ **Recognition does not require knowing where the object is.** A great advantage of this approach is that objects can be recognized without knowing where they are and which size they have. Thus object recognition can be potentially faster than if normalization were required.

— **Usually information is lost.** The invariant features being extracted will be typically tailored to the natural visual environment. Thus if unnatural visual stimuli are presented, the object representation may be insufficient and invariant recognition may degrade. Another type of information loss results from the lack of spatial information which can potentially cause confusion between objects composed of the same local features in different spatial arrangements (but see Ullman and Soloviev 1999; Mel and Fiser 2000) and interference between different objects in the visual field.

+ **Processing towards recognition.** Extracting invariant features is already an important step towards object recognition. Thus invariance and recognition are achieved largely simultaneously.

Notice that the four pros and cons listed in this section are complementary to those listed in the previous section and that the first two and the last two pros and cons in each list form pairs related to the same property. Because of this complementarity it may be an appealing idea to combine these two approaches; this issue will be touched upon in "Open Questions."

Dynamic Routing Circuit Model

The most prominent neural model for shift and size invariant recognition in the visual system based on a normalization is probably the dynamic routing circuit model by Olshausen, Anderson, and Van Essen (1993). This model implements the normalization in a rather direct fashion; see figure 16-2. The connectivity between two successive layers is controlled by routing control units, which can turn on or off certain subsets of connections. If the appropriate connections are activated, a region in the input layer, referred to as the window of attention, is projected to the output layer in a standardized size. This provides a normalized representation of the attended region, based on which recognition can be performed. The connections between the different layers are organized such that small shifts and rescalings can be realized at the lower stages while larger shifts and rescalings are realized at the higher stages. The input layer is associated with the LGN and the output layer is associated with AIT. A closely related model has been presented by Postma, van den Herick, and Hudson (1997).

Several aspects of the visual system listed in "Neurobiological Constraints" can be easily accounted for by the routing circuit model. The network has a layered structure and achieves shift and size invariance (although in principle the invariances do not depend on the type of stimulus; compare "Evidence for

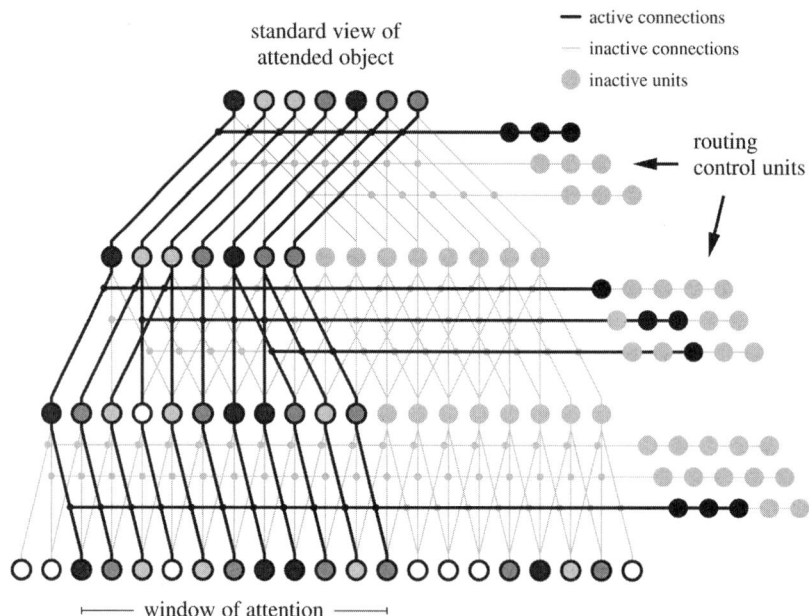

Figure 16-2. Schematic illustration of a routing circuit. The activity of units represents a feature value, such as local light intensity, and is indicated by different gray values. The same type of feature is used in the whole network (no feature hierarchy). Most of the existing connections between two successive layers are disabled (gray lines) through inhibitory mechanisms by the routing control units. The remaining active connections (black lines) establish a mapping between a region in the input layer (*bottom layer*), referred to as window of attention, and the output layer (*top layer*). This provides a normalized view of the attended object.

Shift and Size Invariance in the Visual System"). Units have increasing receptive field sizes and potentially increasing invariances from bottom to top. There is a clear split between *what-* and *where-*information represented in the main network and the routing control units, respectively. Finally, attention is naturally implemented in the network (although only in its spatial form; compare "Neurobiological Constraints"). However, there are also some open issues here. Feedback is only used in an indirect fashion through the routing control units; there is no role for direct feedback yet. The network does not show a feature hierarchy. So far only local light intensities are represented in the network from bottom to top. The network could easily be adapted to other features such as Gabor-wavelet responses, but this would not introduce a feature hierarchy. Speed is also an open issue. Since the network in its present form requires the control of the routing prior to recognition, it is unclear how the model could account for rapid recognition within 150 milliseconds. Finally, no ideas as to how a routing circuit network could be setup by self-organization have been worked out yet. Some of these open issues will be discussed in greater detail later in this chapter.

Invariant Feature Networks

There are a great number of neural models based on the idea of extracting invariant features. Prominent examples are the neocognitron (Fukushima, Miyake, and Ito 1983), higher-order neural networks (e.g., Reid, Spirkovska, and Ochoa 1989), and the weight-sharing backpropagation network (LeCun et al. 1989). Invariant features are typically extracted in two steps. First, features are extracted, then invariance is achieved by pooling over those units whose features are related by a specific transformation. For shift invariance, for instance, each feature is first extracted in a location specific manner by a set of units with identical receptive fields distributed over the whole input layer. In neural network models this is often achieved by a weight-sharing constraint. Pooling over a neighborhood of these units sensitive to the same local feature at different locations yields a feature specific response that is invariant to local shifts. A neural module extracting shift invariant features is illustrated in figure 16-3. The extracting and pooling might also be performed on a dendritic tree rather than by a layer of units (Mel, Ruderman, and Archie 1998). Size invariance can be treated analogously to shift invariance if in the first step common features of different size are extracted (Gochin 1994). By combining many of these modules at different locations and different levels, one obtains a hierarchical network extracting invariant features of increasing complexity and increasing invariance. Figure 16-4 shows such a hierarchical network, which is similar to a neocognitron (Fukushima, Miyake, and Ito 1983).

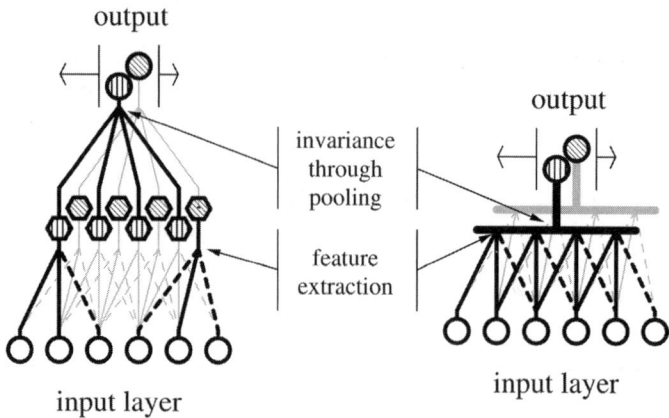

Figure 16-3. Modules for extracting shift invariant features. Black lines highlight the receptive fields of some selected units (solid lines indicate excitatory connections; dashed lines indicate inhibitory connections). *Left*, in a first step arrays of units (hexagons) with identical receptive fields extract features (indicated by different textures) at any given location. In a second step the activity of all units sensitive to the same feature is pooled. The pooling units (textured circles) then respond to this feature invariant to local shift within the receptive field (indicated by the arrows to the left and right of the pooling units). *Right*, the same computation could also be performed on the dendritic tree (thick lines) rather than by an explicit neural layer. However, for clarity the left type of illustration will be used in this chapter.

The basic architecture has been extended in several ways. To include top-down attention to a location Salinas and Abbott (1997) have added so-called gain fields to allow selecting a local region and enable feature extracting units only there. One can also imagine top-down attention to objects or features if the facilitation acts on different sets of units sensitive to a common feature rather than location as illustrated in figure 16-5 (compare Koch and Ullman 1985). These attentional control mechanisms are similar to those in the routing circuit model in that they work top-down and require indirect feedback. In Fukushima (1986) saliency driven attention based on direct feedback was implemented. Assume that at the top level there are units that respond to objects (single units are considered here for simplicity and may actually correspond to larger groups of neurons). If several objects are present in the visual field, several of these units will be active. Through a winner-take-all mechanism, one of these units can be selected and feedback connections can be used to trace back which of the feedforward units and connections gave rise to the response of the winning unit. By facilitating those connections and units and suppressing others, the system can attend to the most salient object. This is also illustrated in figure 16-5.

The detailed connectivity of an invariant feature network is too complicated to be determined genetically. It is therefore interesting to see that invariances

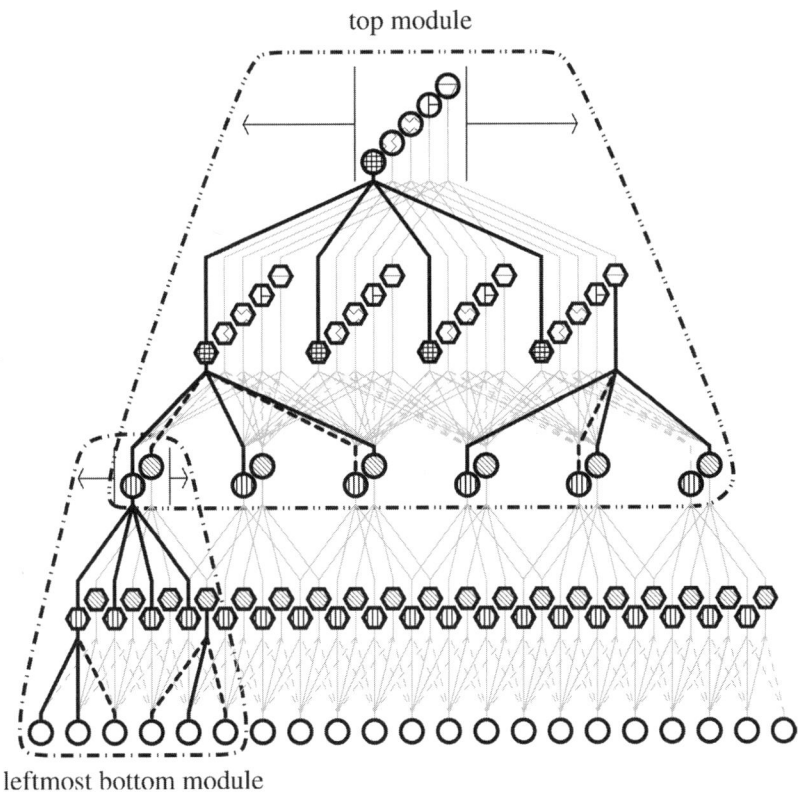

top module

leftmost bottom module

Figure 16-4. A hierarchical network for extracting invariant features can be built by replicating the module depicted in figure 16-3 at different locations and different levels. The top module has the same structure as the bottom modules except that more complex features in greater number are combined and larger invariance is achieved. The bottom modules have some overlap. As one proceeds up the hierarchy, spatial specificity is traded for feature specificity.

can be learned in a hierarchical network based on visual experience (Wallis and Rolls 1997; Wiskott 1999), leading to a connectivity like in a typical invariant feature network. The respective learning principle is based on the assumption that the external world changes slowly while the primary sensory signals change quickly, for example, the response of retinal photoreceptors change quickly due to their small receptive field sizes. Unsupervised learning of invariances can then be based on the objective of extracting slowly changing features from the quickly varying sensory input (Földiák 1991), which then leads to a robust and invariant representation of the environment.

Invariant feature networks can account for many aspects of the visual system listed in "Neurobiological Constraints." They have a layered structure with increasing feature complexity as well as increasing shift invariance. Size

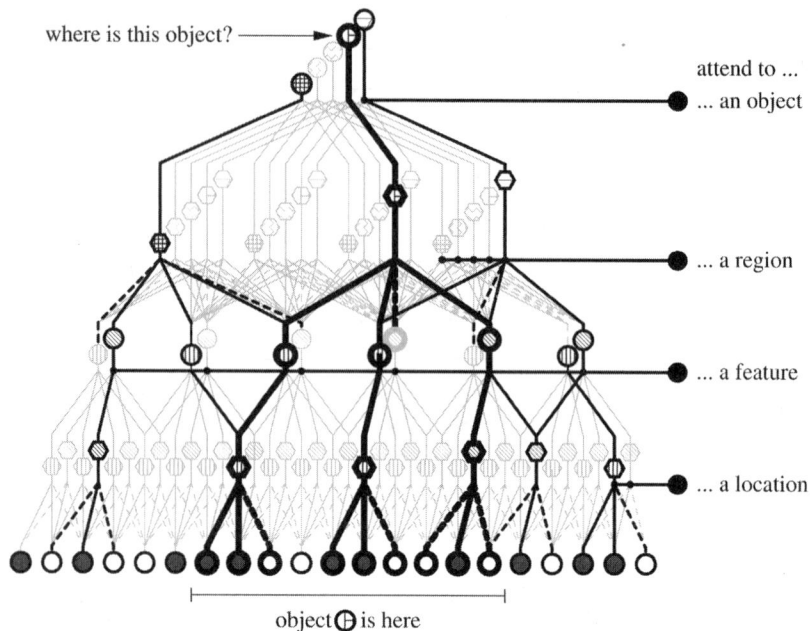

Figure 16-5. The hierarchical network for extracting invariant features with two types of attentional control. First, the control units to the right can facilitate or enable (thin black lines) and disable (connections not shown) certain sets of units and connections so that only information about a particular object, region, feature, or location is being processed. This is a form of top-down attentional control. Second, those connections and units that gave rise to the activity of a particular unit at the top can be facilitated through direct feedback connections (thick black lines and units). This is a form of saliency-driven attention. If only the facilitated part of the network remains active, the location of the attended object can be easily determined by some mechanisms not illustrated here.

invariance could be dealt with analogously. Object recognition can be very fast, since without attentional control all processing is purely feedforward. Attention can be implemented in various ways which also provides a function for direct feedback connections. Finally, simulations have shown that the connectivity of invariant feature networks can be learned based on visual experience. Only one of the major issues considered here remains unclear: How can an invariant feature network process *where*-information separately from *what*-information?

Open Questions

In the preceding sections a short review of the two main classes of neural models for shift and size invariant recognition was given. It became clear that the routing circuit model leaves more questions open than invariant feature networks. However, this is not surprising, since many more researchers have been working on the latter. Thus, at the current stage of the discussion both models should be considered further. I will discuss two open questions regarding the routing circuit model and one open question regarding the invariant feature networks.

How Can Routing Circuits Have a Feature Hierarchy?

Routing circuits only achieve a normalization. No feature extraction or other kind of processing towards object recognition is performed. Thus, routing circuits do not have a feature hierarchy. If the input layer represents gray values, the output layer also represents gray values. This is in conflict with neurophysiological results that indicate that neurons in higher layers are selective for more complex features than those in lower layers. Thus, the question is, how can routing circuits have a feature hierarchy?

The first naive approach would be to strictly alternate a feature extraction stage like the first stage in figure 16-4 and a routing stage. The feature extraction stage would increase the number of units at any location. This is no problem in the invariant feature network since there the density of locations that need to be represented is reduced in the pooling stages. However, this is not the case in the routing stages. In the standard routing circuit model the total number of units is only decreased by reducing the overall size of the represented field but the density of complex features would be the same as that of the simple features at the bottom layer, which would lead to an explosion of the number of units and a very redundant representation.

Thus, it is clear that the spatial density of the representation has to decrease from bottom to top. A second approach would therefore be to include the dynamic routing mechanism into the standard invariant feature network. This is illustrated in figure 16-6. A problem with this solution, however, is that local features can only be represented with a fixed spacing or, alternatively, local distortions and deviations from the regular spacing have to be explicitly represented by the routing control units and possibly also memorized. In any case, this architecture looks very similar to the standard invariant feature network with attentional control. It is therefore an open question whether a feature hierarchy could be included in the routing circuit model in a sensible way while preserving properties that differ significantly from the invariant feature network.

How Can Dynamic Routing Circuits Be Fast Enough?

Another open question with the dynamic routing circuit is its speed. We know that humans can do visual classification tasks, for example, animal versus no

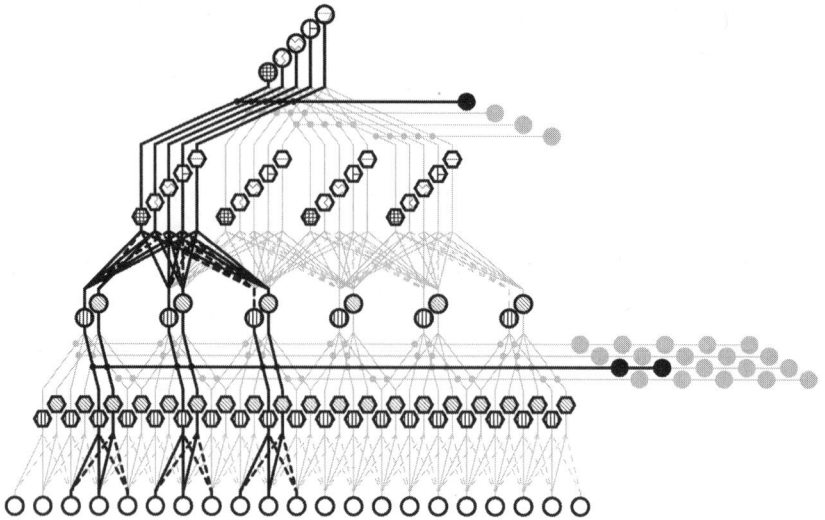

Figure 16-6.   A hierarchical network that is a combination of an invariant feature network and a dynamic routing circuit. The pooling step is replaced by dynamic routing, thereby preserving control over the spatial normalization transformation. With a rigid routing scheme like the one indicated in this figure, local features could be represented only with a fixed spacing. This problem could be solved by a more flexible routing scheme that would also include local distortions. This, however, would make the control more demanding and the architecture similar to the original invariant feature network with top-down attentional control, which does not depend on the detailed control of the distortions.

animal, within about 150 milliseconds regardless of the accurate location of the objects in the visual field. In the general case the standard procedure for invariant object recognition in a dynamic routing circuit would probably be as follows: (1) The window of attention is widely open and the presented image gets propagated through the routing circuit. (2) Some mechanism determines the region in the visual field the system has to attend to. (3) The routing control units are appropriately activated and the dynamic routing is established. (4) The image gets propagated through the routing circuit once again such that the attended object is normalized to standard size and location. (5) Object recognition can take place based on the normalized view. Even if we use very optimistic estimates for the times used by these five steps, it is clear that the whole process cannot be finished within 150 milliseconds. Thus, the question is, how can fast object recognition be explained by dynamic routing?

One simple solution would be to assume that the rapid recognition experiments done so far do not require dynamic routing and can be explained simply with the window of attention widely open (Bruno Olshausen, 2000, personal communication). The task of distinguishing animal pictures from nonanimal pictures (Thorpe, Fize, and Marlot 1996) might be solved just based on simple feature detectors, such as eye- and fur-detectors, which do not require a normalization of the input image. If this solution turns out to be true, it would indicate that the routing circuit can initially work in an invariant feature network mode.

Another solution might be to assume that in simple cases the window of attention can be determined very rapidly on a low level. By modifying an earlier experiment (Subramaniam et al. 1995) Biederman and coworkers have performed rapid recognition experiments in which black line drawings were presented on a white background every 70 milliseconds at one of four different locations in the visual field and the subjects had to respond when they recognized a particular object. Performance was nearly perfect despite the fact that the location of stimulus presentation changed randomly from image to image, that is, every 70 milliseconds (Irving Biederman, 2000, personal communication). In this experiment the location of the objects could have been detected purely on the basis of low-level cues (black lines on white background). Such a low-level based attentional control would bypass the ventral pathway and permit to skip step (1) above and reduce the time used for step (2). The control mechanism proposed in Olshausen, Anderson, and Van Essen (1993) is of this type, but the control dynamics is recurrent and based on gradient ascent, which makes the system slow. However, if a more rapid version of this control mechanism could be developed, it might be possible that the system would be fast enough even if dynamic routing were required for recognition.

A more detailed analysis of existing experimental data and possibly further experiments are required to validate or rule out one or both of these solutions.

How Can Invariant Feature Networks Process Where-*Information?*

The major open question for the invariant feature networks is that of how they process *where*-information. There are some models that address this issue. Hummel and Biederman (1992), for instance, have built a network model called JIM (John and Irv's model) for recognizing three dimensional objects made of simple geometric shapes. Part of JIM extracts invariant features and other parts process information about the location and size of the features. However, this model is incomplete in that the *where*-information is not extracted from the input image but provided separately by hand. Learning to extract invariant features and *where*-information from input images has been demonstrated (Jacobs, Jordan, and Barto 1991; Wiskott 1999), in the former case through supervised learning and in the latter case through unsupervised learning. However, in both cases only one object in front of a blank background was visible at a time so that extracting *where*-information was greatly simplified.

Thus, processing *where*-information in an invariant feature network remains an open issue. However, in context of attentional control some of the necessary machinery has already been developed. Two basic mechanisms are needed. Firstly, communication from the *where*- to the *what*-system requires a mechanism for focusing on a particular location to answer the question, what do I see here? This can be done by the top-down attentional mechanism described in "Invariant Feature Networks." Secondly, communication from the *what*- to the *where*-system requires a mechanism for determining the location of a recognized object to answer the question, where do I see this? This can be done by the saliency driven attentional mechanism also described in "Invariant Feature Networks." Thus, we see that some basic machinery for communication between the *what*- and *where*-system has been developed. What remains to be done is to link together these existing components and to develop a readout mechanism for object location in case of saliency driven attention. A detailed comparison with psychophysical and neurophysiological results would then have to show how plausible such a model would be.

Other Models

There are a number of models for shift and size invariant recognition that have not been considered in the discussion above. Some of these models will now be briefly mentioned.

Several models are based on a fixed global invariance transformation, such as a combination of the log polar and Fourier transform (Cavanagh 1978) or the R-transform (Reitböck and Altmann 1984). These are an extreme case of an invariant feature network. They are not considered here for two reasons. First, since the transformation is global, recognition requires a segmented object; cluttered scenes cannot be processed well. Second, the transformation requires a very precise connectivity and no mechanisms as to how these structures could be learned or self-organize have been presented.

Dynamic link matching (Bienenstock and von der Malsburg 1987; Konen, Maurer, and von der Malsburg 1994; Wiskott and von der Malsburg 1996) is a dynamics which establishes a topology-preserving mapping between similar objects in two different layers. This can be used to find a normalizing transformation to achieve invariant recognition. Since the dynamics requires an interplay between synaptic changes and induced correlations, dynamic link matching is too slow to account for the rapid invariant recognition capabilities of humans and is not considered here. However, dynamic link matching may play an important role in the self-organization of the visual system (Wiskott and von der Malsburg 1996).

An interesting mechanism for extracting invariant features has been proposed by Buonomano and Merzenich (1999). It is based on local interactions generating a temporal code that is shift invariant. It would be interesting to investigate to what extent this mechanism could substitute for the invariant feature module described in "Invariant Feature Networks."

Conclusion

How does our visual system achieve shift and size invariance? I have discussed this question here from a theoretician's point of view by giving an overview over computational models and pointing out some open issues in developing these models further. At the level of the discussion presented here the invariant feature networks seem to be consistent with most of the neurobiological constraints listed in this chapter, while there are more open issues for the dynamic routing circuit model, which represents here the computational approach of normalization. However, this picture might change if the discussion is carried further and more experimental details are taken into account. It may also be that both types of networks need to be combined as briefly considered in "Open Questions." The visual system could work in an invariant feature mode in the beginning and then use mechanisms of dynamic routing for a refined perception. Finally, it may also turn out that neither of the two network models discussed here is realized in the visual system. Some alternative models were mentioned.

The question of shift and size invariance may appear to be too specific to be worth being raised as one of the important questions in systems neuroscience. Wouldn't be the question of how our brain builds invariant representations in general be much more suitable? I think it depends on the answer. Either the brain solves all invariance problems in a similar way based on a few basic principles or it solves each invariance problem in a specific way that is different from all others. In the former case the more general question would be appropriate and one could consider the more specific question of shift and size invariance as a representative example. Solving the problem of shift and size invariance would then provide the key to all other invariance problems. In the latter case, that is, if all invariance problems have their specific solution, the more general question would indeed be a set of questions and as such not appropriate to be raised and discussed here. There is, of course, a third and most likely alternative, and that is that the truth lies somewhere between these two extremes.

Note

Received November 2000, accepted January 2001.

Acknowledgments I am grateful to Bruno Olshausen and Irving Biederman for fruitful discussions and valuable feedback and to Raphael Ritz for critically reading the manuscript. I also wish to thank Andreas Herz for providing excellent working conditions. Financial support was given by HFSP (RG 35–97) and the Deutsche Forschungsgemeinschaft (DFG).

References

Biederman, I. and E. E. Cooper (1991). Evidence for complete translational and reflectional invariance in visual object priming. *Perception 20*, 585–593.

Biederman, I. and E. E. Cooper (1992). Size invariance in visual object priming. *Journal of Experimental Psychology: Human Perception and Performance 18*(1), 121–133.

Bienenstock, E. and C. von der Malsburg (1987). A neural network for invariant pattern recognition. *Europhysics Letters 4*(1), 121–126.

Buonomano, D. V. and M. Merzenich (1999, January). A neural network model of temporal code generation and position-invariant pattern recognition. *Neural Computation 11*(1), 103–116.

Cavanagh, P. (1978). Size and position invariance in the visual system. *Perception 7*, 167–177.

Desimone, R. and J. Duncan (1995). Neural mechanisms of selective visual attention. *Annual Review of Neuroscience 18*, 193–222.

Dill, M. and M. Fahle (1998, January). Limited translation invariance of human visual pattern recognition. *Perception and Psychophysics 60*(1), 65–81.

Felleman, D. J. and D. C. Van Essen (1991, January). Distributed hierarchical processing in the primate cerebral cortex. *Cerebral Cortex 1*, 1–47.

Földiák, P. (1991). Learning invariance from transformation sequences. *Neural Computation 3*, 194–200.

Fukushima, K. (1986). A neural network model for selective attention in visual pattern recognition. *Biological Cybernetics 55*, 5–15.

Fukushima, K., S. Miyake, and T. Ito (1983). Neocognitron: A neural network model for a mechanism of visual pattern recognition. *IEEE Trans. on Systems, Man, and Cybernetics 13*, 826–834. Reprinted in *Neurocomputing*, J. A. Anderson and E. Rosenfeld, Eds., MIT Press, Cambridge, MA, pp. 526–534.

Furmanski, C. S. and S. A. Engel (2000, March). Perceptual learning in object recognition: Object specificity and size invariance. *Vision Research 40*(5), 473–484.

Gochin, P. M. (1994, September). Properties of simulated neurons from a model of primate inferior temporal cortex. *Cerebral Cortex 5*, 532–543.

Hummel, J. E. and I. Biederman (1992). Dynamic binding in a neural network for shape recognition. *Psychological Review 99*(3), 480–517.

Ito, M., H. Tamura, I. Fujita, and K. Tanaka (1995). Size and position invariance of neural responses in monkey inferotemporal cortex. *Journal of Neurophysiology 73*(1), 218–226.

Jacobs, R. A., M. I. Jordan, and A. G. Barto (1991). Task decomposition through competition in a modular connectionist architecture: The what and where vision task. *Cognitive Science 15*, 219–250.

Kobatake, E. and K. Tanaka (1994). Neuronal selectivities to complex object features in the ventral visual pathway of the macaque cerebral cortex. *Journal of Neurophysiology 71*(3), 856–867.

Koch, C. and S. Ullman (1985). Shifts in selective visual attention: Towards the underlying neural circuitry. *Human Neurobiology 4*(4), 219–227.

Konen, W., T. Maurer, and C. von der Malsburg (1994). A fast dynamic link matching algorithm for invariant pattern recognition. *Neural Networks 7*(6/7), 1019–1030.

LeCun, Y., B. Boser, J. S. Denker, D. Henderson, R. E. Howard, W. Hubbard, and L. D. Jackel (1989). Backpropagation applied to handwritten zip code recognition. *Neural Computation* 1(4), 541–551.

Mel, B. W. and J. Fiser (2000). Minimizing binding errors using learned conjunctive features. *Neural Computation* 12(2), 247–278.

Mel, B. W., D. L. Ruderman, and K. A. Archie (1998). Translation-invariant orientation tuning in visual "complex" cells could derive from intradendritic computations. *The Journal of Neuroscience* 18(11), 4325–4334.

Merigan, W. H. and J. H. R. Maunsell (1993). How parallel are the primate visual pathways? *Annual Review of Neuroscience* 16, 369–402.

Nazir, T. A. and J. K. O'Regan (1990). Some results on translation invariance in the human visual system. *Spatial Vision* 5(2), 81–100.

Nowak, L. G. and J. Bullier (1997). The timing of information transfer in the visual system. In K. Rockland et al. (Eds.), *Cerebral Cortex*, Volume 12, Chapter 5, pp. 205–241. New York: Plenum Press.

Olshausen, B. A., C. H. Anderson, and D. C. Van Essen (1993). A neurobiological model of visual attention and invariant pattern recognition based on dynamic routing of information. *Journal of Neuroscience* 13(11), 4700–4719.

Oram, M. W. and D. I. Perrett (1994). Modeling visual recognition from neurobiological constraints. *Neural Networks* 7(6/7), 945–972.

Postma, E. O., H. J. van den Herik, and P. T. W. Hudson (1997). SCAN: A scalable neural model of covert attention. *Neural Networks* 10(6), 993–1015.

Reid, M. B., L. Spirkovska, and E. Ochoa (1989). Simultaneous position, scale, and rotation invariant pattern classification using third-order neural networks. *International Journal of Neural Networks—Research & Applications* 1(3), 154–159.

Reitböck, H. J. P. and J. Altmann (1984). A model for size- and rotation-invariant pattern processing in the visual system. *Biological Cybernetics* 51, 113–121.

Salinas, E. and L. F. Abbott (1997, June). Invariant visual responses from attentional gain fields. *Journal of Neurophysiology* 77(6), 3267–3272.

Subramaniam, S., I. Biederman, P. Kalocsai, and S. R. Madigan (1995, May). Accurate identification, but chance forced-choice recognition for RSVP pictures. In *Proceedings of Association for Research in Vision and Ophtalmology, ARVO'95, Ft. Lauderdale, Florida.*

Thorpe, S., D. Fize, and C. Marlot (1996, June). Speed of processing in the human visual system. *Nature* 381, 520–522.

Tovée, M. J., E. T. Rolls, and P. Azzopardi (1994). Translation invariance in the responses to faces of single neurons in the temporal visual cortical areas of the alert macaque. *Journal of Neurophysiology* 72(3), 1049–1060.

Ullman, S. and S. Soloviev (1999, October). Computation of pattern invariance in brain-like structures. *Neural Networks* 12(7/8), 1021–1036.

Ungerleider, L. G. and M. Mishkin (1982). Two cortical visual systems. In D. J. Ingle, M. A. Goodale, and R. J. W. Mansfield (Eds.), *Analysis of Visual Behaviour*, Chapter 18, pp. 549–586. Cambridge, MA: MIT Press.

Wallis, G. and E. Rolls (1997). Invariant face and object recognition in the visual system. *Progress in Neurobiology* 51, 167–194.

Wiskott, L. (1999). Learning invariance manifolds. *In Proc. Computational Neuroscience Meeting, CNS'98, Santa Barbara*. Special issue of *Neurocomputing* 26/27, 925–932.

Wiskott, L. and C. von der Malsburg (1996). Face recognition by dynamic link matching. In J. Sirosh, R. Miikkulainen, and Y. Choe (Eds.), *Lateral Interactions in the Cortex: Structure and Function*, Chapter 11. http://www.cs.utexas.edu/users/nn/web-pubs/htmlbook96/: The UTCS Neural Networks Research Group, Austin, TX. Electronic book, ISBN 0-9647060-0-8.

PART V

ORGANIZATION OF COGNITIVE
SYSTEMS

17

What Is Reflected in Sensory Neocortical Activity: External Stimuli or What the Cortex Does with Them?

Henning Scheich, Frank W. Ohl, Holger Schulze, Andreas Hess, and André Brechmann

Sensory cortical activity is manifest in spatiotemporal patterns. These patterns are usually conceived as representations of stimuli in maps. Various lines of evidence suggest, however, that these map-based patterns also depend on the specific purpose served by the cortical processing. Therefore, a broader concept of the function of sensory neocortex seems to be required.

The Duality Problem

Sensory cortex maps can be conceived to interface two complementary functions, a phenomenon addressed here as *the duality problem*. On the one hand, cortex maps serve the specification of sensory input information (stimulus attributes or features) by neuronal mechanisms and functional organization. This can be considered as the implicit aspect of pattern recognition (the *bottom-up principle*). Maps of this type are the way in which the cortex copes with the variability of the stimulus world generating orderly spatial representations of what is similar and dissimilar. On the other hand, mechanisms and organization of sensory maps may also serve cognitive functions to make sensory information explicit and available in different contexts, namely by selective attention and discrimination, object constitution, categorization, mental imagery as well as by selective long-term storage and recollection of information of interest (the *top-down principle*). The bottom-up process is the result of polysynaptic and chiefly upstream analysis along the sensory pathway with certain concluding steps specific to cortex. The substrate for top-down processing is much less clear but could in principle occur by feedback interaction in sensory cortex chiefly with other forebrain sources namely with the same polymodal "activation areas," frontal cortex, cingulate cortex areas which have long been identified to maintain feedback systems with the hippocampus formation (Goldman-Rakic, Selemon, and Schwartz 1984; Insausti, Amaral, and Cowan 1987; Jones and Powell 1970; van Hoesen and Pandya 1975; for review see Farah 1997; Posner 1997; Squire, Shimamura, and Amaral 1989).

There is a general tendency to view cognitive processes as separated from bottom-up processing in sensory cortex maps, that is, to allocate relevant mechanisms in nonprimary sensory areas or in higher-order "association areas." But no principal reason would exclude the existence of duality even in primary sensory areas, except that it has proven extremely demanding to experimentally disentangle the interfacing of such processes in sensory cortex maps. Thereby it is only part of the problem that most studies of sensory cortex have been carried out in anesthetized animals, thus revealing merely bottom-up processing. Even in the awake brain of a behaving subject, any analysis of this sort has a unidirectional bias: one cannot exclude or abolish the bottom-up processing to determine the top-down contribution. It is only possible to vary and minimize the top-down demands to have an estimate of the bottom-up prerequisite.

The concept of duality in the sensory cortex and the problem of proof are particularly well exemplified with respect to learning and memory, selective attention, and mental imagery. The study of organization of various types of cognitive learning has generated the view that a number of systems, especially limbic structures and parts of prefrontal and parietal cortex, are essential for selection and intermediate retention of to be stored information. But it has long been believed that "memory is stored in the same neuronal systems that ordinarily participate in perception, analysis and processing" of this information, thus in the sensory cortex (see the introduction of the review by Squire, Mishkin, and Shimamara, 1990, p. 15). The involvement of the sensory cortex, however, has remained largely a conjecture since the basic clues on the respective critical roles of limbic and prefrontal structures, lesion-induced functional impairments, have strong ambiguities in the case of sensory cortex lesions due to this duality (compare discussion in Ohl et al. 1999).

Similarly, in the words of Posner (1997, p. 617), "When attention operates during task performance, it will operate at the site where the computation involved in the task is usually performed. Thus, when subjects attend to color, form, or motion of a visual object they amplify blood flow in various extrastriate areas (Corbetta et al. 1991)." Thus, in contrast to the aforementioned problem in memory research there are already results from brain imaging studies to support duality in the sensory cortex. Results of this sort are also available in the case of mental imagery (for review see Farah 1997).

In a certain sense, the brain may have a problem in distinguishing the results of bottom-up and top-down processing in the same structure. If the two are intermingled in the same neurons, stimulus identity may not be secured in stimulus representations, the flow of fresh information may be interrupted by top-down processing, and the distinction of what is real and what is imagination may be jeopardized, to name only a few problems.

We propose here a concept for the auditory cortex illustrated by several examples by which the duality problem might be solved by the cortex. Key components are:

Orderly spatial organization of neuronal ensembles devoted to top-down recognition needs as superimposed maps.

Representation of different stimulus features in separate maps if top-down demands are different for the recognition of these features.

Temporal sequential order of representations of bottom-up and top-down information by different activity states in the same maps.

It should be clear that these components are questions rather than firm propositions. Furthermore, some of the arguments in this chapter may not be experimentally testable but are rather based on comparative evidence in animal and human cortex, that is, on evolutionary arguments.

Maps and States

At this point, it seems important to make the crucial distinction of how a cortex map is usually constructed from stimulus-response data of many neurons and what it subsequently predicts about the representation (or identification) of a single stimulus in the map. In a given location, a neuron is tested with an array of stimuli providing the response profile (or receptive field) of that neuron. Ideally, variation of a certain stimulus class is made along its different dimensions (features like color, shape, orientation, etc.) separately to explore the tuning to variants of a given feature and the selectivity for different features. The procedure is repeated with multiple locations in a raster across the presumed map. This way one obtains multiple response profiles, from which a response space can be constructed over the raster with the dimensionality of the number of stimulus variations. In practice this is not done, but singularities of response profiles, such as maximum excitation or inhibition, are selectively used to connect locations with comparable response properties in the map (isostimulus contours). Thereby, within a map the information about the global distribution of a response to any given stimulus is lost.

The selection of the seemingly most important response aspects entails only *local views of the response distribution in the map*. It is important to point out that this type of map characterization is useful to estimate where in the map maximal contrasts between feature variants can be expected as long as it is not implied that local views constitute stimulus identities. Since these local views are, in essence, arbitrary classification procedures, it becomes impossible to determine from them the complete representation of a given stimulus and whether the stimulus has a unique representation in the map. Local views of the map similar to units receptive field properties are bottom-up aspects of stimulus processing (Phillips and Singer 1997). They do not by themselves allow addressing of a stimulus or one of its features as an entity, that is, they do not make it explicit.

From a local view, "artificial" problems may arise that are less essential in a global view of the map's activation pattern, namely whether flexible ensembles of feature-selective neurons have to be generated to recognize stimuli (Hardcastle 1994). Advantages of the alternative *global view* of maps (unique stimulus representation by global activation patterns) have long been recognized (for review see Freeman 2000). It has also been recognized by electrophysiological techniques (DeMott 1970; Lilly 1954) and by optical imaging (Arieli et al. 1996; Hess and Scheich 1996; Horikawa et al. 1996; Horikawa, Nasu, and Taniguchi 1998; Hosokawa et al. 1999; Taniguchi and Nasu 1993; Uno, Murai, and Fukunishi 1993) that local response features in maps show spatial shifts during the time course of the response to any given stimulus. This strongly argues for the relevance of a global view of maps. The global spatiotemporal activation pattern of a complete map is called here a state of the map.

Global Views, Map Superposition, and Top-Down Demands

A global view of a given map would produce a unique representation of any stimulus if a sufficient diversity of unit response properties (receptive fields) was provided. When top-down demands require sorting out (making explicit) a particular stimulus feature (e.g., color, tone frequency, sound periodicity) and its variation, a separate map in the cortex would be a strategy to provide a global view of this feature (Kaas 1982, 1993). Alternatively, a global view can still be provided if a further, correspondingly specialized, map is superimposed on the given map (superposition map) (e.g., Cohen and Knudsen 1999; Schreiner 1995, 1998). It should be pointed out that the process of making a particular feature explicit by the top-down demand may imply putting the recruited global activity state (actual state) into relation to the nonrecruited activity states (potential states) represented in the map. Thus a map in the global view, may be the specific frame for the cognitive construction of an explicit feature. It should also be noted that a global view of the activity distributions within the frame of a map offers relatively simple ways to "read out" information about stimuli.

The Auditory Cortex: A Test Ground for Views on Cortical Maps

Before embarking on this topic a few principles of auditory cortex organization and open questions related to this organization, which become relevant during the course of this chapter, are described. One of the most conservative organizational principles in the auditory pathways of mammals are tonotopic maps reflecting the place code for frequency channel analysis in the cochlea up to the cortex (Merzenich and Brugge 1973; Merzenich, Knight, and Roth 1975; Scheich 1991; Scheich, Heil, and Langner 1993; Thomas et al. 1993). The geometry of these bottom-up tonotopic maps basically consists of a tonotopic gradient, along which neurons show an orderly change of highest sensitivity (characteristic frequency; CF) and of highest discharge rate (best frequency; BF) in response to systematically varied pure tone frequencies. Orthogonal to this tonotopic gradient neurons have roughly the same CF or BF. This isofrequency dimension of the maps obtained with pure tones does not reflect the analytical potential of neurons but rather a common dominant subcortical input from the tonotopic frequency channel. Neurons along an isofrequency dimension have highly varied sensitivities for additional single or complex acoustic parameters (Heil, Rajan, and Irvine 1992, 1994; Ohl and Scheich 1997b; Schreiner 1995, 1998).

A characteristic property of the auditory cortex in all mammalian species is the multiplicity of maps distinguished with tone stimuli. Usually several of these maps are tonotopically organized. (The failure to show this so far in some species is presumably due to insufficient mapping.) Size and geometries of these tonotopic maps (spatial relationship between tonotopic and isofrequency dimension) and spatial characteristics of nontonotopic maps, which still show tone responsiveness of most neurons, are species specific. Geometries are also characteristically different among the multiple maps of a species.

The reasons for multiple representations of cochlear tonotopic information have been sought in the separate brain analysis of various complex acoustic parameter spaces (stimulus features such as amplitude modulations [AM], frequency modulations [FM], noise bandwidths, segmentations, harmonic structures, and spectral envelopes). But in the best-studied cases with this bottom-up concept, cat, monkey, and gerbil auditory cortices, mapping experiments have shown that representation of any feature is not exclusive to any particular map (Eggermont 1998; Rauschecker, Tian, and Hauser 1995; Schreiner and Urbas 1988; Schulze et al. 1997). Thus, these differences between maps may only be gradual.

On the other hand, there is a detailed analysis of the highly specialized maps in auditory cortex of the mustache bat, which suggests indeed the principle of parametric representation of different features of echo sounds in separate maps (e.g., Suga 1994). It is interesting that this apparent counterexample to cats, monkeys, and gerbils may prove the case for a subtle yet important distinction of what multiple maps in the auditory cortex represent. Variations of echo features from the bat's own voice reflected from obstacles and prey are not simply stimulus features serving pattern recognition. They manifest at the same time detailed behavioral meaning for the hunter, namely, information about size, surface structure, relative speed, and direction of objects. The information that is derived from such features subsequently entails adequate specific behavioral acts. Consequently, in the mustache bat, stimulus features represent qualities equivalent to behavioral categories. For the special case of the bat a cognitive generalization for a particular behavior can be made from one or very few stimulus features. For less specialized animals this cognitive specialization should be possible from numerous stimulus features, presumably in an opportunistic fashion. This may be the reason why their different maps appear to be much less feature specific.

In summary, the specialized case of the bat auditory cortex, which presumably evolved on the basis of self-generated "predictable" sound patterns (in contrast to individually unpredictable acoustic environments of most other animals) may still imply a common principle with multiple maps in auditory cortex of other animals: their significance may chiefly lay in cognitive demands of different behavioral tasks. The consequences of this distinction will become clear from examples in the following sections taken from Mongolian gerbil and human auditory cortices. There it will be shown (1) that neurons organized in a given map may derive information of comparable behavioral relevance from different stimulus features and (2), conversely, that different tasks executed on the same stimulus material lead to activation of different maps. Such functional aspects of maps are not easily reconciled with a strict bottom-up concept of stimulus specification but rather point to a decisive influence of top-down processing in constitution and use of such maps.

Examples of Superimposed Maps in the Auditory Cortex

Evidence for Map Superposition in the Auditory Cortex

The primary auditory cortex (AI) map, because of its rigid tonotopic input organization, is a suitable example for demonstrating the principle of top-down influences on this framework. One useful aspect is its parcelation into multiple ensembles of task-dedicated neurons or submaps. An early example described in cat are ensembles of neuron types related to binaural processing (excitatory-excitatory, EE, or excitatory-inhibitory, EI), which were originally thought to form multiple stripes orthogonal to tonotopic organization (Middlebrooks, Dykes, and Merzenich 1980; Schreiner 1991) but which may be irregular patches of similar neuron types (Phillips and Irvine 1983; for review see Clarey, Barone, and Imig 1991). Furthermore, Schreiner and Sutter (1992) described in cat multiple representations of complex features related to frequency tuning overlaying the best frequency (BF) organization. This organization related to specification of stimulus features may consist of patches with different spatial order in individual animals (Heil, Rajan, and Irvine 1992). These organizational principles already suggest a tendency in AI to map and thereby separate the results of multiple filter properties of neurons that are generated stepwise along the auditory pathway. For instance, space maps of auditory sources are found in inferior and superior colliculi (Knudsen, du Lac, and Esterly 1987), but spectral aspects of the localized sounds do not seem to be distinguished in detail. Similarly, periodicity of sounds is mapped in the inferior colliculus, presumably serving complex pitch discrimination (Langner 1992), but this processing step is not yet sufficient for the use of periodicities for object constitution or foreground-background discrimination (compare the section, "Foreground-Background Decomposition").

Therefore, it is of considerable interest that map principles are found in AI that can be considered as strategies to disentangle separable stimulus features, that is, to make them more explicit. This is the case in the following examples in which a superimposed map is either independent of the tonotopic organization (the section, "Superimposed Periodicity Maps") or exists in a specific relation to the tonotopic gradient (the section, "Superimposed Vowel Map").

Superimposed Periodicity Maps

Using optical imaging of intrinsic signals as well as electrophysiological mapping of the primary auditory cortex (AI) of the gerbil, we recently described a representation of sound periodicity with a continuous, almost cyclic functional gradient superimposed on the tonotopic organization (figure 17-1; Schulze et al. 2002). As pointed out by Nelson and Bower (1990), the geometry of a topographic stimulus representation has implications for the neuronal

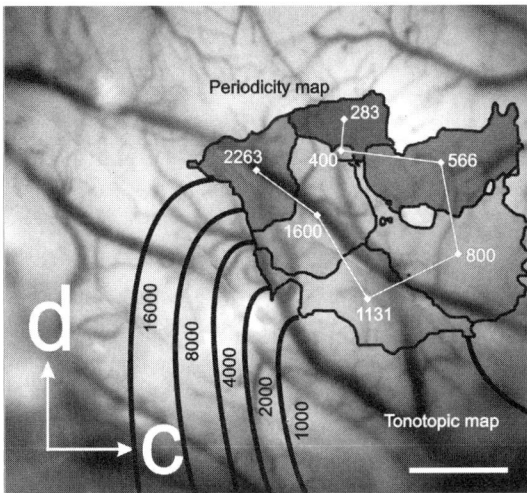

Figure 17-1.   Topography of periodicity map in gerbil AI. See color insert.

computations that can be carried out within a map, in other words, different types of computations require different optimal maps (compare Scheich 1991). According to Nelson and Bower (1990), continuous maps are those in which the computationally relevant parameter is continuously represented along a functional gradient. Computations carried out in such maps are characterized by predominantly local interactions in the problem space, for example, lateral inhibition.

With reference to these postulates we suppose that the cyclic geometry of the functional gradient in this map is the result of computational requirements beyond simple local interactions: the cyclic arrangement of stimulus representations, such as periodicities, allows for equivalent interconnection across the center of the map of neurons that represent a certain value of a stimulus with neurons representing all other values of that parameter. Such a pattern of interconnections facilitates computational algorithms for which global synchronous interactions between arbitrary distances in the parameter space are critical. Thus, a cyclic map could provide a pattern of interconnections that facilitates the extraction of a particular value of the represented parameter from a mixture of other values by a neuronal implementation of a "winner-take-all" algorithm (Haken 1991; Schmutz and Banzhaf 1992; Waugh and Westervelt, 1993) in addition to the local lateral interaction mechanism thought to operate in linear topographic maps. In such linear maps, for example, tonotopic maps, only the contrast between neighboring neurons (and stimulus representations) is enhanced. There are also other examples of cyclic maps: the map of stimulus amplitudes composed of two facing half-cycles of amplitude representation in the auditory cortex of the mustached bat (Suga 1977) and the pinwheel-like arrangement of the iso-orientation domains in the visual cortex (Bonhoeffer and Grinvald 1991) may also provide a substrate for a winner-takes-all computation.

The identification of individual signals from a mixture is a common task for the auditory systems of all vocalizing species (Bodnar and Bass 1999). Gerbils are social animals living as family groups in elaborate burrows, with many such groups in close vicinity in a given area (Ågren, Zhou, and Zhong 1989; Thiessen and Yahr 1977). They have a rich repertoire of vocalizations, of which a particular group of alarm calls (Yapa 1994) is characterized by harmonic spectra with periodicities in the range represented in the cortex map reported here. Gerbils may therefore use a winner-takes-all algorithm in their periodicity map for selecting the vocalizations of individuals within a group. It remains to be seen whether an area in rostral human auditory cortex activated by fore-ground-background decomposition tasks (Scheich et al. 1998) contains a map for extracting the periodicity (fundamental) of a specific speaker's voice in noisy situations such as cocktail parties (von der Malsburg and Schneider 1986).

Interestingly, sound periodicities in the range below about 100 Hz, which are perceptually different form sounds with high periodicities (see below), are not represented within the described cyclic map but are represented via a temporal code (phase locking) in AI. Neurons coding for sound periodicity via this temporal code are not continuously distributed across AI but instead form a scattered, nontopographic map instead (Schulze and Langner 1997). Obviously, there are two different maps for one sound parameter (periodicity) in the auditory cortex, one that is continuous and represents high periodicities that elicit the percept of certain pitches, and one that is scattered and represents low periodicities, which are perceived as rhythm or roughness (compare Schulze and Langner 1997). While from the point of view of the continuous parameter space of periodicities (the bottom-up view), there would be no need to generate separate representations, this parcelation becomes understandable from a top-down point of view, since the two submaps represent stimuli from different perceptual categories and serve different purposes.

Superimposed Vowel Map

A further example of superposition of maps shedding new light on the issue of local versus global views of maps is given by the case of vowel representation in the auditory cortex (Ohl and Scheich 1997b; Ohl and Scheich 1998). This study aimed to solve the problem that on the one hand psychophysical experiments have shown that the discrimination of human vowels chiefly relies on the frequency relationship of the first two peaks of a vowel's spectral envelope, the so-called formants F1 and F2 (Peterson and Barney 1952) but on the other hand all previous attempts to relate this two-dimensional (F1,F2)-relationship to the known one-dimensional topography of the tonotopic gradient had failed.

A first linear hypothesis would have predicted that in a tonotopic map the activation pattern evoked by a two-formant vowel would resemble the superposition of the activation patterns produced by narrowband signals centered around the frequencies F1 and F2, respectively. This view was plausible because single units' activities in the peripheral auditory system (as measured by rate or synchrony codes; Young and Sachs 1979) when plotted over their characteristic frequency give a profile similar to the power spectrum of a vowel. Instead, we found a different unique activation pattern that held no obvious relationship to patterns produced by single formants. We first showed that in a part of gerbil primary auditory cortex field AI, single units showed nonlinear spectral interaction mechanisms as a function of varied F2-F1 spectral distance. These response properties formed a new map, superposed on the tonotopic map, in which a nonlinear transformation of the psychophysically determined (F1,F2)-relationship was represented as a spatial gradient. The transformed relationship was isomorphic to the original (F1,F2)-relationship with respect to identification and classification of vowels (Ohl and Scheich 1997b).

This finding helped to identify an interpretational bias inherent in earlier attempts, namely the a priori assumption that single units in a map contribute to the coding predominately of those parts of the vowel spectrum that corresponded to their characteristic frequencies (*feature extraction principle*). Consequently, this view entails the problem that those individually coded features subsequently have to be put into relation to each other again (*feature binding problem*) by yet-unknown mechanisms. Alternatively, it has been argued (Ohl and Scheich 1998) that the availability of a global activity pattern in the above-described transformed map eliminates the need for feature extraction and rebinding processes during vowel recognition and classification. By emphasizing the relevance of the activity state over the map geometry, even the requirement for isomorphism between activity patterns and perception has been disputed (Pouget and Sejnowski 1994; Freeman 2000).

Top-Down Processing

Same Stimuli, Different Tasks

It was recently shown in the Mongolian gerbil that the right (but not left) auditory cortex is essential for the distinction of rising versus falling direction of frequency modulations (FM). Gerbils trained in a shuttle box to discriminate directionally mirror-imaged frequency sweeps lose the capability after lesion of right but not of left auditory cortex and are no longer capable of acquiring such discriminations (Wetzel et al. 1998b). Gerbils are capable, however, of learning pure tone discriminations after auditory cortex lesions (Ohl et al. 1999).

These results gave rise to the question whether human auditory cortex might be specialized in a similar way. Right human auditory cortex specializations for certain aspects of FM discrimination might be suspected also because there are numerous reports on deficits of right brain lesioned patients with respect to the distinction of prosodic information in speech (e.g., Ackermann, Hertich, and Ziegler 1993; Joanette, Goulet, and Hannequin 1990; Johnsrude, Penhune, and Zatorre 2000). Prosody is the dimension of speech that chiefly conveys information about emotions, intentions, and various other social and biological attributes of individual speakers independent of the linguistic content of speech. Thereby, slow modulation of voice fundamental frequency (speech contour) as well as some variations of spectral motion of vowel formants seem to play an important role.

The question of FM directional discrimination was addressed in human auditory cortex by low-noise functional magnetic resonance imaging (fMRI) using essentially the same selection of FM stimuli as in the gerbil experiments. The first specific task (the control task) was to listen to an unknown stimulus collection and to determine any commonality among the stimuli. Subjects easily identified the concept of the stimuli, describing them as whistles with changing pitch. Responses in auditory cortex were measured in a so-called block design, integrating activation across the whole period of stimulus presentation. This experiment generated either balanced activation in right and left auditory cortices or a left dominance of activation, depending on the individual, and produced a left-lateralized activation across our cohort of subjects (figure 17-2, left panel).

Subsequently, the same stimulus selection was presented again, but subjects were informed that they had to identify the rising FM samples (or the falling FM samples) by key pressing. The result was different from the previous experiment in spite of hearing the same stimuli. Activation of the auditory cortex was lateralized to the right side (figure 17-2, right panel; Brechmann and Scheich 2005). There were also special redistributions of activity in some territories of the right auditory cortex that will not be described here. The shift of activation upon distinction of FM direction could also be demonstrated with a stimulus set in which long and short as well as rising and falling FM sweeps were present (Brechmann and Scheich 2005). Discrimination of the stimulus-duration aspect generated balanced activation of right and left auditory cortexes, while discrimination of the FM directional aspect again generated dominant right auditory cortex activation.

In light of the gerbil experiment showing that the rodent right auditory cortex is indispensable for the discrimination of FM direction, the fMRI results

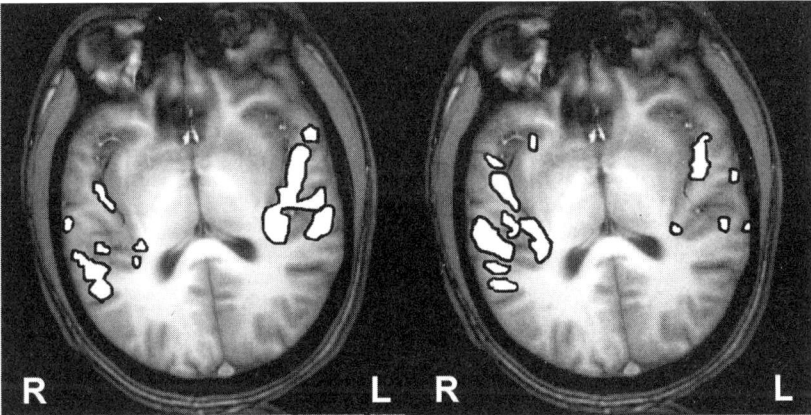

Figure 17-2. Task-dependent fMRI activation in a subject during listening to the same linear FM sweeps varied in frequency range and direction of sweep. *Left panel*, left-lateralized activation: Identification of the common denominator of the unknown stimuli, that is, FM sweeps. *Right panel*, right-lateralized activation: Identification of the sweeps with downward direction.

suggest at least a special role of right auditory cortex in humans in this respect. The most important point to make here is that mechanisms related to FM directional processing are only engaged when this stimulus feature is to be explicitly distinguished, not simply upon stimulus presentation. This demonstrates that top-down influences can dominate the activation in auditory cortex to the extent that even a change of hemispheric lateralization can result.

Different Cues, Same Task
Foreground-Background Decomposition

Periodicity analysis provides one possible cue by which certain classes of auditory objects that coincide in time may be distinguished. In the section, "Superimposed Periodicity Maps," a superposition map was described in gerbil AI, which not only allows the systematic mapping of periodicity but also to select among coinciding objects with different periodicities. While these mechanisms presumably allow discriminating among certain classes of competing auditory objects, the fundamental concept of object constitution in a mixture of sounds is broader and more demanding. In its best-known form it has been addressed as the "cocktail party effect" (Cherry 1953; von der Malsburg and Schneider 1986). While Cherry (1953), who coined the name, was under the assumption that the effect is a sound localization problem, it is clear now that the problem of foreground-background decomposition can also be solved monaurally (Yost and Sheft 1993). Several other mechanisms of spectral and temporal discrimination besides periodicity analysis may be used, presumably in an opportunistic fashion, a matter of ongoing research.

Some lines of experiments with low-noise fMRI in the human auditory cortex have led to the concept that a rostral area on the dorsal surface of the temporal lobe, anterior to Heschl's gyrus, may be an area of central importance in foreground-background decomposition. Several mechanisms relevant in this context may converge in this area previously named territory TA (Scheich et al. 1998).

One of our experiments addressed the question of how, in a series of complex tones (notes of different musical instruments), matching pairs (instrument and note identical) are identified in the presence of a continuous background. This background consisted of a loud sawtooth frequency modulation of a tone which masked fundamental and lower harmonic frequencies of the instrument notes (Scheich et al. 1998). This constellation is a simplified version of the task that a conductor has while monitoring different instruments in an orchestra. Note that instrument notes but not the background in this case contained pitch-related periodicity information.

In the first experiment, the total effect of instrument notes, background, and discrimination task was determined in the auditory cortex as referred to interval periods without any stimulation. This revealed strong bilateral activation in all previously known primary and nonprimary subdivisions of auditory cortex (territories TA, T1, T2, T3).

The second experiment served to isolate the effect of the foreground-dependent task. The described situation was the same, but the background alone continued though the reference periods. Thereby the FM background had been calibrated in such a way that it maximally activated primary-like areas in the auditory cortex. Consequently, because foreground plus background had been referred to the background alone, it was expected that much of the primary-like activation in auditory cortex would cancel out. This was indeed the case. Of the original activation determined in the first experiment, merely the activity in the rostral area TA was maintained in the second experiment and was not significantly different. This suggests that in contrast to the other areas, the background alone had very little direct influence on TA and did not appreciably influence the effect of the foreground task in this area.

There the properties of TA do not depend on particular stimuli. This became apparent during the course of a study on level-dependent activation of human auditory cortex with a large set of different rising and falling FM sweeps (Brechmann, Baumgart, and Scheich 2002). TA, in contrast to other auditory cortex areas, showed little activation and no level-dependent change of activation for intermediate and high levels of stimuli. Only when the FM level (35 dB SPL) fell below the level of the background noise of the MRF scanner (48 dB SPL) TA activation was strongly increased. Thus, activation of this area increased when the targets had to be retrieved from a louder yet distinguishable scanner noise. This is a result similar to the previous experiments and provides the additional information that mechanisms in TA are challenged especially if the background strongly interferes with the foreground targets.

A Motion-Selective Map in the Auditory Cortex

If principles of a map organization strongly relate to the constitution of behavioral meaning of stimuli (top-down demands) in addition to the analysis of a particular stimulus feature, one should expect that neurons in this map derive information of comparable behavioral relevance from different stimulus features. This seems to be the case in a recently discovered motion-selective area in human auditory cortex (Baumgart et al. 1999). This small area on the lateral planum temporale, dominantly or sometimes exclusively activated on the right side, was identified in further studies with fMRI using motion percepts generated by time-variant interaural cues through headphones. One motion cue is a changing interaural level relationship of stimuli (interaural intensity difference, IID). In this case, a slow amplitude modulation of a carrier sound was generated, which in the case of the control stimulus had the same phase of the modulation cycle at the two ears. This sound was not perceived as having any specific location in space. Conversely, if amplitude-phase cycles of the identical stimuli were slowly shifted interaurally the percept was that of a sound source slowly moving back and forth in azimuth in front of the listener.

With fMRI these two stimuli generated very strong and spatially similar activations in primary and nonprimary areas of the auditory cortex. The subtraction of the two activation patterns, however, yielded a reliable signal intensity increase with the motion stimulus laterally on the right planum temporale (figure 17-3, left panel). It is interesting that a very similar activation increase was obtained with a different interaural cue and a different type of carrier stimulus (figure 17-3, right panel). These were short tone bursts of constant frequency presented with simultaneous onset at the two ears (control) or with successively changing onset between the two ears (motion). The latter contains interaural time differences (ITD) as a motion cue. While the control condition did not generate any location percept of the sound source, the ITD cue led to the percept of small azimuthal jumps of source location.

The similarity of the spatial location of signal increase for IID and ITD cues in the right planum temporale is relevant for assumed hemispheric specializations for extracorporal space analysis by the right hemisphere. Obviously not only visual space cue analysis and multimodal space cue processing is lateralized to the right hemisphere (Bisiach and Berti 1997; Bisiach and Vallar 1988) but also auditory space cue analysis. Whether this relates in any way to a proposed "dorsal stream" of auditory space analysis in the auditory cortex (Romanski et al. 1999) remains to be determined.

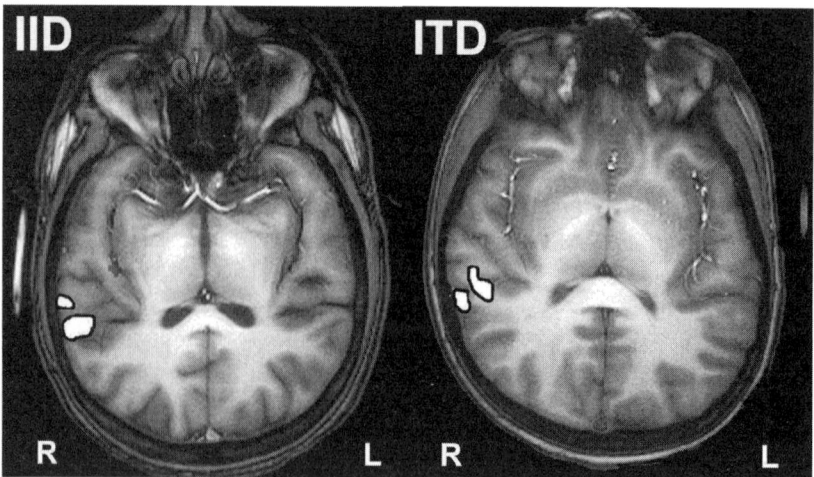

Figure 17-3. Parameter-independence of movement-selective fMRI activation in human auditory cortex. See color insert.

The immediate relevance to the present subject is that the motion-selective area in the right auditory cortex is neither stimulus specific nor motion-cue specific and thus does not fulfill the criterion of an area specialized for specific acoustic features in a bottom-up concept. Rather, the generation of explicit motion percepts seems to be the common denominator. This is further underlined by fMRI experiments using a third motion cue, namely head-related transfer functions using a moving sound source (obtained by a twin microphone array in an artificial human head), which generates a vivid space percept of a movement all around the head. In this case the motion area on the planum temporale is even more strongly activated but on both sides with a dominant activation on the right side (Warren et al. 2002).

Categorization

The concept of multiple overlaying maps in the same neuronal substrate raises the question of ways in which activity states coexisting in this substrate can be organized (1) to avoid confusion by interference effects and (2) to allow the relation of a given activity state to any particular map. We propose that the *sequential temporal order of recruitment* of neural activity states is one mechanism for the disentanglement of maps. While the types of overlaying maps discussed in the section, "Global Views, Map Superposition, and Top-Down Demands," are typically obtained in a particular and widely used experimental approach to uncover regularities in map organization ("mapping experiment"), we will argue in this section that in more natural circumstances activity states emerge to serve particular needs of the behaving organism or subject.

Numerous studies in human auditory cortex have substantiated the view of succession of bottom-up and top-down processing (for review see Kraus and McGee 1991). A particularly interesting field concerned with the temporal interplay of bottom-up and top-down aspects of information processing is the analysis of *selective attention*. Here, a subject focuses attention on a selected subset of the environmental (or proprioceptive) inputs at the expense of less relevant inputs (for review see Posner 1997). Clearly, the attribution of relevance during selection and maintenance of a focus constitutes a top-down modulation of bottom-up information processing. There has been some debate about whether this modulation occurs only at "late" stages (several hundreds of milliseconds) of information processing or already affects "early" stimulus processing. Advances in neuroimaging studies over the last decade have revealed that selective attention can affect both early and late components of information processing (Hillyard et al. 1995). While there is evidence that selective attention involves cortical (rather than more peripheral) mechanisms, these become effective already twenty milliseconds after stimulus onset, that is, when the bottom-up–evoked neural activity arrives in the cortex.

Although this demonstrates that top-down modulations *become effective* at the earliest possible time of cortical stimulus processing, it should not be overlooked that they are the result of preceding experiences, that is, they involve events that happened in the past before the onset of selective attention and reoccurred or continued. In other words, in addition to the time scale in which selective attention becomes effective, we have to draw into consideration the temporal sequential organization of neuronal activity processing the events that gave rise to a particular attentional focus.

A clear example of such temporal separation between bottom-up and top-down aspects of neuronal representation is given by recent findings on the neuronal correlates of auditory categorization learning. In single-trial analyses of high-resolution electrocorticograms, "early" (20–50 ms) neural activity states representing stimulus-related, mainly bottom-up–relayed information could be discerned from "late" (several hundred milliseconds to a few seconds) states representing category-related information (Ohl et al. 2003a, 2003b). In this experiment, Mongolian gerbils, a rodent species especially well suited for complex learning tasks (Ohl and Scheich 1996, 1997a, 2004; Schulze and Scheich 1999) were trained to sort novel, previously unknown, frequency-modulated (FM) sounds into the categories "rising" or "falling," respectively, depending on the direction of the frequency modulation (low pitch to high pitch or vice versa) of the sound (Wetzel et al. 1998a). Discrimination of modulation direction has been demonstrated to be cortex-dependent (Ohl et al. 1999).

First, in behavioral analyses (Wetzel et al. 1998a) it was shown that gerbils can learn this categorization upon training a sequence of pairwise discriminations with a number of sound prototypes (rising vs. falling). In each training block of the sequence, the discrimination of a particular pair of a rising and a falling FM tone was trained until criterion. This was followed by a subsequent training block with another stimulus pair. During this so-called discrimination phase animals showed (1) a gradual increase in discrimination performance for each new prototype pair of FM tones and (2) a gradually declining generalization gradient for gradually varying FM tones taken from the stimulus continuum between the learned prototypes. After training continued for a few blocks, a sudden transition in behavior occurred after which novel stimuli were immediately identified as belonging to either the rising or falling category. This so-called categorization phase is characterized by (1) immediate stimulus identification and (2) categorical perception (sigmoid psychometric functions) instead. The transition from the discrimination phase to the categorization phase occurred abruptly at a point in time during the training history that was specific to the individual animal (figure 17-4).

This behavioral state transition was paralleled by a transition in the organization of a particular type of transient spatiotemporal activity pattern in AI that we called *marked states*. These states were identified by tracking over time a measure of dissimilarity between the spatial activity pattern in a given

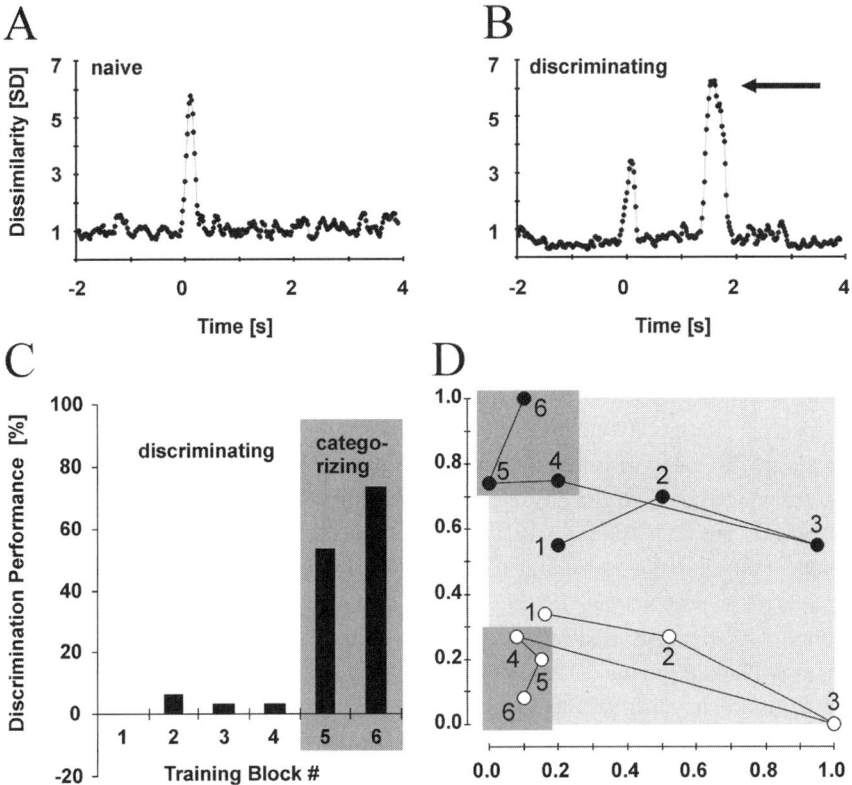

Figure 17-4. Cortical dynamics during categorization learning. (A, B) Dissimilarity functions in a typical trial of a naive and a discriminating animal, respectively. "Marked states" are distinguished by transient maxima in these functions. (C) Discrimination performance attained in the first training sessions of six consecutive training blocks in one animal. In the first sessions of a training block the FM tone pair to be discriminated were novel to the animal. Note that in training blocks 1–4 discrimination performance was negligible in the first session (and built up gradually in later sessions, not shown) (discrimination phase), whereas in blocks 5 and 6, performance was already high in the first sessions (categorization phase). (D) Similarity relations between spatial activity patterns associated with marked states in the same animal. For each training block (numbers) the pattern for the category "rising" (filled circles) and the pattern for the category "falling" (open circles) is depicted. The diagram is so arranged that distances between points are proportional to the dissimilarity of the corresponding activity patterns. Note that when categorization begins between training blocks 4 and 5 (compare C), activity patterns during marked states start to remain similar to the patterns found during the previous training block leading to a clustering in the diagram (darker rectangles). Before the clustering, interpoint distances within and between categories were of similar magnitude.

trial to be analyzed and the mean pattern obtained by averaging over all trials involving stimuli of the respective other category (figure 17-4A, B). This measure peaked after stimulus arrival in the cortex independent of the training state (that is, in both naive and trained animals) reflecting the perturbation of ongoing activity by the incoming stimulus (Freeman 1994; Arieli et al. 1996). Specifically, the spatial activity pattern found in the early marked states reflected the physical stimulus characteristics represented in the tonotopically organized map of primary auditory cortex (Ohl, Scheich, and Freeman 2000; Ohl et al. 2000). With training, additional peaks emerged from the ongoing activity tagging additional marked states occurring at variable later latencies up to four seconds after stimulus presentation (figure 17-4B). The spatial activity patterns associated with these later peaks showed an interesting change at the transition from discrimination learning to categorization: with categorization, the patterns reflected the belonging to the formed category rather than the physical characteristics of the stimuli. The similarity relations between spatial activity patterns during the marked states can be visualized in two-dimensional plots, in which the dissimilarity between two patterns is proportional to the distance between the corresponding points (figure 17-4D). During discrimination learning (training blocks 1–4, compare figure 17-4C), dissimilarities within and between categories were similar in magnitude. In contrast, the dissimilarities after the transition to categorization (blocks 4–6, compare figure 17-4C) were much smaller within a category than between categories.

It is presently unknown how the later activity states forming the superimposed map of category-belonging might form. It is currently being investigated whether they dynamically spread with approximately circular isophase contours over cortical areas starting from point-like centers of nucleation, which would be indicative of a self-organized process in cortex (Freeman and Barrie 2000).

Conclusions

This chapter summarizes experimental evidence from animal and human auditory cortices in favor of the hypothesis that sensory cortex is not simply the head stage of "passive" stimulus analysis, but is also the locus of "active" processes, which make stimulus features cognitively explicit and available for tasks of variable demand. Several organizational principles are held responsible for this bottom-up/top-down interfacing within the sensory cortex. For instance, there is evidence from this and numerous other studies that cognitive processing of incoming stimuli in cortex follows the initial descriptive processing in maps and, as judged from analysis of "cognitive potentials," proceeds in steps.

The new hypothesis suggested here is that top-down cognitive processes create new states within maps that can still be described as spatiotemporal activation patterns but may use coordinates of cognitive similarity rather than similarity at the initial stimulus-descriptive level. Furthermore, the organizational principles of multiple separate maps in the auditory cortex and of superposition maps for several stimulus features within the same area are considered in the light of bottom-up/top-down interfacing. Both types of maps may be regarded not simply as a way to orderly analyze distinguishable stimulus features in parallel but as principles to independently address and make explicit such features for various cognitive demands.

Local views of activity states in maps provide not more than peak activities of local neural ensembles for respective features (traditional feature analysis). We propose instead that global views of maps in terms of activation patterns more fully represent a given stimulus, that is, a neuronal space with both active and nonactive subregions within the frame of a map may be relevant to characterize a stimulus. Similarly, more complex activation patterns generated by map superposition may be resolved better in a global view than in local views. Consequently, we propose that later cognitive states of given maps also provide such global views of maps yet with a cognitive frame of reference.

References

Ackermann, H., Hertich, I., and Ziegler, W. (1993). Prosodische Störungen bei neurologischen Erkrankungen—eine Literaturübersicht. *Fortschritte der Neurologie und Psychiatrie*, 61, 241–253.

Ågren, G., Zhou, Q., and Zhong, W. (1989). Ecology and social behaviour of Mongolian gerbils, Meriones unguiculatus, at Xilinhot, Inner Mongolia, China. *Animal Behavior*, 37, 11–27.

Arieli, A., Sterkin, A., Grinvald, A., and Aertsen, A. (1996). Dynamics of ongoing activity: explanation of the large variability in evoked cortical responses. *Science*, 273, 1868–1871.

Baumgart, F., Gaschler-Markefski, B., Woldorff, M.G., Heinze, H.J., and Scheich, H. (1999). A movement-sensitive area in auditory cortex. *Nature*, 400, 724–726.

Bisiach, E. and Berti, A. (1997). Consciousness and dyschiria. In M.S. Gazzaniga (Ed.), *The cognitive neurosciences* (pp. 1331–1340). Cambridge, MA: MIT Press.

Bisiach, E. and Vallar, G. (1988). Hemineglect in humans. In F. Boller and J. Grafman (Eds.), *Handbook of neurophysiology*, Vol. 1 (pp. 195–222). Amsterdam: Elsevier.

Bodnar, D.A. and Bass, A.H. (1999). Midbrain combinatorial code for temporal and spectral information in concurrent acoustic signals. *Journal of Neurophysiology*, 81, 552–563.

Bonhoeffer, T. and Grinvald, A. (1991). Iso-orientation domains in cat visual cortex are arranged in pinwheel-like patterns. *Nature*, 353, 429–431.

Brechmann, A., Baumgart, F., and Scheich, H. (2002). Sound-level-dependent representation of frequency modulation in human auditory cortex: a low-noise fMRI study. *Journal of Neurophysiology*, 87, 423–433.

Brechman, A. and Scheich, H. (2005). Hemispheric shifts of sound representation in auditory cortex with conceptual listening. *Cerebral Cortex*, 15, 578–587.

Cherry, E.C. (1953). Some experiments on the recognition of speech, with one and with two ears. *Journal of the Acoustic Society of America*, 25, 975–979.

Clarey, J.C., Barone, P., and Imig, T.J. (1991). Physiology of thalamus and cortex. In A.N. Popper and R.R. Fay (Eds.), *The mammalian auditory pathway: Neurophysiology* (pp. 232–334). New York: Springer.

Cohen, Y.E. and Knudsen, E.I. (1999). Maps versus clusters: different representations of auditory space in the midbrain and forebrain. *Trends in the Neurosciences*, 22, 128–135.

Corbetta, M., Miezin, F.M., Dobmeyer, S., Shulman, G.L., and Petersen, S.E. (1991). Selective and divided attention during visual discriminations of shape, color, and speed: functional anatomy by positron emission tomography. *Journal of Neuroscience*, 11, 2383–2402.

DeMott, D.W. (1970). *Toposcopic studies of learning*. Springfield, IL: Thomas Books.

Eggermont, J.J. (1998). Representation of spectral and temporal sound features in three cortical fields of the cat. Similarities outweigh differnces. *Journal of Neurophysiology*, 80, 2743–2764.

Farah, M.J. (1997). The neural basis of mental imagery. In M.S. Gazzaniga (Ed.), *The cognitive neurosciences* (pp. 963–975). Cambridge, MA: MIT Press.

Freeman, W.J. (1994). Neural mechanisms underlying destabilization of cortex by sensory input. *Physica D*, 75, 151–164.

Freeman, W.J. (2000). *Neurodynamics. An exploration of mesoscopic brain dynamics.* Springer, London.

Freeman, W.J. and Barrie, J.M. (2000). Analysis of spatial patterns of phase in neocortical gamma EEGs in rabbit. *Journal of Neurophysiology*, 84, 1266–1278.

Goldman-Rakic, P.W., Selemon, L.D., and Schwartz, M.L. (1984). Dual pathways connecting the dorsolateral prefrontal cortex with the hippocampal formation and parahippocampal cortex in the rhesus monkey. *Neuroscience*, 12, 719–743.

Haken, H. (1991). *Synergetic computers and cognition*. Berlin: Springer.

Hardcastle, V.G. (1994). Psychology's binding problem and possilbe neurobiological solutions. *Journal of Consciousness Studies*, 1, 66–90.

Heil, P., Rajan, R., and Irvine, D.R.F. (1992). Sensitivity of neurons in cat primary auditory cortex to tones and frequency-modulated stimuli. II: Organization of response properties along the "isofrequency" dimension. *Hearing Research*, 63, 135–156.

Heil, P., Rajan, R., and Irvine, D.R.F. (1994). Topographic representation of tone intensity along the isofrequency axis of cat primary auditory cortex. *Hearing Research*, 76, 188–202.

Hess, A. and Scheich, H. (1996). Optical and FDG-mapping of frequency-specific activity in auditory cortex. *NeuroReport*, 7, 2643–2647.

Hillyard, S.A., Mangun, G.R., Woldorff, M.G., and Luck, S.J. (1995). Neural systems mediating selective attention. In M.S. Gazzaniga (Ed.), *The cognitive neurosciences* (pp. 665–681). Cambridge, MA: MIT Press.

van Hoesen, G.W. and Pandya, D.N. (1975). Some connections of the hippocampus. In J.H. Byrne and W.O. Berry (Eds.), *Neural models of plasticity* (pp. 208–239). New York: Academic Press.

Horikawa, J., Hoskowa, Y., Kubota, M., Nasu, M., and Taniguchi, I. (1996). Optical imaging of spatiotemporal patterns of glutamatergic exictation and

GABAergic inhibition in the guinea-pig auditory cortex in vivo. *Journal of Physiology*, 497, 620–638.

Horikawa, J., Nasu, M., and Taniguchi, I. (1998). Optical recording of responses to frequency-modulated sounds in the auditory cortex. *NeuroReport*, 9, 799–802.

Hosokawa, Y., Horikawa, J., Nasu, M., and Taniguchi, I. (1999). Spatiotemporal representation of binaural difference in time and intensity of sound in the guinea pig auditory cortex. *Hearing Research*, 134, 123–132.

Insausti, R., Amaral, D.G., and Cowan, W.M. (1987). The entorhinal cortex of the monkey: II. Cortical afferents. *Journal of Comparative Neurology*, 264, 356–395.

Joanette, Y., Goulet, P., and Hannequin, D. (1990). In Y. Joanette, P. Gouldet, and D. Hannequin (Eds.) *Right hemisphere and verbal communication* (pp. 132–159). New York: Springer.

Johnsrude, I.S., Penhune, V.B., and Zatorre, R.J. (2000). Functional specificity in the right human auditory cortex for perceiving pitch direction. *Brain*, 123, 155–163.

Jones, E.G. and Powell, T.P.S. (1970). An anatomical study of converging sensory pathways within the cerebral cortex of the monkey. *Brain*, 93, 793–820.

Kaas, J.H. (1982). A segregation of function in the nervous system: Why do sensory systems have so many subdivisions? In W.P. Neff (Ed.), *Contributions to sensory physiology* (pp. 201–240). New York: Academic Press.

Kaas, J.H. (1993). Evolution of multiple areas and modules within neocortex. *Perspectives in Developmental Neurobiology*, 1, 101–107.

Knudsen, E.I., du Lac, S., and Esterly, S.D. (1987). Computational maps in the brain. *Annual Reviews in Neuroscience*, 10, 41–65.

Kraus, N. and McGee, T. (1991). Electrophysiology of the human auditory system. In A.N. Popper and R.R. Fay (Eds.), *The mammalian auditory pathway: Neurophysiology* (pp. 335–404). New York: Springer.

Langner, G. (1992). Periodicity coding in the auditory system. *Hearing Research*, 60, 115–142.

Lilly, J.C. (1954). Instananeous relations between the activities of closely spaced zones on the cerebral cortex—electrical figures during responses and spontaneous activity. *American Journal of Physiology*, 176, 493–504.

von der Malsburg, C. and Schneider, W. (1986). A neural cocktail-party processor. *Biological Cybernetics*, 54, 29–40.

Merzenich, M.M. and Brugge, J.F. (1973). Representation of the cochlear partition on the superior temporal plane of the Macaque monkey. *Brain Research*, 50, 275–296.

Merzenich, M.M., Knight, P.L., and Roth, G.L. (1975). Representation of the cochlea within primary auditory cortex in the cat. *Journal of Neurophysiology*, 38, 231–249.

Middlebrooks, J.C., Dykes, R.W., and Merzenich, M.M. (1980). Binaural response-specific bands in primary auditory cortex (AI) of the cat: Topographical organization orthogonal to isofrequency contours. *Brain Research*, 181, 31–48.

Nelson, M.E. and Bower, J.M. (1990). Brain maps and parallel computers. *Trends in Neuroscience*, 13, 403–408.

Ohl, F.W., Deliano, M., Scheich, H., and Freeman, W.J. (2003a). Early and late patterns of stimulus-related activity in auditory cortex of trained animals. *Biological Cybernetics*, 88, 374–379.

Ohl, F.W., Deliano, M., Scheich, H., and Freeman, W.J. (2003b). Anaysis of evoked and emergent patterns of stimulus-related auditory cortical activity. *Reviews in the Neurosciences*, 14, 35–42.

Ohl, F.W. and Scheich, H. (1996). Differential frequency conditioning enhances spectral contrast sensitivity of units in auditory cortex (field AI) of the alert Mongolian gerbil. *European Journal of Neuroscience*, 8, 1001–1017.

Ohl, F.W. and Scheich, H. (1997a). Learning-induced dynamic receptive field changes in primary auditory cortex of the unanaesthetized Mongolian gerbil. *Journal of Comparative Physiology A*, 181, 685–696.

Ohl, F.W. and Scheich, H. (1997b). Orderly cortical representation of vowels based on formant interaction. *Proceedings of the National Academy of Sciences USA*, 94, 9440–9444.

Ohl, F.W. and Scheich, H. (1998). Feature extraction and feature interaction. *Behavioral and Brain Sciences*, 21, 278.

Ohl, F.W. and Scheich, H. (2004). Fallacies in behavioural interpretation of auditory cortex plasticity. *Nature Reviews Neuroscience*, published online, doi: 10.1038/nrn 1366-c1.

Ohl, F.W., Scheich, H., and Freeman, W.J. (2000). Topographic analysis of epidural pure-tone-evoked potentials in geril auditory cortex. *Journal of Neurophysiology*, 83, 3123–3132.

Ohl, F.W., Schulze, H., Scheich, H., and Freeman, W.J. (2000). Spatial representation of frequency-modulated tones in gerbil auditory cortex revealed by epidural electrocorticography. *Journal of Physiology (Paris)*, 94, 549–554.

Ohl, F.W., Wetzel, W., Wagner, T., Rech, A., and Scheich, H. (1999). Bilateral ablation of auditory cortex in Mongolian gerbil affects discrimination of frequency modulated tones but not of pure tones. *Learning and Memory*, 6, 347–362.

Peterson, G.E. and Barney, H.L. (1952). Control methods used in a study of the vowels. *Journal of the Acoustic Society of America*, 24, 175–184.

Phillips, D.P. and Irvine, D.R. (1983). Some features of binaural input to single neurons in physiologically defined area AI of cat cerebral cortex. *Journal of Neurophysiology*, 49, 383–395.

Phillips, W.A. and Singer, W. (1997). In search of common foundations for cortical computation. *Behavioral and Brain Sciences*, 20, 657–722.

Posner, M.I. (1997). Attention in cognitive neuroscience: An overview. In M.S. Gazzaniga (Ed.), *The cognitive neurosciences* (pp. 615–624). Cambridge, MA: MIT Press.

Pouget, A. and Sejnowski, T.J. (1994). Is perception isomorphic with neural activity? *Behavioral and Brain Sciences*, 17, 274.

Rauschecker, J.P., Tian, B., and Hauser, M. (1995). Processing of complex sounds in the Macaque nonprimary auditory cortex. *Science*, 268, 111–114.

Romanski, L.M., Tian, B., Fritz, J., Mishkin, M., Goldman-Rakic, P.S., and Rauschecker, J.P. (1999). Dual streams of auditory afferents target multiple domains in the primate prefrontal cortex. *Nature Neuroscience*, 2, 1131–1136.

Scheich, H. (1991). Auditory cortex: comparative aspects of maps and plasticity. *Current Opinion in Neurobiology*, 1, 236–247.

Scheich, H., Baumgart, F., Gaschler-Markefski, B., Tegeler, C., Tempelmann, C., Heinze, H.J., Schindler, F., and Stiller, D. (1998). Functional magnetic resonance imaging of a human auditory cortex area involved in foreground-background decomposition. *European Journal of Neuroscience*, 10, 803–809.

Scheich, H., Heil, P., and Langner, G. (1993). Functional organization of auditory cortex in the Monglian gerbil (Meriones unguiculatus): II. Tonotopic 2-deoxyglucose. *European Journal of Neuroscience*, 5, 898–914.

Schmutz, M. and Banzhaf, W. (1992). Rubust competitive networks. *Physical Reviews A*, 45, 4132–4145.

Schreiner, C.E. (1991). Functional topographies in the primary auditory cortex of the cat. *Acta Otolaryngology Supplement*, 491, 7–15.

Schreiner, C.E. (1995). Order and disorder in auditory cortical maps. *Current Opinion in Neurobiology*, 5, 489–496.

Schreiner, C.E. (1998). Spatial distribution of responses to simple and complex sounds in the primary auditory cortex. *Audiology and Neuro-Otology*, 3, 104–122.

Schreiner, C.E. and Sutter, M.L. (1992). Topography of excitatory bandwidth in cat primary auditory cortex: Single-neuron versus multiple-neuron recordings. *Journal of Neurophysiology*, 68, 1487–1502.

Schreiner, C.E. and Urbas, J.V. (1988). Representation of amplitude modulation in the auditory cortex of the cat. II. Comparison between cortical fields. *Hearing Research*, 32, 49–64.

Schulze, H., Hess, A., Ohl, F.W., and Scheich, H. (2002). Superposition of horseshoe-like periodicity and linear tonotopic maps in auditory cortex of the Mongolian gerbil. *European Journal of Neuroscience*, 15, 1077–1084.

Schulze, H. and Langner, G. (1997). Periodicity coding in the primary auditory cortex of the Mongolian gerbil (*Meriones unguiculatus*): Two different coding strategies for pitch and rhythm? *Journal Comparative Physiology A*, 181, 651–663.

Schulze, H., Ohl, F.W., Heil, P., and Scheich, H. (1997). Field-specific responses in the auditory cortex of the unanaesthetized Mongolian gerbil to tones and slow frequency modulations. *Journal of Comparative Physiology A*, 181, 573–589.

Schulze, H. and Scheich, H. (1999). Discrimination learning of amplitude modulated tones in Mongolian gerbils. *Neuroscience Letters*, 261, 13–16.

Squire, L.R., Mishkin, M., and Shimamara, A. (1990). *Learning and memory: Discussions in neuroscience*. Geneva: Elsevier.

Squire, L.R., Shimamura, A.P., and Amaral, D.G. (1989). Memory and the hippocampus. In J.H. Byrne and W.O. Berry (Eds.), *Neural models of plasticity* (pp. 208–239). New York: Academic Press.

Suga, N. (1977). Amplitude spectrum representation in the Doppler-shifted-CF processing area of the auditory cortex of the mustached bat. *Science*, 196, 64–67.

Suga, N. (1994). Multi-function theory for cortical processing of auditory information: implications of single-unit and lesion data for future research. *Journal of Comparative Physiology A*, 175, 135–144.

Taniguchi, I. and Nasu, M. (1993). Spatio-temporal representation of sound intensity in the guinea pig auditory cortex observed by optical recording. *Neuroscience Letters*, 151, 178–181.

Thiessen, D. and Yahr, P. (1977). *The gerbil in behavioral investigations; Mechanisms of territoriality and olfactory communication.* Austin, University of Texas Press.

Thomas, H., Tillein, J., Heil, P., and Scheich, H. (1993). Functional organization of auditory cortex in the Mongolian gerbil (*Meriones unguiculatus*): I. Electrophysiological mapping of frequency representation and distinction of fields. *European Journal of Neursocience*, 5, 822–897.

Uno, H., Murai, N., and Fukunishi, K. (1993). The tonotopic representation in the auditory cortex of the guinea pig with optical recording. *Neuroscience Letters*, 150, 179–182.

Warren, J.D., Zielinski, B.A., Green, G.G., Rauschecker, J.P., and Griffiths, T.D. (2002). Perception of sound-source motion by the human brain. *Neuron*, 34, 139–148.

Waugh, F.R. and Westervelt, R.M. (1993). Analog neural networks with local competition. I. Dynamics and stability. *Physical Reviews E*, 47, 4524–4536.

Weinberg, R.J. (1997). Are topographic maps fundamental to sensory processing? *Brain Research Bulletin*, 44, 113–116.

Wetzel, W., Wagner, T., Ohl, F.W., and Scheich, H. (1998a). Categorical discrimination of direction in frequency-modulated tones by Mongolian gerbils. *Behavioural Brain Research*, 91, 29–39.

Wetzel, W., Ohl, F.W., Wagner, T., and Scheich, H. (1998b). Right auditory cortex lesion in Mongolian gerbils impairs discrimination of rising and falling frequency-modulated tones. *Neuroscience Letters*, 252, 115–118.

Yapa, W.B. (1994). Social behaviour of the Mongolian gerbil *Meriones unguiculatus*, with special reference to acoustic communication. Dissertation, University of Munich.

Yost, W.A. and Sheft, S. (1993). Auditory perception. In W.A. Yost, A.N. Popper, and R.R. Fay (Eds.), *Human psychophysics* (pp. 193–236). New York: Springer.

Young, E.D. and Sachs, M.B. (1979). Representation of steady-state vowels in the temporal aspects of the discharge patterns of populations of auditory-nerve fibers. *Journal of the Acoustic Society of America*, 66, 1381–1403.

18

Do Perception and Action Result from Different Brain Circuits? The Three Visual Systems Hypothesis

Giacomo Rizzolatti and Vittorio Gallese

Introduction

Traditionally, perception and action have been considered separate domains. Individuals perceive, then act. The two processes are independent and do not interfere with one another. To this psychological separation corresponds, according to this view, separate anatomical and functional brain sectors. Some sectors mediate perception, others mediate action.

The sensory modality that has been most studied in primates is vision. An unexpected result of these studies was the discovery that in primates the cytoarchitectonic area 19 and the temporal and parietal areas surrounding it are composed of a multiplicity of functionally distinct areas, each containing a more or less complete representation of the contralateral field of vision. Of these "extrastriate" visual areas, some are mostly connected with the temporal lobe, others with the parietal lobe.

A very important and influential attempt to give an explanation of this complex organization was made by Ungerleider and Mishkin (1982) on the basis of their lesion studies in monkeys. According to these authors, visual cortical areas are organized in two separate streams of visual information. A dorsal stream, which includes visual areas V3A, V6, V5/MT, MST, and FST and culminates in the inferior parietal lobule, and a ventral stream, which includes visual areas V3 and V4 and culminates in the inferior temporal cortex. The dorsal stream is responsible for perception of space, while the ventral stream for perception of object.

A different view was advanced by Milner and Goodale (Goodale and Milner 1992; Milner and Goodale 1995). In accord with Ungerleider and Mishkin (1982) they maintain that there is a fundamental functional difference between the dorsal and ventral streams. They deny, however, that the difference is in the resulting percept (space vs. object). According to Milner and Goodale

(1995), the difference is in the output characteristics of the two cortical visual streams. The ventral stream is fundamental for perception. The dorsal stream, in contrast, processes visual stimuli to provide high-order visual information for control of action. It is, therefore, not involved in perception.

A similar view was proposed independently by Jeannerod (1994, 1997). According to him, the ventral stream is responsible for the "semantic mode" of object representation, while the dorsal stream is responsible for the "pragmatic mode" of stimulus processing. The semantic mode of object representation refers to object analysis described in object-centered coordinates. The pragmatic mode indicates the type of processing that stimuli have to undergo for action organization. Although the distinction between a semantic and a pragmatic system proposed by Jeannerod (1994) appears to be more cautious than that of Milner and Goodale (1995), the essence of the two proposals is very similar.

The aim of this chapter is to show that the separation of the cortical visual processing into two streams is insufficient and, in the version where perception and action are kept separated, leads to a misunderstanding of the true nature of perceptual processes. We will show that the processing carried out in the inferior parietal lobule is different from that performed in the inferior parietal lobe and that the so called dorsal stream is in fact formed by two streams: the dorsodorsal stream (D-D) and the ventrodorsal stream (V-D). The D-D stream has basically the characteristics suggested by Milner and Goodale (1995) and Jeannerod (1994) for their dorsal stream, while the V-D stream is responsible not only for the organization of actions requiring space computation but also for space and action perception.

At the end of the chapter we will briefly discuss the relation between action and perception as it emerges from neurophysiological data on the V-D stream. We will propose that both action perception and space perception derive from a preceding motor knowledge based on self-generated actions.

The Organization of the Parietal Lobe: The Dorsodorsal (D-D) and Ventrodorsal (V-D) Streams

Figure 18-1 shows lateral and mesial views of the macaque monkey brain. The intraparietal sulcus is unfolded. The frontal and parietal functional and cytoarchitectonic areas most relevant for the discussion in this chapter, included those buried inside the intraparietal sulcus, are indicated.

A fundamental landmark of the parietal lobe of primates is the intraparietal sulcus. This evolutionarily ancient sulcus (it is present already in prosimians) subdivides the posterior parietal lobe into two main sectors: the superior parietal lobule (SPL) and the inferior parietal lobule (IPL). These two sectors receive different cortical inputs from sensory cortices and have different connections with the motor cortex and frontal lobe.

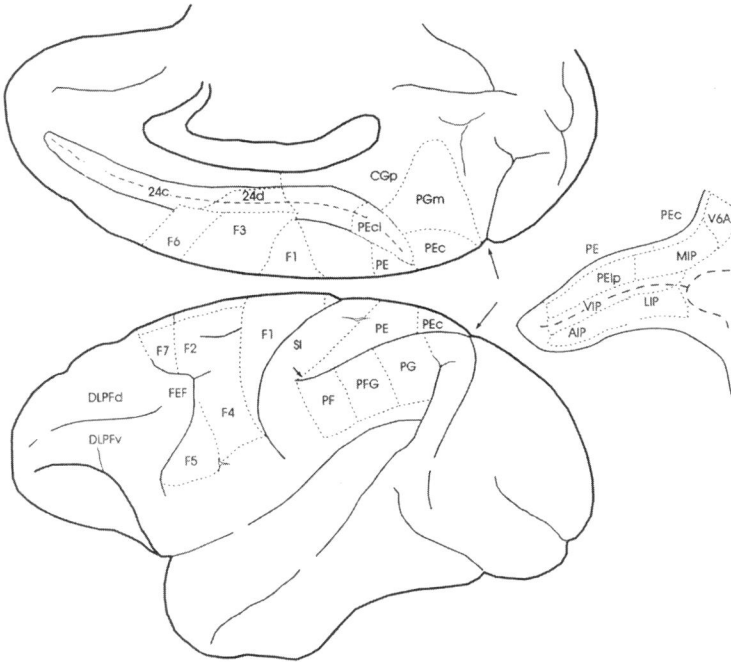

Figure 18-1. Lateral and mesial views of the macaque monkey brain showing the cytoarchitectonic parcelation of the motor cortex and of the posterior parietal cortex. Motor areas are defined according to Matelli, Luppino, and Rizzolatti (1991). The parietal areas, except those of the intraparietal sulcus, are defined according to Pandya and Seltzer (1982). The areas located within the intraparietal sulcus are defined according to physiological data (see text) and are shown in an unfolded view of the sulcus in the right part of the figure. Modified from Rizzolatti et al. (1998).

Early experiments carried out on monkeys showed that SPL is related to the somatosensory system. It receives its main cortical input from the primary somatosensory cortices and in particular from those which code proprioception. It sends its output to the primary motor cortex—area F1—and to the dorsal premotor cortex—area F2—(see figure 18-2). Recent neurophysiological experiments showed that SPL receives also a visual input. Neurons responding to visual stimuli were described in its caudal areas and, in particular, in areas V6A and MIP (Galletti et al. 1996; Caminiti, Ferraina, and Johnson 1996). Both these areas are connected with the dorsal premotor cortex—area F2—(Caminiti, Ferraina, and Johnson 1996; Matelli et al. 1998).

Like SPL, IPL (especially its rostral sector) receives somatosensory afferents. In addition, IPL is the main site of convergence of the pathways from the visual areas of the dorsal stream. IPL projects to the areas of the ventral premotor cortex—areas F4 and F5 (see figure 18-2)—and to the prefrontal

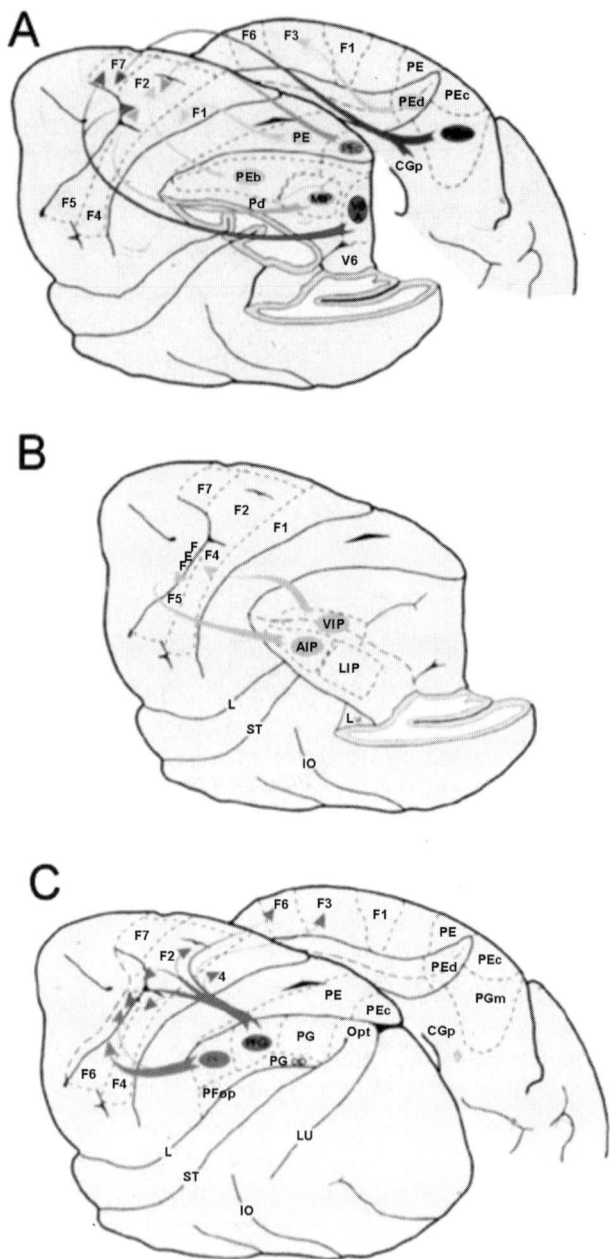

Figure 18-2. Schematic diagrams showing the extrinsic afferent connections of the parietodependent motor areas. See color insert.

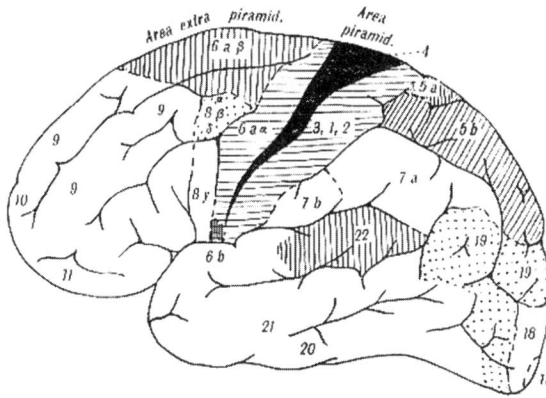

Figure 18-3. Lateral view of the human brain showing the cytoarchitectonic parcellation of Foerster. Modified from Foerster (1936).

lobe. The functional properties of IPL are in accord with the pattern of its anatomical connections. IPL neurons are often bimodal responding both to visual and somatosensory stimuli (for more details see the section, "The Ventrodorsal [V-D] Stream").

The anatomical data reviewed so far indicate that both SPL and IPL receive somatosensory and visual information, but this information is conveyed subsequently to different frontal areas. Before reviewing the functional organization of SPL and IPL in detail, an important point should be clarified, namely that of the homology between the parietal lobe organization of humans and other primates. The human posterior parietal lobe, as that of the monkey, is formed by two lobules—SPL and IPL—separated by the intraparietal sulcus. As shown by Foerster (1936) (see figure 18-3), contrary to the original parcellation of Brodmann, SPL is basically coextensive with area 5 (5a + 5b), while IPL with area 7 (7a + 7b). That is, the cytoarchitectonic organization of the posterior parietal lobe is similar in monkeys and humans. This view was confirmed by von Bonin and Bailey (1947) who, adopting the terminology of von Economo (1929), found that in monkeys as in humans, SPL is formed in large part by area PE (area 5) and IPL by areas PF (7b) and PG (7a). Thus, when monkey data on area 7 are used to discuss functional properties of human parietal lobe, they should be used in reference to human IPL and not SPL as one might be tempted to do on the basis of the (wrong) Brodmann map. Similarly, the data on monkey area 5 should be used in reference to human SPL.

The Dorsodorsal (D-D) Stream

There are two groups of areas that one can distinguish in SPL. One group is formed by areas PE, Pec, and PEip, the other by areas MIP and V6A (figure 18-1).

The areas of the first group are involved in the control of movements based on somatosensory information. Most neurons of these areas respond to somatosensory stimuli and in particular to proprioceptive stimuli (Sakata et al. 1973; Mountcastle et al. 1975; Lacquaniti et al. 1995; Iwamura and Tanaka 1996). Many of them discharge in association with body movements (e.g., arm movements, Kalaska et al. 1990). Area PE projects to area F1 (area 4), while areas PEc and PEip project mostly to area F2 and in particular to the sector of F2 located around the superior frontal dimple (figure 18-2).

The areas of the second group process visual information in addition to the somatosensory one. A large number of neurons responding both to visual and somatosensory stimuli are located in area MIP (Colby and Duhamel 1991). Recent data by Snyder, Batista, and Andersen (1997) showed that in an SPL sector very likely corresponding to area MIP, there are many neurons that respond to visual stimuli. The response of these neurons is potentiated when stimulus presentation is followed by an arm reach movement directed toward the stimulus. This potentiation is lacking in the case of eye movements directed toward the same target.

Neurons responding to visual stimuli are present also in are V6A (Galletti et al. 1996, 1997; Ferraina et al. 1997). According to Galletti et al. (1997), about half of V6A neurons have this property. The remainder discharge mostly in association with eye or arm movements. It is interesting to note that in contrast to what is observed in most visual areas, in V6A there is no magnification of the foveal representation (Colby et al. 1988).

The neurophysiological data on SPL suggest that this region intervenes in the control of arm (and other body parts) movements on the basis of somatosensory and visual information. Clinical studies confirmed this role: lesions centered on SPL produce reaching disorders (optic ataxia). Unilateral lesions typically determine the deficit in the contralesional half field in the absence of hemianopia (Ratcliff and Davies-Jones 1972).

Perenin and Vighetto (1988) carefully studied the anatomical locations of lesions causing optic ataxia and hemispatial neglect. Their results showed that the two syndromes result from of lesions of different posterior parietal sectors. SPL damage determines optic ataxia, while IPL damage brings about unilateral neglect. Similar findings were reported by Vallar and Perani (1987) (see figure 18-4).

Taken together, physiological data and clinical reports strongly support the stance of Milner and Goodale (1995), that the dorsal stream is not involved in perception but only in the control of action. Note, however, that this statement is valid only for SPL. As we will show in the following section, it is not valid for IPL. The absence of perceptual deficits is not observed following damage to ventrodorsal stream.

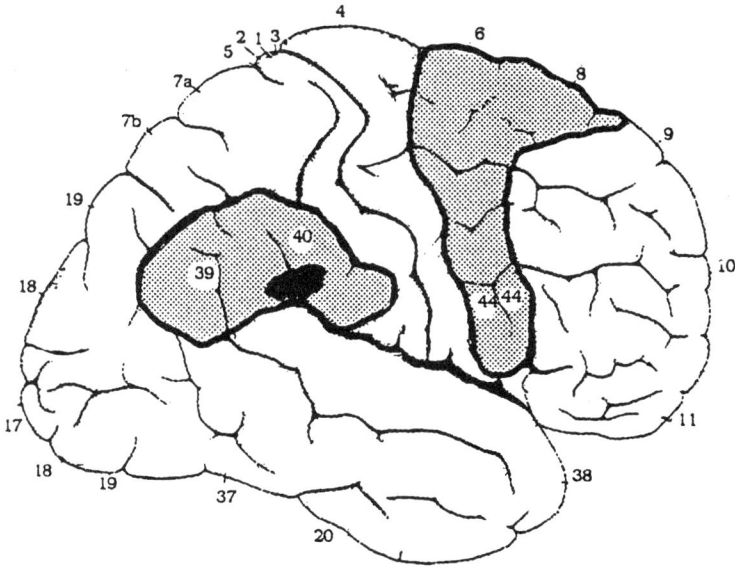

Figure 18-4.   Lateral view of the right human brain showing the anatomical location of lesions (shaded areas) producing visual unilateral neglect. The black area indicates the most frequent lesion site. Modified from Bisiach and Vallar (2000).

The Ventrodorsal (V-D) Stream

The V-D stream consists of IPL and the visual areas that project to it. Anatomically, the areas forming IPL can be subdivided into three groups, areas located on the cortical convexity (PF, PFG, and PG), areas located on the lateral bank of the intraparietal sulcus (AIP, VIP, and LIP), and the opercular areas (PFop and PGop). The connections of IPL with the agranular frontal cortex are shown in figure 18-2.

Areas VIP and LIP are involved in space-directed action and in space representation, whereas areas AIP and PF are involved in action on objects and in action perception. The properties of these areas will be dealt with in the next sections.

The V-D Stream: Action in Space and Space Perception

The cortical circuit that organizes saccadic eye movements in primates is formed by area LIP and its frontal target, the frontal eye fields (FEF). Single-neuron studies showed that LIP neurons respond to the presentation of visual stimuli. Their receptive fields are typically large. They are neither directionally nor orientation selective. LIP receptive fields are coded in retinal coordinates. Many

LIP neurons discharge before saccadic eye movements. Pure movement-related neurons are, however, rare. The functional properties of FEF neurons are in many aspects similar to those of LIP neurons. The percentage, however, of movement-related neurons is much larger in FEF than in LIP. Furthermore, the intensity of neuron discharge associated with saccadic eye movements is typically much stronger in FEF than in LIP.

The effect of reversible inactivation (by muscimol injection) of LIP on oculomotor behavior was recently studied by Li, Mazzoni, and Andersen (1999). As far as motor behavior is concerned, the data showed that after the inactivation the latency of saccades towards the contralesional target increased and displayed a higher variance than before injection. The duration of saccades remained virtually unmodified, while the metrics of saccades was affected but only in memory-related saccades. Reversible inactivation of area LIP (Li, Mazzoni, and Andersen 1999) doesn't produce spatial neglect. However, when two stimuli are simultaneously presented, the monkey consistently prefers the stimulus ipsilateral to the inactivation site. Similar results were found following lesion of LIP (Lynch and McClaren 1983).

Lesion of FEF produces effects that are more dramatic (Latto and Cowey 1971; Rizzolatti, Matelli, and Pavesi 1983; van der Steen, Russell, and James 1986). Immediately following the lesion there is a marked decrease of spontaneous and evoked saccadic eye movements towards the side contralateral to the lesion. Stimuli presented contralateral to the lesion are frequently neglected. They are virtually always neglected when the animal's attention is already engaged on another stimulus. The deficit is more severe when the stimuli are presented far from the animal (extrapersonal space) than when they are close to it (peripersonal space) (see Rizzolatti, Matelli, and Pavesi 1983). Neither LIP nor FEF lesions produce tactile deficits (personal neglect). It is important to note that bilateral lesions of FEF accompanied by bilateral lesions of the superior colliculus produce a complete lack of saccadic eye movements (Schiller, True, and Conway 1980).

In conclusion, it is clear from both electrophysiological and lesion experiments that the primary function of the circuit formed by areas LIP and FEF is that of programming ocular saccades to specific space locations. Nevertheless, in spite of the fact that the whole circuit machinery is "constructed" for this purpose, the lesion of this oculomotor circuit does not produce only oculomotor deficits but also deficits in space perception. This clearly indicates that space perception is not a separate function, located in areas dedicated to this purpose, but derives from operations necessary for action organization.

In order to reach an object in space, it is necessary to "know," in addition to the initial position of the arm, where the object is located with respect to the individual that makes the action. The transformation of object location into head and arm reaching movements is performed by a circuit formed by areas VIP and F4 (figure 18-2).

Single-neuron recordings showed that in monkey area VIP there are two main classes of neurons activated by sensory stimuli: visual neurons and bimodal, visual and tactile neurons. Purely visual neurons are strongly selective for the direction and speed of the stimuli. Their receptive fields are typically large. At difference with LIP and FEF neurons, a considerable percentage of VIP neurons code space in reference to the monkey's body (Duhamel et al. 1997).

Bimodal VIP neurons respond to both visual and tactile stimuli. Tactile receptive fields are located predominantly on the face. Tactile and visual receptive fields are usually in register, that is, the visual receptive field encompasses a three-dimensional spatial region around the tactile receptive field (peripersonal space). Some bimodal neurons are activated preferentially or even exclusively when three-dimensional objects are moved towards or away from the tactile receptive field.

Consistent with the single-neuron data are the results of lesion studies. Selective electrolitic lesion of area VIP in monkeys determines mild but consistent contralesional neglect for peripersonal space. No changes were observed in ocular saccades, eye pursuit, or optokinetik nistagmus. Tactile stimuli applied to the contralesional side of the face failed to elicit orienting responses (J.-R. Duhamel, personal communication).

While data on this point are lacking for VIP neurons, there is evidence that most F4 neurons discharge in association with a monkey's active movements (Godschalk et al. 1985; Gentilucci et al. 1988). The movements most represented are head and arm movements, such as head turns and reaching. Most F4 neurons discharge in response to sensory stimuli. According to their sensory responses, F4 neurons were subdivided into two classes: somatosensory neurons and bimodal, somatosensory and visual neurons (Gentilucci et al. 1988; Fogassi et al. 1992, 1996). Recently, trimodal neurons, responding also to auditory stimuli, were described (Graziano et al. 1999). Tactile receptive fields, typically large, are located on the face, chest, arm and hand. Visual receptive fields are also large. They are located in register with the tactile ones, and similarly to VIP, confined to the peripersonal space (Gentilucci et al. 1983, 1988; Fogassi et al. 1992, 1996; Graziano, Yap, and Gross 1994).

Studies of the visual properties of F4 neurons showed that in most F4 neurons the receptive fields do not change position when the eyes move (Gentilucci et al. 1983; Fogassi et al. 1992, 1996; Graziano, Yap, and Gross 1994). This indicates that the visual responses of F4, unlike those of LIP and FEF neurons, do not signal positions on the retina but positions in space relative to the observer. The spatial coordinates of the receptive fields are anchored to different body parts and not to a single reference point, as suggested by some motor theorists on the basis of psychological experiments. Visual receptive fields located around a certain body part (e.g., the arm) move when that body part is moved (Graziano et al. 1997). Allocentric coding was also tested and contrasted with egocentric coding: in all tested neurons the receptive field organization was found to be coded in egocentric coordinates (Fogassi et al. 1996).

Unilateral lesion of the ventral premotor cortex that includes area F4 produces two series of deficits: motor deficits and perceptual deficits (Rizzolatti, Matelli, and Pavesi 1983; see also the inactivation studies by Schieber 2000 and Fogassi et al. 2001). Motor deficits consist of a reluctance to use the contralesional arm, spontaneously or in response to tactile and visual stimuli, and a failure to grasp with the mouth food presented contralateral to the side of the lesion. Perceptual deficits concern the contralesional peripersonal space and the personal (tactile) space. A piece of food moved in the contralesional space around the monkey's mouth does not elicit any behavioral reaction. Similarly, when the monkey is fixating a central stimulus, the introduction of food contralateral to the lesion is ignored. In contrast, stimuli presented outside the animal's reach (far space) are immediately detected.

As shown in figure 18-4, neglect in humans occurs after lesion of the IPL and, less frequently, following damage of the posterior part of the frontal lobe (see Bisiach and Vallar 2000). The most severe neglect in humans occurs after lesion of the right IPL. In full-fledged unilateral neglect, patients may show a more or less complete deviation of the head and eyes towards the ipsilesional side. Routine neurological examination shows that patients with unilateral neglect typically fail to respond to visual stimuli presented in the contralesional half field and to tactile stimuli delivered to the contralesional limbs.

As in monkeys (Rizzolatti, Matelli, and Pavesi 1983), in humans neglect may selectively affect the extrapersonal or the peripersonal space. In humans, this dissociation was first demonstrated by Halligan and Marshall (1991). They reported the case of a patient with neglect who, when asked to mark the center of a line in the near space, displaced the midpoint mark to the right, as typically neglect patients do. However, when the test was made in the far space the neglect dramatically improved or even disappeared. A similar dissociation was reported more recently by Berti and Frassinetti (2000). Other authors described the opposite dissociation: severe deficits in tasks carried out in the extrapersonal space but slight or no deficit for tasks performed in the peripersonal space (see Cowey, Small, and Ellis 1994, 1999; Shelton, Bowers, and Heilman 1999).

The lesions causing neglect in humans are always very large; thus, while the findings of separate systems for peripersonal and extrapersonal space are robust and convincing, any precise localization of the two systems in humans is at the moment impossible.

In conclusion, it is clear that in both monkeys and humans lesions of IPL and its frontal targets determines spatial perceptual deficits. These data are in agreement with the notion that the dorsal stream plays a fundamental role in space perception (Ungerleider and Mishkin 1982). However, there is no "space center." In IPL, there are several circuits that process sensory information for organizing eye and arm movements. It is their lesion that determines a deficit in space perception.

The V-D Stream and Object Awareness

It is interesting to note that, contrary to the view of Milner and Goodale (1995), IPL not only appears to play a fundamental role in space perception, but is also necessary for object awareness.

Initial evidence in favor of this comes from a work of Volpe, Ledoux, and Gazzaniga (1979) that showed that patients with right parietal lesion and clinical evidence of extinction denied seeing the stimuli presented in the affected field when two stimuli were simultaneously presented to the right and left of the fixation point. Yet, despite this fact, the patients were still able to make a same-different judgement when forced to do so by the examiner. The authors interpreted their data as evidence for the existence of implicit knowledge of the stimuli, even when the patients were unaware of their presence.

A capacity for processing object information without having explicit knowledge of the process was subsequent described by Marshall and Halligan (1988) in a patient with a severe neglect. This patient was repeatedly presented with two drawings of a house, one above the other. The two houses were identical on the right side but were different on the left side. The patient denied any difference between the pairs of houses even when one them was drawn as if burning. Nonetheless, when forced to choose the house in which she would prefer to live, the patient always chose the one that was not burning.

Berti and Rizzolatti (1992) confirmed these findings in a systematic way. In their experiments, patients with severe unilateral neglect were required to respond as fast as possible to target stimuli presented to the normal field by pressing one of two keys according to the category of the target. The presented stimuli were pictures of animals and fruits. Before showing these stimuli, pictures of animals and fruits were presented to the neglected field as priming stimuli. The patients denied of seeing these stimuli (primes), yet the results showed that the primes facilitated the patients' responses to the stimuli shown in the normal field. This was observed not only in "highly congruent conditions," that is, when the prime stimulus and the target were physically identical, but also when prime and stimulus represented two elements of the same category but were physically dissimilar.

Taken together these findings indicate that patients with neglect are able to process stimuli presented to the neglected field up to a categorical level of representation. Yet these patients are not aware of the result of the processing in the case of lesion of IPL. This implies that individuals must have the parietofrontal sensorimotor circuits intact in order to achieve object awareness. This is required even for those stimuli, such as fruits or animals, that are analyzed in the ventral stream. The ventral stream's processing is not sufficient to get perception without parietal spatial processing.

The V-D Stream: Action on Objects

The V-D stream, in addition to organizing movements requiring space computation, is also involved in the organization of grasping and manipulation hand movements. These movements, in order to be executed, require the computation of the size and shape of objects rather than space (Arbib 1981; Jeannerod 1986).

The parietal area that plays a central role in sensory-motor transformations for grasping is area AIP (Taira et al. 1990; Sakata and Taira 1994; Murata et al. 2000). AIP neurons fall into three main classes: visual-dominant neurons, visual-and-motor neurons, and motor-dominant neurons. Visual-dominant neurons discharge during object fixation and during grasping in light, but not during grasping in dark. Visual-and-motor neurons discharge both during grasping in light and grasping in dark, but their response is stronger during grasping in light. They also discharge during object fixation. Motor-dominant neurons discharge during grasping in both light and dark. They are silent during object fixation.

Area AIP is reciprocally connected with area F5, and specifically with a sector of it (F5ab) located in the posterior bank of the arcuate sulcus (Luppino et al. 1999). Electrical stimulation studies showed that area F5 contains a representation of hand movements (Rizzolatti et al. 1981, 1988; Kurata and Tanji 1986; Hepp-Reymond et al. 1994).

Single-neuron recordings demonstrated that most F5 neurons code specific actions (Rizzolatti et al. 1988). Among them, the most represented are "grasping," "holding," "tearing," and "manipulating" neurons. Neurons of a given class respond weakly or not at all when similar movements are executed in a different context. Many neurons in each class code specific types of hand shaping, such as, for example, precision grip, whole hand prehension, and finger prehension.

About 20 percent of F5 neurons located in F5ab discharge in response to the visual presentation of 3D objects (Murata et al. 1997). These neurons have been called "canonical neurons." The majority of them are activated selectively by the presentation of objects of a certain size, shape, and orientation. The visual specificity is typically congruent with the motor specificity. It is important to note that the neuron response to object presentation occurs even when no action upon the presented object is requested and the monkey has simply to release a lever following object presentation (Murata et al. 1997).

What can be the explanation for this behavior? The observed neuron discharge is certainly not related to motor preparation because the response is also present when no response toward the object is required. Similarly, the specificity of the response for certain objects rules out unspecific factors, such as attention or intention to act, which were equal for all the presented objects. The interpretation that is left is that object presentation produces a representation of the observed stimulus in motor terms: a potential action. When an appropriate stimulus is presented, canonical F5 neurons code a potential action. This action may be executed or not. We will see the important implications of this concept for understanding another class of F5 visuomotor neurons, the "mirror neurons," and more generally for action driven perception.

Taken together, these neurophysiological data strongly suggest that IPL is involved in transforming size and shape of objects (coded into AIP) into the appropriate motor schema for acting upon them (coded in F5). Strong support for this view comes from inactivation studies. Inactivation of area AIP produces a dramatic deficit in the capacity of the monkey to shape its hand based on visual information on the object size and shape. The deficit is not due to pure motor deficit or to reaching problems (Gallese et al. 1994). A similar deficit in visuomotor transformations for grasping was recently described following inactivation of F5ab (Fogassi et al. 2001).

Deficits in visuomotor transformation for hand movements were reported also in humans following lesion of the parietal lobe (Jeannerod 1986; Pause et al. 1989). Until recently, however, little was known about the location of the finger-movement representation in the parietal cortex of humans. Recently, it was reported that a selective deficit in the coordination of finger movements required for object grasping occurs following damage to the anterior part of the lateral bank of the intraparietal sulcus (Binkofski et al. 1998). The same intraparietal site becomes active when normal subjects perform prehension movements (Binkofski et al. 1998, 1999). These data strongly suggest that homologous parietal structures mediate visuomotor transformation for grasping in human and monkeys. The homology is further supported by the demonstration that manipulation in humans determines the activation of area 44 (Binkofski et al. 1999; Ehrsson et al. 2000), which is the area considered the human homologue of the monkey's area F5.

In conclusion, IPL, in addition to being involved in space computation and perception, is also crucially involved in the organization of object-directed hand movements. Contrary to space circuits, the circuit formed by AIP-F5 does not appear to be involved in perception. It is possible, however, that it could play a role in providing semantic value to objects during development (see Rizzolatti and Gallese 1997).

The V-D Stream: Action Perception

In F5, in addition to canonical neurons there is a second class of visuomotor neurons: "mirror neurons." The defining functional characteristic of these neurons, which are mostly located on the convexity of F5 (F5c), is that they discharge both when the monkey makes a particular action and when it observes another individual (monkey or human) making a similar action. Mirror neurons do not respond to object presentation, even when the object (e.g., food) is of interest to the monkey. Most of them do not respond either to the sight of a mimed action (Gallese et al. 1996; Rizzolatti et al. 1996).

The observed actions most effective in triggering mirror neurons are grasping, holding, manipulating, and placing. The majority of mirror neurons are active during the observation of one action only. The others respond to two or, rarely, three actions. Some mirror neurons are selective not only to the general goal of the observed action, (e.g., grasping an object), but also to the way in which the observed action is performed (e.g., observation of precision grip).

Mirror neurons show generalization in response to visual stimuli. Thus, the distance at which an action is executed does not influence their response. As far as the object target of the observed action is concerned, its meaning also does not appear to matter. The response to food is typically the same as that to three-dimensional solids.

A comparison between mirror neurons' motor and visual properties indicates that almost all neurons show congruence between the effective observed and executed actions. For some neurons this congruence is extremely strict, that is, the effective motor action (e.g., precision grip) coincides with the action that, when seen, triggers the neurons (e.g., precision grip). For other neurons, the congruence is broader. These broadly congruent neurons are of particular interest, because they generalize the goal of the observed action across many instances of it.

F5c is heavily connected with IPL and especially with area PF (see figure 18-2). As described by Hyvärinen (1981), this area contains a large variety of functionally different neurons. Many of them respond to passive somatosensory stimuli, while others are bimodal or respond predominantly to visual stimuli. About one third of PF neurons fire during monkey's active movements (Leinonen et al. 1979; Hyvärinen 1981).

The functional properties of area PF were recently reinvestigated (Fogassi et al 1998; Gallese et al. 2002). In agreement with previous reports, it was found that the majority of PF neurons responded to somatosensory stimuli, to visual stimuli, or to both. Many visual neurons became active only when the monkey observed biological action. Some of these neurons were active also during action execution, and their motor properties matched the visual properties. This finding indicates that a set of PF neurons has characteristics similar to those of F5 mirror neurons.

PF receives an important input from the cortex located in the rostral part of the superior temporal sulcus (STSa). In an important series of experiments, Perrett et al. (1989) showed that in this region there is a variety of neurons that discharge when the monkey observes biological movements. Some of STSa neurons respond to head movements, others to body movements, others to eye movements (see for review Carey et al. 1997). Particularly interesting is the observation that some STSa neurons respond selectively to the observation of hand-object interactions (Perrett et al. 1990). STSa neurons, however, do not appear to have the motor properties that characterize F5 and PF mirror neurons.

What may be the functional role of the STSa-PF-F5c circuit? Two answers to the question were given. The first is that this circuit is involved in action recognition, the second that it plays a crucial role in imitation. There is little doubt that when an individual makes a voluntary action, he or she knows its consequences. This knowledge is the result of an association between the internal representation of that action and its consequences. Thanks to mirror neurons this knowledge can be extended to actions performed by others. When the observation of an action performed by another individual activates neurons that represent motorically the same action, the action is recognized because the representation evoked by observation corresponds to that internally generated during action programming (Rizzolatti et al. 1996; Gallese et al. 1996; see also Gallese and Goldman 1998; Gallese 2001).

The second interpretation of mirror neurons is that they play a role in imitation. There is consensus that monkeys are unable to imitate hand gestures. F5 mirror neurons cannot be therefore involved in this function. Yet there is growing evidence that, in humans, the same cortical areas that show mirror properties (activation during action execution and action observation) are also involved in imitation (Iacoboni et al. 1999; Nishitani and Hari 2000). It is reasonable to submit, therefore, that a mechanism like that found in the monkey's F5 is at the basis also of imitation. In order to have imitation there should be, however, an additional mechanism that allows one to use intentionally the representation of the observed action. Imitation appears therefore to have the same neural basis as action understanding, but requires also some additional mechanisms that apparently developed only in highest primates and in humans.

At present there are neither lesion nor inactivation studies specifically addressing the role of F5 mirror neurons in action understanding in monkeys. In humans, however, there is evidence that frontal aphasic patients are frequently impaired in pantomime recognition (Brain 1961; Gainotti and Lemmo 1976; Duffy and Watkins 1984; Bell 1994). This evidence supports the view that responsible for the deficit in pantomime recognition is a lesion of Broca's area and the premotor cortex adjacent to it, and, within it, that part where hand actions are represented. Such a deficit is exactly what one would predict if Broca's area and adjacent area 6 were endowed with a mechanism for action recognition as we have proposed.

Another possible piece of evidence supporting a link between the mirror matching system and action perception in humans comes from studies of apraxic patients. It has been shown (Heilman, Rothi, and Valenstein 1982; Heilman and Rothi 1993) that some patients with ideomotor apraxia and cortical lesions centered on the left IPL have difficulty in recognizing and discriminating between correctly and incorrectly performed meaningful actions. Brain imaging experiments in humans have shown an activation of IPL during action observation (Grafton et al. 1996; Grèzes, Costas, and Decety 1998; Buccino et al. 2001) as well as during action execution (Binkofski et al. 1998, 1999; Ehrsson et al. 2000).

In conclusion, IPL appears to be involved, together with STSa and F5, in action recognition and imitation. Thus, the stance that IPL is not involved in perception does not appear to be convincing.

The Ventral Stream

The ventral stream, as defined by Ungerleider and Mishkin (1982), consists of a series of areas that conveys information from area 17 (V1) to the inferotemporal lobe. These areas include area V2, the ventral portion of V3, and V4.

Single-neuron recordings showed that moving from V1 to V2 and then to V3 and V4, the coding of visual features becomes more and more complex. The complexity becomes extremely high in the inferotemporal cortex. As shown first by Gross, Rocha-Miranda, and Bender (1972), many inferotemporal neurons respond best to complex visual stimuli such as faces and hands. Recent studies confirmed these findings showing that neurons in the most anterior part of the inferotemporal cortex are selective to specific biological stimuli such as faces or part of faces (Tanaka et al. 1991; Kobatake and Tanaka 1994). Other recent studies showed that neurons sensitive to similar features (e.g., mouths) are clustered together in columns through the depth of the inferotemporal cortex (Fujita et al. 1992).

In 1939 Kluver and Bucy described a syndrome that appears in rhesus monkeys following a bilateral temporal lobectomy. This syndrome consists of "psychic blindness" (inability to recognize familiar objects), tendency to orally examine available objects, unusual tameness, and an increase in sexual activity (Kluver and Bucy 1939). Further studies demonstrated that affective and sexual symptoms were due to lesion of the limbic system. When lesions were confined to the inferior temporal cortex, the deficit concerned only the monkey visual capabilities. The monkeys were unable to discriminate between visual stimuli but performed well when they could rely on spatial cues as, for example, in the "landmark" task (Mishkin 1954; Mishkin and Pribram 1954). The effect of lesions of the inferotemporal areas was interpreted as evidence that the ventral stream is responsible for the analysis of the quality of objects. It enables the visual system to categorize visual inputs as visual objects, regardless of the visual conditions in which objects are presented.

In humans, since the study of Lissauer (1890) it is known that damage to the inferior temporal cortex (or its disconnection from visual areas) leads to a failure to recognize objects ("visual form agnosia"; see Benson and Greenberg 1969). The best know agnosic patient is probably D. F., extremely well studied by Milner and Goodale (1995). D. F. suffered an extended lesion of areas 18 and 19 due to carbon monoxide poisoning. Following lesion, D. F. showed severe impairment in recognizing the size, shape, and orientation of visual objects, while preserving the capacity to interact with them. Milner and Goodale (1995) stressed the dissociation between the perceptual and action-oriented behaviour of D. F. and used this evidence to support their notion that "there is a clear dissociation between the visual pathways supporting perception and action in the cerebral cortex."

There is no doubt that agnosic patients, such as D. F., are severely impaired in recognizing and discriminating objects. The point is, however, that their perceptual deficits do not encompass the whole spectrum of perception, as epitomized by their intact spatial skills. Agnosic patients, in contrast to neglect patients, and similarly to hemianopic patients, are aware of the existence of space and utilize this knowledge in a fully conscious way. Furthermore, they are able to perceive motion, as recognized by Milner and Goodale themselves (1995). Finally, a case of semantic visual agnosia was recently reported in which recognition of objects that mostly involved their visual representation was severely impaired, while recognition of objects that was based also on their sensorimotor representation was reasonably intact. Most interestingly, the patient was able to recognize observed gestures almost normally (Magnié et al. 1999).

Although Milner and Goodale (1995) acknowledged that the adaptive behavior of primates, humans included, relies on the integration of both their visual streams, the action/perception dichotomy they advocate is very rigid and unable, in our view, to account for the spared perceptual abilities above reviewed, especially if one considers that the integration between the two visual systems they advocate is essentially unidirectional, namely from the ventral stream to the dorsal stream and not vice versa. In the next section we will present several lines of evidence pointing to an important involvement of the motor system in supporting processes traditionally considered to be "perceptual," such as space perception and action recognition.

Perception and Action Are Not Independent Functions: Action Comes First

The data we have reviewed indicate that perception and action do not derive from the activity of separate streams of cortical areas. The most fundamental perceptual abilities—space perception and recognition of actions made by others—are intermingled with action organization. As we will argue more specifically in this section, perception can be explained only if one takes into consideration the relationship between the agent and his or her environment. Action is fundamental for building a meaningful description of the visual world. Only through action is visual experience "validated" so it can acquire a meaning for the individual.

Strong evidence in favor of this account of perceptual processes is provided by the properties of mirror neurons (Gallese et al. 1996; Rizzolatti et al. 1996). The issue here is how to build a meaningful account of actions made by others. We certainly do not deny that there is a system that analyzes and describes the pictorial and kinematic aspects of the observed actions. What we challenge is the view that such an analysis is sufficient to provide an understanding of the observed action. Without a reference on an existent knowledge, the pictorial description of the external events will be devoid of any meaning for the observing individual. This impasse is solved if it is admitted that the motor system generates a basic knowledge resulting from experience and action validation and that there is a neural system that matches actions made by others on this motor knowledge. The properties of mirror neurons show that such a system indeed exists. The activity of mirror neurons represents actions (actions whose meaning is known to the subject) and allows their recognition when an action made by another agent matches this representation.

A further argument in favor of the primacy of action is the organization of space perception. The conventional view on space is that space perception is mediated by the activity of a brain center specifically devoted to space representation. This putative multipurpose space area should be located in the parietal lobe (Critchley 1953; Hyvärinen 1982).

This view is wrong (Rizzolatti et al. 1997, 2000; Colby and Duhamel 1996; Graziano and Gross 1995). There is no evidence whatsoever that a "space center" exists. On the contrary, a large amount of experimental evidence, which we have reviewed in this chapter, shows that lesion of sensorimotor circuits whose primary function is that of controlling action in space also produces deficits in space awareness. It is rather surprising that lesion of one of these circuits is sufficient, at least immediately following the lesion, to determine neglect. A possible interpretation of this finding is that space awareness requires, to be achieved, the "vote" of all circuits that organize actions in space. Regardless, however, of how this occurs, the notion that space awareness derives from the activity of a multiplicity of brain circuits is not only supported by lesion data but also by the recovery that occurs in neglect following even large brain lesions. Without a multiple space representation this phenomenon could not occur.

The notion that space perception depends on movement is by no means new. Historically, particularly interesting was the attempt of von Helmoltz (1896) to substitute for the Kantian notion that space is necessary a priori to organize sensory experience, the notion that this "a priori" is generated by exploration behavior. In our opinion, a strong support for the notion that space derives from motor activity is the demonstration of the existence of peripersonal space.

Peripersonal space is a sort of visual extension of our body surface. It is anchored to the body, and it moves when the body and its parts move. Peripersonal space does not make much sense in physical terms. From a sensory point of view, there is no reason whatsoever that eyes with normal refraction should select light stimuli coming exclusively from a space sector located around the body of the perceiver. Light stimuli arriving from far or near should be equally effective when there is a normal refractory system and a normal accommodation.

The studies of the functional properties of F4 neurons provided interesting data on what these neurons code and therefore on the nature of peripersonal space. In principle, there are two main possibilities. The first is that F4 neurons code space "visually." If this were the case, given a reference point (e.g., the cutaneous receptive field) F4 neurons, when activated, should indicate specific spatial locations, regardless of the stimulation's temporal dimension. For example, a locus fifteen centimeters from the tactile receptive field should remain fifteen centimeters from it regardless of how an object reaches this position. The spatial map, determined by receptive field organization, should be basically static. The alternative possibility is that the discharge of F4 neurons codes potential motor actions directed towards particular spatial locations. These potential motor actions create a motor space around the individual. When a visual stimulus is presented in the peripersonal space, it evokes automatically one of these potential motor actions that, regardless of whether the action is subsequently executed or not, maps the spatial stimulus position in motor terms.

Arguments in favor of the visual hypothesis are the tight temporal link between stimulus presentation and the onset of neural discharge, the response constancy, and the presence of what appears to be a visual receptive field. If, however, there is a strict association between motor actions and stimuli that elicit them, it is not surprising that stimulus presentation would determine the effects just described. Evidence in favor of a motor hypothesis comes from the properties of F4 neurons in response to moving stimuli. Fogassi et al. (1996) moved three dimensional objects toward a monkey at different velocities. The stimulus trajectory was such that the stimulus started to move outside of the receptive field and then entered inside it. The results showed that the receptive fields of the majority of the neurons expanded in depth with the increase of stimulus velocity. Two examples are shown in figure 18-5. These results are not compatible with the view that the peripersonal space is coded in geometrical terms, because this view implies that spatial map should be static. In contrast, if space is motor, since time is inherent to movement, the spatial map should have dynamic properties and vary, therefore, according to the change in time of the object spatial location.

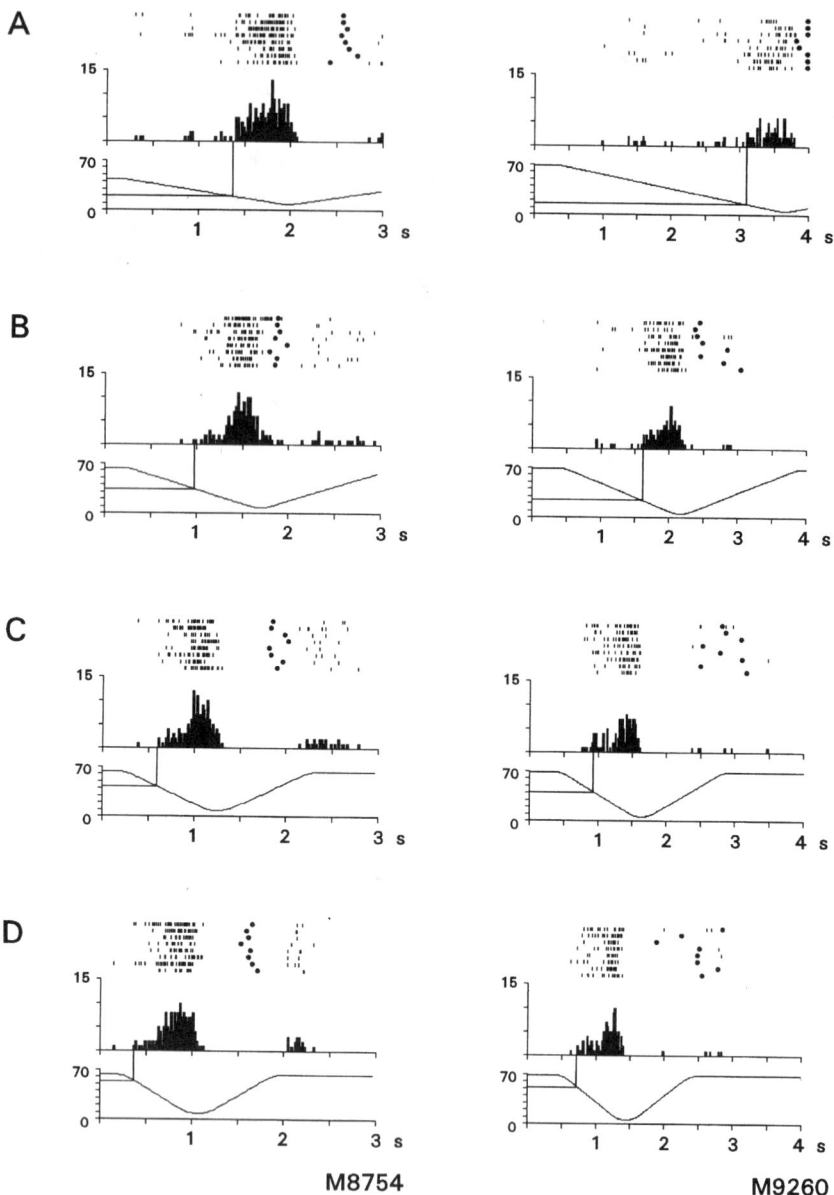

Figure 18-5. Examples of two F4 neurons with somatocentered visual receptive fields expanding in depth with increasing velocity of the approaching stimulus (a robot arm). Four different velocities were used: 20 cm/s (A), 40 cm/s (B), 60 cm/s (C), and 80 cm/s (D). Each panel indicates rasters illustrating the neural activity during individual trials (large dots, fixation point dimming), response histograms (abscissae, time; ordinates, spikes per bin; binwidth, 20 ms), and variation in time of the distance between the robot arm and the monkey (abscissae, time; ordinates, cm). The descending part of the curves indicates movement of the stimulus toward the monkey, while the ascending part indicates the movement away from the

In conclusion, the data reviewed here show that the motor system creates internal representations of actions. These representations may be used for various purposes. One is action generation, but others are action understanding and space perception. These functions are unlikely to be the only cognitive functions that the motor system performs. Its capacity to "validate" experience renders it unique for acquiring knowledge, even abstract knowledge (see Lakoff and Johnson 1980, 1999) about the external world. The role of the motor system in semantics is largely unexplored, but although this issue is complicated by several variables including verbal mediations, we would not exclude the idea that some basic aspects of object semantics may be linked to and derive from the motor system.

Acknowledgments   We thank G. Buccino and L. Fogassi for their critical reading of the manuscript. Supported by MURST and HFSPO.

References

Arbib, M.A. (1981) Perceptual structures and and distributed motor control. In Handbook of Physiology—The Nervous System, II, part 1, edited by V.B. Brooks, pp. 1449–1480. Bethesda, Md.: American Physiological Society.

Bell, B.D. (1994) Pantomime recognition impairment in aphasia: An analysis of error types. Brain Lang.: 47; 269–278.

Benson, D.F., and Greenberg, J.P. (1969) Visual form agnosia: A specific deficit in visual discrimination. Arch. Neurol.: 20; 82–89.

Berti, A., and Frassinetti, F. (2000) When far becomes near: Re-mapping of space by tool use. J. Cogn. Neurosci.: 12; 415–420.

Berti, A., and Rizzolatti, G. (1992) Visual processing without awareness: Evidence from unilateral neglect. J. Cogn. Neurosci.: 4; 345–351.

Binkofski, F., Buccino, G., Posse, S., Seitz, R.J., Rizzolatti, G., and Freund, H.-J. (1999) A fronto-parietal circuit for object manipulation in man: Evidence from an fMRI-study. Eur. J. Neurosci.: 11; 3276–3286.

Binkofski, F., Dohle, C., Posse, S., Stephan, K.M., Hefter, H., Seitz, R.J., and Freund, H.-J. (1998) Human anterior intraparietal area subserves prehension. A combined lesion and functional MRI activation study. Neurology: 50; 1253–1259.

Bisiach, E., and Vallar, G. (2000) Unilateral neglect in humans. In Handbook of Neuropsychology, 2nd edition, vol. I, edited by F. Boller, J. Grafman, and G. Rizzolatti, pp. 459–502. Amsterdam: Elsevier Science B.V.

von Bonin, G., and Bailey, P. (1947) The Neocortex of *Macaca mulatta*. Urbana: University of Illinois Press.

Brain, W.R. (1961) Speech Disorders: Aphasia, Apraxia and Agnosia. Washington, D. C.: Butterworth.

---

monkey. Vertical lines under histograms: beginning of the neuron discharge. The intersection of these lines with the curves representing the distance between the robot arm and the monkey corresponds to the outer border of the neuron's receptive field at different velocities. From Fogassi et al. (1996).

Buccino, G., Binkofski, F., Fink, G.R., Fadiga, L., Fogassi, L., Gallese, V., Seitz, R.J., Zilles, K., Rizzolatti, G., and Freund, H.-J. (2001) Action observation activates premotor and parietal areas in a somatotopic manner: An fMRI study. Eur. J. Neurosci.: 13; 400–404.

Caminiti, R., Ferraina, S., and Johnson, P.B. (1996) The sources of visual information to the primate frontal lobe: A novel role for the superior parietal lobule. Cereb. Cortex: 6; 319–328.

Carey, D.P., Perrett, D.I., and Oram, M.W. (1997) Recognizing, understanding and reproducing action. In Handbook of Neuropsychology, vol. XI, edited by F. Boller and J. Grafman, pp. 111–129. Amsterdam: Elsevier.

Colby, C.L., and Duhamel, J.-R. (1991) Heterogeneity of extrastriate visual areas and multiple parietal areas in the macaque monkeys. Neuropsychologia: 29; 517–537.

Colby, C.L., and Duhamel, J.-R. (1996) Spatial representations for action in parietal cortex. Cogn. Brain Res.: 5; 105–115.

Colby, C.L., Gattass, R., Olson, C.R., and Gross, C.G. (1988) Topographical organization of cortical afferents to extrastriate visual area PO in the macaque: A dual tracer study. J. Comp. Neurol.: 269; 392–413.

Cowey, A., Small, M., and Ellis, S. (1994) Left visuo-spatial neglect can be worse in far than near space. Neuropsychologia: 32; 1059–1066.

Cowey, A., Small, M., and Ellis, S. (1999) No abrupt change in visual hemineglect from near to far space. Neuropsychologia: 37; 1–6.

Critchley, M. (1953) The Parietal Lobes. New York: Hafner Press.

De' Sperati, C., and Stucchi, D. (1997) Recognizing the motion of a graspable object is guided by handedness. Neurorep.: 8; 2761–2765.

Duffy, J.R., and Watkins, L.B. (1984) The effect of response choice relatedness on pantomime and verbal recognition ability in aphasic patients. Brain Lang.: 21; 291–306.

Duhamel, J.-R., Bremmer, F., Ben Hamed, S., and Graf, W. (1997) Spatial invariance of visual receptive fields in parietal cortex neurons. Nature: 389; 845–848.

von Economo, C. (1929) The Cytoarchitectonics of the Human Cerebral Cortex. Oxford: Oxford University Press.

Ehrsson, H.H., Fagergren, A., Jonsson, T., Westling, G., Johansson, R.S., and Forssberg, H. (2000) Cortical activity in precision- versus power-grip tasks: An fMRI study. J Neurophysiol: 83; 528–536.

Ferraina, S., Garasto, M.R., Battaglia-Mayer, A., Ferraresi, P., Johnson, P.B., Lacquaniti, F., and Caminiti, R. (1997) Visual control of hand-reaching movement: Activity in parietal area 7m. Eur. J. Neurosci.: 9; 1090–1095.

Foerster, O. (1936) Motorische felder und bahnen. Sensible corticale felder. In Handbuck der neurologie, vol. 6, edited by O. Bumke and O. Foerster, pp. 1–448. Berlin: Springer.

Fogassi, L., Gallese, V., Buccino, G., Craighero, L., Fadiga, L., and Rizzolatti, G. (2001) Cortical mechanism for the visual guidance of hand grasping movements in the monkey: A reversible inactivation study. Brain: 124; 571–586.

Fogassi, L., Gallese, V., Di Pellegrino, G., Fadiga, L., Gentilucci, M., Luppino, G., Matelli, M., Pedotti, A., and Rizzolatti, G. (1992) Space coding by premotor cortex. Exp. Brain Res.: 89; 686–690.

Fogassi, L., Gallese, V., Fadiga, L., Luppino, G., Matelli, M., and Rizzolatti, G. (1996) Coding of peripersonal space in inferior premotor cortex (area F4). J. Neurophysiology: 76; 141–157.

Fogassi, L., Gallese, V., Fadiga, L., and Rizzolatti, G. (1998) Neurons responding to the sight of goal-directed hand/arm actions in the parietal area PF (7b) of the macaque monkey. Society of Neuroscience Abstracts: 24; 257.5.

Fujita, I., Tanaka, K., Ito, M., and Cheng, K. (1992) Columns for visual features of objects in monkey inferotemporal cortex. Nature: 360; 343–346.

Gainotti, G., and Lemmo, M.S. (1976) Comprehension of symbolic gestures in aphasia. Brain Lang.: 3; 451–460.

Gallese, V. (2001) The "shared manifold" hypothesis: From mirror neurons to empathy. Journal of Consciousness Studies: 8; 33–50.

Gallese, V., and Goldman, A. (1998) Mirror neurons and the simulation theory of mind-reading. Trends in Cognitive Sciences: 12; 493–501.

Gallese, V., Fadiga, L., Fogassi, L., and Rizzolatti, G. (1996) Action recognition in the premotor cortex. Brain: 119; 593–609.

Gallese, V., Fadiga, L., Fogassi, L., and Rizzolatti, G. (2002) Action representation and the inferior parietal lobule. In Common Mechanisms in Perception and Action: Attention and Performance, vol. 19, edited by W. Prinz and B. Hommel, pp. 334–355. Oxford: Oxford University Press.

Gallese, V., Murata, A., Kaseda, M., Niki, N., and Sakata, H. (1994) Deficit of hand preshaping after muscimol injection in monkey parietal cortex. Neuroreport: 5; 1525–1529.

Galletti, C., Fattori, P., Battaglini, P.P., Shipp, S., and Zeki, S. (1996) Functional demarcation of a border between areas V6 abd V6A in the superior parietal gyrus of the macaque monkey. Eur. J. Neurosci.: 8; 30–52.

Galletti, C., Fattori, P., Kutz, D.F., and Battaglini, P.P. (1997) Arm movement-related neurons in visual area V6A of the macaque superior parietal lobule. Eur. J. Neurosci.: 9; 410–413.

Gentilucci, M., Fogassi, L., Luppino, G., Matelli, M., Camarda, R., and Rizzolatti, G. (1988) Functional organization of inferior area 6 in the macaque monkey: I. Somatotopy and the control of proximal movements. Exp. Brain Res.: 71; 475–490.

Gentilucci, M., Scandolara, C., Pigarev, I.N., and Rizzolatti, G. (1983) Visual responses in the postarcuate cortex (area 6) of the monkey that are independent of eye position. Exp. Brain Res.: 50; 464–468.

Godschalk, M., Lemon, R.N., Kuypers, H.G.J.M., and van der Steen, J. (1985) The involvement of monkey premotor cortex neurones in preparation of visually cued arm movements. Behav. Brain Res.: 18; 143–157.

Goodale, M.A., and Milner, A.D. (1992) Separate visual pathways for perception and action. Trends Neurosci.: 15; 20–25.

Grafton, S.T., Arbib, M.A., Fadiga, L., and Rizzolatti, G. (1996) Localization of grasp representations in humans by PET: 2. Observation compared with imagination. Exp. Brain Res.: 112; 103–111.

Graziano, M.S.A., and Gross, C.G. (1995) The representation of extrapersonal space: A possible role for bimodal visual-tactile neurons. In The Cognitive Neurosciences, edited by M.S. Gazzaniga, pp. 1021–1034. Cambridge, Mass.: MIT Press.

Graziano, M.S., Hu, X.T., and Gross, C.G. (1997) Coding the locations of objects in the dark. Science: 277; 239–241.

Graziano, M.S.A., Reiss, L.A.J., and Gross, C.G. (1999) A neuronal representation of the location of nearby sounds. Nature: 397; 428–430.

Graziano, M.S.A., Yap, G.S., and Gross, C.G. (1994) Coding of visual space by premotor neurons. Science: 266; 1054–1057.

Grèzes, J., Costes, N., and Decety, J. (1998) Top-down effect of strategy on the perception of human biological motion: A PET investigation. Cogn. Neuropsychol.; 15; 553–582.

Gross, C.G., Rocha-Miranda, C.E., and Bender, D.B. (1972) Visual properties of neurons in the inferotemporal cortex of the macaque. J. Neurophysiol.: 35; 96–111.

Halligan, P.W., and Marshall, J.C. (1991) Left neglect for near but not far space in man. Nature: 350; 498–500.

Heilman, K.M., and Rothi, L.J. (1993) Apraxia. In Clinical Neuropsychology, 3rd edition, edited by K.M. Heilman and E. Valenstein, pp. 141–163. New York: Oxford University Press.

Heilman, K.M., Rothi, L.J., and Valenstein, E. (1982) Two forms of ideomotor apraxia. Neurology: 32; 342–346.

von Helmoltz, H. (1896) Die tatsache der wahrnehmung. In Vorträge und Reden, vol. II. Braunschweig.

Hepp-Reymond, M.-C., Hüsler, E.J., Maier, M.A., and Qi, H.-X. (1994) Force-related neuronal activity in two regions of the primate ventral premotor cortex. Can. J. Physiol. Pharmacol.: 72; 571–579.

Hyvärinen, J. (1981) Regional distribution of functions in parietal association area 7 of the monkey. Brain Res.: 206; 287–303.

Hyvärinen, J. (1982) Posterior parietal lobe of the primate brain. Physiol. Rev.: 62; 1060–1129.

Iacoboni, M., Woods, R.P., Brass, M., Bekkering, H., Mazziotta, J.C., and Rizzolatti, G. (1999) Cortical mechanisms of human imitation. Science: 286; 2526–2528.

Iwamura, Y., and Tanaka, M. (1996) Representation of reaching and grasping in the monkey postcentral gyrus. Neurosci. Lett.: 214; 147–150.

Jeannerod, M. (1986) The formation of finger grip during prehension. A cortically mediated visuomotor pattern. Behav. Brain Sci.: 19; 99–116.

Jeannerod. M. (1994) The representing brain: neural correlates of motor intention and imagery. Behav. Brain Sci.: 17; 187–245.

Jeannerod, M. (1997) The Cognitive Neuroscience of Action. Oxford: Blackwell.

Kalaska, J.F., Cohen, D.A.D., Prud'homme, M., and Hyde, M.L. (1990) Parietal area 5 neuronal activity encodes movement kinematics, not movement dynamics. Exp. Brain Res.: 80; 351–364.

Kluver, H., and Bucy, P.C. (1939) Preliminary analysis of functions of the temporal lobes in monkeys. Arch. Neurol. Psychiat.: 42; 979–1000.

Kobatake, E., and Tanaka, K. (1994) Neuronal selectivities to complex object features in the ventral visual pathway of the macaque cerebral cortex. J. Neurophysiol.: 71; 856–867.

Kurata, K., and Tanji, J. (1986) Premotor cortex neurons in macaques: Activity before distal and proximal forelimb movements. J. Neurosci.: 6; 403–411.

Lacquaniti, F., Guigon, E., Bianchi, L., Ferraina, S., and Caminiti, R. (1995) Representing spatial information for limb movement: Role of area 5 in the monkey. Cereb. Cortex: 5; 391–409.

Lakoff, G., and Johnson, M. (1980) Metaphors We Live By. Chicago: University of Chicago Press.

Lakoff, G., and Johnson, M. (1999) Philosophy in the Flesh: The Embodied Mind and Its Challenge to Western Thought. New York: Basic Books.

Latto, R., and Cowey, A. (1971) Visual field defects after frontal eye-field lesions in monkeys. Brain Res.: 30; 1–24.

Leinonen, L., Hyvärinen, J, Nyman, G., and Linnankoski, I. (1979) Function properties of neurons in lateral part of associative area 7 in awake monkeys. Exp. Brain Res.: 34; 299–320.

Li, C.-S.R., Mazzoni, P., and Andersen, R.A. (1999) Effect of reversible inactivation of macaque lateral intraparietal area on visual and memory saccades. J. Neurophysiol.: 81; 1827–1838.

Lissauer, H. (1890) Ein fall von seelenblindheit nebst einem beitrag zur theorie derselben. Arch. Psychiat.: 21; 222–270.

Luppino, G., Murata, A., Govoni, P., and Matelli, M. (1999) Largely segregated parietofrontal connections linking rostral intraparietal cortex (areas AIP and VIP) and the ventral premotor cortex (areas F5 and F4). Exp. Brain Res.: 128; 181–187.

Lynch, J.C., McLaren, J.W. (1983) Optokinetic nystagmus deficits following parieto-occipital cortex lesions in monkeys. Exp. Brain Res.: 49; 125–130.

Magnié, M.N., Ferreira C.T., Giusiano B., and Poncet M. (1999) Category specificity in object agnosia: Preservation of sensorimotor experiences related to objects. Neuropsychologia: 37; 67–74.

Marshall, J.C., and Halligan, P.W. (1988) Blindsight and insight in visuo-spatial neglect. Nature: 336; 766–767.

Matelli, M., Govoni, P., Galletti, C., Kutz, D.F., and Luppino, G. (1998) Superior area 6 afferents from the superior parietal lobule in the macaque monkey. J. Comp. Neurol.: 402; 327–352.

Matelli, M., Luppino, G., and Rizzolatti, G. (1991) Architecture of superior and mesial area 6 and of the adjacent cingulate cortex. J. Comp. Neurol.: 311; 445–462.

Milner, D., and Goodale, M.A. (1995) The Visual Brain in Action. Oxford: Oxford University Press, 1995.

Mishkin, M. (1954) Visual discrimination performance following partial ablations of the temporal lobe. II. Ventral surface vs. hippocampus. J. Comp. Physiol. Psychol.: 47; 187–193.

Mishkin, M., and Pribram, K.H. (1954) Visual discrimination performance following partial ablations of the temporal lobe. I. Ventral vs. lateral. J. Comp. Physiol. Psychol.: 47; 14–20.

Mountcastle, V.B., Lynch, J.C., Georgopoulos, A., Sakata, H., and Acuna, C. (1975) Posterior parietal association cortex of the monkey: Command functions for operations within extrapersonal space. J. Neurophysiol.: 38; 871–908.

Murata, A., Fadiga, L., Fogassi, L., Gallese, V., Raos, V., and Rizzolatti, G. (1997) Object representation in the ventral premotor cortex (area F5) of the monkey. J. Neurophysiol.: 78; 2226–2230.

Murata, A., Gallese, V., Luppino, G., Kaseda, M., and Sakata, H. (2000) Selectivity for the shape, size and orientation of objects in the hand-manipulation-related neurons in the anterior intraparietal (AIP) area of the macaque. J. Neurophysiol.: 83; 2580–260.

Nishitani, N., and Hari, R. (2000) Temporal dynamics of cortical representation for action. Proc. Natl. Acad. Sci. USA: 97; 913–918.

Pandya, D.N., and Seltzer, B. (1982) Intrinsic connections and architectonics of posterior parietal cortex in the rhesus monkey. J. Comp. Neurol.: 204; 196–210.

Pause, M., Kunesch, E., Binkofski, F., and Freund, H.-J. (1989) Sensorimotor disturbances in patients with lesions of the parietal cortex. Brain: 112; 1599–1625.

Perenin, M.-T., and Vighetto, A. (1988) Optic ataxia: a specific disruption in visuomotor mechanisms. I. Different aspects of the deficit in reaching for objects. Brain: 111; 643–674.

Perrett, D.I., Harries, M.H., Bevan, R., Thomas, S., Benson, P.J., Mistlin, A.J., Chitty, A.K., Hietanen, J.K., and Ortega, J.E. (1989) Frameworks of analysis for the neural representation of animate objects and actions. J. Exp. Biol.: 146; 87–113.

Perrett, D.I., Mistlin, A.J., Harries, M.H., and Chitty, A.J. (1990) Understanding the visual appearance and consequence of hand actions. In Vision and Action: the Control of Grasping, edited by M.A. Goodale, pp. 163–180. Norwood, N.J.: Ablex.

Ratcliff, G., and Davies-Jones, G.A.B. (1972) Defective visual localization in focal brain wounds. Brain: 95; 49–60.

Rizzolatti, G., Berti, A., and Gallese, V. (2000) Spatial neglect: neurophysiological bases, cortical circuits and theories. In Handbook of Neuropsychology, 2nd edition, vol. I, edited by F. Boller, J. Grafman, and G. Rizzolatti, pp. 503–537. Amsterdam: Elsevier Science B.V.

Rizzolatti, G., Camarda, R., Fogassi, M., Gentilucci, M., Luppino, G., and Matelli, M. (1988) Functional organization of inferior area 6 in the macaque monkey: II. Area F5 and the control of distal movements. Exp. Brain Res.: 71; 491–507.

Rizzolatti, G., Fadiga, L., Gallese, V., and Fogassi, L. (1996) Premotor cortex and the recognition of motor actions. Cogn. Brain Res.: 3; 131–141.

Rizzolatti, G., Fogassi, L., and Gallese, V. (1997) Parietal cortex. From sight to action. Curr. Op. Neurobiol.: 7; 562–567.

Rizzolatti, G., and Gallese, V. (1997) From action to meaning. In Les Neurosciences et la Philosophie de l'Action, edited by J.-L. Petit. Paris: Librairie Philosophique J. Vrin.

Rizzolatti, G., Luppino, G., and Matelli, M. (1998) The organization of the cortical motor system: new concepts. Electroencephalogr. Clin. Neurophysiol.: 106; 283–296.

Rizzolatti, G., Matelli, M., and Pavesi, G. (1983) Deficits in attention and movement following the removal of postarcuate (area 6) and prearcuate (area 8) cortex in macaque monkeys. Brain: 106; 655–673.

Rizzolatti, G., Scandolara, C., Matelli, M., and Gentilucci, M. (1981) Afferent properties of periarcuate neurons in macaque monkeys. I. Somato-sensory responses. Behav. Brain Res.: 2; 125–146.

Sakata, H., and Taira, M. (1994) Parietal control of hand action. Curr. Opin. Neurobiol.: 4; 847–856.

Sakata, H., Takaoka, Y., Kawarasaki, A., and Shibutani, H. (1973) Somatosensory properties of neurons in the superior parietal cortex (area 5) of the rhesus monkey. Brain Res.: 64; 85–102.

Schieber, M. (2000) Inactivation of the ventral premotor cortex biases the laterality of motoric choices. Exp. Brain Res.: 130: 497–507.

Schiller, P.H., True. S.D., and Conway, J.L. (1980) Deficits in eye movements following frontal eye-field and superior colliculus ablation. J. Neurophysiol.: 44; 1175–1189.

Shelton, P.A., Bowers, D., and Heilman, K.M. (1990) Peripersonal and vertical neglect. Brain: 113; 191–205.

Snyder, L.H., Batista, A.P., and Andersen, R.A. (1997) Coding of intention in the posterior parietal cortex. Nature: 386; 122–123.

van der Steen, J., Russell, I.S., and James, G.O. (1986) Effects of unilateral frontal eye-field lesions on eye-head coordination in monkey. J. Neurophysiology: 55; 696–714.

Taira, M., Mine, S., Georgopulos, A.P., Murata, A., and Sakata, H. (1990) Parietal cortex neurons of the monkey related to the visual guidance of hand movement. Experimental Brain Research: 83; 29–36.

Tanaka, K., Saito, H.A., Fukada, Y., and Moriya, M. (1991) Coding visual images of objects in the inferotemporal cortex of the macaque monkey. J. Neurophysiol.: 66; 170–189.

Ungerleider, L. G., and Mishkin, M. (1982) Two visual systems. In Analysis of visual behavior, edited by D.J. Ingle, M.A. Goodale, and R.J.W. Mansfield, pp. 549–586. Cambridge, Mass.: MIT Press.

Vallar, G., and Perani, D. (1987) The anatomy of spatial neglect in humans. In Neurophysiological and Neuropsychological Aspects of Spatial Neglect, edited by M. Jeannerod, pp. 235–258. Amsterdam: North-Holland Co./Elsevier Science Publishers.

Volpe, B.T., Ledoux J.E., and Gazzaniga M.S. (1979) Information processing of visual stimuli in an "extinguished" field. Nature: 282; 722–724.

19

What Are the Projective Fields of Cortical Neurons?

Terrence J. Sejnowski

The evolutionary expansion of the cerebral cortex in humans (Allman 1999) underlies our extraordinary flexibility in interacting with the world and each other (Quartz and Sejnowski 2002). Despite a century of research on the cerebral cortex and intense study of the properties of single cortical neurons during the last fifty years, we are still far from understanding how the cortex makes this possible (Sejnowski 1986). The goal of this chapter is to propose a line of research that could help us uncover new principles of cortical function.

The traditional way to study the properties of cortical neurons is to measure their responses to sensory stimuli or to observe their activity during the performance of actions. The inputs to a neuron can be explored by carefully choosing the sensory stimulus, but the receptive field properties only provide part of the information needed to characterize the neuron. The impact of a neuron on other neurons, which is called its projective field, is equally important to its function (Lehky and Sejnowski 1988). Experiments that reveal the projective fields of cortical neurons could provide missing information needed to interpret their function.

Cortical neurons do not act individually but in concert with other neurons in widely separated areas of the cortex (Churchland and Sejnowski 1992). Moreover, the cortex has reciprocal connections with the thalamus, which filters all sensory information that reaches the cortex and is essential in understanding its function (Sherman, chapter 4, this volume). Communication between different parts of the cortex involves the transmission as well as the reception of information (Laughlin and Sejnowski 2003). Thus, the goal of uncovering the projective fields of tightly interacting neurons would complement the knowledge we now have of their receptive fields.

What Is the Receptive Field of a Neuron?

The concept of receptive field is central to understanding the response properties of sensory neurons. It grew out of early studies on sensory receptors by Sherrington (1906), in which he defined an "adequate" stimulus as one that causes the receptor to respond in a specific manner, such as light for photoreceptors, and the receptive field of the receptor as the spatial region of the sensory surface over which the adequate stimulus generates a response. The receptive field of a central neuron is an extension of this concept and is defined as the region of the sensory surface which when stimulated by an adequate stimulus produces an excitatory response. The stimulus that elicits the strongest response is used to define the trigger feature of the cell, which has led to the view of cortical cells as feature detectors.

The receptive field continues to be highly relevant for current experimental studies of the cortex, but it has evolved over the years as we have learned more about cortical neurons. First, the response to the presentation of a sensory stimulus is not purely passive as previously assumed: the response can be modulated by attention and reward expectation (Fries et al. 2001; Richmond, Liu, and Shidara 2003). Moreover, cortical neurons have intrinsic properties that allow neurons to generate activity without an external sensory stimulus (Llinas 1988). Second, there are regions outside the classically defined receptive field that can modulate the response of the cell (Allman, Miezin, and McGuinness 1985; Movellan et al. 2002). These surround influences take into account the context of the stimulus, such as the modulation of perceived color by the color of the surround, which is important for color constancy under varying illumination (Wachtler, Albright, and Sejnowski 2003).

The receptive field, as modified by recent discoveries, is still used today to experimentally characterize cortical neurons. Knowing the receptive field properties of a neuron helps to understand its possible function in the brain. For example, a neuron that has a visual receptive field and prefers vertical edges presumably has a function that is visual and is related in some way to analyzing vertical edges. However, knowing only the receptive field of a neuron is not sufficient to conclude what its function is in the brain.

What Is the Projective Field of a Neuron?

The focus of this chapter is on properties of a neuron that are complementary to those that characterize the receptive field. In addition to knowing what stimulus excites a neuron, it is equally important to know what the impact that neuron has on downstream targets. This is called the "projective field" of a neuron (Lehky and Sejnowski 1988), which will be defined more precisely later, but first we need to motivate the need for the new concept.

Over a hundred years ago, neurologists had determined the projection pathways between different cortical areas and had begun to speculate about their functions (Wernicke 1900). Pathways were traced from the periphery to the cortex and between cortical areas. This was a form of "connectionism" in which the information stored in different parts of the cortex could be linked by direct pathways joining them. More recently, single-unit recording techniques have been used to establish the properties of the projection cells and to show how their properties are transformed along the pathways (Hubel and Wiesel 1962).

In 1988 Sidney Lehky and I developed a neural network model based on the early stages of processing in the visual system whose function was to compute the shape, or more precisely the curvature, of a curved surface from its shaded image (Lehky and Sejnowski 1988). This is called the "shape-from-shading" problem in computer vision. The details about how the network was constructed are not important here; what is interesting is that many of the model neurons in the network had receptive field properties that were similar to the simple cells in the primary visual cortex of monkeys. Simple cells have excitatory and inhibitory subregions that form an oriented spatial filter.

In addition to testing each of the neurons in the model with the same stimuli used by visual physiologists to characterize the receptive fields of real neurons, we were able examine how each neuron affected the output layer, which re-presented the curvature of the shaded object in the input image. Some of the model neurons that resembled simple cells were indeed behaving like a filter and provided information about the orientation of the curved surface; however, others had an entirely different function and instead served as detectors for the direction of the illumination and sign of the curvature. They tended to have a bimodal distribution of firing rates, at either a low rate or a high rate of firing.

The shape-from-shading network model demonstrated first that the visual system was capable of performing this computation and second that it was not possible to deduce the function of a neuron solely from its receptive field prop-erties. In addition, the connections of the neuron to the output were also needed, which we called the projective field of the neuron. In retrospect, it is obvious that a neuron without any output cannot have a computational function and that the same neuron can have more than one function depending on where it projects and what that information is used for. This raises the interesting question of how to experimentally determine the projective fields of neurons in different parts of the brain and whether this will provide further insight into the functions of single neurons. This chapter is an exploration of this question.

How Can the Projective Field of a Neuron Be Measured?

The receptive field of a neuron depends on its synaptic inputs; in contrast, the projective field of a neuron depends on its axonal arborization. Thus, one source of evidence about the projective field of a neuron is its axonal targets. For example, a clue to the function of a cortical neuron may be found if it projects to a motor structure, such as the ventral horn of the spinal cord. However, in addition to knowing where a neuron projects, it is equally important to know the impact that the firing of the neuron has on the downstream target neurons. This aspect of neural function can be approached with stimulation techniques.

In an ideal experiment, a single neuron would be stimulated to assess its impact by comparing a behavioral measure with and without stimulation. This is a technically difficult experiment, but several groups now have used patch recording of cortical neurons in vivo, so it is at least feasible. When a single neuron in the deep layers of rat barrel cortex was stimulated with a train of stimulus, it produced a sequence of small coordinated movements in multiple whiskers that outlasted the stimulus (Brecht et al. 2004). The effectiveness of the stimulus was enhanced during waking states. This suggests that minimal size of a group of neurons that can support a complex behavioral state could be on the low side, perhaps as low as a few dozen.

With extracellular stimulation more than one neuron is activated, but it is not known how many or how strongly they are activated. For example, microstimulation of the motor cortex, designed to produce a minimal muscular contracture, was used to map out the different body regions on the surface of the cortex (Asanuma, Arnold, and Zarzecki 1976). It is interesting that extracellular stimulation of neurons in the visual system, such as the frontal eye fields in prefrontal cortex, and even the primary visual cortex, can stimulate eye movements to the corresponding region of the visual field that was stimulated (Pouget, Fisher, and Sejnowski 1993).

More recently, extracellular stimulation of parts of the visual cortex of awake monkeys has been used to alter their perceptual judgment. In visual area MT, with neurons that respond selectively to the direction of motion, electrical stimulation can bias the monkey to respond in the preferred direction of the neurons in the column that is being stimulated (Salzman et al. 1992). It is not known whether the perception or the action plan of the monkey is altered by this stimulation, since we do not have access to a monkey's visual awareness (Crick and Koch, chapter 23, this volume).

Perhaps the most dramatic example of the impact of stimulating cortical neurons occurs in humans during brain surgery to prevent epilepsy (Penfield and Roberts 1959). Electrical stimulation was used to map out the cortical regions that need to be spared during surgical resection, such as the language areas (figure 19-1). Stimulation of cortical areas in awake patients sometimes evoked sensory percepts and vivid recollections of past events:

> When electrical stimulation recalls the past, the patient has what some of them have called a "flash-back." He seems to re-live some previous period of

time and is aware of those things of which he was conscious in that previous period. It is as though the stream of consciousness were flowing again as it did once in the past. (p. 45)

These early studies used a crude surface electrode and high currents to stimulate the cortex. This almost certainly resulted in current spread, so that the actual neurons that were stimulated could not be determined. At high currents, cortical neurons directly under the electrode are blocked from

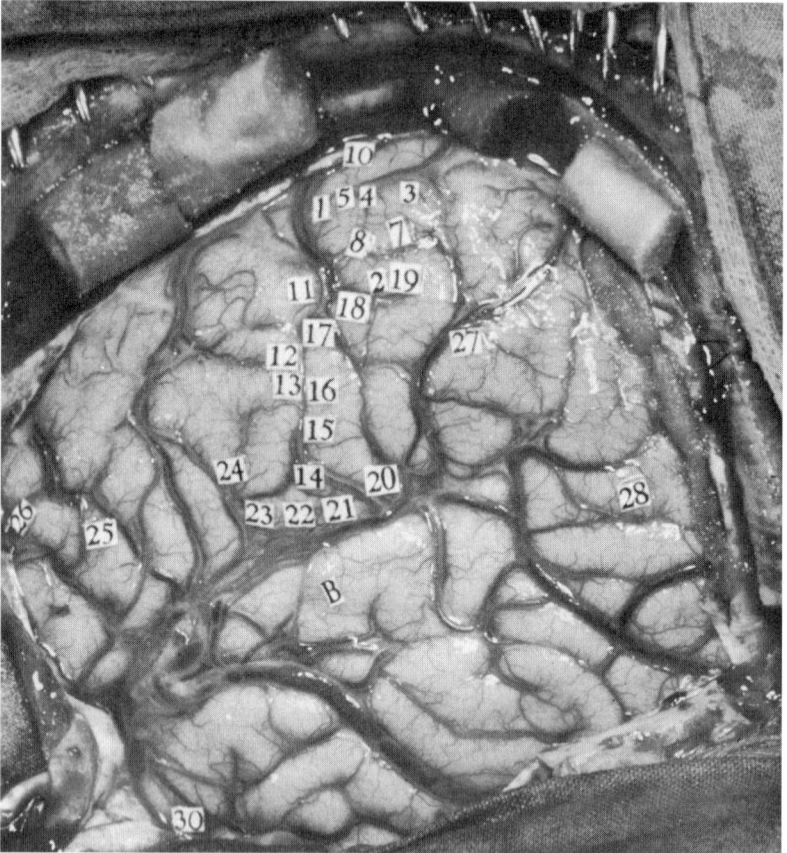

Figure 19-1.   During an operation to cut out the epileptic focus, different parts of the cortex are stimulated to identify speech areas. In this patient, the brain is exposed over the left temporal lobe and the numbered tickets, dropped onto the surface of the cortex, indicate points of positive responses to electrical stimulation. Some of the responses to the stimulation were: 16, tingling in the tip of the tongue; 21, opening of the jaw; 27, the patient said, "Oh, I know what it is. That is what you put in your shoes." After withdrawal of the electrode he said, "foot"; 30, patient tried to talk but could not. Reproduced with permission from Penfield (1959).

firing, and the activated neurons were probably located in a large penumbra surrounding the site of stimulation. At best, these experiments provide only hints about the consequences of stimulating smaller populations of neurons.

The Projective Fields of Networks of Neurons in the Motor Cortex

Ideally, the stimulus strength and duration should not be too low or too high in order to recruit assemblies of neurons typical of those that are active during normal brain states. In a study of neurons in the motor cortex, Graziano, Taylor, and Moore (2002) explored the impact of stimulating neurons in the motor cortex of monkeys with trains of stimuli similar in frequency and duration to the activity recorded in the motor cortex during a limb movement. This level of stimulation produced coordinated contractions in a set of muscles. When the arm region was stimulated, complex arm movements were observed that resulted in a change in the posture of the arm relative to the body of the monkey. When the same location was stimulated, regardless of initial arm position the arm of the monkey ended in the same posture, as shown in figure 19-2. The map of the motor cortex using this pattern of stimulation was a map of body postures rather than a map of individual muscles.

Because many neurons in the same column were stimulated at the same time, it is likely that many other neurons were recruited from brain areas to which these neurons projected. These experiments have been criticized because of the uncertainty about which neurons were being stimulated, but from the perspective of the projective field of a neuron, this is precisely what we want to discover. Just as it is important to discover the "adequate" sensory stimulus to map out the receptive field of a neuron, it is also important to discover the "adequate" stimulation pattern to map out the projective field of a neuron. A train of stimuli corresponding to its natural pattern of activity is more likely to provide adequate stimulation than minimal stimulus. After all, the stimulus used to map the receptive field is not a minimal stimulation, but one that produces the maximal response of the neuron.

Another issue that these experiments raise is the question of whether we should be thinking in terms of individual neurons or populations of neurons in defining the projective field. It is rare that a single neuron will fire in a column of neurons in response to sensory stimulus or during a motor action, so it might be best to focus on the projective field of a network of interacting neurons that typically act together. One of the advantages of the columnar architecture that has been observed throughout the cortex is that neurons with similar properties are often located nearby.

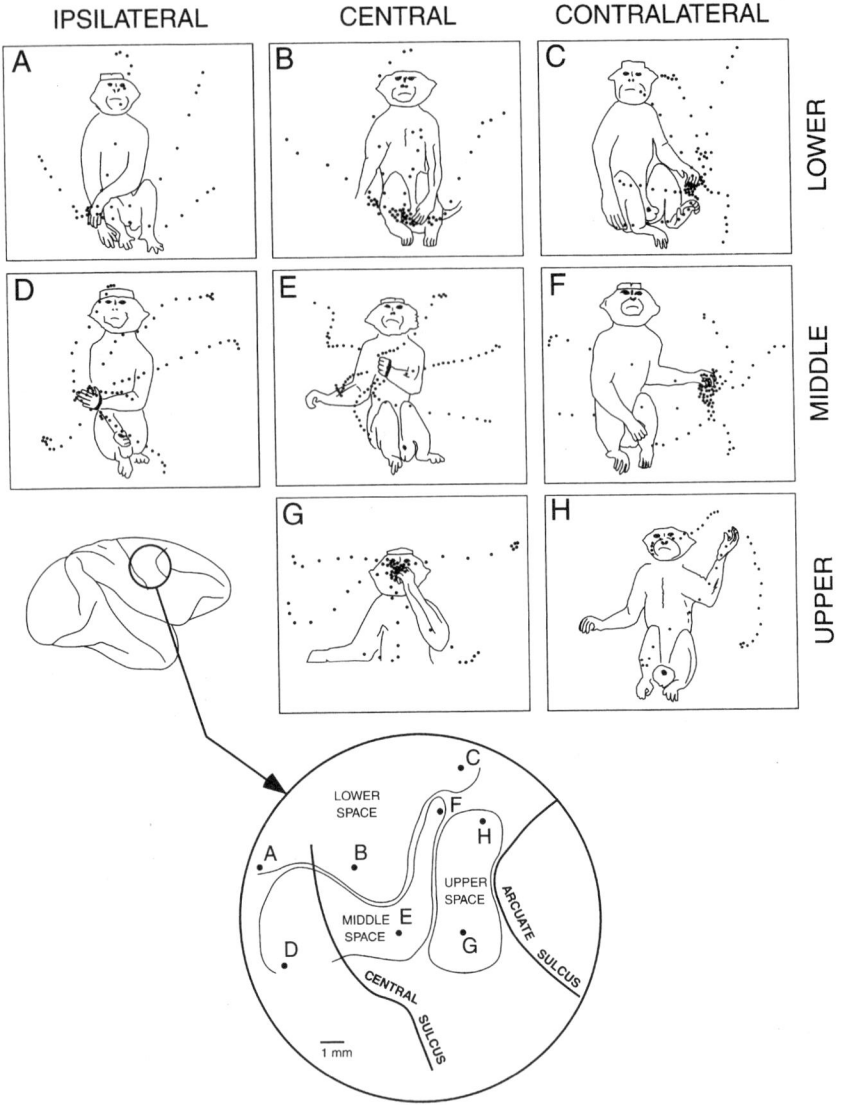

Figure 19-2. Characteristic postures produced by stimulating different regions of the precentral cortex of a monkey. The magnified view at the bottom shows the locations of the stimulation sites. The area to the left of the lip of the central sulcus represents the anterior bank of the sulcus. Stimulation on the right side of the brain caused movements mainly of the left side of the body. For all sites, stimulation trains were presented for 500 ms at 200 Hz. Adapted from Graziano, Taylor, and Moore (2003).

The Penfield Project

In view of this recent work in the motor cortex, the pioneering experiments by Wilder Penfield take on a renewed interest. Could it be that Penfield had been able to stimulate the projective fields of cortical networks and that his anecdotal reports of complex streams of thought were "cognitive postures" similar to the muscular posture elicited by a train of stimuli in the monkey motor cortex? It is not possible to come to any conclusion based on the few extant reports, but similar experiments could be undertaken today under more controlled conditions. It is now common practice to implant an array of electrodes on the surface similar to those used by Penfield and also depth electrodes in the hippocampus and amygdala. Recordings from single cortical neurons in humans are possible (Kreiman, Fried, and Koch 2002) and electrical stimulation also could be undertaken at the same time.

Similar explorations could be undertaken in monkeys, where much more is known about the properties of neurons in different cortical areas and the projections between them. The goal would be to identify which patterns of stimulation produce complex behaviors or can influence the performance of a monkey in a cognitive task. At the same time it should also be possible in monkeys to record simultaneously from several cortical areas to determine the extent to which the pattern of stimulation in one area spreads to other areas of cortex. The overall goal of characterizing the projective fields of neurons in this way could be called the "Penfield Project."

Mirror Neurons

Is it possible for the receptive field and projective field of a neuron to match each other? That is, what if the sensory stimulus that a neuron responds to is in some way related to the motor output that would be elicited by stimulating this neuron? Such neurons have been found in the prefrontal cortex and other brain areas. These neurons respond when a monkey observes complex actions, such as a precision grip of an object, and are also activated when the monkey makes the same action (Rizzolatti, Fogassi, and Gallese 2002). They are called mirror neurons because both their sensory responses and motor correlates match. Although stimulation experiments have not yet been performed for brain areas that contain mirror neurons, similar to those in motor cortex (Graziano, Taylor, and Moore 2002), it is likely that an adequate stimulus would evoke actions similar to those that make the neurons respond.

Mirror neurons appear to solve the inverse problem: given a desired goal defined in terms of a sensory state, such as reaching for an object, what motor commands should be given to achieve that state? Another example is the problem of pronouncing words in a foreign language. Phonemes that are not present in a native language are especially difficult to mimic. If there are mirror neurons in the language areas of the human brain that respond to specific phonemes and when activated produce the same phoneme, then it is possible that a mirror system for language can be used to bootstrap the basic elements of speech and communication (Rizzolatti and Arbib 1999). There are of course many other aspects of language that need to be explained, but the complexity of the articulation should not be underestimated, and a solution to the inverse problem for speech articulation goes a long way toward explaining some aspects of language acquisition.

The existence of mirror neurons raises the question of what aspects of human behavior may be learned by example. Mimicry is a form of supervised learning that does not require a detailed feedback for each muscle. With an internal mirror system, it is possible to learn by observing, which opens up a powerful way to gain skills through social interactions.

Autonomy

We are not simple stimulus-response machines, but many research protocols put subjects into response paradigms that do not allow for the flexibility that may be the most important feature of the cortex. To better understand the contribution of the cortex to behavior during unconstrained conditions, we need to go beyond the artificial dichotomy between the sensory and motor systems that is often imposed by experimental paradigms (Churchland, Ramachandran, and Sejnowski 1994).

It is now possible to record reliably from hundreds of neurons simultaneously from the brains of freely moving animals (Wilson and McNaughton 1993). When a rat is allowed to explore an environment, neurons in its hippocampus fire specifically to locations in the environment, called the "place fields" of the neurons. However, the responses of these neurons are not simple sensory responses, since the place fields are preserved in the dark. Furthermore, the activity patterns of a hippocampal neuron can be shifted by altering the environment or changing the task that the rat is expected to perform (Gothard et al. 1996). This suggests that neurons in the hippocampus are driven as much by internal states of the animal as by the outside world.

Ultimately we need to develop a more sophisticated way to understand the dynamics of the brain's internal states that are not dominated by either sensory inputs or motor actions, but by internal states that are autonomously generated. Autonomous brain activity, sometimes called spontaneous activity, has been known for over a century from electroencephalographic (EEG) recordings from the scalp. Autonomous brain activity is characterized by brain oscillations over a wide range of frequencies, which depend on the behavioral state of the animal (Destexhe and Sejnowski 2001).

Recordings of field potentials in monkeys suggest that some brain oscillations observed in the cortex may be related to attention and expectation (Fries et al. 2001). In humans, new methods for analyzing EEG recordings have revealed that the ongoing EEG may interact with sensory input, which can alter the phases of internal oscillations (Makeig et al. 2002). These recent discoveries about brain dynamics suggest that an important function of the thalamocortical system is to regulate the flow of information between brain areas, making it possible for the brain to rapidly respond to changing contingencies in the world and to anticipate these changes (Salinas and Sejnowski 2001). Rhythmic activity in the cortex continues during sleep states, whose function we are just beginning to appreciate (Sejnowski and Destexhe 2000).

Conclusion

As new methods are devised for recording from thousands of neurons simultaneously in widespread areas of primate brains (Carmena et al. 2003; Hoffman and McNaughton 2002) and as new methods are developed to analyze these signals (Zhang et al. 1998; Abbott and Sejnowski 1999), it should be possible to observe how the flow of information between brain areas is regulated by the behavioral state of the animal. In addition to characterizing the receptive field properties of neurons, it is also important to study their projective fields. The projective fields of interacting assemblies of neurons provide the basic "keyboard" for the brain's interaction with the world and with itself. By combining information about the receptive and projective fields of cortical neurons, an overall picture should emerge of how autonomous behaviors arise from dynamic brain states (Sejnowski 2003).

References

Abbott, L. F., and Sejnowski, T. J. (eds.). (1999) Neural Codes and Distributed Representations: Foundations of Neural Computation. Cambridge, Mass.: MIT Press.

Allman, J. (1999) Evolving Brains. New York: Scientific American Library.

Allman, J., Miezin, F., and McGuinness, E. (1985) Direction- and velocity-specific responses from beyond the classical receptive field in the middle temporal visual area. Perception 14: 105–126.

Asanuma, H., Arnold, A., and Zarzecki, P. (1976). Further study on the excitation of pyramidal tract cells by intracortical microstimulation. Exp. Brain Res. 26, 443–461.

Brecht, M., Schneider, M., Sakmann, B., and Margrie, T. W. (2004) Whisker movements evoked by stimulation of single pyramidal cells in rat motor cortex. Nature 427: 704–710.

Carmena, J. M., Lebedev, M. A., Crist, R. E., O'Doherty, J. E., Santucci, D. M., Dimitrov, D., Patil, P. G., Henriquez, C. S., and Nicolelis, M. A. (2003) Learning to control a brain-machine interface for reaching and grasping by primates. PLoS Biol. 1(2): 193–208.

Churchland, P. S., and Sejnowski, T. J. (1992) The Computational Brain. Cambridge, Mass.: MIT Press.

Churchland, P. S., Ramachandran, V. S., and Sejnowski, T. J. (1994) A critique of pure vision. In: C. Koch and J. Davis (eds.), Large-Scale Neuronal Theories of the Brain, 23–60. Cambridge, Mass.: MIT Press.

Destexhe, A., and Sejnowski, T. J. (2001) Thalamocortical Assemblies: How Ion Channels, Single Neurons and Large-Scale Networks Organize Sleep Oscillations. Oxford: Oxford University Press.

Fries, P., Reynolds, J. H., Rorie, A. E., and Desimone, R. (2001) Modulation of oscillatory neuronal synchronization by selective visual attention. Science 291: 1560–1563.

Gothard, K. M., Skaggs, W. E., Moore, K. M., and McNaughton, B. L. (1996) Binding of hippocampal CA1 neural activity to multiple reference frames in a landmark-based navigation task. J. Neurosci. 16(2): 823–835.

Graziano, M.S.A., Taylor, C.S.R., and Moore, T. (2002) Complex movements evoked by microstimulation of precentral cortex. Neuron 34: 841–851.

Hoffman, K. L., and McNaughton, B. L. (2002) Coordinated reactivation of distributed memory traces in primate neocortex. Science 297: 2070–2073.

Hubel, D. H., and Wiesel, T. N. (1962) Receptive fields, binocular interaction and functional architecture in the cat's visual cortex. J. Physiol. 160: 106–154.

Kreiman, G., Fried, I., and Koch, C. (2002) Single neuron correlates of subjective vision in the human medial temporal lobe. Proc. Natl. Acad. Sci. USA 99: 8378–8383.

Laughlin, S. B., and Sejnowski, T. J. (2003) Communication in neural networks. Science 301: 1870–1874.

Lehky, S., and Sejnowski, T. J. (1988) Network model of shape-from-shading: Neural function arises from both receptive and projective fields. Nature 333: 452–454.

Llinas, R. R. (1988) The intrinsic electrophysiological properties of mammalian neurons: Insights into central nervous system function. Science 242: 1654–1664.

Makeig, S., Westerfield, M., Jung, T.-P., Enghoff, S., Townsend, J., Courchesne, E., and Sejnowski, T. J. (2002) Dynamic brain sources of visual evoked responses. Science 295: 690–694.

Movellan, J. R., Wachtler, T., Albright, T. D., and Sejnowski, T., (2002) Naive Bayesian coding of color in primary visual cortex. Adv. Neural Inf. Process. Syst. 15: 221–228.

Penfield, W., and Roberts, L. (1959) Speech and Brain-Mechanisms. Princeton: Princeton University Press.

Pouget, A., Fisher, S. A., and Sejnowski, T. J. (1993) Egocentric spatial representation in early vision. J. Cogn. Neurosci. 5: 150–161.

Quartz, S., and Sejnowski, T. J. (2002) Liars, Lovers and Heroes: What the New Brain Science Has Revealed About How We Become Who We Are. New York: Harper-Collins.

Richmond, B. J., Liu, Z., and Shidara, M. (2003) Neuroscience: Predicting future rewards. Science 301: 179–80.

Rizzolatti, G., and Arbib, M. A. (1999) From grasping to speech: imitation might provide a missing link: reply. Trends Neurosci. 22(4): 152.

Rizzolatti, G., Fogassi, L., and Gallese, V. (2002) Motor and cognitive functions of the ventral premotor cortex. Curr. Opin. Neurobiol. 12(2): 149–154.

Salinas, E., and Sejnowski, T. J. (2001) Correlated neuronal activity and the flow of neural information. Nat Rev Neurosci 2: 539–550.

Salzman, C. D., Murasugi, C. M., Britten, K. H., and Newsome, W. T. (1992) Micro-stimulation in visual area MT: effects on direction discrimination performance. J. Neurosci. 12: 2331–2355.

Sejnowski, T. J. (1986) Open questions about computation in cerebral cortex. In: J. McClelland and D. Rumelhart (eds.), Explorations in the Microstructure of Cognition, Volume 2: Applications, 372–389. Cambridge: MIT Press.

Sejnowski, T. J. (2003) The computational self. Annals of the New York Academy of Sciences 1001: 262–271.

Sejnowski, T. J., and Destexhe, A. (2000) Why do we sleep? Brain Res. 886(1–2): 208–223.

Sherrington, C. (1906) The Integrative Action of the Nervous System. Cambridge: Cambridge University Press.

Wachtler, T., Sejnowski, T. J., and Albright, T. D. (2003) Representation of color stimuli in awake macaque primary visual cortex. Neuron 37: 681–691.

Wernicke, C. (1900) Grundriss der Psychiatrie in klinischen. Leipzig, Ed Thieme.

Wilson, M. A., and McNaughton, B. L. (1993) Dynamics of the hippocampal ensemble code for space. Science, 261: 1055–1058.

Zhang, K., Ginzburg, I., McNaughton, G. L., and Sejnowski, T. J. (1998) Interpreting neuronal population activity by reconstruction: Unified framework with application to hippocampal place cells. J. Neurophysiol., 79: 1017–1044.

20

How Are the Features of Objects Integrated into Perceptual Wholes That Are Selected by Attention?

John H. Reynolds

Introduction

One of the central observations that has driven modern thinking about vision is that despite the largely parallel architecture of much of early visual processing, the brain is surprisingly limited in its capacity to process visual information. Many of the stimuli that activate the retinae are, in fact, never perceived. Simons and Chabris (1999) recently provided an amusing illustration of this by showing that when attending to a group of people passing a ball to and fro, 58 percent of observers failed to see a man in a gorilla suit standing in plain view at the center of the group, thumping his chest (see figure 20-1). Limited capacity is also strikingly illustrated by the phenomenon of change blindness. In one paradigm, two photographs of real-world scenes are presented successively, separated by a brief blank interval. The two scenes are identical except for a blatant difference such as the presence of an engine under the wing of an airplane at the center of one image and its absence in the next. Despite the large scale of such changes, observers are often unable to tell the difference between the two images, even after many repetitions (for a review, see Rensink 2002).

In both of these examples, stimuli subtending many degrees of visual arc and thus activating hundreds of thousands of retinal cones and tens of millions of rods go unnoticed until attention is directed to them. Thus, there is an enormous amount of information present in the retinal input of which the observer is completely oblivious. This striking reduction in the amount of information from what is implicitly present in the retinal input reflects the limited capacity of some stage (or stages) of sensory processing, decision making, or behavioral control. Somewhere between stimulating the retinae and generating a behavioral response, vast amounts of information are filtered out of the

Figure 20-1. A remarkable example of our limited capacity to process visual information. Observers were asked to monitor the progress of a game in which players passed a ball around. The majority of viewers were unaware of a man in a gorilla suit standing in plain view at the center of the display. If the visual system were unlimited in capacity, observers could have monitored the game while fully processing information about other elements in the scene, including the gorilla. From Simons and Chabris (1999); used by permission of Pion.

visual stream. A key challenge facing visual neuroscientists is to understand the nature of this computational bottleneck and the neural mechanisms that are in place to ensure that behaviorally relevant information is selected to pass through it.

Single unit recording studies in awake, behaving monkeys have characterized competitive selection mechanisms in the extrastriate cortex, and the role they play in selecting out behaviorally relevant stimuli for decision making and control of behavior. These studies have shown that when attention is directed to a location in space, this boosts the responses of neurons that are selective for the stimulus appearing at the attended location, while suppressing neurons that would otherwise respond to stimuli appearing at unattended locations. There is accumulating evidence, described later in the chapter, that this selection process arises from the resolution of competition among mutually suppressive populations of neurons, which is biased in favor of stimuli appearing at the attended location.

The single-unit recording studies that have helped characterized these circuits have focused exclusively on local interactions among feature-selective neurons within a single visual area. However, some studies have found that the visual system is limited in terms of the number of objects that can be processed simultaneously, not the number of individual features. And there is mounting evidence from psychophysical studies of visual attention in humans that competition is more global in nature, spanning multiple visual areas that encode the different features of objects and involving stimulus representations in which different features of one stimulus are selected together while suppressing features of unattended stimulus. These psychophysical studies have found that when an observer attends to one feature of an object (say its color or its orientation) in order to discriminate it, other features of the same object (such as its texture or its motion) can be discriminated at no additional cost. In contrast, performance is severely impaired when making the identical feature judgments about two different objects. Thus, attention selects whole objects for processing, and in so doing suppresses processing of the features of nonselected objects.

Taken together, these psychophysical and neurophysiological observations raise the question of how the different features of an object, encoded by neurons in many different areas of visual cortex, become linked together so that attending to and discriminating one feature of an object causes other features of that object to be selected, while suppressing processing of the features that make up competing objects. Understanding, at the neuronal level, how the features of objects are linked together and selected as wholes promises to enrich our understanding of the relationship between the responses of feature-selective neurons and our experience of features as being integrated into perceptual wholes.

Competitive Circuits in Extrastriate Cortex Mediate Attentional Selection of Behaviorally Relevant Stimuli

Single-unit recording studies of visual processing in alert monkeys have found evidence that selection of behaviorally relevant stimuli depends, in part, on neuronal circuits in extrastriate visual cortex. Moran and Desimone (1985) found that when two stimuli appear together within the receptive fields (RFs) of neurons in the later stages of the ventral visual processing stream, the neuronal response to the pair depends on which of the two stimuli is attended. For each cell, they identified two stimuli: one that elicited a large response when presented alone (the cell's preferred stimulus) and another that elicited a small response (a poor stimulus for the cell). The response to the pair was larger when the monkey attended to the preferred stimulus than when attention was directed to the poor stimulus. Similar changes in response have also been observed in the dorsal stream (Treue and Maunsell 1996).

This is precisely the pattern of results one would expect if multiple stimuli appearing simultaneously in the visual field activate populations of neurons that mutually inhibit one another, and the resulting competition is biased in favor of the behaviorally relevant stimulus (Desimone and Duncan 1995). Reynolds, Chelazzi, and Desimone (1999) tested this model using a behavioral paradigm that allowed them to isolate automatic competitive mechanisms from attentional modulation. Like Moran and Desimone, they recorded neuronal responses when stimuli appeared alone or in pairs within the RF, and examined the effect of directing attention to one of the stimuli. Consistent with earlier results, they found larger responses to the pair when the preferred stimulus was attended than when the poor stimulus was attended. They reasoned that if these changes in firing rate depend on automatic competitive mechanisms that are biased by goal-directed feedback signals, these competitive interactions should be evident even when attention was directed away from the RF of the neuron.

They found that with attention directed away from a neuron's RF, the response to the preferred stimulus was typically suppressed by the addition of the poor stimulus. The magnitude of this suppression was determined by the neuron's selectivity for the two stimuli, such that a very poor stimulus was typically more suppressive than a stimulus that elicited an intermediate response. Further, they found that changes in firing rate with attention depend on these competitive interactions. Directing attention to the poor stimulus magnified its suppressive effect, causing the neuron to respond at a rate that was nearer to the response elicited by the poor stimulus alone (figure 20-2A). Directing attention to the preferred stimulus strongly reduced the suppressive effect of the ignored poor stimulus (see figure 20-2B).

Thus, with two stimuli in the RF, attention to one of them can cause either an increase or a decrease in response, and this depends on the neuron's selectivity for the two stimuli. Recanzone, Wurtz, and Schwarz (1997) and Recanzone and Wurtz (2000) have found a very similar pattern of results in areas MT and MST in the dorsal stream, and Chelazzi et al. (1993) found similar patterns in the inferior temporal cortex, suggesting that selection of behaviorally relevant stimuli is mediated by a canonical biased competitive circuit. These effects are qualitatively different from the effect of attention to a single stimulus appearing alone in the RF, which typically leads to an increase in firing rate (Bushnell, Goldberg, and Robinson 1981; Mountcastle et al. 1987; Spitzer, Desimone, and Moran 1988; Gottlieb, Kusunoki, and Goldberg 1998; Treue and Martinez-Trujillo 1999; McAdams and Maunsell 1999). Such increases are observed regardless of whether the stimulus is a poor or a preferred stimulus for the cell (McAdams and Maunsell 1999; Reynolds, Pasternak, and Desimone 2000). Such increases in response to a single stimulus in the RF may reflect a goal-directed bias in favor of behaviorally relevant stimuli. Other evidence for a goal-directed feedback signal comes from studies conducted by Luck et al. (1997), who found that when attention

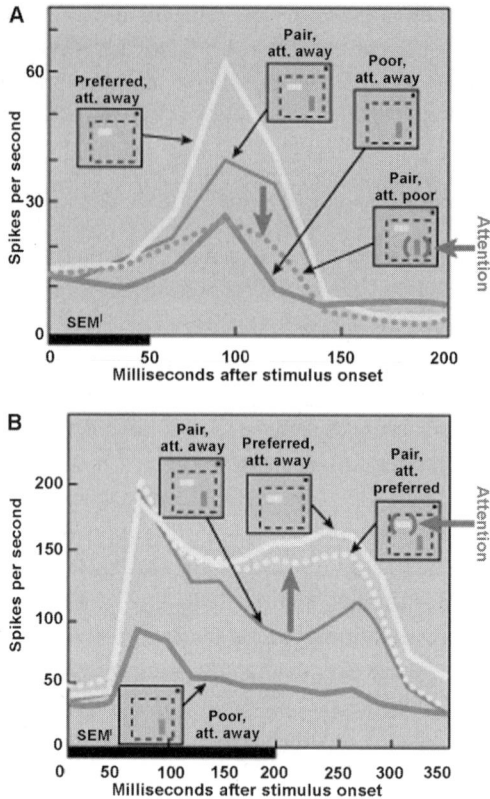

Figure 20-2. Attention to one stimulus of a pair filters out the effect of the ignored stimulus. See color insert.

is directed to a location within a neuron's RF *prior to the appearance* of a stimulus, this causes a spatially selective increase in the neuron's baseline firing rate. Feature-selective increases in baseline activity have also been reported in neurons that are selective for a stimulus that must be stored in memory in order to perform a search task (Chelazzi et al. 1993, 1998) or a match to sample task (Fuster and Jervey 1981; Miyashita and Chang 1988). Consistent with the idea that the elevated baseline activity reflects feedback that biases competition between stimuli, when multiple stimuli subsequently appear in the RF, competition tends to be resolved in favor of the neurons that showed elevated baseline activity during the delay interval.

Reynolds, Pasternak, and Desimone (2000) found evidence that the increase in response that is observed with attention to a single stimulus reflects an increase in the effective strength of a stimulus, which could help put it at a competitive advantage against unattended stimuli. They measured contrast response functions of V4 neurons when monkeys either attended to a location inside the RF or away to a location far from the RF. As illustrated in figure 20-3, they found that attention causes neurons to respond to stimuli that are too faint to elicit a response when unattended. That is, when attention was directed to a stimulus, V4 neurons respond as though the physical contrast of the stimulus had increased. In a related study, Fries et al. (2001) have identified a mechanism that may explain the observed increase in stimulus strength with attention. Recording in area V4, they found that when attention is directed to a stimulus, this increases high-frequency synchronization among different neurons that respond to the attended stimulus. This study used spikes on one electrode to trigger the selection of brief (300-millisecond) segments of local field potential (LFP) data recorded on a separate electrode. The LFP is a measure of neuronal activity averaged over several hundred neurons near the electrode tip, so if the neurons that give rise to the LFP were uncorrelated with the neurons recorded on the other electrode, averaging together many such samples would result in a flat line. Instead, Fries and colleagues found that the spike-triggered average LFP signal often contained two dominant frequency bands: one in the gamma frequency range, and a second, lower frequency component. When attention was directed to the stimulus that drove the spike and LFP channels, the gamma band power often increased, reflecting an increase in the degree to which neurons tended to synchronize at the gamma frequency. An increase in high frequency synchronization would be expected to cause spikes generated by multiple neurons to arrive at target neurons within a narrow time window, magnifying their impact.

The observation that attention to a single stimulus increases the effective contrast of a stimulus, possibly through an increase in high frequency synchronization, coupled with the observation that competition is resolved in favor of the attended stimulus, suggest a simple model to explain how attention biases the competition between multiple stimuli in the visual field. Specifically, cortical cells are "hard wired" to respond preferentially to the highest-contrast stimulus in their RF, and attention capitalizes on this mechanism by increasing the effective contrast of the attended stimulus.

Reynolds and Desimone tested this proposal by varying the relative contrast of two stimuli within V4 neurons' RFs (2003). They found that when attention was directed away from the RF, competitive interactions were biased in favor of the stimulus with the greatest luminance contrast. They also found that attention to the lower-contrast stimulus enabled it to overcome this competitive disadvantage and control the neuronal response. These results suggest that a relatively simple cortical circuit can explain how attentional feedback signals can select behaviorally relevant stimuli, even in the presence of more salient distracters.

In summary, single-unit recording studies in monkeys trained to perform attentionally demanding tasks show that competitive circuits in extrastriate cortex can be biased in favor of behaviorally relevant stimuli, enabling the neurons they activate to elicit a response while suppressing the responses of

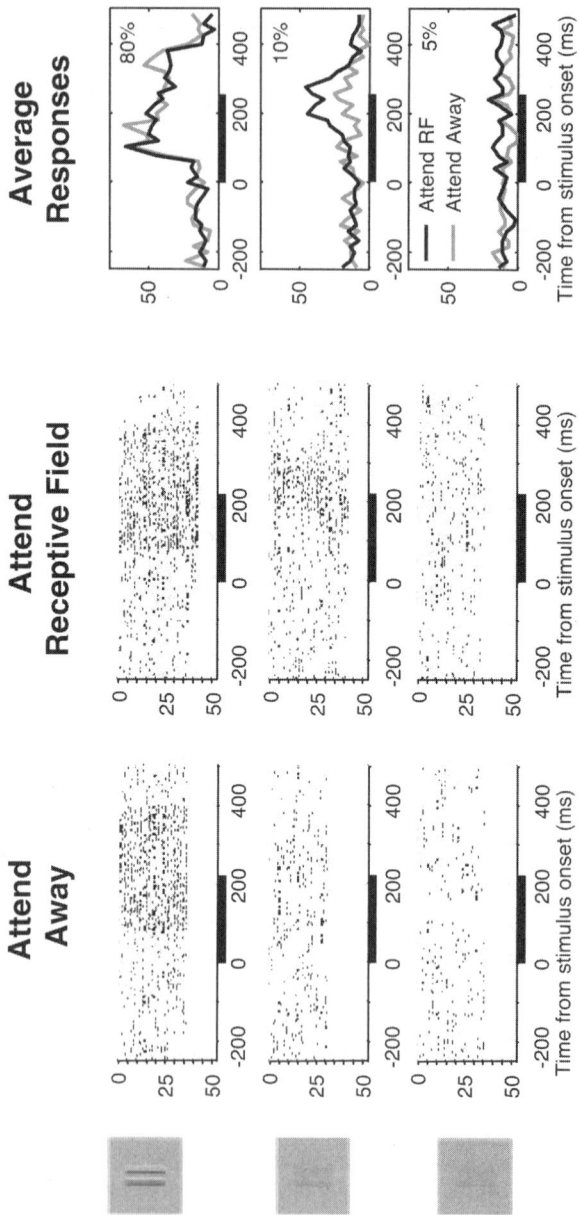

Attend Away — **Attend Receptive Field** — **Average Responses**

Time from stimulus onset (ms)

80%

10%

5%

Attend RF
Attend Away

Time from stimulus onset (ms)

neurons that are activated by ignored stimuli. When an observer is asked to detect the appearance of a target at a particular location, this is believed to generate spatially selective feedback signals in parts of parietal cortex that provide task-appropriate spatial reference frame (Gottlieb et al. 1998), which appears to bias competition within extrastriate cortex in favor of stimuli appearing at the cued location. Similarly, when an observer searches for an object in a cluttered scene, feature-selective feedback biases competition in favor of stimuli that share features in common with the target of the search. As a result, neurons that encode these features are put at a competitive advantage over neurons that do not encode features of the target, enabling them to suppress the responses of neurons that would otherwise be activated by irrelevant distractor stimuli.

Evidence That Competition Occurs among Coherent Objects

While these studies provide compelling evidence that attentional selection occurs through the resolution of competition in favor of the attended stimulus, they have not brought into focus what type of stimulus representation engages in competition. In all these studies, competing stimuli appeared at separate locations, and this confounds selection of locations with selection of objects. The observed competitive interactions could therefore reflect either competitive

---

Figure 20-3. Single neuron example: improved sensitivity with attention. Each of the three rows show responses of a V4 neuron that were elicited by the stimulus when it was presented at different levels of contrast, indicated by the gratings on the left of each row. Contrast increases from 5 percent contrast in the bottom row to 10 percent in the middle row, up to 80 percent contrast in the top row, illustrated by the grating stimuli on the left. The raster grams in the first two columns show trial-by-trial responses to the identical stimulus when it was ignored (*left column*) and when it was attended to (*middle column*). Each row of tick marks corresponds to an individual trial, and each tick mark indicates the time of an individual action potential relative to the onset of the stimulus. The dark bar at the bottom of each panel indicates the time during which the stimulus was present (250 ms duration). The panels in the right column show responses from the first two columns averaged over trials. The two lines in each panel show the mean response when attention was either directed away from the receptive field (gray line) or directed toward the receptive field (black line). The stimulus elicited a robust response when it was presented at 80 percent contrast, whether or not it was attended to. The faintest stimulus (5 percent contrast) was too faint to be detected by the neuron, even when the monkey was attending to its location. However, attention enabled the cell to detect a 10 percent contrast stimulus, which it did not detect when attention was directed elsewhere. That is, attention shifted the neuron's response threshold, making it more sensitive to stimuli appearing at the attended location. Adapted from Reynolds, Pasternak, and Desimone (2000), copyright 2000, with permission from Elsevier.

interactions among locations or competition between coherent objects. However, psychophysical and functional imaging paradigms developed to distinguish between object-based and spatial attention show that attention can select out whole objects for processing, and that this suppresses processing of other objects.

These psychophysical studies can be interpreted as reflecting the resolution of competition among coherent object representations. This section provides a selective review of the psychophysical evidence that attention can and does select out coherent object representations.

Early psychophysical studies showed that observers can attend to one stimulus while suppressing another, even when the two stimuli are superimposed and cannot therefore be selected by spatial attention alone (Neisser and Becklen 1975; Rock and Gutman 1981). Subsequent studies have found that when an observer makes a judgment about one feature of an object (e.g., its color) simultaneous judgments about other features of the same object (e.g., its orientation and motion) do not interfere with the first judgment. This suggests that when attention is directed to one feature of an object, all of the features that make up the object are automatically selected together (Duncan 1984; Valdes-Sosa, Cobo, and Pinilla 1998, 2000; Blaser, Pylyshyn, and Holcombe 2000). Consistent with this intepretation, O'Craven, Downing, and Kanwisher (1999) have found that discriminating one feature of an object results in increased cerebral blood flow in cortical areas that respond to the irrelevant features of the attended object, but not in areas that respond to features of an unattended overlapping object.

Despite this automatic selection of multiple features of a single object, it is difficult to simultaneously monitor features of different objects. For example, Blaser, Pylyshyn, and Holcombe (2000) found that observers can easily report the color and orientation of one object. They cannot, however, simultaneously make accurate judgments about the color of one object and the orientation of another, overlapping object. On trials in which subjects made correct judgments about one object they were severely impaired in their judgments about the other object. Similar results have been reported using overlapping line drawings (Duncan 1984) and overlapping surfaces (Valdes-Sosa, Cobo, and Pinilla 1998, 2000). Vecera and Farah (1994) have found that this inability to simultaneously make accurate judgments about two different objects occurs regardless of whether the stimuli are overlapping or at separate locations. This suggests that attention can select spatially invariant object representations. Consistent with the notion that objects can be the target of attention, studies of inhibition of return (IOR), the impaired processing of a recently attended stimulus, have found IOR for objects even when they move from one location to another (Tipper, Driver, and Weaver 1991). Related results show that attention selects out objects, even when their parts are distributed in space. Baylis and Driver (1993) presented ambiguous displays that could be perceived either as containing one or two objects and found that judging the relative location of two contours was more difficult when the contours were perceived as belonging to two objects rather than one. Egly, Driver, and Rafal (1994) found that attention to one part of an object improves an observer's ability to detect a change in another part of the same object, relative to when the change appeared in an equidistant part of an unattended object. He and Nakayama (1995) have also provided evidence that attention to one part of an object improves processing of other parts of the same object. They found that attention could be directed efficiently to elements that were arranged to form a single surface, relative to when the elements did not cohere into a common surface.

In addition to these studies that show that attention can be allocated to whole objects or surfaces, a number of studies have shown that during visual search, attention can be directed to preattentively defined objects appearing among distracters. For example, Wolfe (1994) found that whereas search for targets defined by conjunctions of two colors is typically inefficient (see Cave and Wolfe 1990), attention can easily select a target defined by the same two colors if the target is preattentively grouped as an object composed of two differently colored parts. In a related study, Rensink and Enns (1998) showed that attention selects preattentively grouped objects by showing that efficient search for an irregularly shaped object among squares can be made ineffi- cient if the irregular shape of the target is perceived as resulting from occlusion. Studies of attentional deficits among patients with right parietal lesions provide further evidence that objects can receive substantial proces- sing before being selected by attention. In extinction, which is most often observed after damage to the right parietal lobe, patients can detect stimuli on either side of the visual field but are often unaware of a contralesional stimulus if it is presented concurrently with a stimulus presented in the un- damaged hemifield. In addition to this spatial attentional deficit, patients studied by Arguin and Bub (1993) and Driver et al. (1994) were also impaired in processing the contralesional side of objects, indicating that objects were processed to some degree despite the failure of attentional control mechanisms in the parietal lobe. Driver (1995) provided further evidence that substantial object processing occurs in neglect patients by showing that their perceptual impairments depend on how parts of objects are grouped into perceptual wholes.

Competitive Interactions among Surfaces

The above-mentioned experiments show that attention can select objects for processing even when two objects appear at the same location in space and that selection of one object impairs processing of the other object. Valdes-Sosa,

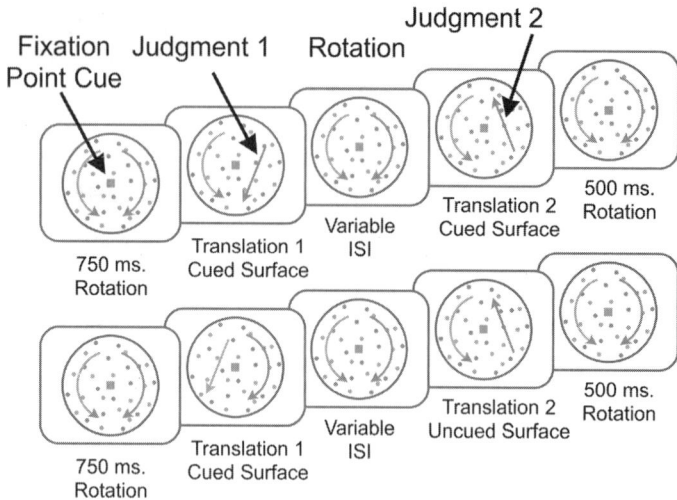

Figure 20-4. Object-based attention task introduced by Valdes-Sosa and colleagues. See color insert.

Cobo, and Pinilla (2000) have identified one of the clearest examples of such selection using an ingenious paradigm designed to examine surface-based attention isolated from the influence of feature-based or spatial attention. Their paradigm is illustrated in figure 20-4. On each trial, observers viewed two random dot patterns (one red, one green) that rotated around a common center in opposite directions. The fixation point color (red or green) acted as an endogenous cue that directed subjects to attend to the surface of the corresponding color. After a brief delay, the cued surface translated briefly in one of eight directions while the uncued surface continued to rotate. After this translation, both surfaces rotated until one of the two surfaces underwent a second brief translation. On each trial, observers reported the directions of the two shifts. The endogenous cue indicated which surface would shift first so observers could ignore one surface in order to reliably report the first shift.

Figure 20-5, adapted from a study that replicated the results of Valdes-Sosa and colleagues, shows average performance in this task. Observers were able to report the first translation accurately and could also report the second translation of the *cued* surface accurately even when two successive translations occurred with an interstimulus interval (ISI) as short as 150 milliseconds. If the *uncued* surface translated second, however, judgments were severely impaired, and this impairment lasted approximately 600 milliseconds.

These results can be interpreted as arising from the limited capacity of the visual system to process information about multiple objects coupled with the slow dwell time of attention (Valdes-Sosa, Cobo, and Pinilla 2000). According to this interpretation, the endogenous cue, which was 100 percent valid, enabled observers to attend to one of the two surfaces and their performance

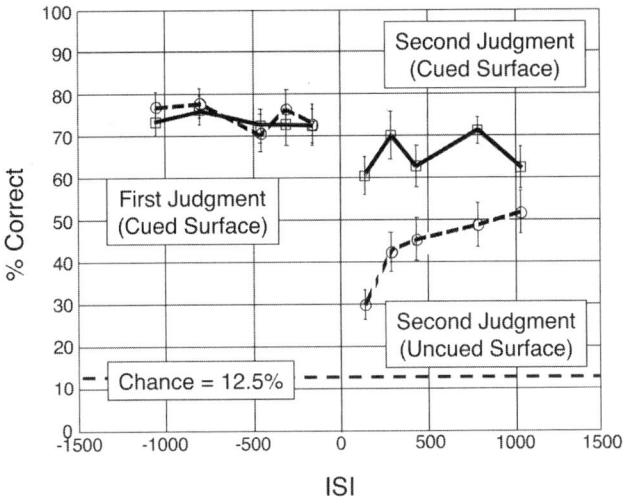

Figure 20-5. Mean accuracy across eight subjects in reporting the direction of two successive translations averaged across trials in which the fixation point color indicated which surface would translate first. Chance performance, indicated by dashed horizontal line, was 12.5 percent. ISI indicates the duration of the interval between the offset of the first translation and the onset of the second translation. By convention, negative ISIs correspond to judgments of the first translation, and positive ISIs correspond to judgments of the second translation. Lines indicate whether the first and second translations occurred on the same surface (solid line) or different surfaces (dashed line). Error bars indicate standard errors of mean performance across subjects. Observers correctly reported the direction of the first translation (which was always on the surface cued by the fixation point), regardless of whether the second translation also occurred on the cued surface (solid line) or occurred on the other surface and regardless of how soon after the first translation the second translation occurred. Subjects reported the second translation accurately if it occurred on the cued surface. However, subjects were severely impaired in making judgments about the second translation when it occurred on the uncued surface. This impairment was greatest at the shortest ISIs tested (150 milliseconds) and gradually diminished over time. Adapted from Mitchell et al. (2003) copyright 2003, with permission from Elsevier.

was therefore high. However, either surface could undergo the second translation with equal probability, and therefore, observers had to divide attention between the two surfaces. The extra cost of attending to two objects caused their performance on the second judgment to be poorer, on average, than their performance on the first judgment. The observation that this reduction in performance occurred primarily when judgments were of the surface that was not endogenously cued can be explained as resulting from the initial allocation of attention to the endogenously cued surface.

Previous studies show that attention remains locked on a cued stimulus for a similar period of several hundred milliseconds, during which time judgments of other stimuli are impaired (Duncan, Ward, and Shapiro 1994).

The biased competition model offers an alternative explanation for these behavioral effects, which also explains why the impairment in the second judgment is so tightly time locked to the first translation. As noted above, competitive circuits in extrastriate cortex can be biased by internally generated feedback signals that favor an endogenously cued stimulus. They can also be biased by stimulus-driven biases such as differences in the relative salience of competing stimuli. For example, as noted by Reynolds and Desimone (2003), an increase in the luminance contrast of one stimulus can cause it to suppress responses elicited by lower-contrast stimuli nearby, even when attention has been endogenously cued to a location far from the RF to perform an attention-demanding task. The first translation causes a momentary increase in the salience of the translating surface, and this increase in salience might cause competition to be resolved in favor of the surface that underwent the first translation. If so, the resulting suppression of the responses of neurons that encode the nontranslating surface would be expected to impair processing of information about the uncued surface.

To test this hypothesis, Reynolds, Alborzian, and Stoner (2003) repeated the original experiment but removed the endogenous cue by replacing the colored fixation point with a noninformative gray fixation point. If the surface-dependent difference in the accuracy of second translation judgments were due to the slow withdrawal of endogenously cued attention, then this performance difference should disappear when the endogenous cue was removed. However, if the impairment simply reflected the automatic resolution of competition between the two surface representations, then a translation of one surface should impair judgment of the subsequent translation of the other surface, even in the absence of the endogenous cue.

The results of this experiment are presented in figure 20-6, which shows accuracy of translation judgments following removal of the endogenous cue. Removing the endogenous cue caused a small reduction in the accuracy of judgments of the first translation, indicating that observers used the endogenous cue when present. However, judgments of the first translation were still quite accurate, indicating that observers could easily attend to both surfaces. Removing the endogenous cue did not cause a statistically significant change in the accuracy of second translation judgments, indicating that the impairment in judging the uncued surface was not the result of the endogenous cue. Successive translations of the same surface were judged accurately, even though the endogenous cue was absent. However, translation of one surface strongly impaired the accuracy of judgments of the other surface. In a related experiment, Reynolds, Alborzian, and Stoner (2003) found that when the onset of one surface was delayed, which would be expected to cause a transient increase in response among neurons selective for the features of the delayed onset surface, this also resulted in a marked impairment in observers' ability to report brief translations on the old surface, as compared to the new. While

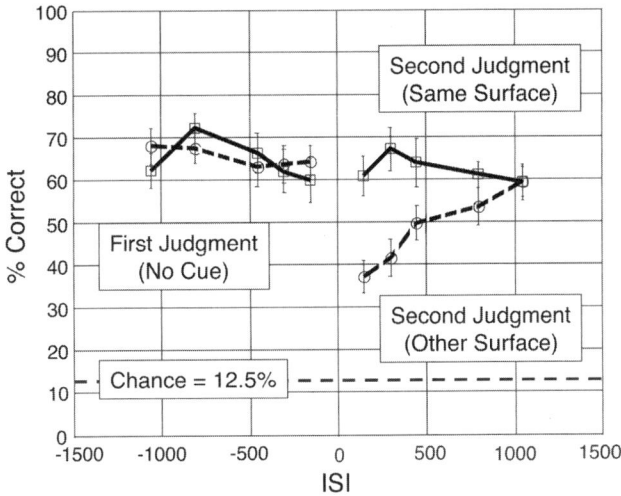

Figure 20-6. Mean accuracy across eight subjects in reporting the direction of two successive translations for trials in which the fixation point was gray. Conventions are identical to those used in figure 20-5. Despite the absence of the endogenous cue, observers were able to report the direction of the first translation on 65.1 percent of trials, and their performance did not depend on whether the second translation also occurred on the cued surface (solid line) or occurred on the other surface (dashed line) and did not depend on ISI. Subjects reported the second translation accurately if the same surface translated twice. However, if one surface translated, this severely impaired the observers' ability report a shift of the other surface. As was the case in the cued condition, this impairment was greatest at the shortest ISIs tested (150 milliseconds) and gradually diminished over time. Adapted from Mitchell et al. (2003) copyright 2003, with permission from Elsevier.

these results are compatible with the predictions of the biased competition model, the most direct test of the model is to record the responses of neurons activated by the two surfaces and see whether neurons that respond to one of two superimposed stimuli are suppressed under conditions that lead to impaired behavioral performance. In a probe experiment, Fallah et al. (2002) have recorded the responses of V4 neurons using the same stimuli employed by Reynolds, Alborzian, and Stoner (2003). One set of dots was of the neuron's preferred color and the other was of an isoluminant nonpreferred color. The response elicited by the preferred-color surface was consistently suppressed by the addition of the nonpreferred surface. Moreover, the suppression caused by the nonpreferred surface was greater when it, rather than the preferred surface, had a delayed onset. Remarkably, the duration of this differential suppression lasted for hundreds of milliseconds with suppression being maximal immediately following the appearance of the poor stimulus, and falling off over a time course that was similar to the recovery time of the behavioral impairment observed by Valdes-Sosa, Cobo, and Pinilla (2000) and Reynolds, Alborzian, and Stoner (2003). These results show that competitive circuits in area V4 are not limited to mediating competition between spatial locations and may play a role in object-based attention. The similarity of the time courses of the neuronal and behavioral impairments lends support to this proposal.

Conclusions

This chapter has outlined recent progress in our understanding of how the brain compensates for the limited capacity of the visual system and has emphasized limitations of this understanding. Competitive selection mechanisms have been identified in extrastriate visual cortex that can be biased in favor of behaviorally relevant stimuli. These mechanisms subserve selection by ensuring that neurons that encode features of behaviorally relevant stimuli remain active and are, therefore, able to transmit information about the selected stimulus to other areas of the brain associated with behavioral control. Despite progress in our understanding of how these competitive selection mechanisms mediate selection of spatially separate stimuli, relatively little is known about how coherent objects are selected for processing. Mounting psychophysical evidence shows that attention can and does select out coherent objects for processing. When attention is directed to one feature of an object, other features of the same object can be evaluated at no additional cost, but judgments of feature of unattended objects are impaired. Discovering where objects are represented as coherent wholes that can be selected by attention for processing is a major open question now confronting systems neuroscience.

One exciting possibility is that the competitive circuits that mediate spatial selection may also play a role in competitive selection of whole objects. Consistent with this, single-unit recordings in area V4 show that competitive interactions are observed among stimuli that are superimposed at the same location and cannot, therefore, be selected by a purely spatial selection mechanism. Moreover, when one of two superimposed stimuli is selected, neurons that encode the other stimulus are suppressed, and the time course of this neuronal suppression is similar to the duration over which human observers are impaired in judging the suppressed stimulus. These results appear to suggest that competitive circuits in extrastriate cortex may mediate competition among whole objects. It remains to be seen how features of objects are integrated in the first place so that attending to one feature of an object causes other features of the same object to be selected, while suppressing neurons that encode features of unattended objects.

Acknowledgments   Funding provided by the Fritz B. Burns Foundation, the F. M. Kirby Foundation, Inc., and the George Hoag Family Foundation.

References

Arguin M, and Bub DN (1993) Evidence for an independent stimulus-centered spatial reference frame from a case of visual hemineglect. *Cortex* Jun;29(2): 349–357.

Baylis GC, and Driver J (1993) Visual attention and objects: evidence for hierarchical coding of location. *J Exp Psychol Hum Percept Perform* Jun;19(3): 451–470.

Blaser E, Pylyshyn ZW, and Holcombe AO (2000) Tracking an object through feature space. *Nature* Nov 9;408(6809):196–199.

Bushnell MC, Goldberg ME, and Robinson DL (1981) Behavioral enhancement of visual responses in monkey cerebral cortex. I. Modulation in posterior parietal cortex related to selective visual attention. *J Neurophysiol* Oct;46(4):755–772.

Cave KR, and Wolfe JM (1990) Modeling the role of parallel processing in visual search. *Cognit Psychol* Apr 22;22(2):225–271.

Chelazzi L, Duncan J, Miller EK, and Desimone R (1998) Responses of neurons in inferior temporal cortex during memory-guided visual search. *J Neurophysiol* Dec;80(6):2918–2940.

Chelazzi L, Miller EK, Duncan J and Desimone R (1993) A neural basis for visual search in inferior temporal cortex. *Nature* 27:363(6427):345–347.

Desimone R and Duncan J (1995) Neural mechanisms of selective visual attention. *Annu Rev Neurosci* 18:193–222.

Driver J (1995) Object segmentation and visual neglect. *Behav Brain Res* 1995 Nov;71(1–2):135–146.

Driver J, Baylis GC, Goodrich SJ, and Rafal RD (1994) Axis-based neglect of visual shapes. *Neuropsychologia* Nov;32(11):1353–1365.

Duncan J (1984) Selective attention and the organization of visual information. *J Exp Psychol Gen* 113(4):501–517.

Duncan J, Ward R, and Shapiro K (1994) Direct measurement of attentional dwell time in human vision. *Nature* May 26;369(6478):313–315.

Egly R, Driver J, and Rafal RD (1994) Shifting visual attention between objects and locations: evidence from normal and parietal lesion subjects. *J Exp Psychol Gen* 123(2):161–177.

Fallah M, Stoner GR, and Reynolds JH (2002) Competitive selection of superimposed stimuli in V4. *Society for Neuroscience Abstracts* 57:6.

Fries P, Reynolds JH, Rorie AE, and Desimone R (2001) Modulation of oscillatory neuronal synchronization by selective visual attention. *Science* Feb 23;291 (5508):1560–163.

Fuster JM and Jervey JP (1981) Inferotemporal neurons distinguish and retain behaviorally relevant features of visual stimuli. *Science* May 22;212(4497): 952–955.

Gottlieb JP, Kusunoki M, and Goldberg ME (1998) The representation of visual salience in monkey parietal cortex. *Nature* Jun 29;391(6666): 481–484.

He ZJ and Nakayama K (1995) Visual attention to surfaces in three-dimensional space. *Proc Natl Acad Sci USA* Nov 21;92(24):11155–11159.

Luck SJ, Chelazzi L, Hillyard SA, and Desimone R (1997) Neural mechanisms of spatial selective attention in areas V1, V2, and V4 of macaque visual cortex. *J Neurophysiol* Jan;77(1):24–42.

McAdams CJ and Maunsell JHR (1999) Effects of attention on orientation-tuning functions of single neurons in macaque cortical area V4. *J Neurosci* Jan 1;19(1):431–441.

Miyashita Y and Chang HS (1988) Neuronal correlate of pictorial short-term memory in the primate temporal cortex. *Nature* Jan 7;331(6151):68–70.

Moran J and Desimone R (1985) Selective attention gates visual processing in the extrastriate cortex. *Science* Aug 23;229(4715):782–784.

Mountcastle VB, Motter BC, Steinmetz MA, and Sestokas AK (1987) Common and differential effects of attentive fixation on the excitability of parietal and prestriate (V4) cortical visual neurons in the macaque monkey. *J Neurosci* Jul;7(7):2239–2255.

Neisser U and Becklen R (1975). Selective looking: Attending to visually significant events. *Cognitive Psychology* 7:480–494.

O'Craven KM, Downing PE, and Kanwisher N (1999) fMRI evidence for objects as the units of attentional selection. *Nature* 401(6753):584–587.

Recanzone GH and Wurtz RH (2000) Effects of attention on MT and MST neuronal activity during pursuit initiation. *J Neurophysiol* 83(2):777–790.

Recanzone GH, Wurtz RH, and Schwarz U (1997) Responses of MT and MST neurons to one and two moving objects in the receptive field. *J Neurophysiol* Dec;78(6):2904–2915.

Rensink RA (2002) Change detection. *Annu Rev Psychol* 53:245–277.

Rensink RA and Enns JT (1998) Early completion of occluded objects. *Vision Res* Aug;38(15–16):2489–2505.

Reynolds JH, Alborzian S, and Stoner GR (2003) Exogenously cued attention triggers competitive selection of surfaces. *Vision Res* Jan;43(1): 59–66.

Reynolds JH, Chelazzi L, and Desimone R (1999) Competitive mechanisms subserve attention in macaque areas V2 and V4. *J Neurosci* Mar 1;19(5):1736–1753.

Reynolds JH and Desimone R (2003) Interacting roles of attention and visual salience in V4. *Neuron* Mar 6;37(5):853–863.

Reynolds JH, Pasternak T, and Desimone R (2000) Attention increases sensitivity of V4 neurons. *Neuron* Jun;26(3):703–714.

Rock I and Gutman D (1981) The effect of inattention on form perception. *J Exp Psychol Hum Percept Perform* Apr;7(2):275–285.

Simons DJ and Chabris CF (1999) Gorillas in our midst: sustained inattentional blindness for dynamic events. *Perception* 28:1059–1074.

Spitzer H, Desimone R, and Moran J (1988) Increased attention enhances both behavioral and neuronal performance. *Science* Apr 15;240(4850):338–340.

Tipper SP, Driver J, and Weaver B (1991) Object-centered inhibition of return of visual attention. *Q J Exp Psychol A* 43(2):289–298.

Treue S and Martinez-Trujillo J (1999) Feature-based attention influences motion processing gain in macaque visual cortex. *Nature* 399(6736):575–579.

Treue S and Maunsell JH (1996) Attentional modulation of visual motion processing in cortical areas MT and MST. *Nature* Aug 8;382(6591):539–541.

Valdes-Sosa M, Cobo A, and Pinilla T (1998) Transparent motion and object-based attention. *Cognition* 66(2):B13–B23.

Valdes-Sosa M, Cobo, A and Pinilla T (2000) Attention to object files defined by transparent motion. *J Exp Psychol Hum Percept Perform* 26(2):488–505.

Vecera SP and Farah MJ (1994) Does visual attention select objects or locations? *J Exp Psych: General* 123:146–160.

Wolfe JM (1994) Guided Search 2.0: A revised model of visual search. *Psychonomic Bulletin and Review* 1(2):202–238.

21

Where Are the Switches on This Thing?

L. F. Abbott

Introduction

Controlled responses differ from reflexes because they can be turned off and on. This is a critical part of what distinguishes animals from automatons. How does the nervous system gate the flow of information so that a sensory stimulus that elicits a strong response on some occasions, evokes no response on others? A related question concerns how the flow of sensory information is altered when we pay close attention to something as opposed to when we ignore it. Most research in neuroscience focuses on circuits that directly respond to stimuli or generate motor output. But what of the circuits and mechanisms that control these direct responses, that modulate them and turn them off and on?

Self-regulated switching is vital to the operation of complex machines such as computers. The essential building block of a computer is a voltage-gated switch, the transistor, that is turned off and on by the same sorts of currents that it controls. By analogy, the question of my title refers to neural pathways that not only carry the action potentials that arise from neural activity, but are switched off and on by neural activity as well. By what biophysical mechanisms could this occur?

In the spirit of this volume, the point of this contribution is to raise a question, not to answer it. I will discuss three possible mechanisms— neuromodulation, inhibition, and gain modulation—and assess the merits and short-comings of each of them. I have my prejudices, which will become obvious, but I do not want to rule out any of these as candidates, nor do I want to leave the impression that the list is complete or that the problem is in any sense solved.

Neuromodulators versus Neurotransmitters

Neuromodulators can dramatically alter the responsiveness of neurons and the transmission properties of synapses (Marder and Calabrese 1996). They also have profound impacts on behavioral responsiveness—wakefulness and sleep being a prime example. However, neuromodulators are thought to work on a rather coarse scale, both temporally and spatially. Thus, while they might be able to activate large numbers of neurons on a seconds time scale, they may not be able to target small enough groups of specific neurons rapidly enough to fulfill the switching role we seek. As a result, neuromodulation is not generally considered to be a candidate mechanism for rapid and precise switching of complex neural circuits and responses. Nevertheless, it is good to keep in mind that this standard wisdom may be wrong (see Sherman and Guillery 1998), and neuromodulation may play a bigger role in neuronal switching than we currently suspect.

Neurons or Synapses

The state of a neural circuit and the information that it represents is generally associated with the level and pattern of activity of its neurons. Following this conventional view, as I do here, switching in neural circuits corresponds to modifying that neural activity. However, it is worth mentioning an alternative way of thinking. Synapses are remarkably plastic over a large range of time scales (see Abbott and Nelson 2000, for example). This raises the possibility that a neural circuit might more appropriately be characterized by the state of its synapses rather than by the state of its neurons. Neural activity might then play a switching role by putting synapses into an appropriate functional state, and switching might be accomplished primarily by modifying the synapses within a neural circuit. In the examples I discuss later in this chapter this is not the case; switching is accomplished by modifying neurons not synapses.

Hard versus Soft Switches

Attention can have both modulatory and gating effects on neuronal responses. For some neurons, attention modulates response amplitude while leaving selectivity unaltered (Connor et al. 1996; McAdams and Maunsell 1999; Treue and Martínez-Trujillo 1999). For other neurons, attention has a more dramatic gating effect, making it difficult to evoke any response at all in the absence of attention (Gottlieb, Kusunoki, and Goldberg 1998; Seidemann,

Zohary, and Newsome 1998). These two types of modification correspond to what we might call soft and hard switching.

Due the existence of a threshold for action potential generation, hard switching can be accomplished by strong inhibition. In other words, a neuron can be switched from a responsive to a nonresponsive state by hyperpolarizing it below threshold so it cannot fire any action potentials. Such a mechanism has been proposed in the context of shifter circuits, an interesting discussion of and proposal for switching in neural circuits by Anderson and Van Essen (1987) and expanded upon by Olshausen, Anderson, and Van Essen (1993). An observed correlate of this form of switching may be the up- and down-state behavior seen in intracellular in vivo recordings (see Stern, Jaeger, and Wilson 1998, for example).

Using hard, on-off switching through strong inhibition requires us to postulate that inhibitory neurons play a much more active and precise role in cognitive processing than they are generally given credit for. The general picture that emerges has circuits of excitatory neurons responding to stimuli and generating motor responses, with a network of inhibitory neurons controlling these excitatory networks and the responses they generate. A puzzle here might be the relatively small fraction of inhibitory neurons in cortex given that these are supposed to be responsible for controlling and switching sensory pathways and motor responses in a precise manner.

An alternative to the hard switching provided by strongly inhibiting neurons is a form of soft switching produced by modulating the gain of neurons. Gain modulation appears to be a primary mechanism by which cortical neurons nonlinearly combine input signals (reviewed by Salinas and Thier 2000). It shows up in a wide range of contexts including the gaze-direction dependence of visual neurons in posterior parietal cortex (Andersen and Mountcastle 1983; Andersen, Essick, and Siegel 1985), and the effects of attention on neurons in areas V4 (Connor et al. 1996; McAdams and Maunsell 1999) and MT (Treue and Martínez-Trujillo 1999). Gain modulation has been proposed as a mechanism for generating a variety of "nonclassical" receptive field properties of neurons in primary visual cortex (Heeger 1992), and for the neural computation of coordinate transformations relevant for tasks ranging from visually guided reaching (Zipser and Andersen 1988; Salinas and Abbott 1995; Pouget and Sejnowski 1997) to invariant object recognition (Salinas and Abbott 1997).

Although gain modulation is clearly associated with attentional effects, it is not obvious how it could be used to generate switching in neuronal circuits. Gain modulation is a more subtle effect, sometimes modify responses by only 10–20 percent, than the on-off switching we seek. Thus, for gain modulation or any other soft switching mechanism to be a viable candidate for neuronal switching, additional mechanisms must be introduced to amplify the effects of modest gain modulations to provide all-or-none, on-off switching. The remaining sections are devoted to this issue.

Amplifying Gain Modulation

How can a small amount of gain modulation lead to dramatic changes in behavior? Here, in work done in collaboration with Jian Zhang (Zhang and Abbott 2000), I discuss two possible answers to this question. These involve the circuits shown in figure 21-1. In the model of figure 21-1A, responses in a downstream neuron (the upper neuron in figure 21-1A) are normally suppressed by a rather precise balance between excitatory and inhibitory input. Gain modulation of the network neurons driving this downstream neuron (the starred neurons in figure 21-1A) disrupts this balance allowing strong excitatory input to drive the downstream neuron. Amplification arises due to the cancellation of strong excitatory drive by an equally strong balanced inhibitory input (see Shadlen and Newsome 1994; Troyer and Miller 1997, for example).

The second example uses gain modulation to alter the effective recurrent connections of a neural circuit. In the circuit of figure 21-1B, the synaptic connections among the bottom row of neurons are funneled through the upper row of neurons (such an architecture was studied in a different context by Hahnloser et al. 1999). Neurons in the upper row are subject to gain modulation (as indicated by the stars). Modifying the gain of the upper row of neurons changes the synaptic connectivity among the lower row of neurons, which can dramatically alter the selectivity and response amplitude of the network activity evoked by feedforward input.

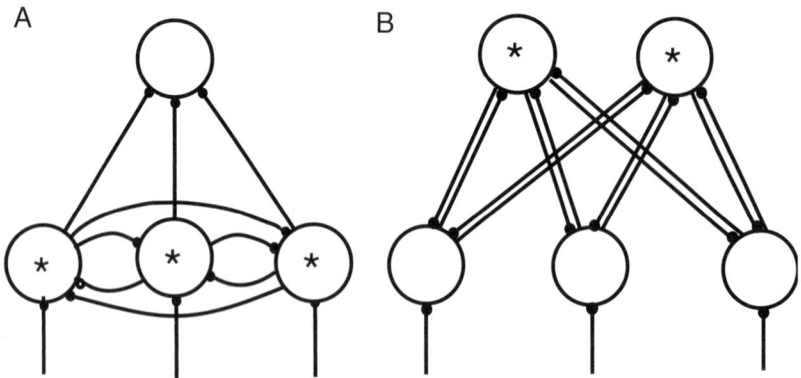

Figure 21-1. Two networks that amplify gain modulation effects. In both panels, neurons subject to gain modulation are denoted by stars. Feedforward input arising from a hypothetical stimulus enters the network through the afferents indicated at the bottom of the circuit diagrams. (A) A network of recurrently connected, gain-modulated neurons drives a downstream neuron. (B) A network of neurons (*lower row*) is interconnected by pathways that pass through a set of gain-modulated neurons (*upper row*).

Switching through Modulation of Balanced Synaptic Input

In our implementation of the circuit of figure 21-1A, hundred recurrently connected integrate-and-fire neurons with gain modulation (represented by the starred units in figure 21-1A) drive a single downstream integrate-and-fire neuron (the unstarred neuron in figure 21-1A). Feedforward inputs to the recurrently coupled network neurons were chosen so that each of them is tuned to a parameter characterizing the sensory input, which we refer to as an image orientation angle in subsequent figures. The connections between the recurrent network and the downstream neuron were developed by an anti-Hebbian learning rule, which established a balance between excitatory and inhibitory inputs to the downstream neuron (Zhang and Abbott 2000).

Gain modulation is applied to the neurons in the recurrent network by adjusting the effective membrane time constant of the integrate-and-fire model. Figure 21-2A shows the effect of the gain modulation on the response of one network neuron to different stimulus angles. This is a typical gain-modulated response tuning curve similar to those seen experimentally (see McAdams and Maunsell 1999, for example). The modulation is multiplicative, and the modulated neurons retain their selectivity when their response amplitudes are modified. When the network is gain modulated, different neurons are modulated differently. Figure 21-2B shows the effect of modulation on the entire population response of the network to a single stimulus orientation. The amount of modulation is small for all network neurons.

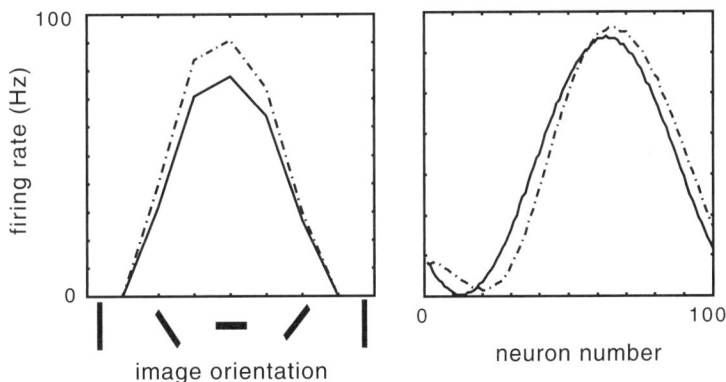

Figure 21-2. Gain modulation of the network neurons of figure 21-1A. *Left*, each network neuron is tuned to the orientation of the stimulus as indicated by its firing rate in response to different image orientations. Solid curve is without gain modulation and dashed curve is with gain modulation. *Right*, the entire population response to a given stimulus orientation is plotted by graphing the firing rate of each neuron as a function of its identifying index. Gain modulation shifts the responses indicated by the solid curve to responses indicated by the dashed curve.

Figure 21-3 shows that modest gain modulation of the network neurons has a large impact on the response of the downstream neuron. Without modulation, the downstream neuron responds weakly, if at all, to the stimulus (upper panel of figure 21-3). When the population activity of the network neurons is slightly modified by gain modulation, the cancelation of balanced excitatory and inhibitory inputs responsible for this weak response is disrupted, and the downstream neuron responds to the stimulus vigorously and in a selective manner (lower panel of figure 21-3).

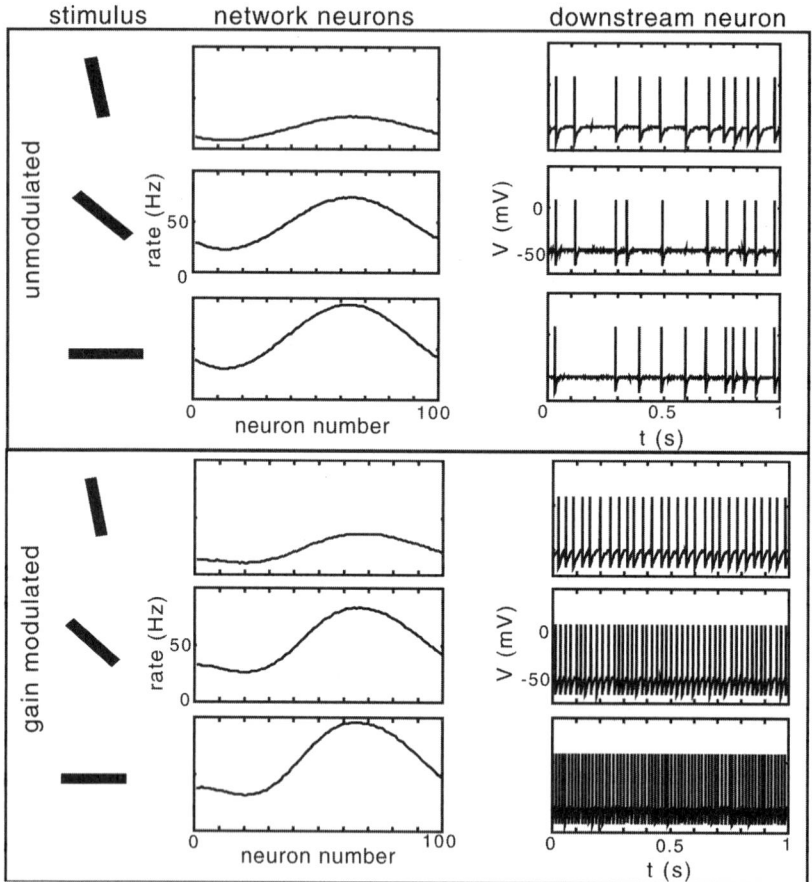

Figure 21-3. Responses of the network and downstream neurons to different stimulus orientations without and with gain modulation. Stimulus orientations are indicated at left. Population responses of the network neurons are plotted in the left column of graphs in the same format as in the right panel of figure 21-2. The membrane potential of the downstream neuron is shown in the right column of plots. *Top*, without gain modulation, the downstream neuron responds weakly to the stimulus. *Bottom*, with gain modulation, the downstream neuron shows a strong, tuned response to the stimulus.

Any system for attention-based switching must distinguish attended stimuli from large-amplitude (i.e., high-contrast) stimuli. In another words, we want to pay attention to stimuli because they are significant, not simply because they are intense. This is a potential problem with switching achieved by hyperpolarizing a neuron below its action potential threshold. For a given level of hyperpolarization, an intense enough stimulus might evoke a response. In the model with balanced input, increasing image intensity (contrast) raises the level of both excitatory and inhibitory input to the downstream neuron, so the net effect is small (left column of figure 21-4). A strong, contrast-dependent response is only generated when gain modulation throws off this balance (right column of figure 21-4).

Switching through Gain Modulation of Recurrent Pathways

A recurrent network can selectively amplify specific aspects of the input it receives, with selectivity determined by the synaptic connections within the network (Abbott, 1994; Douglas et al., 1995). When recurrent excitation is strong (near the limit of instability), the network amplifies by a large factor, and it becomes highly selective. The amplitude of the network response in this case is very sensitive to the overall level of excitation within the network.

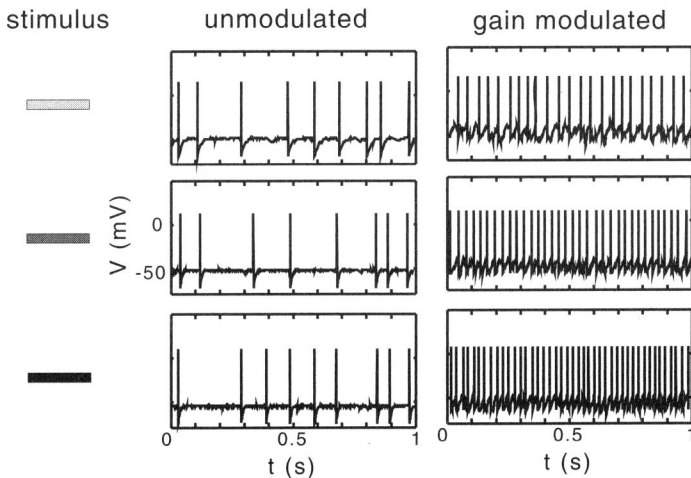

Figure 21-4. Effect of contrast on the responses of the downstream neuron. The level of contrast of the stimulus is indicated at left. The left column of plots shows the membrane potential of the downstream neuron in the absence of gain modulation of the network neurons. The response is weak and independent of contrast. The right column of plots shows that the responses of the downstream neuron are strong and sensitive to contrast when the network neurons are gain modulated.

In the architecture of figure 21-1B, the recurrent connections among the network neurons are affected by the gain of the gain-modulated neurons. Suppose that we label the neurons in the lower row of figure 21-1B with indices i or j and those in the upper row with an index a. Let M_i^a denote the strength of a synapse from gain-modulated neuron a to network neuron i and N_j^a denote the strength of a synapse from network neuron j to gain-modulated neuron a. In a linear approximation, the matrix of synaptic weights for the network neurons is proportional to $\sum_a M_i^a N_j^a$. If neuron a is gain modulated by a factor g_a, this connectivity matrix is modified to $\sum_a g_a M_i^a N_j^a$. Thus, gain modulation modifies the connectivity of the network. This can have a large effect on the response of the network, especially if it is operating with a high degree of amplification.

Discussion

To extend our understanding of neural circuits from the representation of information to cognitive processing, we must face complex issues of neural control and switching of neural circuits. At the present time, we are more in a gathering than a ruling-out situation, and any reasonable candidate switching mechanism for controling sensory and motor circuits is worth studying and analyzing.

References

L.F. Abbott. "Decoding neuronal firing and modeling neural networks," *Quarterly Review of Biophysics* **27** (1994): 291–331.

L.F. Abbott and S.B. Nelson. (2000) "Synaptic plasticity: Taming the beast," Nature Neuroscience **3** (2000): 1178–1183.

R.A. Andersen, G.K. Essick, and R.M. Siegel. "Encoding of spatial location by posterior parietal neurons," *Science* **230** (1985): 450–458.

R.A. Andersen and V.B. Mountcastle. "The influence of the angle of gaze upon the excitability of light-sensitive neurons of the posterior parietal cortex," *Journal of Neuroscience* **3** (1983): 532–548.

C.H. Anderson and D.C. Van Essen. "Shifter circuits: A computational strategy for dynamic aspects of visual processing," *Proceedings of the National Academy of Sciences of the United States of America* **84** (1987): 6297–6301.

D.E. Connor, J.L. Gallant, D.C. Preddie, and D.C. Van Essen. "Responses in area V4 depend on the spatial relationship between stimulus and attention," *Journal of Neurophysiology* **75** (1996): 1306–1308.

R.J. Douglas, C. Koch, M. Mahowald, K.A.C. Martin, and H.H. Suarez. "Recurrent excitation in neocortical circuits," *Science* **269** (1995): 981–985.

J.P. Gottlieb, M. Kusunoki, and M.E. Goldberg. "The representation of visual salience in monkey parietal cortex," *Nature* **391** (1998): 481–484.

R. Hahnloser, R.J. Douglas, M. Mahowald, and K. Hepp. "Feedback interaction between neuronal pointers and maps for attentional processing," *Nature Neuroscience* **8** (1999): 746–752.

D.J. Heeger. "Normalization of cell responses in cat striate cortex," *Visual Neuroscience* **9** (1992): 181–198.

E. Marder and R.L. Calabrese. "Principles of rhythmic motor pattern generation," *Physiological Reviews* **76** (1996): 687–717.

C.J. McAdams and J.H.R. Maunsell. "Effects of attention on orientation-tuning functions of single neurons in macaque cortical area V4," *Journal of Neuroscience* **19** (1999): 431–441.

B.A. Olshausen, C.H. Anderson, and D.C. Van Essen. "A neurobiological model of visual attention and invariant pattern recognition based on dynamical routing of information," *Journal of Neuroscience* **13** (1993): 4700–4719.

A. Pouget and T.J. Sejnowski. "Spatial transformations in the parietal cortex using basis functions," *Journal of Cognitive Neuroscience* **9** (1997): 222–237.

E. Salinas and L.F. Abbott. "Transfer of coded information from sensory to motor networks," *Journal of Neuroscience* **15** (1995): 6461–6474.

E. Salinas and L.F. Abbott. "Invariant visual responses from attentional gain fields," *Journal of Neurophysiology* **77** (1997): 3267-3272.

E. Salinas and P. Thier. "Gain modulation: A major computational principle of the central nervous system," *Neuron* **27** (2000): 15–21.

M.N. Shadlen and W.T. Newsome. "Noise, neural codes and cortical organization," *Current Opionion in Neurobiology* **4** (1994): 569–579.

S.M. Sherman and R.W. Guillery. "On the actions that one nerve cell can have on another: Distinguishing drivers from modulators," *Proceedings of the National Academy of Science (USA)* **95** (1998): 7121–7126.

E. Seidemann, E. Zohary, and W.T. Newsome. "Temporal gating of neural signals during performance of a visual discrimination task," *Nature* **394** (1998): 72–75.

E.A. Stern, D. Jaeger, and C.J. Wilson. "Membrane potential synchrony of simultaneously recorded striatal spiny neurons in vivo," *Nature* **394** (1998): 475–478.

S. Treue and J.C. Martínez-Trujillo. "Feature-based attention influences motion processing gain in macaque visual cortex," *Nature* **399** (1999): 575–579.

T.W. Troyer and K.D. Miller. "Physiological gain leads to high ISI variability in a simple model of a cortical regular spiking cell," *Neural Computation* **9** (1997): 971–983.

J. Zhang and L.F. Abbott. "Gain modulation in recurrent networks," In Bower, J. ed. *Computational Neuroscience, Trends in Research 2000* (Amsterdam: Elsevier, 2000), 623–628.

D. Zipser and R.A. Andersen. "A back-propagation programmed network that simulates response properties of a subset of posterior parietal neurons," *Nature* **331** (1988): 679–684.

22

Synesthesia: What Does It Tell Us about the Emergence of Qualia, Metaphor, Abstract Thought, and Language?

V. S. Ramachandran and Edward M. Hubbard

And yet there should be no combination of events for which the wit of man cannot conceive an explanation. Simply as a mental exercise, without any assertion that it is true, let me indicate a possible line of thought. It is, I admit, mere imagination; but how often is imagination the mother of truth?

—Sherlock Holmes

Introduction

In this chapter, we would like to propose an approach to the problem of understanding the mind somewhat different from the traditional ones in cognitive psychology and neuroscience (e.g., reaction time measurements, computational modeling, black-boxology, single-unit physiology, conventional electroencephalography). Instead of trying to tackle the problem of the mind problems head-on, we approach it by investigating phenomena that are robust and repeatable yet do not fit the big picture of cognitive science as currently understood. This approach is especially likely to be useful in cognitive science since it still very much in its infancy. It was a commonly used—and immensely successful—strategy in the early days of physics (e.g., Faraday's laws relating magnetism to electricity discovered by tinkering with bits of wire and magnets, radioactivity, X-rays) and biology (e.g., bacterial transformation, the discovery of homeotic mutations by Bateson).

But this approach is not fashionable in psychology partly because of the lingering pernicious effects of behaviorism and partly because psychologists like to ape mature quantitative physics—even if the time isn't ripe. Asking simple questions starting from "first principles" isn't fashionable among psychology

432

Faraday +
Maxwell.

professors or neuroscientists (although, thank God, it still is among students). But surely, every science must go through an earlier Faraday stage before graduating to a mature Maxwellian stage, and psychology, in our view, is now still at the Faraday stage; there's simply no point in trying to jump ahead (although we would love to be proved wrong!).

We are not advocating naive empiricism; every good experiment is guided by unconscious theoretical intuitions and conjectures about what might be true. Instead of replacing more conventional approaches, we merely suggest that an additional source of valuable insights might come from just good old-fashioned phenomenology and from looking at "odd" experiences that have been largely ignored by the establishment. We believe that such an approach will not only inject a sense of fun and adventure into the pursuit of psychology (harking back to the days of Helmholtz and William James) but also suggest completely new lines of investigation and help get young neuroscientists out of the cul-de-sacs of methodology that now dominate the field. Until recently, psychology professors strongly discouraged their students from using the "C word"—consciousness—which is odd considering that this is what psychology was supposed to be about! But thanks largely to Francis Crick and Christof Koch, there has been a great revival of interest in this topic (see chapter 23, this volume). To borrow an analogy used by Horace Barlow in a different context, a neuroscientist or cognitive psychologist studying the functions of the brain without invoking consciousness is like a Martian biologist trying to understand the testicles without invoking sex!

Maybe we are stating the obvious; no colleague of ours would knowingly deny the need to ask questions starting from "first principles" (the kind of seemingly naive but embarrassingly hard-to-answer questions that very bright undergraduates often ask). But what about the other strategy we mentioned: pursuing "oddities" or anomalies (to use Kuhn's term)? We have sometimes felt that one can become a good scientist (or even a great one) simply by pursuing phenomena that more mature scientists have brushed aside as anomalies. This isn't as easy as it seems because most anomalies turn out to be false alarms (polywater, Elvis sightings, telepathy, spoon bending) rather than real ones (bacterial transformation, Uranus deviating from its orbit, ulcers caused by bacteria). The wisdom lies in knowing which ones to pursue and which ones to ignore.

For a scientific phenomenon to gain wide acceptance and attention, three criteria must be fulfilled: (1) it must be real—in the sense of being reliably repeatable, (2) there should be at least some potential "candidate" explanations, and (3) so what? What are the broader implications of the phenomenon in question? Without all three of these criteria in place, you won't succeed in attracting the attention of the scientific community even if you are on the right track. For example, telepathy fulfils criterion 3; it has vast implications. The problem is that it doesn't satisfy criterion 1 and certainly not criterion 2.

3 CRITERIA
FOR PURSUING
A SCIENTIFIC
PHENOMENON

Continental drift (when it was discovered seventy years ago) satisfied 1 and 3 but not 2 (plate tectonics had not been discovered yet), and the same could be said of Avery's discovery of bacterial transformation. Chargaff's base pairing rules satisfied 1 and 2 but not 3 (until Crick and Watson saw their true significance).

One of us (V. S. R.) considered many examples of such "anomalous" phenomena in *Phantoms in the Brain* (Ramachandran and Blakeslee 1998), mostly from the older literature of neurology. In this chapter, we will consider another example, synesthesia. Our goal will be to inspire new interest in this topic so that it gets the attention it deserves. We will try to do so by satisfying the three criteria spelled out above. First, we will show that synesthesia is a genuine sensory effect. Second, we will suggest what the underlying neural mechanisms might be. Third, we will point out that far from being just an oddity, synesthesia might help illuminate some of the most puzzling aspects of the mind such as the evolution of metaphor, language, and even abstract thought in humans. In addition, we will discuss the philosophical riddle of qualia and point out how studying synesthesia and other sensory phenomena (such as the filling-in of scotomas and blind spots) can provide some hints about the evolution, functional significance, and neural correlates of qualia. These will be very speculative ideas, but our goal at this stage is simply to provide a starting point for thinking about these issues, not to provide final answers. In that sense, this essay is written in the same spirit as the others in this volume. (As Sherlock Holmes told Watson, "Moonshine is a better thing than fog.")

Is Synesthesia Real?

In the nineteenth century Francis Galton ([1880] 1997) noticed that certain otherwise normal individuals claim to "see" specific colors when listening to musical notes (e.g., C-sharp might be red) or when looking at printed numbers (e.g., 5 might be red, 7 might be blue). The phenomenon runs in families, suggesting a genetic propensity (Bailey and Johnson 1997). We find that approximately one in two hundred people is a synesthete, an incidence much higher than previously believed. Synesthesia seems so outlandish that there has been a tendency to dismiss it as bogus and it has been treated as an oddity by mainstream neuroscience (although there have been a few notable exceptions; see Baron-Cohen and Harrison 1997; Gray et al. 1997). It is fair to say that the four "usual" explanations for the phenomenon have been:

- These people are just crazy.
- They are acid junkies or potheads (and sure enough, synesthesia is sometimes experienced by LSD users, but to us that makes the phenomenon more interesting, not less).

- They are just remembering early childhood memories (perhaps from having played with colored refrigerator magnets). However, this does not explain why it runs in families (unless the same magnets were passed on) nor why the rest of us nonsynesthetes don't have such experiences (you might think of cold when seeing ice but you certainly don't feel cold).

- They are being "metaphorical" as when you or I say cheese is "sharp." Cheese is soft to touch, so why do you say sharp? You might respond that it is just a metaphor, but why do you use a tactile adjective for a taste sensation? How do we know that synesthetes are not being equally metaphorical when they say "C-sharp is red"—perhaps they are just more gifted in this regard!

The last two explanations are the most viable, but there are problems with them. The idea that "synesthetes are being metaphorical" doesn't help much; it is an example of the classic blunder of trying to explain one mystery (synesthesia) with another mystery (metaphor) and that rarely gets you very far in science. No one has the foggiest idea of how metaphors are represented in the brain. In this essay, we will turn this explanation upside down and suggest the exact opposite; synesthesia is a concrete sensory phenomenon whose neural basis can be pinned down and that in turn can provide a foothold—an experimental lever—for understanding what the neural basis of metaphors might be.

Synesthesia: A Sensory Effect

Subjective reports of synesthetes vary enormously. However, carefully listening to their introspective reports, one finds strong hints that the phenomenon is probably genuine (and probably sensory, at least in one group of synesthetes whom we refer to as "lower synesthetes"). For instance, our first two synesthetes made it clear that the colors were spatially localized on the number or letter; they were not just remembering the color. Indeed the very opposite was true, the colors actually seemed to facilitate their memory! We have encountered synesthetes who said they learnt to type more quickly and learnt musical scales more easily than their classmates because the letters (or musical notes) were "color coded."

These observations were complemented by psychophysical measurements that we made which provided additional evidence that synesthesia (at least in a subset of synesthetes) is a genuine sensory effect rather than a memory or a metaphor. In J. C., the saturation of induced color decreased monotonically with contrast, and the color vanished completely below about 8 percent to 9 percent contrast even when the number was still clearly visible. The color also vanished if two numbers (say 2 and 5) were alternated in time at 4 to 5 Hz—even though the alternation of the numbers themselves was clearly visible at much higher rates (e.g., 15 Hz). Such a high level of sensitivity to the elementary physical parameters defining the grapheme also supports our view that the effect is indeed sensory.

These findings suggest that synesthesia is sensory but do not provide unambiguous proof. To obtain clear-cut proof that the effect is indeed sensory, we devised the stimulus shown in figure 22-1a, a matrix of randomly placed computer generated 2s with some 5s scattered among them. Since the 2s and 5s are made up of identical features (three horizontal bars and two vertical ones) you cannot spot the 5s except by a detailed item-by-item inspection. Eventually you see that they are arranged to form a triangular shape (a non-synesthetic subject typically takes a few seconds). Our two synesthetes, on the other hand, saw this display as colored (figure 22-1b) and the red 5s clearly "popped out" from the background, so they instantly saw a red triangle segregated from a green background. This experiment shows two things. First, synesthetes are not "crazy" or "making up" this effect. If they were crazy, why are they actually much better at this task than "normals"? Secondly, Treisman (1982) and others have shown that this type of pop-out and segregation can occur only for elementary sensory features like motion, color, depth, and so on—but not for more complex features like graphemes (unless the graphemes also happen to differ along some elementary dimension). Not coincidentally, these so-called elementary features as defined by psychophysics (e.g., "pop-out" or grouping) are often the ones that are extracted early in the sensory-visual areas such as V4 (color) and MT (motion).

Indeed, in 1988 one of us (V. S. R.) showed that three-dimensional shape-from-shading could lead to pop-out and segregation and suggested that there might be cells very early in the visual pathways that extract this feature, a suggestion that has now been confirmed by both single-unit physiology and

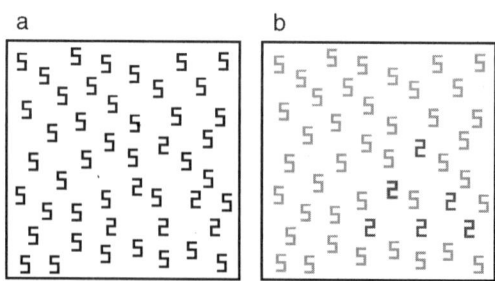

Figure 22-1. Schematic representation of displays used to test whether synesthetically induced colors lead to pop-out. (a) When presented with a matrix of 5s with a triangle composed of 2s embedded in it, control subjects find it difficult to find the triangle. (b) However, because they see the 5s as (say) green (depicted in light gray) and the 2s as red (depicted in dark gray), our synesthetic subjects were able to easily find the embedded shape.

fMRI. So the fact that synesthetes see a red triangle in figure 22-1a strongly suggests that the synesthetically induced colors are genuinely sensory, not some high-level memory association or metaphor. Metaphors and memories don't pop out. These results provided the first clear-cut evidence since the time of Galton that synesthesia was an authentic effect—indeed a genuine sensory phenomenon worthy of further study (Ramachandran and Hubbard 2000, 2001a). This observation has since then been confirmed by other groups.

Another observation we made was also consistent with this. When we showed J. C. and E. R. roman numerals (e.g., V instead of arabic 5) no color was seen! This suggests that in at least our first two "lower" synesthetes it was not the numerical concept (e.g., of ordinality) that induced the color but rather the actual physical appearance of the grapheme.

In the past, it has been claimed that Stroop interference shows that synesthesia is sensory. For example, if the number 5 is seen as red but is presented in the wrong ink color (e.g., green), the synesthete is slower to name the ink color, that is, her reaction time is slightly longer. But this effect merely demonstrates that synesthesia is automatic, not necessarily that it is sensory (MacLeod 1991). Such a slowing can result from interference at almost any stage in processing from sensory input to motor output.

Synesthetically Evoked Colors Can Drive Apparent Motion

In addition to showing that synesthetically evoked colors can lead to pop-out and segregation, we have also found that synesthetically induced colors can drive apparent motion (Ramachandran and Hubbard 2002). As in our first experiment, we began with a matrix of randomly placed 2s with a small cluster of 5s embedded near one corner of the display. This display was flashed briefly (350 msec) and replaced by a similar matrix of 2s that was uncorrelated with the first frame and the cluster of 5s was now in the opposite corner. The two frames were cycled continuously. When nonsynesthetes see this display, they just see random apparent motion. But when J. C. looked at it, he said he saw the red cluster "jumping" or moving back and forth between the two displays. This observation suggests that just as real color can support long-range or form-based apparent motion (Ramachandran and Gregory 1978), so can synesthetically induced colors. This observation is also consistent with our view that synesthesia is mediated relatively early in the sensory processing hierarchy; it may involve the activation of motion neurons in MT by colors evoked in the fusiform. By presenting a continuous sequence of movie frames, we were even able to elicit what appeared to be smooth pursuit eye movements in lower synesthetes (Ramachandran and Azoulai, in press).

More recently we constructed displays to show that even the detection of symmetry (normally thought to be preattentive) can be based on synesthetically

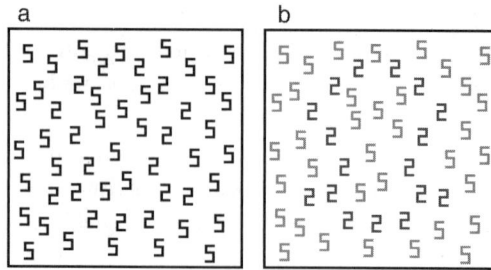

Figure 22-2.   A demonstration of the effect of synesthetically induced colors on symmetry detection. (a) Control subjects find it hard to detect the butterfly-like shape composed of 2s embedded in the display of 5s. (b) However, because he sees 5s and 2s as different colors (depicted in light and dark gray), subject JC was immediately able to spot the symmetric shape in the display. This phenomenological observation needs further experimental confirmation.

induced splotches of color even though the inducing graphemes are not symmetrical (so a nonsynesthete cannot perceive the symmetry; see figure 22-2). The global symmetry of the induced colors overrides the asymmetry of local defining elements (Ramachandran and Hubbard 2002).

What Causes Synesthesia?   Cross activation of the N° grapheme area + V4 (fusiform gyrus).

Our evidence so far strongly supports the notion that synesthesia is a sensory phenomenon, but what causes it? We were struck by the fact that the "number grapheme area," as identified using fMRI by Mauro Pesenti (Pesenti et al. 2000) and Tim Rickard (Rickard et al. 2000), is in the fusiform gyrus and the "color area" of Semir Zeki (Leuck et al. 1989, Zeki and Marini 1998) is in the fusiform gyrus right next to the number area—almost touching it (see figure 22-3). We realized this probably wasn't a coincidence; the most common type of synesthesia is the number (grapheme)-color type. We suggested, therefore, that the phenomenon was caused by a cross-activation of sensory brain maps. By this we mean that the presence of the inducing elements (e.g., numbers) not only evokes activity in the number/form area, as it should, but also automatically evokes activity in another sensory map (e.g., V4 color area) that is just as strong as direct activation by a real sensory stimulus. This creates a vivid phenomenal experience of color that is virtually indistinguishable from real stimulus-driven sensations (but see "Synesthesia, Blindsight, and the Problem of Qualia").

The idea that synesthesia is the result of some kind of "crossed wires" is hardly new and in fact is probably as old as the phenomenon itself. However, this idea is normally couched in vague terms and does not take into account the known physiology. Instead, we attempted a more precise testable formulation of this notion in terms of the known anatomy and physiology of brain areas and

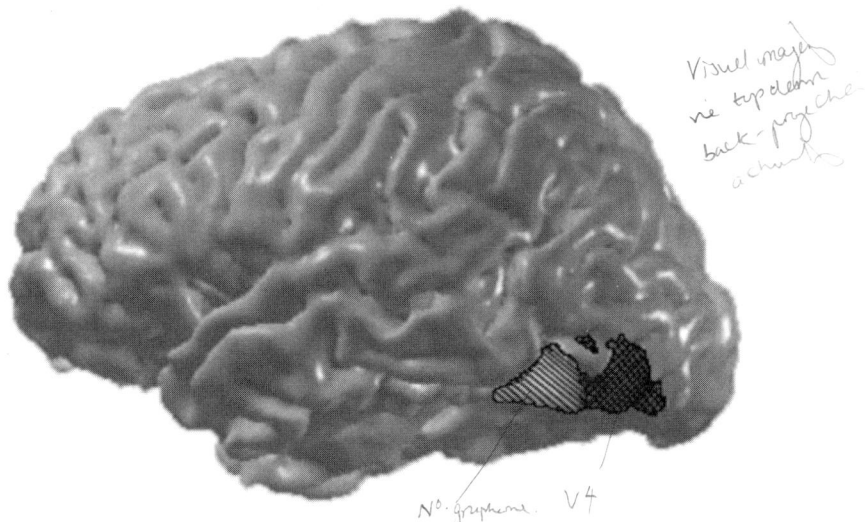

Visul mayer
ne top clen
back-p...
a clu...

No· Grapheme. *V4*

Figure 22-3. Schematic showing that cross-wiring in the fusiform might be the neural basis of grapheme-color synesthesia. Area V4 is indicated by diagonal lines while the number-grapheme area is indicated by cross-hatching.

specifically proposed that synesthesia should lead to activation of color selective regions in the fusiform gyrus via grapheme nodes in the same gyrus (Ramachandran and Hubbard 2000, 2001a). There is now evidence from functional imaging (Nunn et al. 2002) that such activation of V4 does indeed occur. Our own imaging studies (in collaboration with Geoffrey Boynton) have also shown evidence of V4 activation. However, unlike Nunn et al., we have also found activation of "early" areas such as V1 and V2 as well.

How might this cross-activation of sensory maps come about? Recall that synesthesia runs in families and may have a genetic basis. Perhaps some families have a faulty gene (or genes) that cause one of three problems: (1) actual "cross-wiring"—an excess of connections between brain modules that are ordinarily distinct or a failure of "pruning" that is ordinarily produced by that gene, (2) a cross-activation that results from excess activity (but no actual wires) or a disinhibition between brain modules that are ordinarily functionally insulated (this would account for drug-induced synesthesia), or (3) as a variant of (2) an excess activation caused by disinhibition of back-projections from higher to lower visual areas in the hierarchy. If so, synesthesia might provide some clues to understanding visual imagery, which is also thought to require top-down activation through back-projections (see Ramachandran and Hubbard 2001b).

We prefer the more neutral term *cross-activation* to *cross-wiring* because it covers all three potential mechanisms. But two additional questions emerge: (1) if defective genes are involved, why do they affect one brain area and not any others? and (2) does this mean we are implying that synesthesia can occur only between adjacent brain modules, and if so how can we account for more exotic variants like tasting shapes?

The answer to the first question is that the gene may be selectively expressed in certain areas (e.g., fusiform for grapheme-color synesthesia and the temporal-parietal-occipital (TPO)/angular gyrus area for others) due to transcription factors. This would also explain why if you have one type of synesthesia you are also more likely to have another type; perhaps the gene is expressed in several locations in some synesthetes.

Turning to the second question, we have to bear in mind that while it is usually true that adjacent brain modules are more likely to be connected to begin with (and therefore more likely to be involved in cross-activation), even modules remote from each other often have some connections (e.g., Kennedy et al. 1997), and an enhancement of these through the three mechanisms suggested above could therefore mediate synesthesia. One would expect, however, that the likelihood of one map cross-activating another increases with the proximity between the two areas.

An example might be the celebrated case of the "man who tasted shapes" described by Richard Cytowic; he would remark that most tastes evoked distinct shapes (e.g., mint tasted like a cylindrical marble column). We would suggest that this is because the gustatory (taste) cortex is in the insula very close to the hand area of the Penfield map in S1 (Ramachandran and Hubbard 2001b).

Another question is, if cross-activation of brain maps is the correct explanation, why is color the most commonly evoked sensory experience in synesthesia, and why is it much more often number to color rather than color to number? The answer may lie in the manner in which different complex dimensions (sensory or otherwise) are mapped in the brain, a topic about which very little is known. Perhaps the dimensions of color are represented in such a way that it is easier to map them onto other sensory inputs than is possible with (say) touch. This hypothesis would also explain the preferred directionalities and pairings that we see in day-to-day cross-modal metaphors—even in nonsynesthetes.

Higher and Lower Synesthetes

The more outré and grotesque an incident, the more carefully it deserves to be examined, and the very point which appears to complicate the case is, when duly considered and scientifically handled, the one which is most likely to elucidate it.

—Sherlock Holmes

When we began our research on synesthesia in 1999, we were lucky to have stumbled on two subjects who showed all the effects described above; pop-out, reduction of color with lowering contrast, high flicker rates, and so on. This provided, to our knowledge, the first clear, unambiguous proof that we are dealing with a genuine sensory phenomenon. Recall, also, that these subjects saw colors only with arabic numbers but not with roman ones like V, which implies that it is the visual appearance of the grapheme, not the high-level numerical concept (e.g., of sequence or ordinality) that drives the color. This observation is also consistent with the "early" sensory cross-activation model—since the fusiform gyrus represents the visual graphemes—not the numerical concept.

But later we came across synesthetes in whom this wasn't true; in them not just numbers but even days of the week or months of the year were colored (no wonder many scientists thought they were crazy). Monday might be blue, Tuesday red, and Wednesday brown (or December might be yellow and March, blue). What all these have in common is the idea of numerical sequence or ordinality. This ability probably depends on the angular gyrus (it remains to be seen whether that structure mediates both cardinality and ordinality or both). It might be interesting to see if patients with lesions in the left angular gyrus have problems with calendars ("calendar agnosia" or "sequence agnosia"). Our basis for suggesting this is that patients with angular gyrus lesions often have problems with elementary arithmetic (dyscalculia).

We were struck by the fact that some of the "higher" color areas in the color processing hierarchy which receive their input from V4 in the fusiform lie in the general vicinity of the angular gyrus. We suggested, therefore that if the "synesthesia gene(s)" were to be expressed in these higher areas in the vicinity of the angular gyrus it would lead to cross-activation there instead of the fusiform, so the color is driven by the concept rather than visual appearance (Ramachandran and Hubbard 2001a, 2001b). Roughly speaking, depending on whether the gene is expressed mainly in the fusiform or angular gyri you end up with two types of synesthetes, whom we call "lower" and "higher" synesthetes. We are not implying that the distribution is bimodal—that remains to be seen. Indeed, there may be many complex mixed types depending on how widely the gene is expressed. The ratio of incidence of the different types also remains to be determined, but our anecdotal sampling suggests that only about 10 to 20 percent may be true lower synesthetes.

But one prediction we made (Ramachandran and Hubbard 2001a, 2001b) based on this model was that the psychophysical properties of the induced colors should be different in higher and lower synesthetes; for example, in higher synesthetes the induced colors might not lead to pop-out and perceptual segregation. This prediction has been confirmed in an elegant study conducted recently by Merikle, Dixon, and Smilek (2002).

There are two (and only two as far as we know) cases in the literature of patients losing the synesthetically induced colors following brain lesions.

Remarkably, in one of them, reported by Oliver Zangwill (Spalding and Zangwill, 1950) the lesion was in the angular gyrus, and in the second (reported by Sacks et al. 1988) the lesion was in the fusiform, leading to the loss of both color vision and synesthesia. This is perfectly consistent with our speculations about higher and lower synesthetes (see Ramachandran and Hubbard 2001b and Ramachandran 2003 for details).

Lastly, we should point out that in some synesthetes it's the phoneme that seems to evoke the color, not the grapheme. It remains to be seen whether these overlap with the group that we refer to as higher synesthetes. One possibility is that the phoneme-color synesthetes are the higher synesthetes for the lower synesthetes who see colors in printed visual alphabetic graphemes, whereas the number concept/calendar synesthetes are higher synesthetes (as we already noted) for visual number graphemes. The cross-activation for lower synesthetes would be in the fusiform for both types of lower synesthetes and in the vicinity of the TPO junction for both types of higher synesthetes (e.g., involving the angular gyrus for the calendrical ones and the superior temporal gyrus for phoneme-color ones).[2]

The existence of reciprocal connections and cross-activation between nodes that deal with related categories even in nonsynesthetes probably complicates the picture and may account for the variability seen in synesthesia and the existence of mixed types. For example, hearing the word *five* reminds even a nonsynesthete slightly of the grapheme 5, the written word five, the numerical concept of fifth, and the calendar, producing a sort of penumbra of associated neural activation. Such reciprocal coactivation may be even more powerful in synesthetes—blurring the distinction between the node and its penumbra and between different types of synesthesia (and evoking different degrees of activation of the color nodes at different stages of color processing, depending on the particular synesthete you are studying). It's a failure to recognize this that has resulted in some of the controversy about synesthesia in the older literature.

Top-Down Influences on Synesthesia

The fact that synesthesia is the result of cross-activation of sensory maps does not rule out the possibility that "top-down" influences, such as attention and visual imagery can modulate the effect. If you look at figure 22-4 (a large 5 made up of little 3s) the picture is ambiguous and can be seen as either the "forest"—the 5—or the "trees"—the 3s. Intriguingly when we showed it to our two synesthetes, they reported that they saw the color switch from red to green depending on whether they were attending to the forest or the trees (Ramachandran and Hubbard 2001b). This observation implies that even though synesthesia is evoked by the visual appearance alone—not the

3 3 3 3 3
3
3
3 3 3 3
3 3 3 3
3
3
3 3 3 3

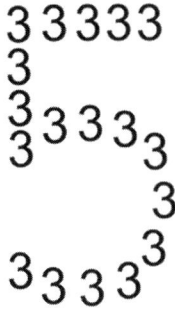

Figure 22-4.   Hierarchical figure demonstrating top-down influences in synesthesia. When our synesthetic subjects attend to the global 5, they report the color appropriate for viewing a 5. However, when they shift their attention to the 3s that make up the 5, they report the color switching to the one they see for a 3.

high-level concept—the manner in which the visual input is categorized, based on attention, is also critical. Similarly in figure 22-5, we have an ambiguous letter in the middle. On its own, it could depict either an H or an A. When flanked on either side by T and E (horizontally) it looks like H, whereas when flanked by C and T (vertically), it looks like an A (as in CAT). Once again, its not the physical appearance of the letter but which "bin" the brain puts it in that determines which synesthetic color is evoked; when our subjects grouped the letters vertically they saw the middle character as A and saw blue, but when they grouped the horizontal letters they saw the middle character as pink (Ramachandran and Hubbard 2001b).

C
TAE
T

Figure 22-5.   Ambiguous stimuli demonstrating further top-down influences in synesthesia. When presented with the ambiguous H/A form in THE CAT, both of our synesthetes reported that they experienced different colors for the H and the A, even though the physical stimulus was identical in both cases.

Hypernormal Stimuli and Art

We were curious as to whether the actual font (e.g., Roman vs. Gothic) of the number or letter can affect the perceived color and in synesthetes (or, at least in lower synesthetes). This does indeed seem to be the case. We expected to find the simplest "prototype" grapheme (e.g., obtained by averaging many different letter fonts) to be most effective and this was often true. Yet, to our astonishment, sometimes a very oddly shaped "deviant" font was actually more effective than fonts that were obviously closer to the original!

The paradox is resolved by considering findings from ethology. Niko Tinbergen found that a seagull chick would beg for food from its mother as soon as it hatches. It does so by pecking at a red spot on the tip of the mothers long brown beak. In fact, you can remove the beak from the mother and wave it in front of the chick and the chick will peck equally vigorously at the disembodied beak. In "recognizing" Mom the chick takes the sensible short cut of assuming that beak = Mom, given that in nature it is not likely to encounter a mutant pig with a beak or a malicious ethologist waving a beak!

Tinbergen found that you don't even need a beak; an oblong piece of cardboard with a red spot on one end will suffice. The chick gets fooled; it shows no preference for a real beak over this cartoon. The reason is that the "beak neurons" are not especially fussy. It is possible to fool them just as a rusty bent key can sometimes fool a badly made lock. But the big surprise came when Tinbergen showed the chick a long thin stick with three red stripes near the end. Amazingly, even though this stimulus doesn't even resemble a real beak, the chick actually prefers this to a real beak! In fact, it goes berserk. Tinbergen doesn't say why this happens but obviously he had stumbled on a super-beak. We suggest that the receptive fields for "beak detection" (in the chick's tectum, rotundum, ectostriatum, or hyperstriatum) are wired up in such a way that they are actually more optimally stimulated by this weird pattern than by a real beak (e.g., embodying the rule "the more red contour, the better" hence the heightened response to three red stripes). We have suggested elsewhere (Ramachandran and Hirstein 1999; Ramachandran 2001) that this is the basis of all art appreciation in humans— especially the appreciation of semiabstract nonrealistic art for which there has hitherto never been an adequate explanation. For example, a cubist portrayal of multiple incompatible views of a face on one plane might hyperactivate "master face cells" in the temporal lobes that are normally wired up to respond to any one of many different views. Since such master cells are constructed out of convergence of axons from "single view" cells earlier in the hierarchy that each respond to only one view, presenting two views in the same receptive field would lead to hyperactivation of the master cell. (This provides a neat neural explanation for Picasso—often touted by social scientists as the ultimate example of that which is irreducible and mysterious in human nature).

Returning to synesthesia, we suggest that the "weird fonts" are hypernormal inputs; that is, they are to your E- or 5-detecting neuronal nodes what the stick with three stripes is for beak neurons. Such hyperactivation of grapheme nodes would produce heightened cross-activation of colors in synesthetes.

Some Additional Phenomenology

It is unfashionable in psychophysics to ask people what they are actually experiencing, but common sense suggests that this might provide useful insights. Here is a tentative list of questions that remain to be answered along with some preliminary observations we have made. Which of these apply only to lower synesthetes, only to higher, or to both remains to be studied carefully (bearing in mind, especially, that there may be more than just two types).

Where exactly is the color perceived? In general, lower synesthetes say they see the color spatially localized on the printed number or letter. Sometimes the color is confined entirely within the margins of the number (or letter) at other times it "bleeds," forming a slight halo around the number. Higher synesthetes, on the other hand, make remarks like, "The color is seen in my mind's eye," and so on. We must bear in mind that this is such an ineffable experience that the subject experiences the same frustration in trying to convey it to us as you might experience in describing colors to a rod-monochromat.

Some synesthetes describe seeing odd "Martian colors" in graphemes that they cannot see in the real world; this has been noticed before but no adequate explanation has been provided in the literature. We would explain it in terms of cross-activation the same way we explained Martian colors seen by our color-anomalous synesthete earlier in this essay. Sometimes synesthetes see "mixed" colors in a number or letter; the number 5 may be blue but in a "patchy" manner, "like a Dalmatian dog." The Martian color effect and the Dalmatian effect are both hard to account for in terms of early memory associations: how can you "remember" something you have never seen in the real world (Ramachandran and Hubbard 2001b)?

Memorability: We have noticed that synesthetic colors are often used as a mnemonic device (Ramachandran and Hubbard 2001b). One of our subjects says she remembers phone numbers by color. When trying to recall the number she conjures up a spectrum—a visual image corresponding to the number in her mind's eye and then proceeds to "read off" the numbers. Another claimed he learned to type much faster than his classmates did because for him the keyboard was color coded. A third subject found it easy to learn musical scales because the piano keys were colored. Each note had a unique color (she was a sound-color rather than grapheme-color synesthete). Merikle et al. (2002) recently reported an ingenious experiment. They showed a random matrix of numbers to a synesthete and found he could more readily memorize the numbers than nonsynesthetes because of the extra cue provided by the color.

Imagery: We found that in some synesthetes, when the subject visualized a number, the color was paradoxically more saturated than when actually seeing the number (Ramachandran and Hubbard 2001b). It is as if when looking at a real number colored (say) black the actual color interfered with or slightly reduced the vividness of the color evoked by cross-activation in (say) the fusiform or higher up, whereas with top-down imagery-driven activation of the grapheme node, there is no bottom-up interference from the real color. We have also seen in one subject that a small amount of real color added to the white grapheme is harder to detect if the color is identical to the synesthetically induced one but not if the color is different; implying that the brain actually "confuses" the synesthetic color for a real color. This observation needs additional verification using more subjects.

Spatial interactions: A double-digit number presented visually is usually seen as the corresponding two colors. If the numbers are too close spatially—in some subjects the colors "clash" and cancel each other out. Intriguingly if the two different numbers or letters evoked the same color the colors actually enhanced each other (one synesthete said "Rama" was extremely red because both R and A were red for her). It remains to be seen whether this is true for grapheme-color synesthetes or only for phoneme-color synesthetes. (or for both).

Spatial precedence rule: If a whole word was presented to some synesthetes, all the letters were always tinged the same color as the first letter. It would be interesting to see if this is true for "double words" or hyphenated words.

Tactile graphemes: Numbers or letters drawn on the palm usually do not evoke colors unless the subject starts visualizing them.

Auditory presentation: If we say "five" to a grapheme-color synesthete, she experiences the color "in her head" (presumably only in higher synesthetes) or automatically evokes the number visually (in lower synesthetes?) along with the corresponding color. In two subjects we found that if we said "two, three, two, three, two, three..." the subject reported seeing a sort of spatially spread-out palette or "spectrum" of corresponding colors in front of her (Ramachandran and Hubbard 2001b).

Silent graphemes: (e.g., "k" in "knock") are still seen vividly colored. It remains to be seen if this is only true for lower synesthetes.

In some bilingual synesthetes (e.g., Chinese, English) one language alone had colored graphemes whereas the other didn't. This requires more intensive study on a larger population of bilinguals. (Remarkably, in one of our synesthetes the colors were evoked only by her second language, Hebrew graphemes, but not her first language, English graphemes).

If a normal person sees your lips pronouncing one phoneme visually on a video tape but hears the wrong sound, he will hear what your lips are "saying" rather than what he is actually hearing—the McGurk effect. In two phoneme-color synesthetes we found that the color evoked corresponded to the perceived phoneme—influenced by visual capture—rather than the actual physical, auditorily presented sound!

Alphanumeric categorization: O can be seen as one of two different colors depending on whether it was seen as the letter O or the number zero.

We presented partially occluded amodally completed letters and numbers (e.g., Bregman's Bs; Bs that are partially covered by an ink blot so only parts of them are visible—see figure 22-6). Without the occluder in place, the B fragments did not produce color, but as soon as the occluder was introduced (so the observer completed the Bs perceptually behind it), the color was clearly visible, although, oddly, it seemed to spread in front of the occluder (and after such "priming" even the fragments started to evoke colors!).

Emotions: Anecdotal observations suggest that synesthetes have a disproportionately strong emotional reactions to their color (or other) associations. For example, if a number is presented in the "right" color it feels harmonious whereas if presented in the wrong color it feels strongly aversive—like discordant notes or "nails scratching on a blackboard" as one synesthete told us. We have suggested (Ramachandran and Hubbard 2001b) that this may occur in those synesthetes in whom the "hyperconnectivity" gene is expressed not only between sensory maps but also between sensory areas and areas that subserve emotions (e.g., between IT and amygdala or nucleus

Figure 22-6. The Bregman Bs. When presented with the B fragments, our synesthete JC reported experiencing no colors. However, when presented with the Bs with the ink spot over them, JC reported experiencing the appropriate colors for the Bs. The color was clearly visible, although, oddly, it seemed to spread in front of the occluder (and after such "priming" even the fragments started to evoke colors).

accumbens). Such individuals may have a disproportionately large emotional reaction (and perhaps galvanic skin response, or GSR) to even trifling discord or harmony. Such heightened emotions could also provide a basis for learning to play an important role in synesthesia—because they would progressively amplify preexisting small biases (given the disproportionately large reinforcement).

The color associations with numbers or letters are not completely random across synesthetes. There is a greater tendency for A to be red, for example, and U to be blue; the general trend for vowels seems to be that front vowels (tongue closer to the front of palette) evoke shorter wavelengths and back vowels, longer wavelengths. This requires detailed quantitative study. It may reflect the way in which phonemes are produced—and therefore mapped—in certain brain areas so that the cross-activation is also systematic along certain dimensions (Ramachandran and Hubbard 2001b).

In view of our "cross-activation of brain maps" hypothesis, it is important to look for systematic trends in the data. There has been a tendency in this field to regard any such observed trends as spurious as soon as an exception crops up, but this is unwise. Bear in mind the lesson to be learnt from Mendeleyev's discovery of the periodic table. Trends were initially noticed (e.g., Newland's law of octaves) but were prematurely given up as silly even though they set the stage for the subsequent development of the periodic table. When Mendeleyev arranged elements according to increasing atomic weights, an overall pattern emerged. He was not in the least dismayed by the fact that many elements didn't "fit" the overall pattern and pointed out that perhaps the experiments were wrong—perhaps their atomic weights had been measured incorrectly to begin with. He was later vindicated on this point, and the periodic table is here with us to stay. (The weights were wrong because isotopes were not known in the time of Mendeleyev).

Most of the observations on this list are very preliminary and need to be confirmed on additional subjects. We have to be especially wary of the fallacy of our brains' tendency to selectively attend to positive instances that confirm our hunches. It is hard to believe, though, that they will not shed light on the basic mechanisms used by the brain to create perceptual categories of objects and events in the world. Our ideas are also broadly consistent with Edelman's important view (1989) that understanding the cross-activation of brain maps that represent different categories in the world holds the key to solving many of the riddles of the human mind. But whereas Edelman places his emphasis on learning, we argue that hardwired preexisting rules of cross-map translations may have also played an equally important role. Also, Edelman tests his conjectures by using computer simulations, but we have tried to show in this essay that we can use the fascinating condition of synesthesia (and other related phenomena) to experimentally probe these interactions.

Synesthesia, Blindsight, and the Problem of Qualia

Consciousness is one of the great, unsolved riddles of science. It is so mysterious that it's not even clear whether it is a philosophical problem or a scientific one (recall that many so-called philosophical problems have been usurped by science during the last century). It's easy to fall into the trap of thinking that you are conscious of everything that goes on in your brain (but not in, say, the liver or the kidney, which obviously have nothing to do with conscious awareness). Actually, the evidence suggests that you are not conscious of even most of what goes on in your brain. Thanks mainly to Freud, we know that most of your behavior is governed by a cauldron of emotions and motives of which you are completely unaware or only dimly aware and that what you call your conscious life may be no more than an elaborate posthoc rationalization to justify your actions. At a very mundane level, think of what happens when you drive a car while talking to a friend next to you. Unless something very salient and unexpected happens (e.g., an actual zebra crosses at a zebra crossing), your consciousness is focused on the conversation, and you are largely unconscious of all the complex computations your visual pathways are engaged in while negotiating traffic and dodging trucks and pedestrians. We may conclude from this that routine activities like driving do not require consciousness, but the kind of "offline" symbol manipulation involved in activity like conversation does require consciousness. (It is interesting that it's very hard to imagine the converse of this scenario, paying all your attention to the driving while having a meaningful conversation unconsciously with your friend.)

Saying that most events in your brain—and the corresponding behaviors—don't reach consciousness is easy; but is surprisingly hard to prove scientifically. We realized that synesthesia might provide some clues because it provides a way of selectively activating a set of visual areas while "skipping" or bypassing others, something you cannot do in nonsynesthetes except by direct brain stimulation. Instead of asking the somewhat nebulous question "what is consciousness" or "what is the self," we can refine our approach to the problem by focusing on just one aspect of consciousness—our awareness of visual sensations—and ask ourselves, "Given our detailed knowledge of the anatomy and function of the thirty visual areas in the brain, how does the activity of neural circuits in these areas allow us to experience (say) a red sensation?" (see Crick and Koch, chapter 23, this volume; Ramachandran and Blakeslee 1998).

Does conscious awareness of red require activation of all or most of these thirty areas in the human brain that have been identified as being involved in vision or only a small subset of them? What about the whole cascade of activity from the retina to the thalamus to the visual cortex (area 17) before the messages get relayed to the thirty areas? Is their activity also required, or can you skip them and directly activate V4 (the color area in the fusiform) and experience an equally vivid red? If you look at a red apple, you would ordinarily activate the visual area for both color (red) and form (apple-like). But what if you artificially stimulate the color area while silencing the cells concerned with form (assuming that's even possible), would you then experience "disembodied" red color floating out there in front of you like a mass of amorphous ectoplasm or other spooky stuff? And lastly, we know that there are many more projections going backward from each area in the hierarchy of visual processing to "earlier" areas than there are going forward, and the function of these back-projections is completely unknown. Is their activity required for conscious awareness of red? What if you could selectively silence them with a chemical while you look at a red apple, would you lose awareness? These questions come perilously close to being the kind of impossible-to-do armchair "thought experiments" that philosophers revel in, but the key difference is that these experiments can be done not just in principle but also in practice—and maybe even within our lifetimes.

Let us return, now, to synesthesia to see how it may actually help us answer some of these seemingly intractable questions. To achieve this goal, we took advantage of a visual phenomenon known as "crowding." If you fixate on the small + in figure 22-7, you will find that it's quite easy to discern the number 5 off to one side—even though you are not looking at it directly. But if we now add four other numbers (like 2s) all around it, then you can no longer identify the middle number (5)—it looks out of focus. Normal volunteers are at chance level identifying this number. This is not due to reduced acuity in the periphery of your vision; after all you could see the 5 perfectly clearly when it wasn't surrounded by 2s. The reason you can't identify it is because of your limited attentional resources; the flanking 2s somehow "distract" your attention away from the central 5 and prevent you from seeing it.

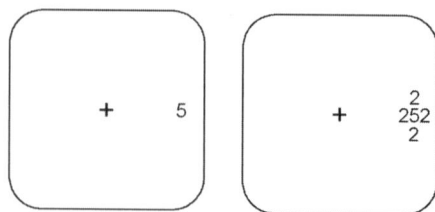

Figure 22-7. A demonstration of the crowding effect. A single grapheme presented in the periphery is easily identifiable. However, when it is flanked by other graphemes, the target grapheme becomes much harder to detect. Synesthetic colors are effective (as are real colors) in overcoming this effect.

The big surprise came when we showed the same display to our two synesthetes. Our question was, if they "see" the middle number as colored, would this somehow make it more conspicuous and thereby "rescue" it from the crowding effect? Imagine our amazement when they looked at the display and made remarks like "I cannot see the middle number—it's fuzzy. But it looks red so I guess it must be a 5"! This is a remarkable observation, for it suggests that even though they were not consciously registering the middle number, it was nonetheless being processed somewhere in the brain at an unconscious level and evoking the appropriate color! And then they could use this color to intellectually deduce what the number was. If our theory is right, the implication would be that the number is processed in the fusiform gyrus and evokes the appropriate color before the stage at which the crowding effect occurs in the brain—indeed even before the number is consciously perceived! This was strong evidence that we are dealing with cross-activation relatively early in sensory processing—not some high-level metaphorical association or memory. Just as you can drive a car negotiating traffic without even being conscious of it (because your attention is on the conversation), the number is registered unconsciously and evokes the appropriate color—even though you can't see the number itself. We have seen this effect in two subjects, but we do not yet know how general it is.

At first, this finding may seem to be at odds with our claim that top-down imagery can influence the induced colors. However, on further reflection, there's really no contradiction. The crowded number is not ambiguous—it allows the full activation of the corresponding cells in the fusiform and its lack of visibility is because of attentional limitations at some later stage. Consequently, it activates corresponding color nodes in V4 even though it is "silenced" later. With the ambiguous display (like CAT), on the other hand, the middle letter partially activates two grapheme nodes equally effectively until top-down imagery or context biases one of the two nodes, thereby allowing it to reach threshold and activate the corresponding color.

Finally, we should point out that the converse of the blindsight effect is also true. We found in two synesthetes that if the number is presented in peripheral vision and scaled for eccentricity, the evoked color vanishes even though the number is still clearly visible (Ramachandran and Hubbard 2001a). This effect has since then been confirmed by Naom Sagiv and Lynne Robertson (personal communication) although they find that the fact that the number getting close to the frame of the monitor seems to be critical—not eccentricity. Whatever the ultimate interpretation of this effect, as we noted originally, it represents a "double dissociation" from the blindsight effect. In blindsight an invisible (or indiscernible) grapheme evokes color whereas the fall-off with eccentricity implies that the very opposite is true under other circumstances—a number that is clearly visible can fail to evoke color. Indeed, in what may be the first report of laterality effects in synesthesia, we found that in one of these subjects graphemes were colored mainly in the right visual field but not in the left; implying asymmetric cross-activation in left versus right fusiform (see Ramachandran and Hubbard 2001a; see also the recent imaging study by Nunn et. al 2002, which has revealed striking asymmetries).

A Color-Blind Synesthete

Color blindness can occur when the color areas in the brain are damaged by a head injury, tumor, or stroke, but the most common cause is an inherited absence of or deficiency in one of the three cone pigments. Such a person is born with an incapacity to perceive the full range of colors that normals can see. Depending on which cone pigment is lost and the extent of loss, he may either be partially or completely color blind. This was the problem experienced by one of our synesthetes, S. S.; he had the inherited form of cone-pigment deficiency and experienced fewer colors in the world than most of us do.

The odd observation he made was that he often saw numbers tinged with colors that he could otherwise never see in the real world. He referred to them, quite charmingly and appropriately, as "Martian colors" that were "weird" and seemed quite "unreal." Ordinarily one would be tempted to ignore such remarks as being crazy, but in this case, the explanation was staring at us in the face: we realized that our theory about cross-activation of brain maps provides a neat explanation for this seemingly incomprehensible phenomenon.

Remember, S. S.'s cone receptors are deficient, so he cannot, under any circumstances, see the full range of colors by looking at anything in the real world. And yet in all likelihood the cortical receiving centers in the brain concerned with colors (such as V4 in the fusiform) were probably programmed by a different set of genes that are not affected in S. S.'s brain, so these areas are perfectly normal even though his retina is unable to process the full range of colors to deliver to the areas. But if you show him a number, the "form" of the number gets processed all the way up to the fusiform and produces cross-activation of cells in V4 much as you could using real colors in a normal person. It's as if you were able to "skip" the retinal processing of color and directly access and stimulate the full range of color-coded cells in V4, depending on which number was presented. Since S. S. has never experienced (and can never experience) these colors in the real world of real objects and can do so only by looking at numbers, he sees them as strange and spooky—nothing like he can ordinarily see—so he calls them "Martian colors."

This striking observation provides a direct empirical answer to the old philosophical question of whether someone can experience visual qualia that they have never experienced before. For example, if someone was born blind and had her sight restored as an adult, would she experience genuine qualia and, if so, how can she tell us? Our answer to this three-hundred-year-old question is "Yes, she will."

A Solution to the Riddle of Qualia

There are two questions: First, why do only some neural events have "qualia" associated with them whereas others do not (e.g., your pupil will contract in response to a light even when you are asleep or semicomatose)? What is the common denominator of those neural events that do versus those that do not? Second, what are the functional characteristics that distinguish qualia-laden events from unconsciously processed ones? Elsewhere one of us (V. S. R.) has suggested that four functional characteristics need to be fulfilled for the emergence of qualia (Ramachandran and Hirstein 1997). We refer to these as the four laws of qualia:

Qualia are irrevocable and beyond dispute; that is, "this is yellow"—not "maybe it is yellow."

Qualia are linked to short-term memory. They are held in a short-term "buffer" to provide enough time for subsequent decisions about processing, that is, to allow the luxury of choice.

Whereas for a spinal reflex arc only one output is possible, the color "yellow' has potentially unlimited implications; yellow flowers, yellow teeth, yellow yolk, and so on (sometimes loosely referred to as "meaning" or semantic reference).

Qualia are always linked to attention.

Criterion number 1 is absolutely crucial in our view. To get a clearer understanding of it, we will illustrate it with a concrete example. Since there is a "hole" or gap in your retina corresponding to the optic disc, if you shut one eye and look at the world there is a region of visual space subtending about five degrees where you are completely blind. If you aim this "blind spot" at a corner of a square, the corner will vanish. But remarkably, if you look at a homogeneous yellow wall, you don't see a gap or dark region corresponding to the gap; it gets "filled in" by the surrounding yellow. Indeed, you can even have a yellow doughnut that surrounds the blind spot (with its inner margin just overlapping the outer edge of the blind spot) and you see a homogeneous yellow disc rather than a ring. Again, the yellow fills in (Ramachandran 1992). This demolishes philosophers' claims that you simply "ignore" what is in your blind spot.

But in what sense, if any, is this different from your failure to notice the huge gap behind your head? If you stand in a yellow room you don't experience a big dark region or vacuum behind your head; you simply assume, infer or deduce that the walls behind you are also yellow. Is this the same kind of filling in as that which occurs in the blind spot?

In some abstract sense, the two are similar, of course; both are examples of interpolation, of the visual system guessing what is in the unsampled region based on what is in the surrounding region. Yet phenomenologically, the two instances are utterly different. In the case of the blind spot you literally see the yellow in the middle of the doughnut, whereas for the blind region behind your back you infer or guess that it must be yellow but do not actually see the color localized there; it has no yellow qualia.

This distinction lies at the heart of the riddle of qualia and provides the vital clue for understanding why it evolved. It takes us back to the question we started with, why do some neural events have sensory qualia associated with them whereas others don't? What does having qualia buy you, in evolutionary terms (Ramachandran and Hirstein 1997)?

Our hunch is that when sensory signals reach a certain preestablished criterion of certainty, the perceptual mechanism labels or "flags" the event as being "beyond reasonable doubt" in order to confer certainty and eliminate hesitation from subsequent decisions about processing. In other words, when something is (say) 98 percent likely to be red, then your brain pretends that it is 100 percent certain it is red and creates a corresponding sensory representation that allows you subsequent hesitation-free decision making and perceptual processing. The "pretend" representation also endures in short-term memory long enough to facilitate subsequent decisions and is "tamper resistant" from top-down influences—that is, it is irrevocable. We suggest that in true instances of qualia-laden filling-in, you use the surrounding information (or other relevant information) to say (in effect) "It is more than 98 percent likely that this line is actually passing through the blind spot (or this yellow color extends right across the blind spot). Therefore, we will assume that the sensory support for it is there in the outside world and will actually activate the corresponding sensory cortical areas (either using lateral connections or using feedback from back-projections) in a manner indistinguishable from genuine bottom-up sensory activation."

Contrast this with the gap behind your head where you infer the presence of yellow or the tail of a cat sticking out under the table causing you to infer the whole cat. In both instances you use top-down influences to partially activate the sensory neurons corresponding to "yellow" or "cat" but you don't literally see the color or the animal (because the level of activation is still far less than what you could achieve with a real input). This is important because it allows you (your visual centers) to say, in effect, "Maybe it's a cat but maybe it's a strange new species—so let me reserve judgment; indeed I can just as readily imagine it to be a mutant pig with a cat's tail."

This explains the biological function of genuine sensory filling-in, in particular and qualia in general—its purpose is to flag an input in sensory areas in order to confer stability on behavior, to eliminate hesitation from subsequent decision-making. For a region like the blind spot where you are 98 percent certain it is yellow inside, it makes evolutionary sense to go ahead and fill it in. For the region behind your head, you don't fill in with qualia because there is a fair chance you are wrong (the wall behind you might be red) and the penalty, too high (there might be a lion behind you!). In short, for the region behind your head you create evanescent fleeting representations that are tentative and you are allowed to change your mind about (like the "hypotheses" of science), but for the qualia of your blind spot, you are stuck with it—you cannot ignore it or change your mind about it.

To see this more clearly, imagine that humans evolve into a rabbit-like creature with eyes on the sides, so that the gap behind the head becomes progressively smaller. Our model would predict that when a certain threshold size (approaching the size of the real blind spot), there would be a shift from "finding out" (Dennett style) to true "filling in" (Ramachandran and Churchland style), and when this happens, the four laws of qualia will be fulfilled (which may require the evolution of new neural mechanisms). Dennett's (1991) notion that you don't need to fill in because there is no homunculus viewing the scene is a red herring. The homunculus, in this case, may just be subsequent decision-making mechanisms that may "require" gap-free representations to avoid distraction during subsequent processing (Ramachandran and Hirstein 1997). You are no longer allowed to say, "Maybe it is yellow, but there's a small chance it is really pink." You say, "It bloody well is yellow and I'll treat it as such and forge ahead."

Returning to synesthesia, we believe that the sensations synesthetes experience fulfill the criteria outlined above and therefore have qualia. For example, for a synesthete 4 is green, irrevocably green, and cannot be seen as pink. This has obvious relevance to the key distinction we have made between finding out on the one hand and filling in on the other. Notice, too, that these anomalous qualia probably depend on differences in the wiring of the fusiform (or angular) gyrus. Whether higher synesthetes also experience true qualia that are equally irrevocable remains to be seen.

Why Is Synesthesia More Common in Artists, Poets, Novelists, and Composers?

We now turn to the question of why synesthesia is more common among artsy types—like painters, poets, composers, and novelists—if, indeed it is. According to one recent survey (Domino 1989), as many as a third of all artists claim to have had synesthetic experiences of one sort or another. But is this simply because artists are often more prone to express themselves in vague metaphorical language, or maybe they are just less inhibited about admitting having had such experiences? Or is the incidence genuinely higher? And if so, why? Do artists and poets seem to have synesthesia because they have a vivid imagination and they are prone to talk as if they do? Or is their creativity linked in some deeper way to their synesthesia? Is synesthesia the result of a vivid imagination or a vivid imagination the result of synesthesia?

One thing that artists, poets, and novelists have in common is that they are especially good at using metaphor ("It is the East and Juliet is the sun"). It is as if their brains are set up to make links between seemingly unrelated domains—like the sun and a beautiful young woman. When you hear "Juliet is the sun" you don't say, "Does that mean she is an enormous glowing ball of fire?" (Actually, schizophrenics, who sometimes take metaphors literally, tend to do this).[1] You say, instead, "She is warm like the sun, nurturing like the sun, radiant like the sun" or "She rises from bed like the sun, dispelling the gloom of the night." Your brain instantly finds the right links highlighting the most salient and beautiful aspects of Juliet. In other words, just as synesthesia involves making arbitrary links between seemingly unrelated perceptual entities like colors and numbers, metaphor involves making links between seemingly unrelated "conceptual" realms. Perhaps this isn't just a coincidence. Perhaps the higher incidence of synesthesia in artists is rooted deep in the architecture of their brains.

The key to this puzzle is the counterintuitive observation that at least some high-level concepts are probably anchored in specific brain regions. There is nothing more abstract than a number. Warren McCollough, in emphasizing the very abstract nature of number, once asked the rhetorical question "What is a number that Man may know it? And what is Man that he may know number?" Yet even such an airy abstraction is represented, as we have seen, in a relatively small brain region—the angular gyrus. Indeed if a stroke is small enough, it can even make a person lose the ability to name specific categories (e.g., tools but not fruits and vegetables). All of these entities are "stored" close to each other in the inferior temporal lobe, but they are sufficiently separated from each other that a small stroke can knock out one and not the other. You might be tempted to think of fruits and tools as perceptions rather than concepts. However, two tools (e.g., a hammer and saw) can be visually as dissimilar from each other as they are from a fruit (e.g., a banana). This suggests that it is the concept of "use" that's critical here, not the visual appearance.

If concepts exist in the form of brain maps, then perhaps we have the answer to our question about metaphor and creativity. If a mutation were to cause excess connections between different brain maps, then depending on where and how widely in the brain the trait was expressed, it could lead to both synesthesia and to a propensity toward linking seemingly unrelated concepts and ideas. This would explain the higher incidence, perhaps, of synesthesia among artists, poets, and creative people in general (Ramachandran and Hubbard 2001b; Ramachandran 2003). Obviously, we are not saying that being a florid synesthete is sufficient or even necessary for poetry and metaphor—only that the same gene(s) predispose people to be more prone to creativity and that's why they survived (just as the sickle cell gene confers protection against sleeping sickness in recessive form but is fatal as a double recessive).

These ideas take us back a full circle to where we started. Instead of saying, "Synesthesia is more common among artists because they are being metaphorical" we should say "Artists are better at metaphors because they are synesthetes." Many scientists have argued in the past that synesthesia is a fundamentally different phenomenon from metaphors (e.g., Cytowic [1989] 2002), but we disagree. If our reasoning so far is correct, then the two are closely related and may be based on a deep similarity of brain mechanisms.

Synesthesia and Metaphor—The Missing Link

The evidence we have considered so far suggests that synesthesia is caused by some sort of glitch or minor flaw in brain circuits. But we have also been trying to move away from this idea to the notion that the phenomenon isn't just an odd quirk in some people and perhaps to the idea that we are all synesthetes to some extent.

To appreciate this, look at the two forms in figure 22-8, one of which looks like an inkblot (b) and the other is a jagged piece of shattered glass (a). If we were to ask you, "Which of these is a bouba and which is a kiki?" you will pick the inkblot as bouba and the other one as kiki, even though you have never been told this before. We tried this in a large classroom recently and 98 percent of the students made this choice. Now you might think that this has something to do with the blob resembling the physical form of the alphabet B (for bouba) and the jagged thing resembling a K (as in kiki). However, if you try the experiment on non-English-speaking Tamillians in India or on speakers of Chinese, you find exactly the same thing even though the

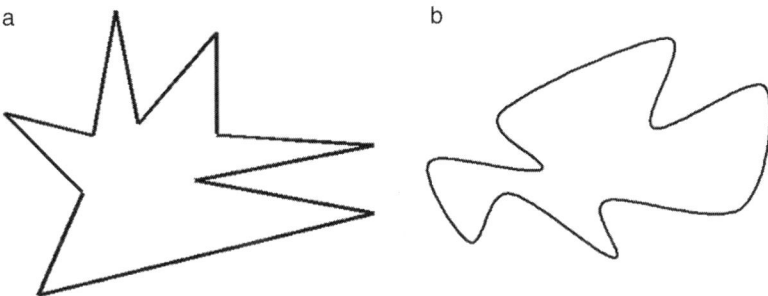

Figure 22-8. Demonstration of kiki and bouba. Because of the sharp inflection of the visual shape, subjects tend to map the name *kiki* onto (a), while the rounded contours of (b) make it more like the rounded auditory inflection of *bouba*.

corresponding alphabets in their language bear no resemblance to these abstract shapes.

So why does this happen? The real reason is that the gentle curves and undulations of contour on the amoeba-like figure metaphorically (one might say) mimic the gentle undulations of the sound "bouba" as represented in the hearing centers in the brain, and the gradual inflection of the lips producing the curved "Booo Baaa" sound, whereas the wave form of the sound "Ki Ki" and the sharp inflection of the tongue on the palette mimic the sudden changes in the jagged visual shape. This experiment proves that there is a sense in which we are all closet synesthetes, although we are in denial about it.

The angular gyrus of the brain receives signals from all the sense modalities—vision, touch, and hearing—and integrates these signals to create an abstract description of the world. We have recently obtained some preliminary evidence that people with damage to this region no longer get the bouba/kiki effect—they are no longer able to match the correct shape with the correct sound. This observation gives as a clue to understanding how abstract thinking might have first evolved in our hominid ancestors.

Let's take another look at the bouba/kiki effect. Clearly, the visual shape is conveyed by light reflecting from the paper and making a spatial pattern of photons dancing on your retina. The sound *kiki*, on the other hand, is conveyed by a time-varying pattern of hair cell movements in your ear. The two have absolutely nothing in common except for the single abstract property of "jaggedness" that is extracted somewhere in your parietal lobes—probably the angular gyrus. So you can think of this structure as performing a very elementary type of abstraction—extracting the common denominator from a set of seemingly dissimilar entities. We don't know how exactly the angular gyrus does this job, but once this ability to engage in cross-modal abstraction emerged it might have paved the way for the more complex types of abstraction that humans excel at. This strategy, the opportunistic "takeover" of one function for a different function is very common in evolution (see "Angular Gyrus, Supramarginal Gyrus, and the Evolution of Abstraction").

One reason this formulation seems counterintuitive at first is that metaphorical associations in ordinary language seem so arbitrary (sharp cheese) but in fact this isn't true; George Lakoff, Mark Johnson, and Mark Turner (e.g., Lakoff 1987; Lakoff and Johnson 1980; Lakoff and Turner 1989) have shown that there are strong directional constraints and some pairings are much more common than others. We have noticed that this is equally true for "genuine" congenital synesthesia and have suggested that in both cases the directionality and preferred pairings reflect anatomical constraints. Indeed, the preferred directions and frequency of associations are probably similar for both (Ramachandran and Hubbard 2001b).

Angular Gyrus, Supramarginal Gyrus, and the Evolution of Abstraction

Earlier we suggested that depending on the stage at which cross-activation occurs, there might be "higher synesthetes" and "lower synesthetes" and that in the former the cross-talk might be in the general vicinity of the angular gyrus, which is known to be involved in high-level numerical concepts. The idea that some types of synesthesia might involve the angular gyrus is also consistent with the old clinical observation that this structure is involved in cross-modal synthesis; information from touch, hearing, and vision is thought to flow together in the angular gyrus to enable the construction of high-level percepts. For example, a cat is fluffy (touch) and it meows (hearing) and has a certain appearance (vision) and sometimes a foul odor (smell), all of which are evoked by the memory of a cat or the sound of the word *cat*. No wonder patients with damage here lose the ability to name things (anomia) even though they can recognize them. Additionally, they have difficulty with arithmetic, which also involves cross-modal integration; you learn to count with your fingers. (Indeed, if you touch the patient's finger and ask him which one it is, he often cannot tell you). All of these bits of clinical evidence strongly suggest that the angular gyrus is a great center in the brain for cross-modal synthesis. So perhaps it is not so outlandish, after all, that a flaw in the circuitry could lead to colors being quite literally evoked by certain sounds.

Intriguingly, we have found that patients with lesions in the angular gyrus are also terrible with metaphors, often taking them literally. Could it be that the angular gyrus, which is disproportionately larger in humans than in apes and monkeys, originally evolved for cross-modal abstraction, but then became coopted for other more abstract metaphors as well? It would be interesting to see, though, whether they are even worse at cross-modal metaphors such as "loud shirt" than at other types of metaphors.

Cross-modal abstraction became increasingly important in mammalian evolution—especially primate evolution. Ancestors of mammals—theraspids—were probably diurnal, but they evolved into nocturnal mammals (perhaps to avoid dinosaurs) that started relying on hearing rather than sight. Later, when they ascended the treetops and eventually became diurnal once again—as primates—they became highly visual creatures with an additional requirement; they had to hook up touch and kinesthesis (proprioception) to vision to allow prehension. Obviously, all of this would have required extensive cross-modal interactions resulting in a disproportionate enlargement of the angular gyrus. The stage was now set for more complex types of cross-modal abstraction, as exemplified by your ability to say which of two irregular shapes you just felt with your hands is identical to the one you are seeing. (After all the two abilities have almost nothing in common; the touch receptors are stimulated serially whereas the visual shape is apprehended in parallel). From this to the kinds of cross-modal synesthesia involved in mediating the bouba/kiki effect was but a short step. The result was the creation of more abstract tokens—the basis of symbol manipulation and thinking in general. Given the way evolution works it is not inconceivable that the angular gyrus (and TPO in general) evolved originally for cross-modal mappings but then became an exaptation for other types of abstraction as well (and perhaps for metaphorical thinking in general). There may be additional division of labor between right and left angular gyri—the latter for abstract and action-related metaphors, the former for spatial and embodied metaphors.

Broadly consistent with our claim that the TPO (including the angular and supramarginal gyri) played a critical role in the evolution of uniquely human abilities is the well-known clinical observation that lesions in this general area lead to deficits in demonstrating tool use—ideomotor apraxia (e.g., the patient cannot mime brushing his teeth or using a hammer even though language comprehension is intact and there is no motor clumsiness). Tool use requires you to conjure up a visual-kinesthetic image and you cannot do this without the kind of cross-modal integration that occurs in the TPO region—especially the supramarginal gyrus. Can it be a coincidence that in addition to mathematics, the "lexicon," and metaphors, another uniquely human ability, skilled tool use, was also made possible through the explosive enlargement of this region of the brain? We have now begun testing patients with ideomotor apraxia from this evolutionary standpoint. Would they have special difficulties with subassembly in tool use? (For example, hafting the head of a hammer to a handle using a string and using the composite to hammer a nail.) And could such subassembly routines have served as an exaptation for the hierarchical structure of language to which it bears a formal resemblance? Could the supramarginal gyrus been originally "pinched off" from the angular gyrus through gene duplication and subsequently developed additional specializations? And finally, the mirror neuron system was originally discovered in monkey ventral premotor area (a homologue of Broca's area in humans), but is it possible that this area receives its major input from the inferior parietal lobule, which may also have cells with mirror-neuron like properties? If so one might speculate that this system—involved in tool use and compromised in apraxia—may have set the stage for the evolution of imitation by children, the basis for culture (Ramachandran 2000). One of us (V. S. R.) found recently that a patient with ideomotor apraxia caused by damage to the left supramarginal gyrus (and hence, possibly, to the mirror neuron system) also had considerable difficulty understanding action metaphors such as "he couldn't put his finger on it" or "a touching remark." This suggests that a sophisticated neuron system is needed not only for comprehension of another person's actions but also for representing action metaphors. This crucial difference in the level of sophistication of the mirror neuron system between humans and lower primates would explain why any monkey can reach for a peanut but only a human can reach for the stars or even understand what that means.

Evolution of Language

> When [Scotland Yard detectives] Gregson and Lestrade and Athelney Jones are out of their depths—which, by the way, is their normal state—the matter is laid before me.
>
> —Sherlock Holmes to James Watson M.D.

The evolution of language is one of the oldest puzzles in psychology. It generated such acrimonious and unproductive debate in the nineteenth century that the French linguistics society introduced a formal ban on all papers dealing with this topic.

But problems cannot be removed by censorship. The question is still unanswered: How did an ability as sophisticated and complex as human language evolve in just two or three hundred thousand years, a mere blink in evolutionary time? How can the blind workings of chance—natural selection—transform the emotional grunts and howls of ape-like ancestors to all the linguistic sophistication of a Shakespeare or a George Bush? More specifically, how did our astonishing lexicon evolve, our vocabulary of thousands of words? And even more puzzling; how did the hierarchic "tree" structure of syntax, which involves the embedding of clauses within clauses and recursivity (e.g., "Is Delhi the capital of India, and IF that is true, THEN would it make sense to focus our troops in that region?") evolve. Did Chomskyan deep structure of syntax evolve as a separate module out-of-the blue as Chomsky implies, or did it coevolve with semantics (or meaning)?

There have been four standard theories:

Alfred Russell Wallace: Language emerged through divine intervention.

Chomsky: God knows what happens when you pack one hundred billion cells into such a tiny space. Maybe new "singularities" emerge, or even new laws of physics. Language is just too sophisticated and has too many interlocking parts to have evolved through plain old natural selection.

Stephen Jay Gould: Language wasn't specifically selected for by natural selection for its function—open-ended communication. On the contrary, language is the specific deployment of a more general-purpose mechanism—thinking—that evolved for other purposes (just as our fingers evolved for climbing but can now be used for counting; which explains why most mathematical systems use base ten).

Steven Pinker: Language is an "instinct" just like coughing or sneezing (albeit more sophisticated), which evolved through conventional Darwinian selection. The reason it seems so mysterious to us is because the intermediate steps are missing.

Who is right? We can discount (1) altogether—it is not testable; (2) also comes close to implying a miracle, although Chomsky doesn't actually invoke divine intervention.

Gould's view, (3), seems reasonable, but we are uncomfortable with it because it merely postpones the problem; it doesn't solve it. Saying language is simply a deployment of thinking doesn't tell us much because we have no idea how thinking evolved—if anything, it is even more complex!

Pinker is on the right track; indeed how else could it have evolved other than through natural selection? But he doesn't go far enough, the idea is too general. It is a bit like answering the question "How does digestion work?" with the answer "By obeying the laws of thermodynamics." Obviously, this must be true, but it doesn't tell us anything about the precise mechanisms of digestion. In the case of language, what we would like to know is, "What was the actual sequence of steps that culminated in language? What was its exact trajectory through the fitness landscape"? For answering this question, Gould's idea of exaptations is much more useful, although he doesn't tell us what the exaptations might have been that resulted eventually in language.

We will replace these four ideas with what we call the synesthetic bootstrapping theory of language (Ramachandran and Hubbard 2001b). To understand this we need to go back to the bouba/kiki example. Why do most people who have never seen these shapes before automatically associate the jagged shape with the sound "kiki" and the undulating shape with "bouba"? We suggested this is an example of a preexisting universal cross-modal synesthesia that involves abstracting the property (or "Fourier components" if you prefer; the terminology isn't important) of jaggedness or undulation in the angular gyrus. Recall that we had seen a patient with a lesion here who no longer made this association.

This observation gives us the vital clue to understanding the emergence of words—of the lexicon of early hominids. For it shows that there is a preexisting nonarbitrary correspondence between word sound and object appearance. Imagine a band of ancestral hominids trying to develop a shared vocabulary. Of course the leader didn't say, "Hey look at this; let's call it a banana—all of you say after me BA-NA-NA"? We would suggest, instead, that there was a built-in bias to associate certain sounds with certain visual shapes and this bias—however small—may have been important in getting you started on a shared vocabulary. This type of cross-modal nonarbitrary synesthesia may be based on cross-activation between visual maps in the fusiform and auditory/phoneme maps in the superior temporal gyrus, although it may also require a "translation" mediated by the angular gyrus (see arrows in figure 22-9).

But this is only one small step; there are also two other key steps required for this to be viable theory of proto-language in particular and language evolution in general. In addition to the visual to auditory cross-activation (bouba/kiki) there are also forms of cross-activation between (1) auditory phoneme areas in the superior temporal gyrus and the visual areas and (2) the motor speech areas in the Broca's area in the frontal lobe which has "maps" of motor programs for moving tongue, lips, palate, larynx, and so on for phonation and vocalization. Such a cross-activation may also occur between the visual appearance of the

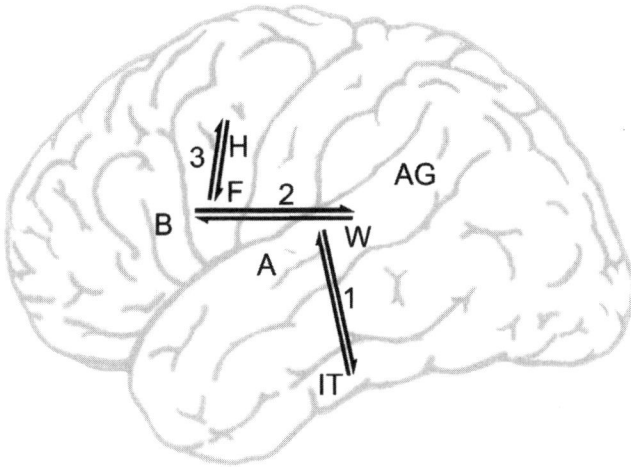

Figure 22-9. A new synesthetic bootstrapping theory of language origins. Arrows depict cross-domain remapping of the kind we postulate for synesthesia in the fusiform gyrus. (1) A nonarbitrary synesthetic correspondence between visual object shape (as represented in IT and other visual centers) and sound contours represented in the auditory cortex (as in our *bouba/kiki* example). Such synesthetic correspondence could be based on either direct cross-activation or mediated by the angular gyrus—long known to be involved in intersensory transformations. (2) Cross-domain mapping (perhaps involving the arcuate fasiculus) between sound contours and motor maps in or close to Broca's area (mediated, perhaps, by mirror neurons). (3) Motor-to-motor mappings (synkinesia) caused by links between hand gestures and tongue, lip and mouth movements in the Penfield motor homunculus (e.g., the oral gestures for "little" or "diminutive" or "teeny weeny" synkinetically mimic the small pincer gesture made by opposing thumb and index finger (as opposed to "large" or "enormous"). The cross-wiring would necessarily require transforming a map of two-dimensional hand gestures into one-dimensional tongue and lip movements (e.g., the flexion of the fingers and palmar crease in "come hither" is mimicked by the manner in which the tongue goes back progressively on the palate). And you pout your lips to say "you," "vous," or "thoo" as if to mimic pointing outward whereas "me," "mois," and "I" mimic pointing inward toward yourself.

object and motor speech maps directly without auditory mediation. The reason for thinking this is three fold: First, think of words like "teeny weeny" or "un peu" or "diminutive." Can it be a coincidence that your lips actually mimic the teeny weeny visual appearance of the object (compare with "large" or "enormous")? Second, even a newborn infant will stick its tongue out if he watches you doing it, so there seems to be a built-in visual appearance to oral movement translation. Third, a class of neurons discovered in the homologue of the Broca's area in monkeys (called "mirror neurons") will fire not only when the monkey performs a manual action, for example, grabbing a peanut; but also when it watches another monkey performing the same action! It seems likely that such neurons exist for vocal movements as well—making it easier for children to imitate parental sounds (partly as a result of the built-in translation from visual to motor and partly from auditory to motor maps). The extent to which such neurons are hardwired versus learnt through Hebbian mechanisms remains to be investigated.

Another telling example is words like "fuDGe," "truDGe," and "sluDGE," in which the movement of the tongue on the palette actually mimics the "stickiness" and viscosity of the seen or felt mud or chocolate (gradual adhesion followed by catastrophic release). One also wonders whether phenomena like dancing are a form on sensory-to-motor cross-activation in normal people. Indeed, there is even an example in the literature of a form of synesthesia in which the subject had an uncontrollable urge to automatically assume certain postures in response to certain sounds (Devereux 1966).

The last piece that completes our theory comes from an effect first noticed by Charles Darwin. He pointed out that when we cut a piece of paper with scissors our jaws clench and unclench unconsciously as if to echo the hand movements. Darwin couldn't have known this, but the effect is probably caused by a sort of "spillover" of motor signals from the hand motor area to the mouth motor area that lies right below it in Penfield's motor homunculus in the precentral gyrus, an effect that we call "synkinesia." So there appears to be a built in translation between manual and oral gestures, and like synesthesia this may rely on cross-activation of brain maps; in this case motor rather than sensory maps.

Many linguists don't like the theory that manual gesturing could have given rise to or at least set the stage for vocal language, but synkinesia suggests that they may be wrong. When you want to depict something tiny you make a tiny pincer like gesture opposing your thumb and forefinger, and notice that your lips do the same when saying "teeny weeny" or "diminutive," synkinetically mimicking your fingers. Or when you beckon someone to come toward you, you stick your arm and hand out, palm up, and then flex your palm and fingers toward yourself; a gesture that the tongue mimics on the palette when it bends gradually backward before hitting the roof—while you make the sound "hither" ("ither" in Hindi). Similarly, you pout your lips outward away from you for "go" and inward toward you when saying "come."

Assume our ancestral hominids communicated mainly through emotional grunts, groans, howls, and shrieks produced by the right hemisphere and anterior cingulate. Later they evolved a rudimentary gestural system that became progressively more elaborate and sophisticated. If the gestures were than "translated" through synkinesia into orofacial movements, and if emotional guttural utterances were channeled through these mouth and tongue movements, you get the first spoken words. The last piece of the puzzle falls in place.

But how did gesturing itself evolve? The most likely scenario is that actual actions became conventionalized into gestures of communicative intent. For example, it is easy to see how the actual hand movement used for pulling someone toward you might have become progressively conventionalized into a "come hither" gesture (and the converse would be true for "push" and "go"). There is no language on earth for which the opposite is true.

So we now have three effects going on; visual to auditory synesthesia as in bouba/kiki mediated by visual-fusiform to auditory-superior temporal cross-activation (with possible angular gyrus mediation); visual plus auditory representations in the back of the brain producing coactivation of corresponding Broca's motor mouth maps ("diminutive," "teeny weeny," etc.); and lastly synkinesia—the mouth mimicking hand gestures (as in teeny weeny, come hither, etc.) which, in turn, evolved through conventionalization and ritualization of real actions. Each of these biases might have been quite small to begin with, but that's all you need in evolution; it is often the initial emergence of a trait that is hard to explain—once it is in place further embellishments are easy. Although the biases may have been small in our ancestors, given the arrangement we have proposed they can progressively bootstrap each other leading to a sort of "avalanche effect" and culminating in modern language.

One objection to this view might be that there are many objects for which words in modern languages are utterly dissimilar. For example, dog is "chien" in French, "kutta" in Hindi, and "nai" in Tamil. But this may be because once protolanguage evolved in ancestral hominids they could have diverged substantially; it's the initial emergence of a trait that's often hard to explain in evolution. Once a trait emerges, though, further embellishments and divergences are easy to evolve (especially if you add "culture" to the equation).

An obvious advantage with our scheme is that we can now look at aphasias from this evolutionary framework (figure 22-9) instead of the classical black boxology of cognitive science and artificial intelligence. (We have already noted that the bouba/kiki effect is compromised in patients with angular gyrus lesions.) The model we present is in fact much more consistent with the central tenets of the neural networks revolution pioneered by Terry Sejnowski, Jeff Hinton, Jay Mclellland, Jeff Ellman, and others, except that we place our emphasis here on evolution rather than learning algorithms. (The two approaches are perfectly compatible, of course.) Another prediction from our scheme is that if "higher" synesthetes have an inherited "flaw" in the angular gyrus region, some of them may exhibit the other components of the Gerstmann's syndrome (acquired angular lesions), such as dyscalculia, left–right confusion. We have recently found strong hints that this might indeed be true (in an informal survey we did of synesthetes attending a conference at UCSD).

We should point out at this stage that our view is different from the now discredited onomatopoeic theory of language. The key difference is that in the latter theory words arose from using sounds that are arbitrarily associated with objects; for example, "bow wow" for dog, whereas in our account the link between shape and sound (bouba/kiki) is not arbitrary—it involves a higher level of cross-modal abstraction.

One odd observation that is consistent with our view comes from comparative linguistics. Berlin (1994) showed that if fifty bird names and fifty fish names from a South American Indian language are given to American college students to classify into the categories of fish and bird names, they do so significantly above chance, even though Indo-European languages don't share an ancestry with this language. We would regard this as an example of the bouba/kiki effect in action.

What we have come up with so far is really a theory of protolanguage including lexicon and (to some extent) of semantics (as in the abstraction performed by the angular gyrus). But how would you import syntax into this scheme, especially the hierarchical structure of the Chomskyan syntax? We would argue that once protolanguage is in place, then, in conjunction with semantics, it becomes somewhat easier to make this transition than previously believed. But, more important, the evolution of tool use by hominids may have also played an important role in making this transition possible, culminating in full-fledged human language (as argued eloquently by Greenfield 1991). In particular, there is a close operational analogy between the subassembly technique used by late (but not early) hominids and the hierarchical structure of syntax. For example, the sequence shape the hammer's head first then attach it to a handle and then chop the meat bears a remarkable resemblance to the "embedding" of clauses within larger sentences. We suggest that this resemblance arises from homology rather than mere analogy. Frontal brain areas that originally evolved for subassembly in tool use may have been duplicated in evolution through gene duplication and then coopted for a completely novel function—language. Linguists are forever complaining that not every subtle feature of modern language is explained by such schemes, but, surely, what we need is an initial framework or scaffolding to get things started because once that's available it is easier to account for the subsequent emergence of subtleties (such as recursiveness, for example).

Notice that this view differs from Gould's idea that language emerged from general-purpose mechanism such as thinking. We agree with Pinker that natural selection alone was the evolutionary "force" that resulted in language, but we disagree about the specific sequence of events. In particular, we suggest that the evolution of language resulted from an opportunistic coopting of multiple exaptations that were previously selected for other functions altogether (e.g., cross-modal abstraction synesthesia between vocal, visual, and auditory maps; synkinesia resulting from the fortuitous adjacency of the mouth and hand cortexes, subassembly in tool use, interactions with "world-based" constraints such as subject-verb-object sequence). Thus in our scheme the fortuitous cooccurrence of multiple exaptations and subsequent equally fortuitous interactions between them was absolutely critical in setting in motion the autocatalytic events that culminated in modern language. In this regard, we would agree with Gould in placing emphasis on exaptations although not on the particular one—thinking—that he picked.

We should add that while some of the specific ideas on language we have put forth here are novel, most of them have been considered and debated extensively in form or another by other researchers outside the neuroscience community. Any novelty lies in our attempt to weave them into a comprehensible and coherent story that will, hopefully, serve as a catalyst for further inquiry.

Some Problems

> There are problems, Watson, but then there are always problems.
> —Sherlock Holmes

Have we solved the riddle of synesthesia? Obviously not. In fact we have barely scratched the surface. But we hope the speculations in this chapter and some of our experiments provide a framework for future research into this fascinating topic.

Some additionl curious observations remain to be explained We list them here, not in any particular order.

1. Is there a sharp distinction between higher or lower synesthetes or do they represent two ends of a continuum? This is important because the group we refer to as "lower" synesthetes constitutes only about 10 to 15 percent of synesthetes.
2. Although synesthesia often involves adjacent brain areas (e.g., grapheme/color in the fusiform) it doesn't have to. Even far-flung brain regions, after all, can have pre-existing connections that could be amplified (e.g., through disinhibition). But, statistically speaking, adjacent brain areas tend to be more "cross-wired" to begin with and therefore synesthesia is likely to involve those more often.

3. Even some of the more exotic forms of synesthesia can be explained on the basis of proximity of brain maps—for example, our subject M. B. who had tastes evoked when he palpated different shapes and textures (proximity of the insula to the Penfield hand region perhaps?). Another of our subjects experienced colors evoked by different types of nausea (caused, perhaps, by links between the insula and color areas): nausea caused by infection—yellow; binge drinking—red; nausea associated with migraine—green; "hunger pangs" nausea—pink.

4. We have noticed another strange type of synesthesia: every letter or number grapheme is "felt" to be either male or remale, good or bad, or (in one subject) even or uneven. (Her daughter had the same associations, but the even/uneven labels were reversed.) This might represent an exaggeration of the universal tendency to "binarize" the world to simplify information processing, but it isn't easy to explain in terms of cross-activation.

5. We observed that chains of associations, which would normally evoke only memories in normal individuals, would sometimes seem to actually evoke qualia-laden sense impressions in some higher synesthetes. So the merely metaphorical can become quite literal. For example, R is red and red is hot so R is hot, etc. One wonders whether the "hyperconnectivity" (either through sprouting or disinhibition) has affected back-projections between different areas in the neural hierarchy.

6. The cross-activation (either through disinhibition of back-projections or sprouting) model can also explain many forms of "acquired" synesthesia that we have discovered. One blind patient with retinitis pigmentosa whom we studied (Armel and Ramachandran 1999) vividly experienced visual phosphenes (including visual graphemes) when his fingers were touched with a pencil or when he was reading Braille. (We ruled out confabulation by measuring thresholds and demonstrating their stability across several weeks; there is no way he could have memorized the thresholds.) A second blind patient whom one of us (V. S. R.) tested with S. Azoulai, could, quite literally, see his hand when he waved it in front of his eyes, even in complete darkness. We suggest that this is caused either by hyperactive back-projections resulting from somatosensory-induced visual imagery or from disinhibition caused by visual loss, so that the moving hand is not merely felt but is also seen. Cells with polymodal receptive fields in the parietal lobes may also be involved in mediating this phenomenon (Ramachandran and Azoulai 2004).

7. While many types of acquired synesthesia can be explained in terms of the known neuroanatomy, others cannot. One of us (V. S. R.) has encountered a patient who, following a pituitary tumor resection and 2 weeks on life support, saw the entire world as being a very specific blue color. After a few weeks this faded, but all things started "tasting blue"—that is, she didn't *see* blue but any taste (other than sweet) was *felt* to be blue and highly aversive. Such observations remind us of how little we really know about the brain.

8. As Galton (1880/1997) pointed out, some synesthetes have complex convoluted "number lines"—any time the subject visualizes a number it seems to always occupy a specific location in space, adjacent numbers being represented on adjacent points on the line. Sometimes the line makes sharp inflections and we have noticed that at these points of inflection simple arithmetic operations are more difficult and error-prone than in other parts of the line. Conversely, some "number savants" have told us that they have number lines that seem to facilitate complex calculations as they wandered around their number landscapes (e.g., the savant Arithmos whom we studied; see "the Brain Man" NOVA 2005). Remarkably, in some individuals, these numer lines occupy "world centered" coordinates; the subject can "inspect" hidden parts of the line by changing his vantage point or moving behind it (the numbers were seen as a mirror image!). In another synesthete studied by one of us (V. S. R.) months of the year formed a semicircle—a sort of half halo—around her head with her head always occupying the current month. For example, if she is in July, then August is to her left and June to her right and December in front. S. Azoulai and I are currently testing this by measuring reaction times (Stroop interference from location) for month names presented either to her left or right. Would these change if she were tested a month later? We have evidence that this is indeed the case. We conclude that in these synesthetes, and perhaps even in normals, time is represented in spatial maps that facilitate computations, a remarkable example of Lakoff-style "embodied cognition."

The introspections of some higher synesthetes are truly bewildering in their complexity. Here is a quotation from one of them: "Most men are shades of blue. Women are more colorful. Because people and names both have color associations the two don't necessarily match." Or "May is red, female, dominant, and round. June is blue, submissive, and a rectangle." Such remarks imply that any simple phrenological model of synesthesia is doomed to fail, although it is not a bad place to start.

In doing science one is often forced to choose between providing precise answers to boring (or trivial) questions (e.g., how many cones are there in the human eye?) or vague answers to big questions such as What is consciousness? or What is a metaphor? Fortunately, every now and then we get a precise answer to a big question (e.g., DNA being the answer to the riddle of heredity). Synesthesia seems to lie half way between these two extremes. While we must not fall into the trap of saying that it explains everything (given its resemblance to classical associationist psychology), it does seem to provide a jumping off point for thinking about certain elusive aspects of the mind that have hitherto been inaccessible to empirical research. Where this trail, with all its twists and turns, will lead, remains to be seen, but meanwhile, as Holmes told Watson, "The game is afoot."

Conclusions

> One's ideas should be as broad as nature if they are to interpret nature.
> —Sherlock Holmes

Synesthesia has been known for over a century, but it has, by and large, been regarded as a curiosity. Only a couple of pioneering researchers, Richard Cytowic ([1989] 2002) and Lawrence Marks (e.g., Marks 1975, 1982) recognized its genuineness and pointed out its importance. But Cytowic was a prophet preaching in a wilderness, and the theories he proposed to explain synesthesia were a bit vague (in our view); he suggested that it was a "throwback" to an earlier stage in phylogeny when the nervous system was undifferentiated. There may be a grain of truth in this, but it doesn't explain the specificity of synesthetic experiences, which is one of its most striking characteristics.

Fortunately, for the neuroscience and psychology community, synesthesia has been rescued from becoming a "fringe" phenomenon. Thanks to the research of several groups (Ramachandran and Hubbard 2000, 2001a; Nunn et al. 2002; Palmeri et al. 2002; Mattingley et al. 2001) there has been a tremendous resurgence of interest in this topic in the last three years, and we can now move from an era of vague phenomenology to the era of experimental research. Such research promises to be exciting for two reasons. First, if a large enough family is identified, we may be able to clone the gene(s). From the gene we can go to the specific brain areas (e.g., fusiform gyrus in lower synesthetes and angular gyrus and TPO junction in higher synesthetes) to detailed psychophysics (e.g., pop-out, blindsight, symmetry, apparent motion, etc.); perhaps all the way to metaphor and Shakespeare—in a single "preparation"! Second, the phenomenon may allow us to probe (through psychophysics, brain imaging, lesion studies, and modeling) the laws of interactions between brain maps (cross-activation). These laws, we believe, hold the key to understanding some of the most mysterious aspects of our minds—such as qualia, metaphor, analogy, and the emergence of abstract thought and language.

Acknowledgments We thank Per Aage Brandt, Geoff Boynton, Francis Crick, Richard Cytowic, Jeffrey Gray, George Lakoff, Diane Rogers-Ramachandran, and Mark Turner for discussions and the NIMH, Charlie Robins, and Richard Geckler for support. We also thank Dr. S.M. Anstis for playing the role of Watson in his many discussions with us. (As Sherlock Holmes told Watson, "Though not luminous yourself, my dear Watson, you are an excellent conductor of light. Some men, though not possessing genius themselves, have a remarkable capacity to inspire it in others.")

Notes

1. We have been struck by the fact that even though schizophrenics are bad at interpreting metaphors and proverbs they are very good at punning ("clang associations"). This seems like a paradox; why are they good with puns but bad at metaphor? The paradox is resolved when you realize that although puns and metaphors are superficially similar, they are the exact opposite of each other. A pun is a superficial or surface similarity masquerading as a deep insight (hence its comic appeal) whereas a metaphor involves illuminating a deep similarity despite surface differences! So perhaps its not altogether surprising that the two are dissociable, sometimes even inversely correlated, in certain disease states. The matter deserves deeper study.

2. In some synesthetes, the sound of the phoneme also seems critical—not just the visual grapheme. This might be based on cross-activation between phonemes in the auditory areas and color areas nearby or back projections to V4. In some of these individuals, the first letter of the word dominates the color. To complicate matters further, in some synesthetes who experience days or months as colored, the color of the first letter tinges the whole word but in many others, it's the numerical placement of the day or month.We have noticed these synesthetes have short identical "runs" of colors for numbers, days and months; Tuesday, Wednesday, and Thursday might be red, yellow, and green but so are March, April, and May or 4, 5, and 6.

References

Armel, K.C., and Ramachandran, V.S. (1999). Acquired synesthesia in retinitis pigmentosa. Neurocase, 54:293–296.

Bailey, M.E.S., and Johnson, K.J. (1997). Synesthesia: Is a genetic analysis feasible? In S. Baron-Cohen and J.E. Harrison (Eds.), Synaesthesia: Classic and Contemporary Readings. (pp. 182–207). Cambridge, MA: Blackwell.

Berlin, B. (1994). Evidence for pervasive synthetic sound symbolism in ethnozoological nomenclature. In L. Hinton, J. Nichols, and J.J. Ohala (Eds.) Sound Symbolism. New York: Cambridge University Press.

Cytowic, R.E. ([1989] 2002). Synaesthesia: A Union of the Senses. New York: Springer-Verlag.

Devereux, G. (1966). An unusual audio-motor synesthesia in an adolescent. Psychiatric Quarterly, 40(3):459–471.

Dennett, D.C. (1991). Consciousness Explained. New York: Little, Brown, and Co.

Dixon, M.J., Smilek, D., Cudahy, C., and Merikle, P.M. (2000). Five plus two equals yellow: Mental arithmetic in people with synaesthesia is not coloured by visual experience. Nature, 406(6794):365.

Domino, G. (1989). Synesthesia and creativity in fine arts students: An empirical look. Creativity Research Journal, 2(1–2):17–29.

Edelman, G.M. (1989). The remembered present: A biological theory of consciousness. New York: Basic Books.

Galton, F. (1880/1997). Colour associations. In S. Baron-Cohen and J.E. Harrison (Eds.), Synaesthesia: Classic and Contemporary Readings. (pp. 43–48). Cambridge, MA: Blackwell.

Gray, J.A., Williams, S.C.R., Nunn, J., and Baron-Cohen, S. (1880/1997). Possible implications of synaesthesia for the hard question of consciousness. In S. Baron-Cohen and J.E. Harrison (Eds.) Synaesthesia: Classic and Contemporary Readings. (p. 173–181). Oxford: Blackwell.

Greenfield, P.M. (1991). Language, tools and brain: The ontogeny and phylogeny of hierarchically organized sequential behavior, Behavioral and Brain Sciences, 4:531–595.

Kennedy, H., Batardiere, A., Dehay, C., and Barone, P. (1880/1997). Synaesthesia: Implications for developmental neurobiology. In S. Baron-Cohen and J.E. Harrison (Eds.), Synaesthesia: Classic and contemporary readings. (pp. 243–256). Oxford: Blackwell.

Lakoff, G. (1987). Women, Fire, and Dangerous Things: What Categories Reveal about the Mind. Chicago: University of Chicago Press.

Lakoff, G., and Johnson, M. (1980). Metaphors We Live By. Chicago: University of Chicago Press.

Lakoff, G., and Turner, M. (1989). More Than Cool Reason: A Field Guide to Poetic Metaphor. Chicago: University of Chicago Press.

Leuck, C.J., Zeki, S., Friston, K.J., Deiber, M.P., Cope, P., Cunningham, V.J., Lammertsma, A.A., Kennard, C., and Frackowiak, R. S. (1989). The colour centre in the cerebral cortex of man. Nature, 340(6232):386–389.

MacLeod, C.M. (1991). Half a century of research on the Stroop effect: An integrative review. Psychological Bulletin, 109(2):163–203.

Marks, L.E. (1975). On coloured-hearing synaesthesia: Cross-modal translations of sensory dimensions. Psychological Bulletin, 82(3):303–331.

Marks, L.E. (1982). Bright sneezes and dark coughs, loud sunlight and soft moonlight. Journal of Experimental Psychology: Human Perception and Performance, 8(2):177–193.

Mattingley, J.B., Rich, A.N., Yelland, G., and Bradshaw, J.L (2001). Unconscious priming eliminates automatic binding of colour and alphanumeric form in synaesthesia. Nature, 410 (6828):580–582.

Merikle, P., Dixon, M.J., and Smilek, D. (2002). The role of synaesthetic photisms on perception, conception and memory. Talk presented at the 12th Annual Meeting of the Cognitive Neuroscience Society, San Francisco, Calif., April 14–16.

Nunn, J.A., Gregory, L.J., Brammer, M., Williams, S.C.R., Parslow, D. M., Morgan, M. J., Morris, R.G., Bullmore, E.T., Baron-Cohen, S., and Gray, J.A. (2002). Functional magnetic resonance imaging of synesthesia: Activation of V4/V8 by spoken words. Nature Neuroscience, 5(4):371–375.

Pesenti, M., Thioux, M., Seron, X., and De Volder, A. (2000). Neuroanatomical substrates of Arabic number processing, numerical comparison, and simple addition: A PET study. Journal of Cognitive Neuroscience, 12(3): 461–479.

Ramachandran, V.S. (1992). Blind spots. Scientific American, 266(5):86–91.

Ramachandran, V.S. (2001). Sharpening Up "The Science of Art." Journal of Consciousness Studies 8(1):9–29.

Ramachandran, V.S., and Azoulai, S. (2004). Blind patients "see" their moving hand in darkness. Paper presented at the Annual Meeting of the Psychonomics Society USA, Minneapolis, Minn., November.

Ramachandran, V.S., and Blakeslee, S. (1998). Phantoms in the Brain. New York: William Morrow.

Ramachandran, V.S., and Hirstein, W. (1997). Three laws of qualia: What neurology tells us about the biological functions of consciousness. Journal of Consciousness Studies, 4(5–6):429–457.

Ramachandran, V.S., and Hirstein, W. (1999). The science of art: A neurological theory of aesthetic experience. Journal of Consciousness Studies, 6(6–7): 15–51.

Ramachandran, V.S., and Hubbard E. (2000). Abstracts of the first international meeting on phantom limbs, Oxford, England.

Ramachandran, V.S., and Hubbard, E.M. (2001a). Psychophysical investigations into the neural basis of synaesthesia. Proceedings of the Royal Society of London B, 268:979–983.

Ramachandran, V.S., and Hubbard, E.M. (2001b). Synaesthesia—a window into perception, thought and language. Journal of Consciousness Studies, 8(12):3–34.

Ramachandran, V.S., and Hubbard, E.M. (2002). Synesthetic colors support symmetry perception, apparent motion, and ambiguous crowding. Talk presented at the 43 Annual Meeting of the Psychonomics Society, November 21–24.

Ramachandran, V.S., and Gregory, R.L. (1978). Does colour provide an input to human motion perception? Nature, 275(5675):55–56.

Rickard, T.C., Romero, S.G., Basso, G., Wharton, C., Flitman, S., and Grafman, J. (2000). The calculating brain: An fMRI study. Neuropsychologia, 38(3):325–335.

Sacks, O., Wasserman, R.L., Zeki, S., and Siegel, R.M. (1988). Sudden color blindness of cerebral origin. Society for Neuroscience Abstracts, 14:1251.

Smilek, D., and Dixon, M. (2002). Towards a synergistic understanding of synaesthesia. Psyche, 8(01).

Spalding, J.M.K., and Zangwill, O. (1950). Disturbance of number-form in a case of brain injury. Journal of Neurology, Neurosurgery, and Psychiatry. 12:24–29.

Treisman, A. (1982). Perceptual grouping and attention in visual search for features and for objects. Journal of Experimental Psychology: Human Perception and Performance, 8(2):194–214.

Zeki, S., and Marini, L. (1998). Three cortical stages of colour processing in the human brain. Brain, 121(9):1669–1685.

23

What Are the Neuronal Correlates of Consciousness?

Francis C. Crick and Christof Koch

Why is this question important? Understanding consciousness is likely to alter our whole view of ourselves as persons. Chalmers (1995) has called this the Hard Problem, but this is a misnomer. There are a host of hard problems in neuroscience, but consciousness is certainly the most mysterious of them and so in some sense the hardest. How is one to explain "qualia"—the subjective aspects of color, the redness of red, or the painfulness of pain—in terms of known science? So far no one has put forward any concrete hypothesis that sounds even remotely plausible. Our strategy is to leave the core of the problem on one side for the time being and instead try first to discover the minimal neural mechanisms jointly sufficient for any one conscious sensation or percept, the neuronal correlates of consciousness, now widely called the NCC. When this is known we may be in a better position to explain qualia in scientific terms.

In this short essay we shall describe our own approach to the problem. We shall do this by asking a series of questions which cover some of the subproblems that are likely to bear on the NCC. In some cases we will sketch possible answers. (For reasons of space we shall not try to cover the approaches of others.) Our main focus will be on the visual system of humans and macaque monkeys. We shall assume that the reader is broadly familiar with most of the relevant psychophysics and neurobiology. One of us (Koch 2004) has written an extensive account of the neurobiology of consciousness and of our attempts to understand consciousness as arising in particular types of brains.

Are We Conscious of All That Goes on in Our Brains?

It is generally agreed that much of the activity of our brains is unconscious. This unconscious activity could be of several kinds. There are regions, such as the enteric nervous system of the digestive tract, of which we are hardly conscious at all, and others, such as the retina, of whose neural activity we are never directly conscious. There is probably considerable activity in the cerebral cortex that is itself unconscious but which leads up to conscious activity. Much of the activity that results from consciousness may be unconscious. And there may be special subsystems that can produce unconsciously a relevant but somewhat stereotyped motor output from a sensory input (more of this later). Thus a key question is: how do the neural activities that correspond to consciousness (the NCC) differ from somewhat similar brain activities that are unconscious?

Our main interest is not the enabling factors that are needed for all forms of consciousness, such as a functioning ascending reticular activating system in the midbrain and intralaminar nuclei of the thalamus (Bogen 1995) but the general nature of the neural and other activity that produces each particular aspect of consciousness, such as the color red, or a middle C, or a sharp pain. We also need to know if there are special mechanisms that bind together different aspects of a particular object/event, such as its color and its shape (the binding problem).

What is consciousness like?

We shall not try at this stage to give an exact definition of consciousness, but it helps to have some ideas of what consciousness is like. There is general agreement that much of consciousness is private. I cannot convey to you exactly how red looks to me, even if experiments show that you and I respond to colors in a very similar way. We have claimed (Crick and Koch 1995a) that this is because at each stage of processing in the cortex, the information symbolized is recoded so that the more internal neural activity is only expressed very indirectly in any motor output, such as speech. On the other hand a person can say whether two similar shades of red appear identical or not and can convey certain other relationships (such as whether B is between A and C in the set ABC). It is not surprising that much of the content of consciousness is largely private. What is mysterious is the exact nature of these internal experiences.

Our main interest is in qualia since that is where the mystery appears to lie. It might be thought that there was general agreement as to what mental processes have qualia, but a recent discussion (Crick and Koch 2000) shows otherwise. Almost everyone would agree that sensations, such as red or pain, have qualia, but Humphrey (in Crick and Koch 2000) has argued that (visual) percepts, such as that of a face or a dog, do not. Others think (as we do) that they do have qualia. Many people believe that their thoughts have qualia, but Jackendoff, Stevens, and Freud all disagree (for references, see chapter 18 in Koch 2004). Dreams seem to have qualia (for example, one can dream in color) but it is less clear whether feelings, emotions and valuations (such as novelty) have qualia of their own or merely produce a particular combination of sensations.

Since there is no general agreement on this topic we must consider all the possibilities. It may be that qualia fall into different classes, with similar but somewhat different properties. For example, it may turn out that to experience qualia the relevant information must be widely distributed in the brain, as Baars (1988, 1997) has suggested, but that "thoughts" are distributed rather differently from "percepts."

Does Consciousness Have a Unity?

It is often claimed that consciousness has a "unity," but this is a somewhat ambiguous term. Focused consciousness does usually appear to have a unity, but there may be somewhat peripheral and rather transient experiences, such as the rather vague consciousness of one's surroundings while driving a car. It is possible that such volatile experiences represents "proto-objects" (Rensink 2000) for which different aspects of the object (such as color, shape, etc.) are only loosely bound together, if at all.

Such experiences may be distinctly different from focused consciousness, which is usually less transient. However it does seem that the brain dislikes contradictory information and often selects a single interpretation, though this may switch over time, as with ambiguous figures or in binocular rivalry.

The unity of consciousness might be thought to imply that all the representations of different aspects of an object (such as its shape, color, motion, etc.) coincided in time in the brain, but there is evidence that this may not be the case (Moutoussis and Zeki 1997a, 1997b; Zeki and Moutoussis 1997).

Can one produce two distinct consciousnesses in one head? Leaving aside multiple personality disorders (since the claims for them are somewhat controversial) it seems that patients (or animals) whose corpus callosum and anterior commissure have been cut do behave as if each side had a separate consciousness, though usually only one side of the brain can speak (Bogen 1993). In humans the speechless nondominant hemisphere can easily perform tactile discriminations, even when a delay is imposed, and appears to remember what it has done. Notice that, at least initially, the relevant sensory information is unable to go from one side to the other subcortically via the brainstem or midbrain. It seems plausible that to join the two split halves together again, to produce a single consciousness, one would have to make very many connections between the two halves and in just the right places. So it would not be easy to join your head to mine by a cable so that I might experience exactly how you see red. Nor is it easy to predict just what the results would be.

Are All Actions Performed Consciously?

Philosophers have invented a creature they call a zombie, which behaves in every way like a normal person but is completely unconscious. We suspect that this hypothetical creature is a contradiction in terms, but there is now highly suggestive evidence that much of our behavior has this zombie character, as advocated by Milner and Goodale (1995). See also a recent short review (Koch and Crick 2001) and Koch (2004). In brief there appear to be systems that can respond very rapidly but in a stereotyped way to relevant visual inputs, for example in reaching for an object, or grasping it, or moving one's eyes. It is only later that one becomes conscious of such acts. There is much anecdotal evidence that this happens in sports. Additional evidence comes from certain patients with complex partial seizures. Such patients make a series of stereotypical actions, often in a way that repeats from one seizure to the next. In some cases they may appear to express emotions by laughing or smiling, but they do not respond to speech or undertake less stereotyped actions. In some cases they have no recollection at all after they recover of what happened during their seizure (Revonsuo et al. 2000).

Additional evidence comes from a patient, D. F., who suffered diffuse brain damage due to carbon monoxide poisoning (Milner et al. 1991). She can see colors and textures but not shapes or orientation. In spite of this she can post an envelope correctly in an oriented slot, but not if a delay of several seconds is imposed between the task and the action. This unconscious system appears to work "online" and leave no immediately accessible memory behind.

Normal people, in making a rapid zombie response, can become conscious of their actions a short time later, though there are cases when what the subject reports does not precisely correspond to what the unconscious zombie system has achieved (for example in the case of eye movements; Goodale, Pélisson and Prablanc 1986).

What would it be like to be a complete zombie? It seems that one would behave somewhat like a sleepwalker, since some sleepwalkers perform many sorts of stereotyped actions without any subsequent recollection of them (Revonsuo et al. 2000; Koch 2004). This might suggest an operational test for consciousness: a subject is conscious when he or she performs reactions which are not stereotyped and has some recollection of them. But how would one decide if an action was stereotyped or not? A possible test would be to impose a delay of several seconds, on the tentative assumption that stereotyped reactions can only be done online, with no delay (Rossetti 1998).

What Is Consciousness For?

This line of thought suggests an answer to a key question: what is consciousness for? Why are we not just a whole series of zombie systems? We have proposed (Crick and Koch 1998) that this would not produce highly intelligent and flexible behavior since we would need far too many such systems to duplicate human behavior. Instead evolution has produced a general-purpose system which constructs the best interpretation of the sensory input based on the animal's experience and the experience of its ancestors (embedded in its genes) and maintains this interpretation for a time sufficient for the thinking-planning parts of the brain to decide what, if anything, to think or do next. This general-purpose system is necessarily slower than the faster, unconscious zombie systems. To be able to respond very quickly, when needed, to certain inputs but more slowly to more complex situations, would be a considerable evolutionary advantage.

Is Attention Necessary for Consciousness?

This is a difficult and complex subject (for a recent review, see Kanwisher and Wojciulik 2000; Lamme 2003; Koch 2004). It is certain that attention can enrich consciousness, but it is less clear that one would be completely unconscious without it. This main problem is that we only have a very partial understanding of attention. We can think of attention being either top-down or bottom-up. Top-down attention occurs, for example, when a subject undertakes a task, such as "look out for the red square." Bottom-up attention occurs if some aspect of the input is especially salient, such as a sudden movement in one's peripheral visual field, causing the eyes to immediately move in that direction. Unfortunately at present the only psychological criterion for saliency is that it attracts attention, so the argument that saliency leads to bottom-up attention is circular. In vision, rapid eye movement is one of the major forms of attention. Subjects can indeed shift their visual attention without moving their eyes but this is a very unnatural situation since the subject has to make an effort to suppress incipient eye movements.

It seems likely that attention can take several different forms and can occur in many places in the brain. A plausible theory (Desimone and Duncan 1995) is that the effects of attention are stronger when they bias competition in the brain so that a relatively weak influence can decide the outcome of the competition, as in some political elections.

Mack and Rock (1998) have carried out an enormous number of experiments in totally naive subjects to demonstrate that with simple visual inputs a subject who is not expecting it may not see an additional brief visual input while his attention is concentrated nearby. On the other hand if a complete scene is unexpectedly flashed very briefly an observer can always report the gist of it.

Because the neural mechanisms that produce attention are not understood, it seems unlikely that attention can do much to help us to find the NCC except as a variable that influences consciousness in particular cases, so one could contrast cases when an object or event is seen due to attention against similar ones in which it is not seen because attention was lacking. That is, one can usefully study the results of attention even if we do not yet understand all the mechanisms that produce it.

Defects in attention can be produced by brain damage (see the relevant chapters in Feinberg and Farah 1997). A patient with damage to his right parietal cortex may not see objects in his left visual field. This is called neglect. In some cases this happens only when there is also a competing object in his right visual field. This is called extinction. Parietal damage on both sides of the cortex can produce Balint's syndrome. A striking feature of this is that the patient sees only one object at a time and has great difficulty in shifting his attention away from it. (They also have other defects that follow from this.) However, all patients with purely parietal damage are able to see something. The most plausible interpretation of these rather varied cases is that attention is needed to select what is seen and that some parts of the parietal cortex have a key role to play in this selection (Driver and Mattingley 1998).

Is Memory Necessary for Consciousness?

It seems certain that some sort of memory is involved, but which sort of memory? For an outline of the various types of memory, see Squire (1987) and Koch (2004). Episodic memory, although it certainly enriches consciousness, is not necessary. Patients who lack it (due to damage to their hippocampal system), though severely handicapped, are certainly conscious.

Other forms of long-term memory, such as semantic memory, are probably embedded in the strength or the dynamics of synapses, or similar parameters. Such "memories" are likely to be employed in many mental activities but consciousness happens so fast that the protein synthesis, which is necessary to lay down new long-term memories, is unlikely to be a major factor in the NCC.

Procedural memories, such as how to swim or ride a bicycle, do not appear themselves to be conscious, though subjects can become conscious of their actions.

The sort of "memory" described as priming may result from conscious or unconscious activity but, by definition, the particular memory that is primed is unconscious.

Since it seems almost certain that consciousness requires some form of rather short-term memory (of the order of a second or so), it seems likely that in the visual system iconic memory will often be involved. Our impression is that iconic memory, as normally understood, is a psychological concept and may not map easily onto what may be a number of distinct forms of rather short-term memories located in various parts of the brain.

The difficult case is working memory. When recalling a telephone number the digits recalled normally produce temporary activation, such as (silent) speech or visual images, but it is not obvious that each digit is continuously in consciousness over the time involved in working memory, such as ten seconds. A person with a very short digit span, though handicapped, appears to be conscious. It rather seems as if working memory continually reactivates some shorter forms of conscious memory. What these shorter forms are is unclear.

We do not nearly remember as much of a visual scene as we think we do, as shown by change blindness (Rensink, O'Regan, and Clark 1997), since the outside world often acts as a form of consultable memory. If we do become conscious of any one particular object in a scene, we normally remember it for at least a short while unless it is masked.

How Does the Timing of Events Affect the Percept?

The timing of the emergence of consciousness has been called microgenesis (see Bachmann 2000).

We can easily see something of a visual input, such as a flash of lightening, even though it lasts for a very short time, provided it is strong enough. Bloch's law states that for stimulus duration less than about a tenth of a second (for a diffuse flash) the brightness of the stimulus appears the same provided the product of its intensity and its duration is constant. In some sense the system is integrating the input over some short time interval.

How bright does a flash of light appear if its intensity is kept the same but its duration varies from trial to trial? This can be estimated by comparing its apparent brightness with the brightness of a similar but constant light. A typical result has been described by Efron (1967). As the duration gets longer the light appears more intense until, for a duration of about forty milliseconds, it reaches a maximum after which the subjective brightness declines to a steady value. This maximum is four or five times the steady value. As Efron has pointed out this description can be misleading. It expresses the results of multiple trials, each one for one particular duration of a flash. It does not show what a person experiences at a single trial. That is, for a flash of length of, say 125 milliseconds, the subject does not see the brightness of the flash increasing rapidly and then decreasing somewhat. On the contrary he reports that he saw a steady brightness. This distinction, which has been widely overlooked, is an important one.

The results suggest that the NCC comes into being abruptly rather than gradually. Once the relevant neural activity reaches some threshold a constant percept of brightness results, at least for a short time.

How long does a short flash of light appear to last? This is estimated by first comparing the onset of two adjacent flashes of light and then the apparent offset of the target flash with the onset of the companion flash. For short durations of the target flash, the (average) apparent duration stays the same, typically about 130 milliseconds (Efron 1970). For flashes longer than 130 milliseconds, the apparent duration equals the real duration. Similar results were found in audition. However, it is important to realize that the apparent brightness of a flash and its apparent duration are probably handled by different parts of the brain. Notice also that we are not asking here about the delay between the onset of the flash and the onset of the percept (which is difficult to measure exactly) but only about the apparent duration of the percept.

Instead of a single flash of light, what do we see if a complex scene, such as people dining in a restaurant, is flashed for different short durations? The general result is that for very short exposures one perceives the general nature of the scene. As the flash is made longer we can report more and more details. Once again, in any one trial we do not see the scene change. We just see more for longer flashes.

This might suggest that some of the higher levels of the visual hierarchy reach the necessary threshold for consciousness before the lower levels do. Possibly the lower levels need some feedback from the higher levels to boost their activity above threshold.

Masking

There is an extensive literature on the subject of masking (Breitmeyer 1984; Bachmann 2000). When two inputs blend, it is sometimes referred to as integrative masking, though the term blending might be preferable. Pattern masking is when two spatially superimposed patterns of contours interfere. When two patterns are adjacent the interference is called metacontrast.

Suppose a red circular disc and an otherwise identical green circular disc are flashed simultaneously onto the same place in the retina. Not surprisingly the subject sees a yellow disc. If the red disc alone is flashed first, for ten milliseconds, immediately followed by the green disc for ten milliseconds, the subject does not see a red disc turning green, but just a yellow disc. The yellow has a greenish tinge compared to the yellow produced when they are simultaneous. If the green disc comes first, the yellow is a little redder. The subject perceives a mixture of the inputs, with a bias towards the later one— this turns out to be a general rule. This suggests some form of integration, with the later signal having a somewhat greater weight. If one disc appears a hundred milliseconds before the other, little blending occurs. This suggests that, in this instance, the integration time is less than a hundred milliseconds.

When masking produces interference this is because the mask is competing with the target. The subject can easily report the target if there is no mask, but in some cases if the mask follows very shortly after the target his responses are at chance—he reports he did not see the target. He does better if there is a delay of a hundred milliseconds or so between the onset of the target and the onset of the mask. The mask interferes with processing of the target in the integration period leading up to consciousness. It is plausible that once some kind of neural activity due to the target has reached above a certain threshold, the following mask cannot interfere with it so easily. This suggests that the conscious activity may show hysteresis (as Libet [1973] has claimed) since the activity is probably held above threshold to some extent by some mechanism, such as loops with positive feedback.

In other cases (and especially in cases of metacontrast) the subject can see the target if the onset of the target and mask are simultaneous but fails to see it if the onset of the mask is delayed by a short period, typically fifty to a hundred milliseconds or so. This is presumably because the target and the mask are initially activating different places in the brain, and it takes time for the activity due to the mask to interfere with the ongoing activity due to the target. It is plausible that in some of these cases the mask is represented in the nonclassical receptive field(s) of the target. Not surprisingly, no masking occurs in metacontrast if attention can be focused on the target location before the target-mask sequence (Enns and DiLollo 2000).

Until recently, visual psychologists have not related their results to the details of the complex organization of the primate visual system, though a recent review (Enns and DiLollo 2000) outlines a first step in this direction. The study of masking in the alert monkey by neurophysiologists has only just begun (Macknik and Livingston 1998; Thompson and Schall 2000; Macknik and Martinez-Conde 2004). A careful study of the neurophysiological effects of masking should throw light on the integration times for the signal to reach above the thresholds for consciousness and the ways interference works. In general the times involved seem to range from fifty to two hundred milliseconds or more. This upper limit approaches the typical time between eye movements.

Are the NCC Local or Global?

First let us state the apparent paradox. The local effects of brain damage can remove from consciousness the content of certain aspects of (visual) consciousness, as in cases of achromatopsia, the inability to consciously perceive colors (Zeki 1993), prosopagnosia or the inability to identify specific faces or to recognize faces as a class (Bauer 1993), or akinetopsia, the inability to experience motion (Zihl, von Cramon, and Mai 1983). On the other hand, functional brain imaging techniques such as PET or fMRI in human subjects reveal that seeing a simple stimulus, such as fringes, moving dots, faces, and so on, leads to widespread and distributed activity in many parts of the brain (as assayed via changes in brain hemodynamics). This is true for any single brain scan. It is only when the peaks of differential responses are compared that one finds highly localized brain regions. In short, brain damage makes consciousness appear somewhat local, whereas scans suggest it can be more global.

There is also experimental evidence from single-electrode electrophysiology and dye imaging in monkeys that any particular part of the visual cortex responds strongly to only certain aspects of the visual scene. In other words, there is a high degree of localization in the brain. Exactly what is being localized is not always completely clear and, moreover, it can change somewhat with experience.

Essential Nodes

The evidence from brain damage, especially in humans, suggests that certain parts of the cortex are essential for a person to be conscious of certain aspects of the visual sensation or percept, such as color, motion, faces, etc. Zeki and Bartels (1999) have, very reasonably, described such a piece of the cortex as an essential node for that aspect of the percept. What should be considered "one aspect" must be decided by experiment. Thus "motion" is not necessarily a single aspect. Indeed there is evidence from brain damage (Vaina and Cowey 1996) that F. D., a patient with rather limited brain damage, is impaired in the detection of second-order motion but not of first-order motion.

The term should not be taken to imply that a person who possessed only the relevant essential node would be conscious of that aspect of the percept. It is highly probable that to produce that aspect of consciousness the node would have to interact with other parts of the brain. The point is that damage to that essential node would specifically remove that particular aspect of the sensation or percept, while leaving other aspects relatively intact.

One aspect of an object or an event is likely to have an essential node in only one or a few places in the cortex. The NCC for a single object or an event has to symbolize several of its distinct aspects (such as color, shape, motion, etc.) so the NCC might better be described as "multilocal." The neural activity leading up to the NCC or resulting from them is likely to be somewhat more global, so the time sequence might be broadly depicted as:

global   →   multilocal   →   global

with the multilocal activity corresponding to the NCC.

The concept of an essential node is an important one. It implies that if there is no essential node for some possible aspect of consciousness the subject cannot be conscious of it. If Zihl's patient (Zihl, von Cramon, and Mai 1983) can see the car and the car is moving, she will not see its movement if the essential nodes for movement are absent. So it is important not to assume that, for example, the brain can necessarily consciously detect some particular change in its firing activity. It will not be able to do so unless there is some essential node to consciously register that type of change.

The key question, then, is what exactly constitutes an essential node? Here we can only make informed guesses. It seems unlikely to be a single neuron. A better guess would be a small patch of pyramidal cells all of the same neural type, in the same cortical sublayer, which all project to roughly the same place in the brain, so that any single neuron in that place is able to receive inputs from most of this set of similar pyramidal cells. This gives us the advantage of a distributed representation.

It seems unlikely that one such set of neurons would be sufficient to form an essential node, since it is plausible that the signal from the essential node should be widely distributed if they are to contribute to consciousness (Baars 1988, 1997). This leads to the idea that the essential node consists of several of these sets of pyramidal cells, stacked one above another and projecting collectively to many brain areas.

This might suggest that the character of an essential node is that of the columnar property at that small place in the cortex, since in every cortical area that has been carefully examined it is found that there is a property (such as orientation, movement in depth, etc.) that many of the neurons in a small column, approximately perpendicular to the cortical surface, have in common (Crick and Koch 1992). This idea might be called the columnar hypothesis. The feature symbolized in this explicit manner by an essential node corresponds to its particular type of columnar activity.

This need not imply that we are conscious of all columnar properties since there may be other necessary conditions, such as exactly where such a column projects (Crick and Koch 1995b).

What Type of Activity Would We Expect the NCC to Show at Some Essential Node?

The psychological evidence discussed above suggests that some kind of activity needs to get above a threshold and to be maintained there for at least a few hundred milliseconds. A first guess as to the neuronal correlate of this might be a very high average rate of firing, such as 300 to 400 Hz. A possible problem is that such a high rate of firing maintained for some time would rapidly exhaust the supply of synaptic vesicles available for release at those synapses. Moreover, so far there is no sign of neurons in the cortex firing at such a high rate for any considerable time.

Another possibility is that the "activity" is really the onset of correlated firing in the set of neurons that make up the NCC; certainly in the neurons of the same and of different types at one place and possibly (to help with the binding problem) at other relevant places (Singer 1999). This correlated firing would produce a stronger effect on recipient neurons than the corresponding uncorrelated firing. It is possible that by this means one can provide an effect appreciably bigger than any possible high average rate of firing, such as 400 Hz.

How precise do the correlations have to be? One might surmise that for pyramidal cells of one type projecting to the same place, the correlations would be rather exact, perhaps to within several milliseconds. For neurons in the same cortical column, which project to widely separated places, the correlations might be somewhat less precise due to the various time delays involved. In some cases only the average rate of firing might be significant. In others the correlations may take the form of oscillatory firing, not necessarily in the 40-Hz range.

What is required to maintain the activity above a threshold? This could depend on a number of mechanisms, and may well be quite complex. It might be something special about the internal dynamics of the neuron, perhaps due to the accumulation of some chemical such as Ca^{++}, either in the neuron itself or in one or more of the associated inhibitory neurons. It might be due to special re-entrant circuits (Edelman 1989). Excitatory feedback loops could, by reiteratively exciting the neuron, push its activity increasingly upwards so that it not only reached above some critical level but was maintained there for a time. It is well known that, speaking loosely, there are many reentrant pathways in the brain, so the problem is to decide which ones might be relevant and exactly what properties they might have. Some of these loops may be within a given cortical area, but there are many loops between regions which are further apart.

The active essential nodes, at any one time, will be activating many other, related nodes, perhaps by re-entrant pathways (see our addendum in Crick and Koch 2000). However, in general, the activity at these related nodes will not reach the cortical level for consciousness, but the activity there may be enough to produce priming effects, for example, by altering some of the synaptic weights there. We suspect that any successful theory of consciousness will entail at least two varieties (or levels) of activation.

The Snapshot Hypothesis

The picture that emerges from these speculations is a rather curious one. It bears some resemblance to Dennett's multiple drafts model (Dennett 1991), although Dennett's ideas, though suggestive, are not precise enough to be considered scientific. The content of consciousness instead of being a continuous ever-changing flux may be more like a rapid series of "static" snapshots. Movement, like other attributes, is signaled in a static way by special neurons. That is, movement is not symbolized by a change in time of the symbols for position but by special neural symbols that represent movement of one sort or another. If the essential nodes for these symbols are lost, as in Zihl's patient, then though she can see the moving car she cannot see it moving. Sacks (2004) describes patients who, during a visual migraine attack, experience what he evocatively calls cinematographic vision. Perception appears staccato-like, in a series of stills, like the movies run in slow motion. Are the mechanisms that mediate the everyday illusion of continuous perception transiently inactivated during the migraine? This bears further investigation.

It is also possible, as Zeki and his colleagues have suggested, that the various attributes of an object or an event do not cross their thresholds simultaneously, though they probably interact somewhat so that these crossings are usually not too far apart in time. Perhaps one is never directly aware of this lack of exact synchrony (for short time intervals) because there are no essential nodes in the brain to signal these small time differences.

Although we have talked of a series of snapshots there is little evidence for a regular clock in the brain with a mechanism that integrates over intervals of constant duration and then starts afresh over the next interval, sometime called a quantized clock. The duration of any snapshot (or of fragments of the snapshot) is likely to vary somewhat, probably depending on various factors, such as off-signals, competition, and perhaps habituation. Purves and his colleagues (Purves, Paydafar, and Andrews 1996; Andrews, White, and Purves 1996) have described several psychological effects, such as a wagon-wheel effect under constant illumination, which hint that there are some clock-like effects in vision. Psychophysical evidence suggests the involvement of a higher-order motion system modulated by visual attention (VanRullen, Reddy, and Koch 2005).

It also seems plausible that the difference in time of two events is symbolized by systems in the brain that are different from those that signal position, movement, or other attributes. This would account for Libet's "backward referral in time" (Libet et al. 1979). Any small set of neurons can do only a limited job. This explains why there are so many different kinds of essential nodes (since there are so many features that need to be symbolized) and why as Dennett (1991) has emphasized that "there is no place in the brain where it all comes together." These ideas are similar to the views expressed by Minsky (1985). For a recent survey of neural theories of consciousness, including an extended description of his own, see chapter 5 of Bachmann (2000).

The brain has a problem with time, since time is needed to express a symbol (by neurons firing, a process spread out over time) so it is not surprising if it symbolized changes in time, such as movements, in a special way, and differences in time in another way.

We are dealing with an intrinsically parallel system to which we have limited access introspectively. This is probably why we find it so hard to understand how it works. This does not mean that we cannot usefully analyze it into smaller parts that interact dynamically, just as the "holistic" properties of a complex organic molecule, such as a protein, can be understood by the interactions of its many amino acids and the atoms of which they are made. In the brain an essential node may be a useful unit of analysis. It may turn out that the best way to describe the NCC for any one percept is the activities at the relevant essential nodes and their dynamic interactions. This suggestion resembles the dynamic core of Edelman and Tononi (2000).

What Future Experiments Might Uncover the NCC?

It is now possible to approach the NCC experimentally. There are many possible experiments, so we will briefly outline only some of them. As a general rule it may be better to study, in the monkey, cases where in humans the distinction between conscious and unconscious is fairly crisp, rather than difficult borderline cases in which one is barely conscious of the percept.

Study cases where the percept differs significantly from the input, such as binocular rivalry or flash suppression and some of the many types of visual illusion, such as the motion after-effect.

Study differences between conscious processes and somewhat related unconscious ones, such as zombie systems.

Study cases in which parts of the brain are inactivated, preferably reversibly, as by anesthetics, by local cooling, by inactivation via injection of GABA analogues or other methods. Eventually it may be necessary to inactivate reversibly all the neurons of one type in one chosen part of the brain, using techniques from molecular biology. It is not out of the question that a few selected mutations might turn a conscious animal into a zombie.

Follow up promising hypotheses, such as the idea that synchronized firing of certain neurons at an essential node may be required for consciousness (though correlated firing might also be used for other purposes). It may be of crucial importance to study the correlated firing of neurons in a single cortical column.

Will the NCC Explain Qualia?

If and when we understand the nature of the NCC, will we be able to explain qualia? This is Chalmer's hard problem. Naturally this will depend on the nature of the NCC. It seems likely that it will involve the processing of many types of information rapidly and in parallel in a way that expresses the significance of the activity—what philosophers call "intentionality" or "aboutness."

This may depend on what activity has been associated in the past with that particular feature and what possible actions such a feature might lead to, even though the subject may not be immediately aware of such associations. The effects of one essential node on another are likely to be of more than one type. For example, some may be driving while others may only be modulating. These relationships between essential nodes in the brain should be isomorphic with the common and significant relationships that we perceive in the outside world.

As Chalmers (1995) has suggested, one will need to propose some sort of bridging principles to relate, say, the subjective redness of red to the relevant complex neural activity. The key question is not whether we can make somewhat arbitrary bridging principles (*this* activity produces *this color*) but whether this principle can be related to the rest of known science. Perhaps we shall need to reformulate our basic physical theories, possibly in informational terms, though this may involve a wider concept of information that the usual Shannon information based on the transmission of signals along noisy channels. Or perhaps no such bridges are possible—only time will tell.

References

Andrews, T.J., White, L.E., and Purves, D. (1996) "Temporal events in cyclopean vision." Proc of Natl Acad Sci USA 93: 3689–3692.

Baars, B.J. (1988) A Cognitive Theory of Consciousness. Cambridge: Cambridge University Press.

Baars, B.J. (1997) In the Theater of Consciousness. New York: Oxford University Press.

Bachmann, T. (2000) Microgenetic Approach to the Conscious Mind. Amsterdam and Philadelphia: John Benjamins.

Bauer, R.M. (1993) "Agnosia." In: Clinical Neuropsychology. Heilman, K.M. and Valenstein, E., Editors, pp. 215–277. New York: Oxford University Press.

Bogen, J.E. (1993) "The callosal syndromes." In: Clinical Neuropsychology. Heilman, K.M. and Valenstein, E., Editors. New York: Oxford University Press.

Bogen, J.E. (1995) "On the neurophysiology of consciousness: I. An overview." Conscious Cogn 4: 52–62.

Breitmeyer, B.G. (1984) Visual Masking: An Integrative Approach. Oxford: Oxford University Press.

Chalmers, D. (1995) The Conscious Mind: In Search of a Fundamental Theory. New York: Oxford University Press.

Crick, F.C., and Koch, C. (1992) "The problem of consciousness." Sci Am 267: 153–159.

Crick, F.C., and Koch, C. (1995a) "Why neuroscience may be able to explain consciousness." Sci Am 273: 84–85.

Crick, F.C., and Koch, C. (1995b) "Are we aware of neural activity in primary visual cortex?" Nature 375: 121–123.

Crick, F.C., and Koch, C. (1998) "Consciousness and neuroscience." Cereb Cortex 8: 97–107.

Crick, F.C., and Koch, C. (2000) "The Unconscious Homunculus." Neuro-Psychoanalysis 2: 3–11.

Dennett, D.C. (1991) Consciousness Explained. Boston: Little, Brown and Co.

Desimone, R., and Duncan, J. (1995) "Neural mechanisms of selective visual attention." Annu Rev Neurosci 18: 193–222.

Driver, J., and Mattingley, J.B. (1998) "Parietal neglect and visual awareness." Nat Neurosci 1: 17–22.

Edelman, G., and Tononi, G.A. (2000) Universe of Consciousness. New York: Basic Books.

Edelman, G.M. (1989) The Remembered Present: A Biological Theory of Consciousness. New York: Basic Books.

Efron, R. (1967) "The Duration of the Present." Annals of the New York Academy of Sciences 138: 713–729.

Efron, R. (1970) "The relationship between the duration of a stimulus and the duration of a perception." Neuropsychologia 8: 37–55.

Enns, J.T., and DiLollo, V. (2000) "What's new in visual masking?" Trends Cogn Sci 4: 345–352.

Feinberg, T., and Farah, M. (Eds.). (1997). Behavioral Neurology and Neuropsychology. New York: McGraw Hill.

Goodale, M.A., Pélisson, D., and Prablanc, C. (1986) "Large adjustments in visually guided reaching do not depend on vision of the hand or perception of target displacement." Nature 320: 748–750.

Kanwisher, N., and Wojciulik, E. (2000) "Visual Attention: Insights from brain imaging." Nat Rev Neurosci 1: 91–100.

Koch, C. (2004) The Quest for Consciousness: A Neurobiological Approach. Denver, Colo.: Roberts and Publishers.

Koch, C., and Crick, F.C. (2001) "On the zombie within." Nature 411: 893.

Lamme, V.A. (2003) "Why attention and awareness are different." Trends in Cogn. Sci. 7: 12–18.

Libet, B. (1973) "Electrical stimulation of cortex in human subjects and conscious sensory aspects." In: Handbook of Sensory Physiology, Vol. II: Somatosensory Systems. Iggo, A., Editors, pp. 743–790. Berlin: Springer Verlag.

Libet, B., Wright, E.W., Feinstein, B., and Pearl, D.K. (1979) "Subjective referral of the timing for a conscious sensory experience." Brain 102: 193–224.

Mack, A., and Rock, I. (1998) Inattentional Blindness. Cambridge, Mass.: MIT Press.

Macknik, S.L., and Livingstone, M.S. (1998) "Neuronal correlates of visibility and invisibility in the primate visual system." Nat Neurosci 1: 144–149.

Macknik, S.L., and Martinez-Conde, S. (2004) "Dichoptic visual masking reveals that early binocular neurons exhibit weak interocular suppression: Implications for binocular vision and visual awareness." J Cog Neurosci 16: 1049–1059.

Milner, A.D., Perrett, D.I., Johnston, R.S., Benson, P.J., Jordan, T.R., Heeley, D.W., Bettucci, D., Mortara, D., Mutani, R., Terazzi, E., and Davidson, D.L.W. (1991) "Perception and action in form agnosia." Brain 114: 405–428.

Milner, D.A., and Goodale, M.A. (1995) The Visual Brain in Action. Oxford: Oxford University Press.

Minsky, M. (1985) The Society of Mind. New York: Simon and Schuster.

Moutoussis, K., and Zeki, S. (1997a) "A direct demonstration of perceptual asynchrony in vision." Proc R Soc B 264: 393–399.

Moutoussis, K., and Zeki, S. (1997b) "Functional segregation and temporal hierarchy of the visual perceptive systems." Proc R Soc Lond B Biol Sci 264: 1407–14.

Purves, D., Paydartar, J.A., and Andrews, T.J. (1996) "The wagon wheel illusion in movies and reality." Proc Natl Acad Sci USA 93: 3693–3697.

Rensink, R.A. (2000) "The dynamic representation of scenes." Visual Cognition 7: 17–42.

Rensink, R.A., O'Regan, J.K., and Clark, J.J. (1997) "To see or not to see: the need for attention to perceive changes in scenes." Psychol Sci 8: 368–373.

Revonsuo, A., Johanson, M., J-E, Wedlund., and Chaplin, J. (2000) "The zombie among us." In: Beyond Dissociation. Rossetti, Y. and Revonsuo, A., Editors, pp. 331–351. Amsterdam and Philadelphia: John Benjamins.

Rossetti, Y. (1998) "Implicit short-lived motor representations of space in brain damaged and healthy subjects." Conscious Cogn 7: 520–558.

Singer, W. (1999) "Neuronal synchrony: a versatile code for the definition of relations?" Neuron 24: 49–65.

Squire, L.R. (1987) Memory and Brain. New York: Oxford University Press.

Thompson, K.G., and Schall, J.D. (2000) "Antecedents and correlates of visual detection and awareness in macaque prefrontal cortex." Vision Res 40: 1523–38.

Vaina, L., and Cowey, A. (1996) "Impairment of the perception of second order motion but not first order motion in a patient with unilateral focal brain damage." Proc R Soc Lond B 263: 1225–1232.

VanRullen, R., Reddy, L., and Koch, C. (2005) "Attention-driven discrete sampling of motion perception." Proc Natl Acad Sci USA 102: 5291–5296.

Zeki, S. (1993) A Vision of the Brain. Oxford: Oxford University Press.

Zeki, S., and Bartels, A. (1999) "Toward a theory of visual consciousness." Conscious Cogn 8: 225–259.

Zeki, S., and Moutoussis, K. (1997) "Temporal hierarchy of the visual perceptive systems in the Mondrian world." Proc R Soc Lond B 264: 1415–1419.

Zihl, J., von Cramon, D., and Mai, N. (1983) "Selective disturbance of movement after bilateral brain-damage." Brain 106: 313–340.

Author Index

Fries, P., 9, 13, 395, 403, 411
Fritzsch, B., 248
Frost, B. J., 35, 36, 38, 223
Fujita, I., 382
Fukunishi, K., 346
Fukushima, K., 329, 330
Funabiki, K., 253
Furmanski, C. S., 323
Fuster, J. M., 410

Gaese, B. H., 33
Gainotti, G., 381
Galaburda, A. M., 44, 51, 53, 55
Gallant, J. L., 18, 194, 195, 198
Gallese, V., 379–81, 384, 401
Galletti, C., 369, 372
Gallistel, C. R., 45
Galton, F., 434, 469
Gannon, P. J., 49
Gardner, S., 249, 257
Garey, M., 220, 235
Gauger, B., 34, 35
Gazzaniga, M. S., 46, 52, 377
Geiger, J. R. P., 249
Geissler, D. B., 49–51
Gelade, G., 120, 122
Gellatt, J. C., 234
Gelperin, A., 9
Gentilucci, M., 375
Georgopoulos, A., 87, 144
von Gersdorff, H., 248
Gerstner, W., 89–93, 97, 141, 253, 271, 274, 278, 297
Geschwind, N., 44, 51, 53, 55
Gewaltig, M. O., 271
Gibbon, J., 279
Giguère, L., 55
Gilbert, C. D., 119, 121, 169, 177, 201, 202
Giles, F., 49
Girard, P., 110, 115, 117–18, 124
Glass, L., 269
Glick, S. D., 55
Gnadt, J. W., 68, 80n.2
Gochin, P. M., 329
Gockel, H., 309
Godschalk, M., 375
Godwin, D. W., 71, 113
Goebel, R., 125
Gold, J. I., 31
Goldberg, J. M., 252–54
Goldberg, M. E., 126, 409, 424
Golding, N. L., 256–58
Goldman, A., 381
Goldman-Rakic, P. W., 344

Golowasch, J., 245
Golubitsky, M., 227
Goodale, M. A., 367–68, 372, 377, 383, 477
Gothard, K. M., 402
Gottlieb, J. P., 409, 413, 424
Goulet, P., 352
Grady, C. L., 52
Grafton, S. T., 382
Gran, B., 55
Gran, L., 55
Grantham, D. W., 39
Grau-Serrat, V., 246, 253, 254, 256
Gray, C. M., 9, 13, 25, 31, 146, 183, 194, 315
Gray, J. A., 434
Graziono, M. S. A., 375, 384, 399–401
Green, D. M., 69, 71
Greenberg, J. P., 383
Greenberg, S., 256, 309
Greenfield, P. M., 466
Gregory, R. L., 437
Grèzes, J., 382
Griffiths, T. D., 52, 303
Grigg, J. J., 250, 253
Grimes, D. B., 205
Grinvald, A., 6, 162–63, 165, 169, 178, 350
Gronenberg, W., 89
Grose, J. H., 309, 312
Gross, C. G., 375, 382, 384
Grothe, B., 253, 316
Grotschel, M., 220
Gruenbein, D., 40
Grünewald, B., 14–16
Guido, W., 68–71
Guillemin, V., 227
Guillery, R. W., 65, 67, 72–76, 113, 116, 424
Guppy, A., 288, 289
Gur, M., 190, 206
Gutman, D., 414
Gutnick, M. J., 178

Haack, B., 47
Haag, J., 248
Hackett, T. A., 285
Haenny, P. E., 121
Hafter, E. R., 246, 287
Haggard, M. P., 310, 312
Hahnloser, R. H., 216, 426
Haken, H., 350
Hall, J. W., 309, 310, 312
Halligan, P. W., 376, 377

McGee, J., 257, 258
McGee, T., 357
McGlone, J., 55
McGuinness, E., 223, 395
McIntosh, A. R., 51, 52
McLaren, J. W., 374
McNaughton, B. L., 187, 402, 403
Mechler, F., 190
Meck, W. H., 279
Meddis, R., 309
Mel, B., 183
Mel, B. W., 327, 329
von Melchner, L., 25
Mendell, L. M., 75
Mendola, J., 125
Meredith, M., 5–7
Merigan, W. H., 324
Merikle, P., 441, 446
Merzenich, M. M., 84, 91, 146, 316, 336, 347, 348
Meskenaite, V., 75, 110, 113
Miall, C., 279
Micheyl, C., 309
Middlebrooks, J. C., 298, 299, 316, 348
Miezin, F., 223, 395
Miklósi, A., 51, 52
Miller, C. T., 306
Miller, D. A., 226, 231, 234, 235
Miller, E. K., 123
Miller, G. L., 33
Miller, J., 140
Miller, K. D., 426
Milner, A. D., 367–68, 477
Milner, D., 367, 372, 377, 383, 477
Minsky, M., 269, 486
Mishkin, M., 324, 344, 367, 376, 382
Mitchison, G., 221, 223
Miyake, S., 329
Miyashita, Y., 410
Mo, J., 248
Moiseff, A., 28, 29, 246, 284
Molfese, D. L., 46
Molfese, V. J., 46
Mombaerts, P., 16
Monier, C., 106, 109, 229
Monsivais, P., 253
Moore, B. C. J., 309–13
Moore, T., 399–401
Moran, J., 122, 408–9
Morel, A., 68, 116, 119, 122
Morest, D. K., 247, 257
Morest, K., 247, 255
Mori, K., 5, 6
Morris, R. D., 52

Mosbacher, J., 249
Moss, C. F., 307
Mossop, J. E., 316
Motokizawa, F., 5, 6
Mott, D. D., 71
Mott, J. B., 311, 312
Motter, B. C., 121, 123, 206
Moulton, D., 6
Mountcastle, V. B., 137, 372, 409, 425
Moutoussis, K., 476
Movellan, J. R., 395
Movshon, J. A., 106, 183
Mukherjee, P., 67, 69
Mumford, D., 117, 120, 124, 192, 202, 204, 216
Murai, N., 346
Murata, A., 378
Murray, S. O., 204

Näätänen, R., 314
Nakanishi, S., 5, 6
Nakayama, K., 202, 415
Naraghi, M., 9, 12
Nash, J. F., 232
Nass, R. D., 46
Nasu, M., 346
Natschlger, T., 222
Nazir, T. A., 324
Nealey, T. A., 113
Neisser, U., 414
Nelken, I., 32, 312
Nelson, J., 223
Nelson, J. L., 140
Nelson, M. E., 349, 350
Nelson, S. B., 424
Nettleton, N., 46, 52
Neuert, V., 311
Newman, E. B., 287
Newsome, W. T., 25, 106, 161, 312, 424–26
Nguyenkim, J. D., 202
Nicoll, R. A., 71
Nieder, A., 36, 312
Niemiec, A., 310
Nishitani, N., 381
van Noorden, L. P. A. S., 304, 307–9
Nottebohm, F., 47
Nowak, L. G., 107, 113, 115–16, 119, 234, 325
Nowlan, S. J., 222
Nunn, J. A., 439, 452, 470

Obusek, C. J., 315
Ochoa, E., 329

Sur, M., 25
Surlykke, A., 307
Sutter, E. E., 107
Sutter, M. L., 304, 348
Swets, J. A., 69, 71
Swindale, N. V., 221
Szentagothai, J., 223

Taira, M., 378
Takahashi, T. T., 28–29, 31, 39, 246, 287, 290, 292, 313–14
Talcott, J., 285
Tamura, K., 55
Tanaka, K., 178, 325, 382
Tanaka, M., 372
Taniguchi, I., 346
Tanji, J., 378
Tank, D. W., 9, 218, 271, 272
Tao, H.-z., 91
Tarr, M. J., 205
Taschenberger, H., 248
Taylor, C. S. R., 399–401
Tempel, B. L., 250, 253
Tenenbaum, J. B., 205
Theunissen, F., 140
Thier, P., 425
Thiessen, D., 350
Thomas, H., 347
Thomas, J. A., 156
Thompson, J. K., 52
Thompson, J. M., 253
Thompson, K. G., 482
Thompson, L. T., 187
Thompson, S., 288
Thomson, M., 178
Thorell, L. G., 187
Thorpe, S., 138, 152, 326, 335
Tian, B., 315, 347
Tinbergen, N., 24
Tipper, S. P., 414
Tollsten, L., 5
Tomita, H., 115
Tononi, G. A., 487
Tovée, M. J., 323
Trachtenberg, J. T., 112
Trahiotis, C., 246, 286
Treisman, A., 120, 122, 434
Trepel, C., 112
Treue, S., 121, 122, 408, 409, 424, 425
Treves, A., 141
Troyer, T. W., 426
True, S. D., 374
Trussell, L. O., 245, 248–50, 257, 316

Tsodyks, M., 109, 148, 162, 165, 170, 172, 176–78, 274, 276
Tsotsos, J. K., 120, 122
Tsukui, T., 55
Tucker, A. W., 234
Tucker, D. M., 52
Turner, M., 458
Turrigiano, G. G., 245, 259

Ullman, S., 122, 124, 178, 201, 204, 216, 327, 330
Ungerleider, L. G., 119, 122, 324, 367, 376, 382
Unnikrishnan, K., 271, 272
Uno, H., 346
Urbas, J. V., 347

Vaid, J., 46
Vaina, L., 483
Valdes-Sosa, M., 414–16, 419
Valenstein, E., 381
Vallar, G., 356, 372, 373, 376
Vallortigara, G., 51, 52, 55
Van Essen, D. C., 77, 78, 104, 105, 178, 193, 194, 324, 425
VanRullen, R., 486
Vassar, R., 16
Vaughan, J. W., 71, 113
Vecchi, M., 234
Vecera, S. P., 414
Verhey, J. L., 310–11
Vidnyánszky, Z., 76
Vighetto, A., 372
Vignes, M., 71
Vinje, W. E., 18, 194, 195, 198
Vogels, R., 119
Volpe, B. T., 377
van Vreeswijk, C., 149, 150, 157, 161

Wachtler, T., 395
Wada, J. A., 49
Wagner, E., 310–12
Wagner, H., 23, 28, 29, 31–36, 38–39, 292, 313, 315
Walker, G. A., 117
Walker, S. F., 51, 53
Wallis, G., 331
Walmsley, B., 248, 249
Walsh, E. J., 257, 258
Walsh, V., 122
Wang, L. Y., 250
Wang, X., 312
Warburton, A. B., 26
Warchol, M. E., 256, 258

Subject Index

barn owl (*continued*)
  stereo vision in, 36–38
  temporal map of, 84, 91–92, 95,
    97–98
    formation of, 93–95
    tuning synapses at a single
      neuron, 92–93
best frequency (BF), 294, 295, 306–7,
  347, 348
biased competition model, 418
bilingual synesthetes, 446
binaural cross-correlation, 297
binaural hearing, 284
  comparison of barn-owl and
    mammalian, 287–90
  duplex theory of, 287
    anatomical basis of, 289
  models of, 286–87
  unique localization problems
    presenting unique problems,
    290–93
binding problem, 146
blindsight and synesthesia,
  449–52
bottom-up and top-down aspects of
  neuronal representation. *See
  also* top-down processing
  separation between, 357–58
bottom-up principle, 343
bottom-up processing, 306, 344,
  357, 478
brain(s). *See also specific topics*
  lateral view of, 373
    showing cytoarchitectonic
    parcellation, 371
  ways in which they are fast,
    138–41
buildup effect, 309
burst mode of firing, 68–70

calcium influx into presynaptic
  terminal, 248
canonical neurons, 378
categorization learning, cortical
  dynamics during, 357–60
categorization phase, 358
characteristic delay (CD), 286, 290
characteristic frequency (CF), 347
circuit dynamics, 6, 18
  as mechanism for recall, 15
classical receptive field (CRF), 193
cliques of neurons, computing with,
  235–36
cochlea, 84
"cocktail party effect of," 354

coding, 137, 158. *See also* neural
  code(s)
  constraints imposed by, 139–40
  optimal, 152–57
cognitive psychology *vs.* neuroscience,
  432–33
coincidence detection, 251–52, 286
  delay line-coincidence detection
    circuits, 252–53
  models of, relate dendritic structure
    to ITD detection, 254–56
columnar computations, 240.
  See also auditory object
    formation, by spatial
    processing; visual cortical
    columns, computation in
comodulation masking release (CMR),
  310–13
competition occurring among
  coherent objects, evidence of,
  413–15
competitive equilibrium. *See* Nash
  equilibrium
competitive interactions among
  surfaces, 415–19
competitive selection, 407, 420. *See
  also* selective attention
complex neurons, 189–90, 205–6
computational map, 84
computational neuroscience.
  See visual cortical columns,
    computation in
connectionism, 396
consciousness, 449, 474, 475.
  See also neural correlates of
    consciousness
  attention as necessary for, 478–79
  as having a unity, 476
  memory as necessary for, 479–80
  purpose of, 477–78
  vs. unconsciousness, 475
    of actions performed, 476–77
    of what goes on in brain, 474–76
contrast normalization model, 200,
  201
control neurons, 205, 206
correlated random-dot stereograms
  (cRDS), 37, 39
cortex. *See also specific topics*
  reason for obsession with, 3–4
cortical activity, spontaneous. *See*
  spontaneous cortical activity
corticocortical communication and
  thalamic relays, 77–79
corticothalamic axons, 113

creativity, 455–57. *See also* art
cross-activation/cross-wiring,
    438–39, 442, 445, 448,
    459–60, 462, 468. *See also*
    synesthesia
crowding effect, 450–51

decorrelation, 6–7, 9, 17
    defined, 7
delay distributions, 267–70, 272,
    275. *See also* temporal
    sequences; time warp
dendritic nonlinearities. *See under*
    coincidence detection
detection (vision), 222. *See also* edge
    detection
discrimination phase, 358
dorsal stream, 367–68
dorsodorsal (D-D) stream, 368,
    371–72
duality problem, 343–45
dye signal, cortical, 160
    what it represents, 163
dynamic link matching, 336
dynamic routing circuit model,
    327–29
dynamic routing circuits, speed of,
    333–35
dynamical patterns, slow
    as features of a code, 14

edge detection, 223–27
endstopping, 225
energy model, 190, 205
epilepsy, surgery to prevent, 397–98
essential nodes, 483–85
extrastriate cortex, competitive
    circuits in
    mediating attentional selection
        of behaviorally relevant
        stimuli, 407–13

F4, area, 374–76, 385, 386
F5, area, 378–82
F1/F0 ratio, 190
feature binding problem, 352
feature extraction principle, 352
feature-selective neurons, 408.
    See also attentional selection of
    behaviorally relevant stimuli;
    competitive selection
feedback connections, 103, 240,
    324–25
    action of, 114
    as driving *vs.* modulating, 113–17

role in global-to-local interactions,
    117–20
role in matching internal models to
    sensory data, 124–26
what is fed back, 112–13
feedback influences, attention as
    mediated by, 120–24
feedforward model, 103–7, 240. *See
    also* feedback connections
    evidence that has led to
        reexamination of, 107–12
feedforward processing scheme in five
    steps, 139
filters and filtering, 222–23
firing neurons, 271
firing rate(s), 137, 144, 158, 484–85
    probability distribution of, 185–86
"flash-back," 397–98
folds (geometry), 227–29
foreground-background
    decomposition, 354–55
"Fourier power" model, 195
frequency modulations (FM), 352–55,
    358
frontal eye fields (FEF), 373, 374
fusiform, cross-wiring in, as basis of
    grapheme-color synesthesia,
    438, 439

GABA (gamma-aminobutyric acid)
    iontophoresis, 108, 109
Gabor functions, 36, 37
gain modulation, 425
    amplifying, 426–29
    of recurrent pathways, switching
        through, 429–30

head turns, azimuthal components
    of, 32
hearing. *See* binaural hearing
"Hebbian" learning, 89
hemispatial neglect, location of lesions
    producing, 372–76
hemisphere dominance. *See* lateralized
    brain functions
heterosynaptic processes, 91
high threshold conductance (HTC),
    250
hippocampal neurons, 402
homosynaptic processes, 91
horizontal connections and global-
    to-local interactions, 117–20
hunting. *See also under* barn owl
    as complex behavior, 26–28
hunting situation, 27, 28

hypercomplex neurons, 189–90
hypernormal stimuli and art, 444–45

imitation, 381. *See also* mimicry
inferior colliculus (IC), 33–35, 313–14
inferior parietal lobule (IPL), 368–69,
    371, 376–77, 382
inhibition of return (IOR), 414
integrate-and-fire model neuron, 271
intentionality, 487
interaural intensity difference (IID),
    355, 356
interaural level differences (ILDs),
    28–32, 97, 288, 289
interaural phase difference (IPD), 295
interaural time difference (ITD)
    discriminator neurons, 254
interaural time difference (ITD) maps,
    83, 271
interaural time differences (ITDs),
    252, 253, 256, 356. *See also*
    binaural hearing; localization
in barn owl, 28–32, 83, 84, 91, 95,
    97–98, 284, 286–88, 290–95,
    297, 298
distribution of peak ITDs as function
    of neuronal best frequency,
    294, 295
encoding of, 251
frequency channels and
    frequency-dependent shifts in,
    31
in guinea pig, 291, 296
ILDs and, 28–30
as main cue for sound localization,
    31
neural basis of sensitivity to,
    289–90
representations for
    in barn owls *vs.* mammals,
    293–95
    developmental considerations
        regarding, 297–98
    different sensory representations
        providing for different coding
        strategies, 298–99
interstimulus interval (ISI), 416–18
invariant feature networks, 329–32
how they can process *where-*
    information, 335–36

Jefress model, 251, 286, 299

Kenyon cells (KCs), 11–16
Konoshi paradox, 92

laminar axon, 93, 94
laminar nucleus (NL), 84, 94.
    See also nucleus laminaris
language
    evolution of, 461–67
    synesthetic bootstrapping theory of,
        462, 463
language functions, 402
    left-hemisphere dominance of,
        47–51
lateral geniculate nucleus (LGN), 72,
    113, 192
lateral horn interneurons (LHIs), 12
lateral occipital complex (LOC), 204
lateralized brain functions, 44–46, 51,
    55–56
    division of properties and labor
        between the hemispheres,
        52–53
    language and vocalization, 46–50
    reasons for, 44–45, 53–56
    which functions are lateralized,
        50–53
learning rule, 90, 92, 269
learning window, 85, 91
left-handedness, 53
Lemke's algorithm, 235
linear complementarity problems,
    234, 240
LIP, area, 373–75
local field potential (LFP), 9–11,
    166, 411
local neurons (LNs), 10, 11
localization. *See also under* binaural
    hearing
    mammalian code for,
        295–97
low threshold conductance (LTC),
    249–50, 254
low-threshold spike (LTS), 65–68

macaque monkey brain, 368, 369
mapping experiment, 357
map(s), 83–85, 96, 456. *See also under*
    barn owl
    auditory cortex as test ground for
        views on, 346–48
    cyclic, 350–51
    defined, 83
    global views of, 346
    local views of response distribution
        in, 345–46
    motion-selective, in auditory cortex,
        355–57
    state of, 346

neurons (*continued*)
spontaneous cortical activity of, 160
ways in which they are slow, 135–37, 139–41
neurophysiological postulate, 89
neuroscience. *See also specific topics*
behavioral neuroscience, history and future of, 23–26, 40–41
integrative, 3
theories in. *See under* visual cortex, primary
neurotransmitter release, 236, 248
neurotransmitters, neuromodulators, and switching, 424
nodes, essential, 483–85
noise, 140, 165, 195
constraints imposed by, 141
masking, 311
the problem with, 15–16
noise reduction, potential mechanisms for, 16
nonclassical surround, 200
normalization, 326–28
recognition as requiring, 326
nucleus angularis (NA), 246, 247, 258
nucleus laminaris (NL), 251–56, 293. *See also* laminar nucleus
nucleus magnocellularis (NM), 246–48, 251, 252

occlusion, 202, 203
olfactory brain/olfactory system, 17–18
circuit dynamics of, as mechanism for recall, 15
oscillatory synchronization and sparse representations in, 9, 17
causes and behavioral relevance of, 10
decoding, 11–12
hidden activity, 10–11
significance of, 12–13
the problem with decorrelation and perceptual clusters, 17
the problem with noise, 15–16
as system to identify rules of potentially general relevance, 5–6
olfactory bulb (OB) and antennal lobe (AL), 5
decoding temporal sequences without explicit sequence decoding, 14–15
as decorrelators, 6–7, 9

decorrelation, 7
mechanisms and possible formal principles, 7, 9
as "encoding machines," 6
slow dynamical patterns, 6–7, 9
as features of a code, 14
olfactory circuits, locus, 8
olfactory pattern recognition, 5, 6
onset responses
have evolved parallel in birds, 258
octopus cells transforming auditory nerve inputs to produce, 256–58
onsets, encoding, 256
optic ataxia. *See also* visual unilateral neglect
location of lesions producing, 372, 373
optic tectum (OT), 35, 36
oscillatory synchronization. *See under* olfactory brain/olfactory system
owl. *See* barn owl

paddlefish (*Polyodon spathula*), 84–85, 96, 98–99
Pandemonium models, 204
pantomime recognition, 381
parietal cortex, 368, 369. *See also* temporal-parietal-occipital (TPO)/angular gyrus area
parietal lobe, 479. *See also* inferior parietal lobule; superior parietal lobule
organization of, 368–71
parietodependent motor areas, extrinsic afferent connections of, 369, 370
parietofrontal network and attention, 122
Penfield Project, 401
perception
action and, 367–68. *See also under* ventrodorsal (V-D) stream
action as coming before, 383–87
timing of events and, 480–81
perceptual grouping, 17
perceptual learning, as occurring over various time scales, 237
peristimulus time histogram (PSTH), 173, 174
PF, area, 380
phase locking, 84
place code, 251

place fields of neurons, 402
plasticity
    models of cortical, 106–7
    synaptic, 268
pointer chain, 15
polymatrix games, 232–35
    defined, 232–33
postsynaptic potential (PSP), 71–73
pragmatic mode of stimulus
    processing, 368
precentral cortex, 400
preferred cortical state (PCS) of
    neuron, 170–78
primary visual cortex. *See* visual
    cortex, primary
primitive processes, 303
processing steps, 139
projection neurons (PNs), 5–7,
    10–14, 16
projective field(s). *See also* mirror
    neurons
    of networks of neurons in motor
        cortex, 399
    of neuron
        measuring, 396–99
        nature of, 394–96
prosody, 352

qualia
    to be explained by neural correlates
        of consciousness, 474, 475,
        487–88
    four laws of, 453
    riddle of
        solution to, 453–55
        synesthesia and, 449–52

random-dot stereograms (RDS),
    36–39
rate code, 145, 146, 158
reaction time, 138
receptive field (RF) of neuron, 394–95,
    409–11. *See also* mirror
    neurons
recognition, 326, 327
    fast, 325–26
    olfactory pattern, 5, 6
routing circuit(s). *See also* dynamic
    routing circuit model
    with feature hierarchy, 333
    schematic illustration of, 328

scorpion, sand (*Paruroctonus
    mesaensis*), neuronal maps of,
    84–89, 96–97

selective attention, 357–58. *See also*
    competitive selection
semantic mode of object
    representation, 368
sensory cortex maps. *See* map(s)
"shape-from-shading" network model,
    396
simple neurons, 189–90
snapshot hypothesis, 485–87
sound localization. *See also* auditory
    object formation
    neural mechanisms underlying,
        early findings on, 28–29
    nonsensory influences on, 34
    virtual auditory worlds for
        studying contribution of
        single parameters to,
        32–34
sound-localization tasks, use of
    interactive virtual auditory
    worlds in, 39–40
space map, 30
space perception, 376–77. *See also*
    perception
space-specific neurons, 30, 31
sparse coding, 237
sparse representations. *See also under*
    olfactory brain/olfactory
    system
    detection of, 13
sparsening, 13, 17
sparsification, 200
spatial interactions, 446
spatial precedence rule, 446
spatial processing, auditory object
    formation by, 313–14
spatial representation of temporal
    sequences, 266–71, 279
speech functions, 314
    left-hemisphere dominance of,
        47–51
spike count, 140
spike trains, 271
    derailed, 271
spikes, 144–45
spontaneous cortical activity,
    178–79. *See also* neurons,
    single cortical
    cortical processing and behavior
        affected by, 167, 168
    distribution in space and time of,
        167
    as noise *vs.* "music," 165
    relative amplitude of, 166–67
spontaneous regime, 168

tonic mode of firing, 68, 69
tonotopic organization, 84
top-down demands, 346, 355
top-down principle, 343
top-down processing, 343–44,
  360–61, 454, 478
  auditory object formation by,
    314–15
  categorization, 357–60
  different cues, same task
    foreground-background
    decomposition, 354–55
  a motion-selective map in
    auditory cortex, 355–57
  same stimuli, different tasks,
    352–54
  succession of bottom-up and,
    357–58
transients in sound. *See* onset
  responses
traveling salesman problem (TSP),
  219–20

unconscious. *See* consciousness, *vs.*
  unconsciousness
unilateral neglect, visual
  location of lesions producing,
    372–76

V1. *See* visual cortex, primary
validity effect, 33
vector code, 87
ventral pathway of visual systems,
  324, 325
ventral stream, 367–68, 382–83
ventrodorsal (V-D) stream, 368, 373
  action in space and space
    perception, 373–77
  action on objects, 378–79
  action perception, 379–82
  object awareness and, 377–78
VIP, area, 373–75
vision, stereo
  in barn owl, 35–37
visual cortex, functional architecture
  of, 221–25
visual cortex, primary (V1), 182–84,
  206
  as "Gabor filter bank," 182–84,
    197
  new theories regarding, 197, 206
    contour integration, 201–2
    dynamic routing, 205–6
    dynamical systems and limits of
      prediction, 198

sparse, overcomplete
  representations, 199–200
surface representation, 202–3
top-down feedback and
  disambiguation, 203–5
Poisson model of, 157
problems with current view of, 197
  biased sampling of neurons,
    184–88
  biased stimuli, 188–89
  biased theories, 189–90, 192
  ecological deviance, 194–97
  interdependence and contextual
    effects, 192–94
requirements for "understanding,"
  206–7
standard model of V1 simple cell
  responses, 182–83
visual cortical columns, computation
  in, 215–18, 240–41
  abstracting the curve inference
    problem, 225
  curve inference and tangent
    maps, 225–27
  encoding a problem instance,
    228–30
  geometry of intercolumnar
    interactions, 226
  position-orientation
    representation, 227–28
  texture, color, and surface
    properties, 230
columnar computations beyond V1,
  237, 240–41
  geometry of stereo
    correspondence, 237–40
  the columnar machine, 230–32
  computing with cliques of
    neurons, 235–36
  learning, codes, and the games
    neurons play, 236–37
  polymatrix games, 232–35
problem and computational
  abstractions, 215–18, 221–25
  complexity of computations,
    219–21
visual evoked response (VER), 166
visual information processing, limited
  capacity for, 406, 407
visual processing, cortical
  separation into two streams, 368.
    See also perception;
    ventrodorsal (V-D) stream
visual scene analysis, mechanisms of,
  315